Virus as Populations

Virus as Populations
Composition, Complexity, Dynamics, and Biological Implications

Esteban Domingo

Centro de Biología Molecular Severo
Ochoa (CSIC-UAM)
University Automona de Madrid
Cantoblanco
Madrid, Spain

AMSTERDAM • BOSTON • HEIDELBERG • LONDON
NEW YORK • OXFORD • PARIS • SAN DIEGO
SAN FRANCISCO • SINGAPORE • SYDNEY • TOKYO
Academic Press is an imprint of Elsevier

Academic Press is an imprint of Elsevier
125 London Wall, London, EC2Y 5AS, UK
525 B Street, Suite 1800, San Diego, CA 92101–4495, USA
225 Wyman Street,Waltham,MA 02451, USA
The Boulevard, Langford Lane, Kidlington, Oxford OX5 1GB, UK

Notices
Knowledge and best practice in this field are constantly changing. As new research and experience broaden
our understanding, changes in research methods, professional practices, or medical treatment may become
necessary.

Practitioners and researchers must always rely on their own experience and knowledge in evaluating and using
any information, methods, compounds, or experiments described herein. In using such information or methods
they should be mindful of their own safety and the safety of others, including parties for whom they have a
professional responsibility.

To the fullest extent of the law, neither the Publisher nor the authors, contributors, or editors, assume any liability
for any injury and/or damage to persons or property as a matter of products liability, negligence or otherwise, or
from any use or operation of any methods, products, instructions, or ideas contained in the material herein.

Library of Congress Cataloging-in-Publication Data
A catalog record for this book is available from the Library of Congress

British Library Cataloguing in Publication Data
A catalogue record for this book is available from the British Library

ISBN: 978-0-12-800837-9

For information on all Academic Press publications
visit our website at http://store.elsevier.com/

Publisher: *Janice Audet*
Acquisition Editor: *Jill Leonard*
Editorial Project Manager: *Halima Williams*
Production Project Manager: *Julia Haynes*
Designer: *Mark Rogers*

Typeset by SPi Global, India
www.spi-global.com

Printed in USA

Working together
to grow libraries in
developing countries

www.elsevier.com • www.bookaid.org

Contents

Foreword

Viruses are minute organisms that, although mostly invisible, rule the world. Scientists may argue whether these obligatory intracellular organisms belong in the tree of life or whether they were the first replicating entities on earth, but all agree that they have decisively influenced evolution. The vast majority of the planet's 10^{31} viruses live in the oceans, whereas the terrestrial viruses make up only a very small fraction. They were first discovered, however, because their "ghosts" were exposed through diseases in plants and mammals (researchers were unable to "see" these ghosts until electron microscopy revealed them in the late 1930s). Following the discovery of tobacco mosaic virus and foot-and-mouth disease virus in 1898, a bewildering number of viruses have been identified many of which cause terrible human diseases, triggering immense fear as well as admiration: How can these minute creatures live so efficiently and, during pandemics, how can they spread so rapidly across an entire globe?

In principle, all viruses have the same, simple architecture: a relatively small piece of nucleic acid (the DNA or RNA genome), containing all information for proliferation, that is surrounded by a proteinaceous shell for protection and cell attachment, often in combination with a membranous layer. Yet the differences in sizes and shapes of virions for numerous different virus families are astounding. Even more mind-boggling is the wealth of genetic information stored in their small genomes, which directs very different strategies of viral proliferation leading to cell transformation, to cell death, or the death of the entire host. Throughout his professional life, Esteban Domingo has studied the highly complex issues of viral genetics, which qualifies him second to none to summarize the entire field.

Domingo begins his narrative with a short course in general virology, worthy of study for anyone involved in teaching or carrying out research in virology. This is followed by an introduction to the overwhelming, complex secrets of genetic information hidden in even the smallest viral genomes. Domingo's text is not dogmatic. Superbly written, it offers facts and hypotheses and encourages the readers to arrive themselves at the "truth," however, short-lived that truth may be. The reader is pushed to ask questions like: What exactly is a virus and where does it come from? How can we understand infection and viral diseases? How can we prevent their terror?

Beyond experimental breakthroughs that have informed our growing understanding of the molecular nature of viruses and their replication in cells and tissues of their respective hosts, the more recent decades have brought forth new concepts in how we perceive viruses. A shocking surprise in virus research was the landmark discovery by Hans Eggers and Igor Tamm and by Esteban Domingo and Charles Weissmann of the high mutation rate in viral replication, particularly of RNA viruses. The realization of error-prone replication of viruses heralded a paradigm shift in understanding molecular evolution. It changed the landscape of all studies regarding viral replication and pathogenesis. Rather than fixed entities operating with a single, unique nucleic acid sequence, individual species and serotypes of viruses have come to be viewed as genetically heterogeneous, complex populations comprised of a consensus sequence genotype, called by Manfred Eigen and John Holland "quasispecies." As a result of the error-prone replication and subsequent selection of mutations that confer virus fitness in response to cell-specific host factors, immune modulation, or environmental factors, viruses have evolved to exist as quasispecies populations. With these observations as a backdrop, Esteban Domingo offers a comprehensive, up-to-date treatise on the mutable nature of viruses and how this property affects virus-host interactions, viral fitness and adaptation, viral pathogenesis, and the ever-changing prospects for antiviral therapies, including the intriguing possibility of making use of what is called an error catastrophe.

The narrative on the error-prone replication of viruses is amplified by bringing Darwinian principles to bear on the generation of variant genomes (and, as a result, viral quasispecies) and how these virus populations behave in terms of random drift, competition, and adaptation/selection for fitness in different environments. This latter topic is addressed in depth via a discussion of the interactions of virus populations with their hosts and how biological parameters such as receptor usage, codon usage/codon pair frequencies, and the host immune response can provide selective pressure and resultant costs to viral fitness. In a subsequent section of this volume, viral fitness is addressed head-on, with a discussion of how genomic sequences of viruses impact the fitness landscape during viral replication and how subpopulations of virus quasispecies may impact the molecular memory of an evolving quasispecies.

In the second half of this volume, Domingo describes different experimental systems and approaches used to analyze changes in virus population composition resulting from perturbations and selective forces applied during infection and replication. For example, persistent viral infections in cell culture allow the study of experimental evolution, while plaque-to-plaque transfers of viruses experimentally recapitulate infectivity bottlenecks that produce low-fitness viruses which may have extreme phenotypes. In addition, these experimental systems provide tractable platforms to test more general principles of genetics, for example, Muller's ratchet or the Red Queen hypothesis, during viral replication. The book then moves to describe the study of virus transmission and evolution in nature and the many parameters that impact evolution rates. Emergence and reemergence of pathogens is discussed in the context of the complexity of behavior of viral populations intertwined with nonlinear events derived from environmental, sociologic, and ecological factors.

The final sections of the book are devoted to a discussion of some of the more tangible implications of viral quasispecies/sequence evolution in considering its impact on disease prevention and strategies for vaccine design and antiviral therapeutics. This latter topic is analyzed at both the theoretical and pragmatic levels as part of what Domingo calls "virus as moving targets," due to the dynamic nature of viral genome sequences and the proteins they encode. Domingo then concludes this illuminating monograph by bringing the discussion of quasispecies and population dynamics full circle to a focus on nonviral systems (cells, cancer, and other infectious agents) that leave the reader with a number of "big-picture" concepts related to genetic variation, random versus selected replication events, and information theory. It is the universal nature of these concepts that makes this volume essential reading for virologists, evolutionary biologists, population geneticists, and any others who wish to be exposed to an intense level of scholarship on a fascinating topic. *Fiat lux!*

Bert Semler, Irvine, California
Eckard Wimmer, Stony Brook, New York

Acknowledgments

This book is a long story, with a lot to be acknowledged. I begin with the most immediate and then go back in time, with some deviations from linearity. The present core team in my laboratory at Centro de Biología Molecular "Severo Ochoa" (CBMSO) in Madrid, composed of Celia Perales, Ana Isabel de Avila, and Isabel Gallego have done an immense job that has permitted the timely completion of the book (photography at the end of the Acknowledgments). In addition to her scientific leadership, Celia has unique skills to convert scientific concepts into images. Hers are all the original figures in the book. Ana and Isabel have diligently complemented my computer age inabilities by keeping track of end-note and professional typing, always alert to mistakes and inconsistencies in my drafts. Celia, Ana, and Isabel have worked on the book while continuing with experiments and acting as a survival brigade in very difficult times for Spanish science due to drastic budget restrictions. I am deeply thankful to the three of them.

I am also indebted to Eckard Wimmer and Bert Semler for taking time to read the book and to write a Foreword. I deeply appreciate their help, not only now but also at different stages of my activity as a picornavirologist. We shared friendship with John Holland, a decisive name in the scientific contents of the book.

Many thanks go also to Elsevier staff for their support and involvement. In particular, to Elizabeth Gibson in the early stages of book proposal, Jill Leonard for her continuous encouragement, Halima Williams for pushing me the right dose (never beyond a catastrophe threshold), and Julia Haynes for an intelligent and positive attitude during the last stages of book production. I am indebted to Elsevier for publishing the book and also for long-time support reflected in the publication of the two editions of "Origin and Evolution of Viruses," and my involvement in the editorial board of Virus Research initially with Brian Mahy, and recently with Luis Enjuanes, Alina Helsloot, and distinguished colleagues.

Our laboratory has had the privilege to engage in multiple collaborations that have broadened the scope of our research while maintaining the focus on implications of virus complexity. We belong to an active network of experts in liver disease (Spanish acronym CIBERehd) thanks to the interest of Jordi Gomez, Jaume Bosch, Juan Ignacio Esteban, and Josep Quer in our work on viral quasispecies and model studies with hepatitis C virus (HCV). Our current cooperation involves liver disease experts of different institutions, particularly Josep Quer, Josep Gregori, Javier Garcia-Samaniego, Aurora Sanchez-Pacheco, Antonio Madejón, Manuel Leal, Antonio Mas, Pablo Gastaminza, Xavier Forns and their teams. Our connection with HCV goes back to the early 1990s when Jordi Gomez and Juan Ignancio Esteban from Hospital Vall d'Hebrón in Barcelona came to our laboratory to discuss the results that became the first evidence of HCV quasispecies in infected patients (the now classic Martell et al., 1992 paper). The cell culture system for HCV was implemented in Madrid with the help of Charles Rice and the efforts of Celia Perales and Julie Sheldon. These ongoing studies are a source of knowledge and inspiration for us.

Before HCV, there were 35 years of research on foot-and-mouth disease virus (FMDV). About 100 students, postdocs, and visitors contributed to the research whose major aim was to explore if quasispecies dynamics applied to animal viruses. I cannot mention all people involved, but most of them are coauthors of publications quoted in the book. Juan Ortín, the now famous influenza virus (IV) expert, and I organized and shared our first independent laboratory ("down at the corner" of C-V block) at CBMSO, an important institution for Spanish biomedical sciences created in the 1970s under the auspices of Eladio Viñuela, Margarita Salas, David Vázquez, Carlos Asensio, Federico Mayor

Zaragoza, Severo Ochoa, and others. Eladio suggested to Juan and I to work on IV and FMDV because of the unknowns underlying antigenic variation of these viruses, and the problem of producing effective vaccines. And so we did. Juan on IV and me on FMDV. Mercedes Dávila joined as laboratory assistant and remained in the FMDV laboratory for 31 years (!), only to leave as a result of an appointment to manage a central facility at CBMSO. The first graduate students were Francisco Sobrino (Pachi) and Juan Carlos de la Torre. They initiated the FMDV quasispecies work at the pre-PCR age. Mauricio G. Mateu was determinant to expand the scope of our research because at the suggestion and with help of Luis Enjuanes we extended the studies of genetic variation to antigenic variation by producing and using monoclonal antibodies against FMDV. In addition to our own antibodies, we received some from Panaftosa, Brasil, and Emiliana Brocchi from Brescia, Italy. Collaborations were established with Panaftosa and Centro de Virología Animal and INTA-Castelar from Argentina. We also began productive projects with Ernest Giralt and David Andreu on peptide antigens, with Ignasi Fita, Nuria Verdaguer on the structure of antigen-antibody complexes, and David Stuart on the structure of the FMDV serotype used in our research. The collaboration with Nuria Verdaguer and her team continues to date on viral polymerases, particularly with the work of Cristina Ferrer-Orta, in a line of work initiated by Armando Arias at the biochemical level.

The first nucleotide sequences of our FMDV reference virus were obtained by Nieves Villanueva and Encarnación Martínez-Salas. In the late 1980s, Cristina Escarmís (my wife and the person who introduced nucleotide sequencing in Spain and in our laboratory) joined our group to expand our sequencing know-how and to work on the molecular basis of Muller's ratchet. Using routinely nucleotide sequences rather than T1-fingerprints (a transition facilitated by John Skehel, with a summer visit to Mill Hill) was as exciting a change as we experience nowadays seeing deep sequencing data. Additional help to the FMDV research was obtained from Joan Plana, Eduardo L. Palma, María Teresa Franze-Fernández, Elisa C. Carrillo, Alex Donaldson, Joaquín Dopazo, Andrés Moya, Pedro Lowenstein, and María Elisa Piccone, among others.

In the 1980s, there were two events that led to the connection of our laboratory with those of John Holland and Manfred Eigen. One was the publication by John Holland and colleagues in 1982 of the *Science* (1982) article on "Rapid evolution of RNA genomes." The paper was brought to me by my colleague from Instituto de Salud Carlos III José Antonio Melero. What I thought was the forgotten Qβ work of Zürich suddenly revived in a *Science* article that proposed a number of biological implications of high mutation rates and rapid RNA evolution. Not without hesitation, some years later I wrote a letter to John. Since he was extremely receptive, we talked on the phone and met for the first time at the International Congress of Virology in Edmonton, Canada in 1987. Juan Carlos de la Torre went to work with John as a postdoctoral student, and then I spent a sabbatical stay in 1988-1989, followed by several other visits to UC San Diego. Our friendship and shared view of how RNA viruses work lasted until his death in 2013. John and his team (which at the time of my visits included David Steinhauer, David Clarke, Elizabeth Duarte, Juan Carlos de la Torre, Isabel Novella, Scott Weaver, and several students, and visits of Santiago Elena and Josep Quer) were pioneers in linking basic concepts of population genetics with viral evolution. Some of the experiments marked the beginning of fitness assays and lethal mutagenesis, so important in current virology. John was a great support to our work in Madrid in those times of incredulity of high mutation rates and quasispecies. I miss his timely and encouraging comments enormously. The visits to San Diego opened also links with the Scripps Research Institute at La Jolla, when Juan Carlos de la Torre joined Michael B. A. Oldstone department. Each visit to Scripps is an important scientific stimulus, and the friendship and support of Juan Carlos and Michael remain unforgettable.

The contact with Manfred Eigen began in a coincidental manner. The Colombian physicist Antonio M. Rodriguez Vargas invited several European and US scientists to participate in the first Latin American School of Biophysics held in Bogotá in 1984. Invited speakers included the members of Sol Spiegelman's team, Fred Kramer and Donald Mills, and Christof Biebricher from Göttingen. This was the first time I met Christof, and our friendship lasted until his death in 2009. He introduced me to Manfred Eigen and I participated in several Max-Planck Winter Seminars that Manfred organizes at Klosters, Switzerland. I will never forget the discussions with Christof on the theory-experiment interphase of quasispecies, at freezing temperatures combined with a hot soup at a mountain restaurant hut. One of the years, John Drake, the pioneer of mutation rates joined us in extremely lively discussions. At Klosters, I met Peter Schuster with whom I have kept contact since. The discussions with Peter on quasispecies and error threshold have been extremely helpful and clarifying. The Winter Seminars were a key in convincing me that our Madrid research was on the right track, and that I could survive giving a talk in front of Nobel Prize awardees which at the time I perceived as an achievement.

The trans-disciplinary flavor of the seminars at Klosters revived years later in Madrid when in the 1990s the physicist Juan Perez Mercader, at the suggestion of Federico Morán, invited me to join in the organization of a new center termed Centro de Astrobiología (CAB), in Torrejón de Ardoz, near Madrid. Exciting discussions with many scientists conformed a new interest in the nascent science of Astrobiology with participation of noted scientists such as Andrés Moya, Ricard Solé, Federico Morán, Ricardo Amils, David Hochberg, Alvaro Giménez, Luis Vázquez, Ricardo García-Pelayo, Ramón Capote, and Francisco Anguita, among others. The science at CAB opened several collaborations with Susanna Manrubia, Ester Lázaro, and Carlos Briones that continue today, with a monthly seminar with Francisco Montero, Cecilio López-Galíndez, and other colleagues as participants.

The view on viral populations that this book conveys had its origin in work with bacteriophage $Q\beta$ in the laboratory of Charles Weissmann in Zürich. I firmly believe that despite the great recognition that Charles has had as a scientist due to many achievements in different fields, the early $Q\beta$ work that contributed the first site-directed mutagenesis protocols, the birth of reverse genetics, and the first evidence of high mutation rates and quasispecies dynamics, is still underappreciated. Perhaps, as the noted molecular biologist and virologist Richard Jackson once put it, the work came 20 years too early. Fundamental groundwork on the early genetics of bacteriophage $Q\beta$ was performed in Zürich by Martin Billeter, Hans Weber, Eric Bandle, Donna Sabo, Tadatsugu Taniguchi and Richard Flavell. Years later, when Weissmann read Holland's 1982 paper in *Science*, he told me that a new field of research had begun.

I come back to the present. The reason to name so many people in previous paragraphs is not to publicize a biosketch that will interest only my family (and minimally). The reason is that each of the persons, institutions, events, and encounters mentioned has had some influence in the contents of this book. The idea of writing a book occurred to me years back when I produced a chapter for Fields Virology that I had to reduce by a factor of five to fit the required length. Other incentives came from what would be usually considered inconsequential episodes. For example, about 10 years ago, before a talk in France, one of the host scientists told me that he was eager to hear my "new" ideas about virus evolution, when I had been working on quasispecies for almost 30 years! Also, at a meeting on antiviral agents somebody told me that he was unable to select a resistant viral mutant, and he found a good idea my suggestion of increasing the viral population size in the serial passages. These and a few other events reinforced in me the thought that perhaps a book could be of some use. The book is intended to be an introduction to fundamental concepts related to the role that viral population numbers play in several features of virus biology. Related aspects of a given concept are explained in different chapters. This is why many boxes and cross-references among chapters are included. In genetics language, I have aimed at complementation among chapters. The specific examples (some of which attain a considerable degree of detail and are understandably biased toward my expertise) are just an excuse to illustrate general points. Hopefully, the reader will be able to apply them to specific cases.

Before ending, I have additional names to thank for their support on different occasions. At the risk of forgetting some names, they are Simon Wain-Hobson, Noel Tordo, Jean Louis Virelizier, Marco Vignuzzi, Carla Saleh, and other colleagues at Pasteur Institute in Paris, Rafael Nájera, Miguel Angel Martínez, Enrique Tabarés, Albert Bosch, Rosa Pintó, Raul Andino, Craig Cameron, Karla Kirkegaard, Olen Kew, Ernst Peterhans, Etienne Thiry, Paul-Pierre Pastoret, Francisco Rodriguez-Frias, Xavier Forns, Luis Menéndez-Arias, Vincent Soriano, Noemi Sevilla, Steven Tracy, David Rowlands, Louis M. Mansky, Roberto Cattaneo, Stefan G. Sarafianos, Kamalendra Singh, Alexander Plyusnin, Margarita Salas, Jesús Avila, Antonio García Bellido, Carlos López Otín, and Pedro García Barreno. Also, my thanks go to board members of the Spanish Society of Virology and of the Royal Academy of Sciences of Spain. Last but not least, colleagues and staff at CBMSO, and funding agencies, that have made our work possible. My deepest appreciation goes to all.

Esteban Domingo
Cantoblanco, Madrid, Summer 2015
For Iker, Laia, Héctor, Jorge

From left to right Isabel Gallego, Ana Isabel de Avila, Celia Perales, and Esteban Domingo during a discussion of a figure for the book at Centro de Biología Molecular Severo Ochoa

Photography by José Antonio Pérez Gracia

INTRODUCTION TO VIRUS ORIGINS AND THEIR ROLE IN BIOLOGICAL EVOLUTION

CHAPTER CONTENTS

ABBREVIATIONS

AIDS	acquired immune deficiency syndrome
APOBEC	apolipoprotein B mRNA editing complex
CCMV	cowpea chlorotic mottle virus
dsRNA	double stranded RNA
E. coli	*Escherichia coli*

eHBVs	endogenous hepatitis B viruses
HBV	hepatitis B virus
HCV	hepatitis C virus
HDV	hepatitis delta virus
HIV-1	human immunodeficiency virus type 1
ICTV	International Committee on Taxonomy of Viruses
Kbp	thousand base pairs
mRNA	messenger RNA
PMWS	postweaning multisystemic wasting syndrome
RT	reverse transcriptase
RdRp	RNA-dependent RNA polymerase
ssRNA	single stranded RNA
T7	bacteriophage T7
tRNA	transfer RNA
UV	ultraviolet

1.1 CONSIDERATIONS ON BIOLOGICAL DIVERSITY

To approach the behavior of viruses acting as populations, we must first examine the diversity of the present-day biosphere, and the physical and biological context in which primitive viral forms might have arisen. Evolution pervades nature. Thanks to new theories and to the availability of powerful instruments and experimental procedures, which together constitute the very roots of scientific progress, we are aware that the physical and biological worlds are evolving constantly. Several classes of energy have gradually shaped matter and living entities, basically as the outcome of random events and Darwinian natural selection in its broadest sense. The identification of DNA as the genetic material, and the advent of genomics in the second half of the twentieth century unveiled an astonishing degree of diversity within the living world that derives mainly from combinations of four classes of nucleotides. Biodiversity, a term coined by O. Wilson in 1984, is a feature of all living beings, be multicellular differentiated organisms, single cell organisms, or subcellular genetic elements, among them the viruses. Next generation sequencing methods developed at the beginning of the twenty-first century, which allow thousands of sequences from the same biological sample (a microbial community, a tumor, or a viral population) to be determined, has further documented the presence of myriads of variants in a "single biological entity" or in "communities of biological entities." Differences extend to individuals that belong to the same biological group, be it *Homo sapiens*, *Droshophila melanogaster*, *Escherichia coli*, or human immunodeficiency virus type 1 (HIV-1). No exceptions have been described.

During decades, in the first half of the twentieth century, population genetics had as one of its tenets that genetic variation due to mutation had for the most part been originated in a remote past. It was generally thought that the present-day diversity was essentially brought about by the reassortment of chromosomes during sexual reproduction. This view was weakened by the discovery of extensive genetic polymorphisms, first in *Drosophila* and humans, through indirect analyses of electrophoretic mobility of enzymes, detected by *in situ* activity assays to yield zymograms that were displayed as electromorphs. These early studies on allozymes were soon extended to other organisms. Assuming that no protein modifications had occurred specifically in some individuals, the results suggested the presence of several different (allelic) forms of a given gene among individuals of the same species, be humans, insects, or bacteria. In the absence of information on DNA nucleotide sequences, the first estimates of heterogeneity from the numbers of electromorphs were collated with the protein sequence information available. An excellent review of these developments (Selander, 1976) ended with the following premonitory sentence on the role of molecular biology in unveiling evolutionarily relevant information: "Considering the magnitude of this effect, we may not be overfanciful to think that future historians will see molecular biology more as the salvation for than, as it first seemed, the nemesis of evolutionary biology."

The conceptual break was confirmed and accentuated when molecular cloning and nucleotide sequencing techniques produced genomic nucleotide sequences from multiple individuals of the same biological species. Variety has shaken our classification schemes, opening a debate on how to define and delimit biological "species" in the microbial world. From a medical perspective it has opened the way to "personalized" medicine, so different are the individual contexts in which disease processes (infectious or other) unfold. Diversity is a general feature of the biological world, with multiple implications for interactions in the environment, and also for human health and disease (Bernstein, 2014).

1.2 SOME QUESTIONS OF CURRENT VIROLOGY AND THE SCOPE OF THIS BOOK

Viruses (from the Latin "virus," poison) are no exception regarding diversity. The number of different viruses and their dissimilarity in shape and behavior is astounding. Current estimates indicate that the total number of virus particles in our biosphere reaches 10^{32}, exceeding by one order of magnitude the total number of cells. Viruses are found in surface and deep sea and lake waters, below the Earth surface, in any type of soil, in deserts, and in most environments designated as extreme regarding ionic conditions and temperature (Breitbart et al., 2004; Villarreal, 2005; Lopez-Bueno et al., 2009; Box 1.1). The viruses that have been studied are probably a minimal and biased representation of those that exist, with at least hundred thousands mammalian viruses awaiting discovery, according to some surveys (Anthony et al., 2013). This is because high-throughput screening procedures have only recently become available, and also because prevention of disease has provided the main incentive to study viruses. Disease-associated viruses are those most described in the scientific literature.

BOX 1.1 SOME NUMBERS CONCERNING VIRUSES IN THE EARTH BIOSPHERE

- Total number of viral particles: ~10^{32}. This is 10 times more than cells, and they are equivalent to 2×10^8 tons of carbon.
- Virus particles in $1\,cm^3$ of sea water: ~10^8.
- Virus particles in $1\,m^3$ of air: ~2×10^6 to 40×10^6.
- Rate of viral infections in the oceans: ~1×10^{23}/s.
- A string with the viruses on Earth would be about ~2×10^8 light years long (~$1.9 \times 10^{24}\,m$). This is the distance from Earth of the galaxy clusters Centaurus, Hydra, and Virgo.

Based on: Suttle (2007), Whon et al. (2012), and Koonin and Dolja (2013).

Current virology poses some general and fascinating questions which are not easily approachable experimentally. Here are some:

- What is the origin of viruses?
- What selective forces have maintained multiple viruses as parasites of unicellular and multicellular organisms? Why a few viral forms have not outcompeted most other forms? Or have they?
- What has been the role, if any, of viruses in our biosphere?
- Have they played essential evolutionary roles, and do such roles continue at the present time? Are viruses mere selfish, perturbing entities?
- Have viruses been maintained as modulators of the population numbers of their host species?
- Does virus variation play a role in the unfolding of viral disease processes?
- Is the behavior of present-day viruses at the population level an inheritance of their origins, a present-day necessity, or both?

This book deals with some of these issues, mainly those that are amenable to experimental testing. This chapter is an exception in that, despite trying to stick to solid evidence, there are some unavoidable excursions to speculations in an attempt to reconstruct ancient events. Although no research issues in biology are totally independent of others, there are some questions whose answers are directly linked to consider viruses as populations. Examples are the response of viruses to selective constraints (be natural or artificial), mechanisms of short- and long-term diversification, or the effect of viral load and genetic bottlenecks in disease progression. These are the types of topics addressed in coming chapters. They have as a common thread that Darwinian natural selection has an immediate imprint on them, observable in the time scale of days or even hours. The capacity for rapid evolution displayed by viruses represents an unprecedented and often underappreciated development in biology: the direct observation of Darwinian principles at play within short times.

Evolution is defined as a change in the genetic composition of a population over time. In this book evolution will be used in its broader sense to mean any change in the genetic composition of a virus over time, irrespective of the time frame involved and the transience of the change. It is remarkable that only a few decades ago virus evolution (or for that matter microbial evolution in general) was not regarded as a significant factor in viral pathogenesis. Evolution was largely overlooked in the planning of strategies for viral disease control. A lucid historical account of the different perceptions of virus evolution, including early evidence of phenotypic variation of viruses, with emphasis on the impact of the complexity of RNA virus populations, was written by J.J. Holland (2006). The present book was partly stimulated by the conviction that the concept of complexity, despite having been largely ignored by virologists, is pertinent to the understanding of viruses at the population level, with direct connections with viral disease.

1.3 THE STAGGERING UBIQUITY AND DIVERSITY OF VIRUSES: LIMITED MORPHOTYPES

Despite pleomorphism in cells and viruses, presence of envelopes, and viruses being spherical or elongated (helical symmetry), the size of viral particles and their host cells tends to be commensurate with the amount of genetic material that they contain and transmit (Figure 1.1). Viruses can be divided in two broad groups: those that have RNA as genetic material, termed the RNA viruses, and those that

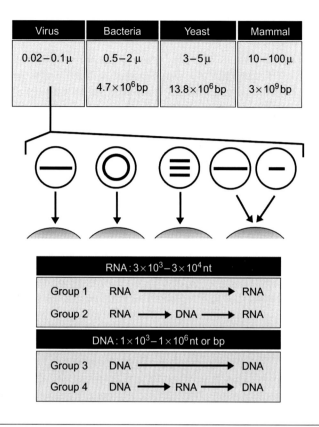

FIGURE 1.1

Representative average diameter values and genome complexity of viruses and some cell types. Diameters are expressed in microns (μ), length of DNA in base pairs (bp), and of RNA in nucleotides (nt). Viral genomes can be linear, circular, segmented, or bipartite (multipartite in general; genome segments encapsidated in separate particles); in the latter case at least two particles, each with one kind of genomic segment, must infect the same cell for progeny production. The bottom boxes describe four groups of viruses according to the type of nucleic acid that acts as replicative intermediate.

have DNA as genetic material, termed the DNA viruses. Both groups, in turn, are subdivided in several orders, families, subfamilies, genera, species, isolates, and multitudes of variants within each isolate. The task of classifying viruses meets with considerable hurdles, and requires periodic revisions by the International Committee on Taxonomy of Viruses (ICTV). ICTV has been essential to provide conceptual order in the vast viral world. One of its objectives is the assignment of newly discovered viruses to the adequate group. A remarkable number of isolates, however, remain unclassified, an echo of the natural diversity of viruses, even among the limited subset that has been isolated and characterized.

According to the structure of their genetic material, RNA viruses can be further subdivided into those that have either single stranded RNA (ssRNA) or double stranded RNA (dsRNA). Both can be either unsegmented RNA (one single piece of RNA) or segmented (two or more pieces or segments of RNA in a single particle or in separate particles). The viral genomic DNA, in turn, can be single stranded DNA (ssDNA) or double stranded DNA (dsDNA), and either linear or circular; in some cases the viral DNA genome is segmented (Figure 1.1).

With regard to the concepts addressed in this book, it is helpful to divide viruses into four groups, depending on whether it is DNA or RNA the type of genetic material which acts as a replicative intermediate in the infected cell (bottom gray shaded boxes in Figure 1.1). The nucleic acids written in the four schemes (two for groups 1 and 3 and three for groups 2 and 4) are those involved in the flow of genetic information. Mistakes in the form of misincorporation of nucleotides during the replication steps indicated by arrows are transmitted to progeny genomes. RNAs produced by transcription to serve solely as messenger RNAs (mRNAs) are essential for gene expression and virus multiplication, but misincorporations in such transcripts are not transmitted to progeny. It could be considered that some mRNA molecule may acquire a mutation relative to the corresponding RNA or DNA template and that this single molecule (e.g., a mRNA encoding a viral polymerase), when expressed, may induce further mutations; we will ignore this possibility since a single mRNA molecule should have a rather limited contribution to the overall genetic variation of a replicating virus population.

Group 1 includes RNA viruses whose genomic replication cycle involves only RNA. They are sometimes called riboviruses. Examples are the influenza viruses, hepatitis A and C viruses, poliovirus, coronaviruses, foot-and-mouth disease virus, or tobacco mosaic virus, among many other important human, animal, and plant pathogens. Their replication is catalyzed by an RNA-dependent RNA polymerase (RdRp) encoded in the viral genome, often organized as a replication complex with viral and host proteins in cellular membrane structures.

Group 2 comprises the retroviruses (such as HIV-1, the acquired immune deficiency syndrome (AIDS) virus, and several tumor viruses) that retrotranscribe their RNA into DNA. Retrotranscription is catalyzed by reverse transcriptase (RT), an RNA-dependent DNA polymerase encoded in the retroviral genome. It reverses the first step in the normal flow of expression of genetic information from DNA to RNA to protein, once known as the "dogma" of molecular biology. This enzyme was instrumental for the understanding of cancer, and in genetic engineering and the origin of modern biotechnology. As a historical account of the impact of H. Temin's work (codiscoverer of RT with D. Baltimore), the reader in referred to Cooper et al. (1995). Retroviruses include a provirus stage in which the viral DNA is integrated into host DNA. When silently installed in cellular DNA, the viral genome behaves largely as a cellular gene.

Group 3 contains most DNA viruses such as herpesviruses, poxviruses and papilloma viruses, and the extremely large viruses (Mimivirus, Megavirus, and Pandoravirus) and their parasitic viruses (La Scola et al., 2008). Their replication is catalyzed by a DNA-dependent DNA polymerase either encoded in the viral genome or in the cellular DNA. Cellular DNA polymerases are involved in the replication of DNA viruses that do not encode their own DNA polymerase.

Finally, Group 4 includes viruses which despite having DNA as genetic material, produce an RNA as a replicative intermediate, the most significant examples being the human and animal hepatitis B viruses (HBVs) and the cauliflower mosaic virus of plants.

Most viruses, from the more complex DNA viruses (i.e., 1200 Kbp (thousand base pairs) for the amoeba Mimivirus, 752 Kbp for some tailed bacteriophages, and up to 370 Kbp for poxviruses, iridoviruses, and herpesviruses), the virophages that are parasites of the giant DNA viruses, the simplest DNA viruses (the circular single stranded 1760 residue DNA of porcine circovirus), RNA bacteriophages (4220 nucleotides of ssRNA for bacteriophage Qβ), or subviral elements (viroids, virusoids, satellites, and helper-dependent defective replicons) show remarkable genetic diversity. However, RNA viruses that replicate entirely via RNA templates; (Group 1 in Figure 1.1), retroviruses; (Group 2); and the hepadnaviruses (Group 4) display a salient genetic plasticity, mainly in the way of a high rate of introduction of point mutations (Chapter 2). Their mutability may be an inheritance of a universal flexibility that probably characterized primitive RNA or RNA-like molecules, thought to have populated an ancestral RNA world at an early stage of life on Earth (Section 1.4.2). Thus, the presence of an RNA at any place in the replicative schemes (Group 1, 2, and 4 in Figure 1.1) implies error-prone replication and the potential of very rapid evolution. "Potential" must be underlined because high error rates do not necessarily result in rapid long-term evolution in nature (Chapter 7). The polynucleotide chain or chains that constitute the viral genome has all the information to generate infectious progeny in a cell, as evidenced by the production of infectious poliovirus from synthetic DNA copies assembled to represent the genomic nucleotide sequence (Cello et al., 2002).

The extent of genetic variation and its biological consequences have been less investigated for DNA viruses than for RNA viruses. The available data suggest that DNA viruses are closer to RNA viruses than suspected only a few years ago, regarding their capacity of variation and adaptation. Evolutionary theory predicts that a high-fidelity polymerase machinery is necessary to maintain the stability of complex genomes (those that carry a large amount of genetic information). This necessity and the various repair activities available to replicative DNA polymerases must be reconciled with the observed diversification of DNA viruses (Chapter 3).

Our current capacity to sample thousands and even millions of viral genomes in relatively short times (a trend that is expanding during the twenty-first century) is revealing an astonishing number of slightly different viral genomes within a single infected host, and even within an organ or within individual cells of an organ! Intrahost diversity of viruses can be the result of coinfection with different viruses (or variants of one virus), of infection that triggers reactivation of a related or unrelated virus from a latent reservoir, diversification within the host, or combinations of these mechanisms. In turn, intrahost virus diversification can result from random sampling events (independent of selection), or from selection acting on variants generated by mutation, recombination or reassortment, or their combined effects (Chapters 2 and 3).

Despite the diversity at the genetic level, viral particles can be grouped in a limited number of morphological types. The more than 7000 bacterial and 150 archeal viruses that have been studied can be assigned to as few as 20 morphotypes. The capsids of nonenveloped (naked) viruses display helical or icosahedral symmetry that determine the architecture of the virion. Variation in size and surface protein distributions can be attained from limited protein folds and the same symmetry principles (Mateu, 2013; Figure 1.2). Divergent primary amino acid sequences in proteins can fold in closely related structures. The "structural space" available to viruses as particles is much more restricted than the "sequence space" available to viral genomes (Abrescia et al., 2012). Sequence space and its mapping into a phenotypic space are key concepts for the understanding of evolutionary mechanisms (Chapter 3).

Three-dimensional structures of entire virions or their constituent proteins can provide an overview of phylogenetic lineages and evolutionary steps in cases in which the information cannot be attained by viral genomics (Ravantti et al., 2013). Yet, minor genetic modifications that do not affect the phylogenetic position of a virus or the structure of the encoded viral proteins in any substantial manner can nevertheless have major consequences for virus behavior. Such consequences include alterations of traits as important as host range and pathogenicity, as discussed in following chapters. How such minor changes in viruses can have major biological consequences may relate to the historical role of viruses in an evolving biosphere. To further address this issue we need to examine how viruses may have originated. This, in turn, begs the question of the origin of life and the possible involvement of viruses in the process.

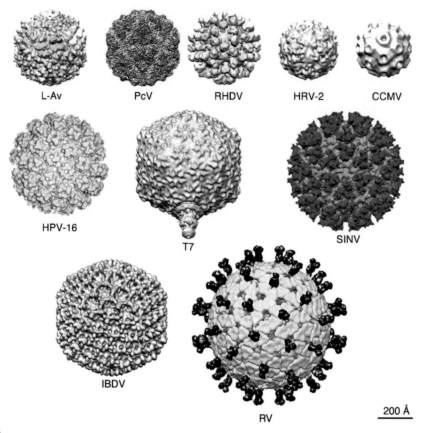

FIGURE 1.2

Examples of spherical bacterial, fungal, plant, and animal virus particles reconstructed from cryo-electron microscopy images. L-Av, *Saccharomyces cerevisae* virus; PcV, *Penicillum Chrysogenum* virus; RHDV, rabbit hemorrhagic disease virus; HRV-2 human rhinovirus type 2; CCMV, cowpea chlorotic mottle virus; HPV-16, human papilloma virus type 16; T7, head of bacteriophage T7; SINV, Sindbis virus; IBVD, Infectious bursal disease virus; RV, rotavirus.

Picture modified from one kindly supplied by J.R. Castón (Castón and Carrascosa, 2013), with permission.

1.4 ORIGIN OF LIFE: A BRIEF HISTORICAL ACCOUNT AND CURRENT VIEWS

An understanding of the mechanisms involved in the origin of life may help penetrating into the origin of previral entities, the possible precursors of the viruses we isolate in modern times. Different notions on the origin of life have been held in human history, often linked to religious debate. Opinions have ranged from a conviction of the spontaneous and easy generation of life from inanimate materials, or its beginning from a unique and rare combination of small prebiotic molecules, or its being the result of a lengthy prebiotic process, or its inevitability as the outcome of the evolution of matter in our universe (or "sets" of universes with the adequate physical parameters, according to some cosmological models).

As little as 150 years ago (not a very distant time from the discovery of the first viruses), there was a general belief in the spontaneous generation of life. This was somewhat paradoxical because chemists of the seventeenth century divided chemistry into mineral chemistry, vegetal chemistry, and animal chemistry. J.J. Berzelious put together animal and plant chemistry and named the resulting discipline "organic" chemistry, which he distinguished from "inorganic" chemistry (Berzelious, 1806). He formulated what was known as the "central dogma of chemistry": "The generation of organic compounds from inorganic compounds, outside a living organism, is impossible." The classical experiments of L. Pasteur provided a definitive proof that, at least under the prevailing conditions on present-day Earth, "life comes from life" (Pasteur, 1861). He established what was considered the "central dogma of biology": "The generation of a whole living organism from chemical compounds, outside a living organism is impossible." The requirement of life to generate life was, however, extended to the belief that "living" and "nonliving" were two totally separate categories in the organization of matter, and that organic compounds, and obviously proteins, could be synthesized only by living cells. This doctrine, called "vitalism" dominated biology for almost a century and in a modified manner it continues today regarding the interpretation of mental activity in humans (matter and spirit as "substance dualism"), and the attitude toward some recent developments in genetic engineering and biotechnology (for a general discussion see Silver, 2007). Aspects of the early views on the origin of life have been addressed in several publications (Rohlfing and Oparin, 1972; Bengtson, 1994; de Duve, 2002; Eigen, 2002, 2013; Lazcano, 2010, among others).

Dogmas are generally not to stay. "Vitalism" was shattered by the chemical synthesis of organic compounds from inorganic precursors (urea by F. Wöhler in 1828, acetic acid by H. Kolbe in 1845, hydrocarbons by D. Mendeleyeev in 1877, and several other compounds by M. Berthelot in the second half of the nineteenth century). The evidence that no "vital force" was needed for such syntheses led F.A. Kekulé to write in his classical textbook on organic chemistry published in 1859-1860: "We have come to the conviction that… no difference exists between organic and inorganic compounds." From then on, organic chemistry became the chemistry of "carbon compounds."

A key experiment to show that components of biological molecules could be obtained from inorganic precursors was carried out in 1953 by S. Miller, working with H.C. Urey. He mimicked the conditions thought to be prevalent in the primitive Earth, and mixed hydrogen (H_2), ammonium (NH_3), and methane (CH_4) in a sealed reactor with an influx of water vapor. Synthesis of a number of organic compounds occurred under the influence of electrical discharges. The *de novo* synthesized chemicals included amino acids (glycine, alanine, aspartic acid, and glutamic acid) formic, acetic, propionic and fatty acids, cyanide, and formaldehyde (Miller, 1953, 1987). Several researchers followed the Miller's approach using other starting chemical mixes, and confirmed that key components of the macromolecules that are associated with living materials (notably purines, pyrimidines, and amino acids) can be made from precursors which were abundant in the primitive Earth or its atmosphere. Today, variant versions of Miller's protocol (including additional starting chemicals, aerosol spread of chemicals, freeze-thaw cycles, different sources of energy, electron beams, etc.) produce interesting information on the synthesis of organic molecules (Dobson et al., 2000; Miyakawa et al., 2002; Bada and Lazcano, 2003; Ruiz-Mirazo et al., 2014). Intense ultraviolet (UV) irradiation may have contributed to the synthesis of compounds relevant to life: ammonia, methane, ethane, carbon monoxide, formaldehyde, sugars, nitric acid, and cyanide. Complex organic compounds (notably aromatic hydrocarbons and alcohols) are also found in interplanetary dust, comets, asteroids, and meteorites, and they can be generated under the effect of cosmic and stellar radiation. Thus, many organic compounds could have been produced within the Earth atmosphere or away from it, and be transported to the Earth surface by meteorites, comets, or rain, to become the building blocks for additional life-prone organic molecules. Places at which peptide bond formation and prebiotic evolution could have been favored are hydrothermal systems and the interface between the ocean and the atmosphere (Chang, 1994; Horneck and Baumstark-Khan, 2002; Ehrenfreund et al., 2011; Parker et al., 2011; Danger et al., 2012; Griffith and Vaida, 2012; Ritson and Sutherland, 2012).

A key issue is the degree of oxidation of the primitive Earth atmosphere. Records of an early surface environment dated 3.8 billion years ago were found in metasediments of Isua, Greenland. These materials suggest that the surface temperature of the Earth was below $100°C$, with the presence of liquid and vapor water, and gases supplied by intense volcanism (CO_2, SO_2, and N_2). The composition of primitive rocks, together with theoretical considerations, suggest a neutral redox composition of the Earth atmosphere (relative abundances: N_2, $CO_2 > CO \gg CH_4$, $H_2O \gg H_2$, $SO_2 > H_2S$) around the time of the origin of primeval forms of life. The possible presence of a reducing atmosphere (N_2, $CO > CH_4 > CO_2$, H_2O, $\sim H_2$, $H_2S > SO_2$) generally or locally is still debated, but increasingly viewed as unlikely. In an oxidative atmosphere yields of amino acids, nucleotides, and sugars would be lower. Either these diminished yields were sufficient or an earlier reduced atmosphere may have accumulated relevant building blocks, among other possible scenarios (Trail et al., 2011). Despite the validity of "life comes from life" in the current Earth environment, the experimental facts suggest that there is no barrier for the generation of life from nonlife, provided suitable environmental conditions are met. In this line, A.I. Oparin proposed that a "primitive soup" could well have been the cradle of life on Earth, as described in his famous treatise on the origin of life (Oparin, 1938; English version 1953), a concept that had already been sketched by C. Darwin.

"Protein first" and "nucleic acid first" as temporal priority for the origin of life is still a contended issue, although in recent decades the preference for nucleic acids due to their superior capacity for self-organization to perpetuate inheritable messages through base pairings has been favored. Paradoxically, however, the building blocks of nucleic acids have been more difficult to obtain from primeval chemicals than the building blocks of proteins. Peptides of about 20 amino acids in length could have been easily formed under prebiotic conditions (Fox, 1988; Fox and Dose, 1992), and the peptides or derived multimers had a potential to display catalytic activities in a protometabolic stage. Short nucleotide and amino acid polymers might have contributed to cross a complexity threshold for self-sustained replication to arise and evolve (Kauffman, 1993). As soon as peptide- or protein-based catalytic activities developed, they had to be coordinated with oligo- or poly-nucleotide replication. This integration of genotypic information with its phenotypic expression may be achieved through hypercyclic couplings, as proposed by M. Eigen and P. Schuster (1979; Eigen, 2013). A synchronous and sharp increase of bacteriophage Qβ RNA and proteins during *E. coli* infection was taken as an experimental proof of the presence of a hypercycle. However, there are multiple molecular mechanisms involved in achieving a necessary increase in viral RNA and proteins during cytolytic infections. The experimental evidence of a general principle of hypercyclic organization should not be taken to mean that Qβ is a remnant of a primitive coupling between replication and translation.

1.4.1 EARLY SYNTHESIS OF OLIGONUCLEOTIDES: A POSSIBLE ANCESTRAL POSITIVE SELECTION

Substantiating the abiotic synthesis of nucleotide- or amino acid-based polymers has been arduous relative to the synthesis of monomeric organic molecules. However, work by L. Orgel and his colleagues documented that polynucleotides could be synthesized from activated nucleotides in the absence of any enzyme (as a review of these early developments, see Miller and Orgel, 1974). Recent work has established that there are multiple chemical pathways for abiotic nucleotide synthesis (reviewed in Ruiz-Mirazo et al., 2014). Mineral surfaces (clay minerals, zeolites, manganates, hydroxides, etc.) have been proposed as key participants in the origin of life, by providing scaffolds for the synthesis of nucleotide and amino acid polymers (Bernal, 1951; Gedulin and Arrhenius, 1994; Ruiz-Mirazo et al., 2014). The relevant features of clays are their adsorption power, ordered structure, capacity to concentrate organic compounds, and their ability to serve as polymerization templates. Adsorption onto mineral surfaces may lower the activation energy of intermolecular reactions. Minerals with excess positive charge on their surfaces might have played a role in the evolution of RNA-like or RNA-precursor molecules. It has been suggested that mineral-organic complexes (chimeras between materials once thought to belong to unmixable categories) could have been the first living organisms endowed with a genetic program (Cairns-Smith, 1965).

Bond formation and chain extension, which are needed to synthesize nucleic acids and proteins prebiotically, are inhibited in liquid water, thus favoring mineral surfaces as potential biogenic sites. Polymer formation could have been facilitated also by heating and drying, with a key involvement of the dried phases in catalysis (reviews in Towe, 1994; Horneck and Baumstark-Khan, 2002). On different clay types, nucleotides and amino acid polymers of dozens of residues have been synthesized, providing a realistic scenario for a transition toward primitive self-replicating entities (Ferris et al., 1996). One of the first types of Darwinian positive selection could operate prebiologically through differential surface binding. We arrive at the paradox that while primitive polymerization reactions might have required dry surfaces, later forms of life evolved in a water-rich environment, being water the major component of cells. "Origin" and "establishment and evolution" of life have been considered either as linked processes that obey similar rules, or distinct events whose investigation requires dissimilar approaches. A distinction between "origin" and "establishment" will become pertinent also when addressing the origin versus the evolution of viruses later in this chapter and the mechanisms of viral disease emergence in Chapter 7.

1.4.2 A PRIMITIVE RNA WORLD

A simplified overview of a course of events that led to the origin of biological systems is depicted in Figure 1.3. The prebiotic synthesis of potential building blocks—which might have been initiated earlier than 5000 million years ago—renders plausible the existence of a pre-RNA era that was then replaced by an RNA world in the late Hadean early Archean periods on Earth. This stage should have been followed by one in which RNA was complemented by DNA as repository of genetic information (Bywater, 2012). Polymers other than DNA and RNA are also capable of encoding evolvable inheritable information (Pinheiro et al., 2012; Robertson and Joyce, 2012). Heterogeneous nucleic acid molecules (including mixtures of ribo- and desoxyribo-polymers) can give rise to functional nucleic acids (Gilbert, 1986; Lazcano, 1994a; Gesteland et al., 2006; Derr et al., 2011; Szostak, 2011; Trevino et al., 2011). RNA enzymes (ribozymes) such as RNA ligases can evolve from random-sequence RNAs (Joyce, 2004; Sczepanski and Joyce, 2014). The critical polymerization reaction involves the formation of a phosphodiester bond and release of pyrophosphate—analogous to the reactions catalyzed by the present-day RdRps—and represent an incipient, primitive anabolism (Eigen, 1992; Lazcano, 1994a; Orgel, 2002, 2004; Dworkin et al., 2003; Joshi et al., 2011). In support of a possible link between catalytic RNA activities and solid mineral surfaces in the origin of life (as summarized in Section 1.4.1), the catalytic activity of the hammerhead ribozyme of the Avocado Sun Blotch viroid was maintained when bound to the clay mineral montmorillonite (Biondi et al., 2007).

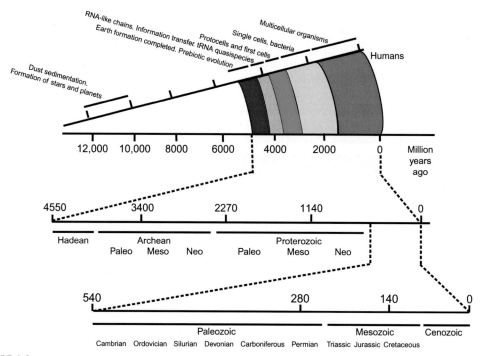

FIGURE 1.3

Schematic representation of geological eons and eras in Earth history, and major prebiological and biological transitions. Time (horizontal lines) is expressed as millions of years before present, counted from the big bang (estimated at ~13,800 million years ago).

Illustration by C. Perales and E. Domingo from information retrieved from references given in the text.

Minimum requirements for an RNA world would be the presence of ribozymes and mechanisms for the intake of energy-rich molecules (Orgel, 2002, 2004). The inherently low copying fidelity of the putative ribozyme polymerases, estimated in 10^{-2} to 10^{-4} errors per nucleotide copied (Wochner et al., 2011) should have ensured genetic variation for selection to act upon variant RNA molecules. Chilarity (the existence of two mirror images or enantiomeres of a molecule) poses a challenge for the chemical origin of biological molecules (Caglioti et al., 2011; Ruiz-Mirazo et al., 2014; Sczepanski and Joyce, 2014). Present-day biological systems use only D-ribose (D from "dextro" or rotation of the plane of polarized light to the right) while chemical condensation reactions produce equal amounts of the D- and L- (levo) forms. Nonenzymatic template-dependent reactions can be inhibited by the incorrect enantiomer. This led to the proposal that analogs devoid of enantiomeric forms, such as glycerol derivatives, could have been the basis of the most primitive genetic systems (Schwartz and Orgel, 1985). Theoretical studies support the notion that initial achiral conditions can evolve toward chilarity, in what has been defined as an extension of punctuated equilibrium to prebiological evolution (Gleiser et al., 2008).

Probably, very little, if anything, remains in our present-day biosphere of a primitive RNA world, let alone traces of the prolonged process that went from chemistry to the first replicating organizations, so profound have been the changes experienced by the Earth and its surroundings over 4 billion years. Contemporary catalytic RNAs (found in the ribosome, as part of some protein complexes, and in some viroids), as well as nucleotide-like coenzymes have been considered as possible molecular remnants of a primitive RNA world (Lazcano, 1994a). Some authors consider the possibility that cells whose genetic material is made of RNA may still hide in some remote sites of our planet (Yarus, 2010). For a transition from a purely RNA (or RNA like) world to an extended scenario with participation of proteins, the presence of a transfer RNA (tRNA) quasispecies and the generation of a genetic code endowed with evolutionary potential must have been critical (Eigen, 1992). Initial theories of how the genetic code might have arisen were put forward by F. Crick, L. Orgel, and C. Woese in the middle of the twentieth century. Main proposals included a stereochemical fit between some amino acids and the corresponding bases (or codons to be), progressive evolution from a one nucleotide to a three nucleotides code, gradual incorporation of amino acids in the coding system, and the frozen accident model of codon universality (for a review of early concepts, see Crick, 1968). New insights on the code origin have come from integrating knowledge of the mechanisms of protein synthesis with likely events in the RNA world. Primitive tRNA quasispecies (around 4000 million years ago, late Hadean, early Archean, Figure 1.3) and tRNA aminoacylating ribozymes evolved to fit the genetic code which was later expanded in coevolution with the translation machinery (Szathmary, 1999; Rodin et al., 2011; Caetano-Anolles et al., 2013). Recent models retain several features of the early proposals. The age of the genetic code has been estimated in 3.8 ± 0.5 thousand million years (Eigen, 2013). The present structure of the genetic code is remarkably redundant, and it minimizes the deleterious effects of mutations, suggesting that protein conservation offers a general selective advantage for most cellular entities. Code redundancy may also contribute to functional mRNA secondary structures and stability (Shabalina et al., 2006; see also Chapter 2).

The advent of DNA as an informational macromolecule that was more stable than RNA, allowed integration of modules to form the first "chromosomes," opening the way to transcriptional regulation prior to protein expression (see also Section 1.5 on the origin of DNA viruses). As in current evolutionary virology, perhaps the most challenging problem to understand early life is to identify the selective constraints that determined (or influenced) the course of events. In contrast to the present-day environmental changes confronted by viruses (multiple and complex, but amenable to experimentation, Chapters 4 and 6), the conditions that permitted primitive genetic entities to acquire expanded coding and signaling capacities defy our imagination. Some aspects are considered next.

1.4.3 LIFE FROM MISTAKES, INFORMATION FROM NONINFORMATION: ORIGIN OF REPLICONS

The most salient attributes of living matter are reproduction, evolvability, energy conversion, and compartmentalization. Two of them are found in viruses: reproduction and evolvability. What selective forces might have led to the integration of these features? Concerning reproduction, there must have been a critical transition from the absence of any inheritable instruction (despite the presence of primitive polymers) to the first molecules endowed with "inheritable" information, for example, a macromolecule capable of producing copies of itself. The transition from "noninformation" to "information" is an essential question for the origin of life that implicates both theory of evolution and information. Current evidence suggests that the process that allowed such critical transformation was slow and inaccurate. Slow because the catalytic RNAs selected in the laboratory are about

10 million-fold slower than most protein enzymes (Jeffares et al., 1998; Yarus, 2010). Inaccurate because pre-enzymatic nucleotide polymerization would rarely display error rates below 10^{-1} to 10^{-2} mutations per nucleotide copied (Inoue and Orgel, 1983). Despite a likely hostile environment, no "predators" (be molecular such as degrading enzymes or others) were present to impede a slow accumulation of variant replicating molecules. Random and unavoidable mistakes allowed polymers to wander in sequence space with reproductive impunity until areas that promoted increased self-organization, replication, and adaptation were encountered. The adaptive potential of mutant distributions is a key concept in the quasispecies theory of the origin of life (Eigen and Schuster, 1979), and a signature of present-day viruses (Chapter 3). Replicative innacuracy and heterogeneity appear as recurrent requirements for the major transitions and adaptedness of the forms generated throughout prebiological and biological evolution. Once a primitive molecular "memory" (signature sequences with some replication ability) was implemented, in the words of M. Eigen: "....information generates itself in feedback loops via replication and selection, the objective being 'to be or not to be'" (Eigen, 1994, 2013). In those times this was the only and simple requirement: to be or not to be.

This singular transition resulted in the first replicating entities that are also termed "replicators" or "replicons." They were selected for replicability, stability, and evolvability with trade-offs (acquisition of benefits for one of the three traits at some cost for another trait) likely playing a role at this stage (see Chapter 4 for trade-offs in virus evolution). Assuming a stage in the absence of peptides, optimization of primitive replicons should have been facilitated by the fact that the same molecule embodied both genotype and phenotype (Eigen, 2013), again a feature of present-day RNA viruses. The genomic RNA by itself determines phenotypic traits, independently of its protein-coding activity. "Priming" of polynucleotide synthesis in the sense we know it today should not have been a limitation since circular RNA or RNA-like molecules could fold partially to prime their own copying. The term "replicon" currently refers to any simple genetic element that encodes sufficient information to be copied (i.e., viruses, plasmids, etc.), even if the copying is carried out by (or in conjunction with) elaborate cell-dependent machineries. Virtual replicons are used in computer simulations, to learn about the dynamics of natural living systems (as reviews, see Adami, 1998; Eigen, 2013).

The environment in which early primitive replicons had to self-organize about 4×10^9 years ago was very different from the environment we have today on Earth. The sun was about 25-30% less luminous than it is today yet it produced more UV light. Due to the absence of oxygen and of an ozone layer, the UV radiation that reached the Earth was 10- to 100-fold more intense than today, the difference being accentuated for the radiation in the 200-280 nm range. Studies of the conversion of UV radiation into DNA-damage equivalents suggest a 2-3 logarithm larger biologically relevant UV radiation during the time of the putative RNA world as compared with today's radiation (Canuto et al., 1982; Chang, 1994; Horneck and Baumstark-Khan, 2002). In such an environment, radiation-related mutational input could have had drastic effects on replicating entities in ways that can be only roughly anticipated from the present-day chemistry. Even the simplest present-day RNA genetic systems, with their small target size, would undergo severe radiation damage. Reconstruction of protein enzyme-free nucleic acid synthesis under the radiation conditions prevalent on Earth during the RNA world development, during late Hadean and early Archean eras offers a fascinating challenge and opportunity of experimental research for the raising field of Astrobiology.

It has been considered that the time elapsed since the Earth attained a life-friendly environment until protocells arose (from about 4500 million to about 3500 million years ago, Figure 1.3) was insufficient for life development. This led to the panspermia theory which proposes that life has an extraterrestrial origin. Panspermia in different forms has been defended by noted scientists such S. Arrhenius in the early twentieth century and later by F. Crick and L. Orgel (discussed by de Duve, 2002). In addition to the time estimates for life generation being arbitrary, our present understanding of how error-prone replication can facilitate evolvability and exploration of novel biological functions (Chapters 2 and 3) converts a 1 million year time period in a long time for life to originate and initiate multiple branches for its development.

1.4.4 UPTAKE OF ENERGY AND A SECOND PRIMITIVE POSITIVE SELECTION

Energy conversion is an essential feature of life. The fact that we have first discussed the origins of inheritable information should not be taken as its being independent from the incorporation of molecules capable of supplying energy for key reactions. The primitive cellular-like organizations might have obtained energy either from organic molecules captured from the external environment (heterotrophy) or from metabolites they synthesized endogenously using external energy (autotrophy). One line of thought considers that it is more likely that the first cells were heterotrophs, and that only later they evolved toward autotrophy, in the form of photosynthesis, which represented a major transition in the repertoire of biosynthetic pathways. According to this model, fermentation reactions were likely the first ones exploited to break energy-rich bonds, as a source of energy for primordial biochemical reactions.

An alternative view is that the first cellular organism was an autotroph, in particular a chemoautotroph (also termed lithotroph) that used inorganic compounds to obtain energy. One of these proposals is that formation of pyrite from hydrogen sulfide was used by primitive cells as an energy source, resulting from reactions such as $FeS + H_2S \rightarrow FeS_2 + 2H^+ + 2e^-$ (Wächtershäuser, 1994). Positive charges on pyrite crystals could accumulate negatively charged molecules (e.g., the products of CO_2 fixation) and undergo reductive reactions. The system might have selected surface-bound polymers rather than monomers, and given rise to biochilarity (selection of one enantiomeric form over another) because of the chiral structure of pyrite (Wächtershäuser, 1988). Then, an evolution toward an Archean carbon-fixation cycle would occur, a precursor of the metabolic cycles found in present-day archeal and bacterial organisms (Ruiz-Mirazo et al., 2014). The picture may have been more complex, as judged by the success of mixotrophic organisms which are capable of switching from phototrophy (light-mediated break down of CO_2 for their metabolism) to heterotrophy in some extreme environments of the present-day Earth (Laybourn-Parry and Pearce, 2007).

The integration of early replication systems and metabolism was probably favored by some compartmentalization of replicative-metabolic units through lipid bilayers (Carrara et al., 2012; Stano et al., 2013; Ruiz-Mirazo et al., 2014). Here a second decisive positive selection event might have entered the scene at the stage of formation of protocells and the first individual cells (3600-3200 million years ago, Figure 1.3; Eigen, 1992). Lipid bilayers endowed with a splitting capacity should have been strongly selected at this stage of life development because of their power to spread. This underlines the concept that Darwinian selection need not be associated exclusively with template-copying processes (Chapter 10). Membrane traffic and reorganizations are essential for the life cycle of many present-day viruses (Huotari and Helenius, 2011). Virus particle stability and capacity to spread, associated with membrane capture and interactions, might derive from the early genomes that exploited membrane-based organelles to achieve functional diversification. Compartmentalization was the key to initiate a cell-based life (Morowitz, 1992).

Despite obvious difficulties in reproducing with our current technology critical physical and chemical processes that were likely involved in the origin of life, there is sufficient evidence to render probable that the most primitive organizations that we would now consider as "living" resulted from the assembly of simple organic compounds that attained a required level of complexity (Kauffman, 1993). The various facets that distinguish living from inanimate matter are recapitulated in the definitions of life that scientists from different backgrounds have proposed, and some of them are listed in Box 1.2. A. Lazcano summarized the current situation in the arduous search on the origin of life: "There are still unsolved problems but they are not completely shrouded in mystery, and this is no minor scientific achievement. Why should we feel disappointed by our inability to even foresee the possible answers to these luring questions? As the Greek poet Konstantinos Kevafis once wrote, *Odysseus should be grateful not because he was able to return home, but on what he learned on his way back to Ithaka*. It is the journey that matters" (Lazcano, 1994b; for a general overview of the history of life on Earth, centered in paleontology, see Cowen, 2005).

The spectacular progress in experimental generation of synthetic life evidences that the assembly of polynucleotide building blocks (of viral, bacterial, and eukaryotic genomes and chromosomes) gives rise to macromolecules that display features of life. Synthetic life is now within reach in years, following assembly principles, that in the absence of man-made technology, took millions of years for the evolution of our young planet.

After this brief survey of origin of life, we can now examine theories on how, when and why viruses arose, and became active actors in our biosphere.

BOX 1.2 SOME DEFINITIONS OF LIFE AND LIVING ORGANISMS

- Life is the property of a system that continuously draws negative entropy (maintains orderliness), and delays decay into thermodynamic equilibrium (Schrödinger, 1944).
- Life is an expected, collectively self-organized property of catalytic polymers (Kauffman, 1993).
- Life is a property of any population of entities possessing those properties that are needed if the population is to evolve by natural selection (Maynard Smith and Szathmáry, 1999).
- Life is what is common to all living beings. This answer is not a tautology, as it allows many attributes to be excluded from the definition of life (de Duve, 2002).
- Life is descent with modification. Replication is life's ultimate chemical and physical survival strategy (Yarus, 2010).
- Life is a self-sustained chemical system capable of Darwinian evolution (NASA Astrobiology Institute).
- Living organisms are metabolic-replicating systems composed of molecules and cells which are subject to spontaneous changes in structure and function due to mutations (Demetrius, 2013).
- Life in three statements: (1) Life is not represented by any fundamental physical structure. (2) Life is an overall organization that is governed by functional rather than by structural principles. (3) In order for life to come about, there must exist some physical principle that controls complexity (Eigen, 2013).
- Life, whatever else it may be, is certainly a regularity among material processes (Eigen, 2013).

1.5 **THEORIES OF THE ORIGINS OF VIRUSES**

Although not in a linear fashion, the number of nucleotides or base pairs in the genetic material—that presumably reflects the amount of genetic information relevant to confer phenotypic traits—increased as evolution led to differentiated organisms. The major theories of the origin of viruses are divided in two opposite categories: those that attribute virus origins to the early development of life, and those that propose that viruses arose when a cellular life was already in place (Figure 1.4). These two views, however, might not be irreconcilable.

From our present understanding of viruses and their genomes, and the likely events in the origin of life discussed in the preceding sections, five main theories—not all totally independent or mutually

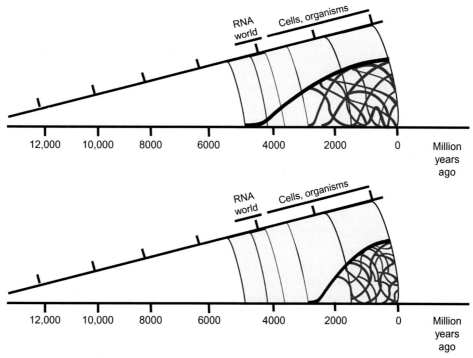

FIGURE 1.4

Two possible course of events regarding when viruses first appeared and participated in the evolution of the biosphere. The scheme of time frames and major biological events (RNA world, first cells and organisms) are those displayed in Figure 1.3. According to the upper diagram, viruses (or previrus-like entities) arose together with the first (precellular) replicating entities. According to the second diagram, viruses (or previrus-like entities) arose when a cellular life had already been established. Presence of virus is generically represented by the external, thick, black curves. The internal red, wavy lines represent generation, dominance and extinction of multiple viral lineages whose numbers and true dynamics will remain unknown.

Illustration by C. Perales and E. Domingo from information retrieved from the different models of virus origin and references included in the text.

exclusive—of the origin of viruses have been proposed. They are summarized next, with some comments that benefit from our current understanding of viruses.

1.5.1 VIRUSES ARE REMNANTS OF PRIMEVAL GENETIC ELEMENTS

• Viruses were involved in the origin of life. Some viruses are the descendants of primitive RNA or RNA-like replicons that preceded cellular forms.

Because of their limited genetic complexity, RNA viruses and subviral RNA elements have been considered possible descendants of the primitive replicating entities that predated cell-based life forms (upper diagram in Figure 1.4). As early as the beginning of the twentieth century, H.J. Muller, L.T. Troland, and J.B.S. Haldane suggested that viruses represented primordial life forms. Influenced by the discovery of bacteriophages by F. d'Herelle, J.B.S. Haldane proposed viruses as intermediates between the prebiotic soup and primitive cells (reviewed in Lazcano, 2010). In those times knowledge of viruses was still superficial from todays' perspective, and lent to daring proposals coherent with viruses being perceived as simple. Despite simplicity and replication being features that could be also attributed to primitive life forms, we have to distinguish the role that virus-like entities might have played in the establishment of early life from the possibility that present-day viruses reflect how early life might have been. A reader versed in the structural and functional complexity of RNA viruses, with all the intricacies of virus-host relationships, will realize how unlikely is that present-day RNA viruses are remnants of an ancestral RNA world, at least as current evidence portrays it. Also, the conditions prevailing in the RNA world did not necessitate that a primitive replicon displays rapid replication—a trait of most present-day viruses—for it to become established, because of the scarcity of predators.

Rather than the RNA viruses, the plant viroids (self-replicating RNAs of 250-400 nucleotides in length) or related genetic elements lodged in the animal world, such as the defective delta agent, also termed hepatitis delta virus (HDV), might be a vestige of primitive genetic elements (Robertson et al., 1992). HDV is dependent on HBV for the completion of its infectious cycle (reviewed in Quer et al., 2008; Taylor and Pelchat, 2010). The HDV genome is a mosaic RNA consisting of a viroid-like RNA and an RNA region whose complementary RNA (antigenomic strand) encodes two forms of a structural protein termed the delta antigen. Both the genomic and antigenomic RNAs possess a strong secondary structure with about 70% paired nucleotides. The delta antigen is encapsidated by the HBV surface antigen as a component of HDV particles. Thus, HDV appears to be the result of an RNA conjunction between a viroid-like RNA and an mRNA-coding region. Such conjoined RNAs might have been the precursors of the modern eukaryotic organization into coding sequences (exons) and intervening sequences (introns) (Sharp, 1985; Robertson, 1992, 1996; Chao, 2007; Taylor and Pelchat, 2010).

The structure of the HDV genome seems to echo processes that originated with primitive RNAs selected for their ability to replicate that incorporated a protein-coding moiety through recombination with other RNAs. Such replication-competent and protein-coding chimeras might have been fully realized in the form of DNA genomes at a later stage of macromolecular evolution. Being circular, the primitive viroid-like RNAs and their conjoined derivatives might have permitted a rolling circle-based replication mechanism that allowed maintenance of the genetic information without the continued presence of primer molecules or other replication-initiation factors.

The question of whether conjoined RNAs or other subviral or viral entities played some key role in early life may never be settled. Adaptation of primitive, replicating virus-like RNAs to an evolved

DNA-RNA-protein world (precellular and cellular), impulsed by the implementation of the genetic code (Section 1.4.2), should have erased genetic signatures of an RNA-only world. The advent of an enzyme that could copy RNA into DNA to carry out reverse transcription should have been instrumental in generating primitive DNA viruses and other DNA-based genomes (Lazcano et al., 1992). Because of its higher stability than RNA, DNA could incorporate different preexistent functional modules to construct increasingly complex genetic systems.

The discovery of the chimeric structure of the HDV genome illustrates how insights into the origin and early evolution of life can be gained from current genomics, despite lacking experimental approaches to recreate episodes that led to virus origins. Even if experiments could be designed, the time frame involved would occupy several generations of scientists, which is not feasible given the current research grant system.

1.5.2 VIRUSES ARE THE RESULT OF REGRESSIVE MICROBIAL EVOLUTION

- Viruses originated from regressive evolution of microbes with a cellular organization, and became parasites of cells.

This theory is quite opposite to the previous one because it presupposes that a cellular world was the source of viruses (lower diagram in Figure 1.4). It was already put forward in the twentieth century when it was evidenced that complex DNA viruses encoded enzymes and immunomodulatory proteins that had cellular counterparts. The virus-generating cells could be either fully functional from the onset or belong to a class of simple cells that parasitized functionally more advanced cells. Now, with the discovery and characterization of giant viruses that probably coexisted with cellular ancestors (Nasir et al., 2012) this theory has acquired new impetus.

The capacity to spread, so inherent to the concept of virus, might have been first attained by cells. At a later stage, the positively transmitted cell could have regressed toward a subcellular transmissible form. Prokaryotic cells can spread effectively among differentiated eukaryotic hosts, and tumor cells have been regarded as transmissible parasites (Banfield et al., 1965; Murgia et al., 2006; Pearse and Swift, 2006). The observation that some "infectious cells" can be disseminated by insects provides a model for an early origin of vector-borne viruses. For a number of viruses, notably HIV-1, it has been recognized that transmission of virus from an infected cell into a recipient cell need not involve a prolonged stay of the virus in the extracellular environment. Infection "synapses" allow intimate cell-to-cell contacts through which virus transmission takes place. It is estimated that synapse-mediated transmission may be 100-fold more efficient than transmission of viral particles released into the extracellular environment. Obviously, acquisition of a capacity for long-range transmission in space and time should have provided a selective advantage to a virus. Perhaps the participation of infection synapses in virus transmission is an evolutionary remnant of an early period in which the infectious entities were cells, with now the "cell-to-environment-to-cell" mode of transmission coexisting with the "cell-to-cell" mode.

The possibility that RNA viruses derive from some "organism" that used RNA as genetic material was suggested initially by D. Baltimore (1980). The evidence that RNA-dependent RNA synthesis is rare in cells suggests that either RNA viruses derived their replicases from a now probably extinct "RNA organism," or that the viral replicases evolved from cellular DNA polymerases (Baltimore, 1980; Forterre, 2005, 2006a,b; Yarus, 2010).

1.5.3 VIRUSES ARE LIBERATED AUTONOMOUS ENTITIES

- Viruses originated from cellular DNA or RNA that evolved to embody autonomous replication, and an extracellular step in their replication cycle.

Related to the regressive evolution model, this theory does not imply an initial cellular nature of the virus-to-be. Rather, some *bona fide* cellular nucleic acids acquired genetic elements that triggered their competence as autonomous, albeit cell dependent elements able to survive transiently outside the cell. This new way of life should have been positively selected if an increased capacity of cell-to-cell transfer conferred an advantage to the cells regarding acquisition of new traits for functional diversification while maintaining a capacity to sustain virus multiplication.

Decades ago, the view that viruses originated from subcellular organelles was a favored one (see e.g., Joklik, 1974). Despite discernible sequence identity between some viral and mitochondrial DNA sequences, no evidence of viruses having functions encoded in cellular organelles and not in chromosomal DNA has been obtained, perhaps reflecting an earlier relationship between viruses and primitive free-living cells rather than between viruses and modern (eukaryotic) cells.

1.5.4 VIRUSES ARE ELEMENTS FOR LONG-TERM COEVOLUTION

- Viruses are as ancient as cells, and coevolved with cells or even with precellular genomic organizations, with which they shared functional modules.

Current genomics of viruses and their host organisms (Bushman, 2002; Mount, 2004; Hacker and Dobrindt, 2006) tends to favor this theory. From the catalogs of regulatory sequences and protein-coding genes from cellular organisms and viruses whose functions have been identified, some viral genes, notably those encoding the viral capsid and the genome packaging machinery, appear to be devoid of a cellular counterpart. Such typical viral genes constitute what has been referred to as the "conserved innate viral self" (Krupovic and Bamford, 2007), that embraces essential functions for the phenotypic traits shared by all viruses. Other viral functions do not deviate in any salient way from cellular functions. They include viral proteins involved in genome replication, and in proteolytic processing of proteins and polyprotein precursors. Comparative genomics suggests that the exchange of functional and structural modules through lateral gene transfers, together with fine adjustments mediated by mutation, have contributed to the coadaptation of cells and autonomous replicons over ancient evolutionary periods (Gorbalenya, 1995; Holland and Domingo, 1998; Jalasvuori and Bamford, 2008; Villarreal, 2008). Even mechanisms that prompt viral variation in cell tropism (Chapter 4) have parallels in differentiated organism. As an example, a two amino acids insertion into ectodysplasin—a member of the tumor necrosis-binding family—alters its receptor specificity, and the differential expression of the two protein versions plays a role in epidermal morphogenesis (Yan et al., 2000). Furthermore, as the number of three-dimensional structures for viral a cellular enzymes has increased to reach thousands, structural similarities between key cellular and viral enzymes (polymerases, proteases) have become apparent. Arguments in favor of either ancestral cells or ancestral viruses being the main source of what we now identify as functional modules may never be settled.

Baltimore (1980) proposed that a limited number of "archetypal" proteins could be responsible for RNA virus function. He named as "archetypal" a "positive virus polymerase," a "negative virus polymerase," a set of "surface" proteins, and several "proteases," among other proteins and regulatory elements. The argument that "archetypal" modules could spread among positive and negative strand

RNA viruses was based on features and mechanisms now recognized as much more profuse than in 1980: the multifunctionality of viral proteins, their capacity to diversify by mutation, and the existence of RNA recombination (Chapter 2). Regulatory strategies were also likely shared by cells and viruses. Small micro-RNAs that now populate the cellular world can act as molecular switches for RNA viral genomes to modulate their replication and gene expression (van Rij and Andino, 2008; Diaz-Toledano et al., 2009). A long coevolution of protocells and primitive virus-like elements that gradually yielded the continuously coevolving cells (at the individual and organismal level) and viruses we see today is a likely course of events.

1.5.5 VIRUSES FROM VESICLES

- "Protoviruses" might have originated in primitive vesicles.

Early protocellular communities probably lacked a cell wall or other compartmentalization barriers, an absence that allowed fluid transfers of metabolites and genetic material (Woese, 2002). The most primitive vesicles might have evolved to contain self-replicating macromolecules (Jalasvuori and Bamford, 2008; Adamala and Szostak, 2013). Vesicles located either in hydrothermal vents, or in other favorable microenvironments of a primitive Earth, had the potential to exchange small molecules between the inside and outside, and materials could reach other vesicles.

Depending on their composition, lipid vesicles could form and remain stable at temperatures of about 100 °C. Their transfer to lower temperatures might have modulated their permeability prior to the stage at which peptide or protein transporters were inserted as membrane components. Here again, "heterogeneities" became important: membranes made of mixtures of amphiphiles display increased thermostability, permeability, and tolerance to divalent cations. Budding vesicles endowed with traits such as growth, division, and permeability should have been positively selected for their ability to spread favorable replicating molecules and protocellular functions. Selected protoviral vesicles became gradually dispensable for the spreading of beneficial genes, and viruses evolved from being solely beneficial entities into displaying also a parasitic behavior that exploited cell resources (compare with Sections 1.7 and 1.8). According to this model, when cells became independent units surrounded by a lipid membrane, viruses promoted selection of cells that expressed peptidoglycan molecules on their surface to decrease or prevent virus infection. This transition could mark the onset of an arms race behavior that has been associated with a survival strategy of many present-day viruses. One of the early outcomes might have been the formation of a cell wall, that allowed vertical transmission of genetic information, and rendered the cell metabolically independent. In addition, it provided an osmotic environment suitable for energy production. Genetic information and molecular devices for energy capture, storage, and use were equally important for a sustainable cellular organization. Thus, according to this theory, viruses originated from protoviral elements whose main function was to spread useful genes horizontally. The evolved viruses acted as selective agents to promote microbial evolution and then became established in an increasingly differentiated cellular world (Villarreal, 2005, 2008; Hendrix, 2008; Jalasvuori and Bamford, 2008).

Geological studies indicate that several mass extinctions of a brief duration (less than 100,000 years, a short geological time!) occurred at several points when multicellular organisms populated the Earth. The end-Permian, end-Triassic, and end-Cretaceous extinctions rank among the most drastic, resulting in profound environmental perturbation (Burgess et al., 2014 and references therein). As depicted in the form of red upward and downward curved lines in Figure 1.4, perturbations associated with viral extinctions (due to massive host extinctions) and severe bottleneck events have periodically blurred traces of many viruses we will never know anything about.

Mechanisms implied by each of the five theories summarized here might have had some participation in the origin of viruses as we know them today, and any model will remain speculative for several and obvious reasons. Viruses have not left a fossil record amenable to analysis with current technology. Moreover, viral genomes can evolve at very high rates in response to environmental necessities (Chapter 7), and reconstructions of how early Earth environments might have looked like are at best imprecise. For these reasons, viruses, independently on when they became active actors in the biosphere (Figure 1.4), are unlikely to have maintained molecular signatures that could shed light on their remote past.

1.6 BEING ALIVE VERSUS BEING PART OF LIFE

An often asked question is whether viruses are alive or not. The answer is debated, as reflected in the various definitions of virus, some of which are listed in Box 1.3. Two opposite views coexist. One of them considers viruses as macromolecular aggregates, that is, viruses are regarded as "chemicals." A second proposal is championed by the "ribovirocell" concept of P. Forterre which implies that a viable, virus-infected cell contains two different organisms that coexist symbiotically: the cell that produces virus and the virus itself. Another facet of the same duality is manifested when viruses are considered merely as perturbing chemicals with no participation in the tree of life versus viruses being actually important players in the tree of life. The dual character of viruses as "alive" during intracellular replication and "not alive" outside the cell was stressed in some early literature (e.g., Davis et al., 1968).

As discussed in Section 1.4, if life is best defined as a conglomerate of complementary features, viruses display two of these features: the capacity to replicate and to evolve. Thus, although it is debatable whether viruses qualify as "alive," they are (and most likely have historically been) an integral part of life. Viruses have influenced the tree of life as we know it, with lateral

---

BOX 1.3 SOME DEFINITIONS OF VIRUS

- Viruses are strictly intracellular and potentially pathogenic entities with an infectious phase and (1) possessing one type of nucleic acid, (2) multiplying in the form of their genetic material, (3) unable to grow and to undergo binary fission, and (4) devoid of a Lipmann system (Lwoff, 1957).
- Viruses are entities whose genomes are elements of nucleic acid that replicate inside living cells using the cellular synthetic machinery and causing the synthesis of specialized elements that can transfer the viral genome to other cells (Luria et al., 1978).
- Viruses are replicating microorganisms that are among the smallest of all life forms (first two editions of *Fields Virology*).
- Viruses are transmissible deoxyribonucleic acid (DNA) or ribonucleic acid (RNA) genetic elements that require a cell for multiplication (Domingo and Perales, 2014).

gene transfers—some mediated by viruses—being a key element in its construction and architecture (Ciccarelli et al., 2006). Because of the difficulties of defining "life" unambiguously, the reader might have noticed that in previous sections the question addressed has been "how is life" rather than "what is life." The question of "why" have viruses persisted in the biosphere is addressed next.

1.7 ROLE OF VIRUSES IN THE EVOLUTION OF THE BIOSPHERE

Two general models have been proposed to explain the maintenance of viruses in the biological world:

- Viruses have persisted because they parasitized opportunistically any cellular niche which was compatible with their replication apparatus while the cells remained viable.

In this view, viruses are "selfish" replicating elements that became successful when increasingly efficient polymerization activities became part of their life cycles. Several observations indicate that what was once considered purely "selfish" (also referred to as "junk") DNA may not be that useless after all. In line with these findings, and the intimate connections between cells and viruses, the view of the latter being purely blind, selfish entities appears as less and less tenable. A different issue is that under some particular environments virus display a "selfish" element-like behavior, as one of the outcomes of their having been positively selected.

- The presence of viruses was positively selected because they promoted cellular variation and functional diversification.

This proposal relates to some of the theories of virus origins (Section 1.5) and the early evolution of life that implied a role for virus-like entities. According to this view, viruses, together with other subcellular genetic elements (plasmids, retroelements, etc.), penetrated the genetic material of ancestral cell forms, acted as agents of lateral gene transfers, and modified the expression profiles of the recipient cells. Probably, there has been (and there still is) a constant flow of genes between cells and viruses and other mobile genetic elements. The abundance of endogenous retroviruses in the mammalian genomes is a clear symptom of such a genetic flow. A nonfunctional viral infectivity factor (Vif) (the HIV-1 protein that can counteract the mutagenic activity of the apolipoprotein B mRNA editing complex (APOBEC) proteins; see Section 2.7 in Chapter 2) was found in the remnant of a rabbit endogenous retrovirus termed rabbit endogenous lentivirus type k (Katzourakis et al., 2007). About 8% of the human genome is made of retroviral-like elements. Present-day human endogenous retroviruses probably contribute to pluripotency of human cells (Santoni et al., 2012).

In addition to the promotion of gene transfers to construct key cellular components, viruses probably acted as selective agents for cells to evolve defense mechanisms against viruses, and this may have originated new cellular functions. Also, viruses could favor survival of some cell types over others, based on differential cell susceptibility to virus infection, thus contributing to cellular diversification. The need to escape viral infection may have furnished novel cell surface receptor proteins through selection of cellular escape mutants (Buckling and Rainey, 2002; Saren et al., 2005). Some experimental systems consisting of persistently infected cells in which the cells and the resident virus coevolve (Chapter 6) illustrate how viruses could act as selective agents to promote cellular variation. Such variation would not necessarily involve exchanges of genetic material between the virus and the cells, provided sufficient genetic variation of cells took place.

Multicellular organisms devoid of viral entities should have endured a long-term disadvantage over an alternative scenario with the coexistence of cells and viruses. Selection by viruses need not be restricted to cells, and it can be extended to entire host organisms and their populations. Abundance of hosts may promote viral epidemiological fitness, which is a factor in viral disease emergence (Chapters 5 and 7). Host subpopulations may be selected by their resistance to epidemic outbreaks by highly pathogenic viruses such as in the 1918 influenza pandemics or currently with the AIDS pandemics and Ebola outbreaks in some parts of Africa. Traditionally, plagues decimated the human population and acted as selective agents for differential survival of individuals.

Selfish-opportunistic and selected-functional are not incompatible models of virus maintenance. Once the instruction to replicate had been positively selected, selfish elements could ensue. As we learn about viral and cellular genomics, the current promiscuity and diversity of viruses (Section 1.3) appear as complementary agencies to promote general biological evolution following Darwinian mechanisms. Viruses might have contributed to the DNA replication machinery of cells, to the formation of the eukaryotic cell nucleus, and to a number of developmental processes (Baranowski et al., 2001; Bushman, 2002; Bacarese-Hamilton et al., 2004; Mallet et al., 2004; Villarreal, 2005, 2008; Forterre, 2006a). Cells are a necessity for viruses and viruses are promoters of cell diversity and, as a consequence, of cellular differentiation (compartmentalization and functional specialization).

1.7.1 CURRENT EXCHANGES OF GENETIC MATERIAL

Present-day viruses reveal several mechanisms of exchange of genetic material that might have roots in early cellular evolution. Temperate bacteriophages (the prototypic example being *E. coli* phage λ) integrate their genomic DNA in the DNA of their host bacteria. The uptake of cellular genes by viruses has been amply documented in transducing bacteriophages (those that can transfer DNA from one bacterium to another), as well as in RNA and DNA tumor viruses. Even RNA viruses that are not known to include a reverse transcription step in their replication cycle can incorporate host RNA sequences. Replication-competent, cytopathic variants of bovine viral diarrhea virus (a type species of the genus Pestivirus of the important family of pathogens *Flaviviridae*) can acquire cellular mRNA sequences in their genome, via nonhomologous recombination (Meyers et al., 1989). Insertion of 28S ribosomal RNA sequences into the hemagglutinin gene of influenza virus increased its pathogenicity (Khatchikian et al., 1989). Some defective-interfering particles of Sindbis virus included cellular tRNA sequences at their 5′-ends (Monroe and Schlesinger, 1983). Sequences related to some flaviviruses can persist in an integrated form into the DNA of the insect vectors *Aedes albopictus* and *Aedes aegypti* (Crochu et al., 2004). Endogenous hepatitis B viruses (eHBVs) have been identified in the genomes of birds and land vertebrates (amniotes), crocodilians, snakes, and turtles. The evidence is that eHBVs are more than 207 million years old, and that ancient HBV-like viruses infected animals during the Mesozoic Era (Suh et al., 2014; Figure 1.4). The existence of alternative mechanisms for the integration of viral genetic material into cellular DNA suggests an ancient origin and a selective advantage of exchanges of genetic information in shaping a diverse and adaptable cellular world (Eigen, 1992, 2013; Gibbs et al., 1995; Villarreal, 2005, 2008).

1.7.2 SYMBIOTIC RELATIONSHIPS

Symbiotic or mutualistic interactions involving viruses are frequent in the present-day biosphere. Human endogenous retroviruses can protect human tissues and the developing fetus against infection by some exogenous retroviruses (Ryan, 2004). Some bacteria require bacteriophage to express virulence determinants (Tinsley et al., 2006). Symbiosis can be established between bacteriophages and animals (Barr et al., 2013). The presence of a dsRNA mycovirus in a fungus is essential for the latter to confer heat tolerance to some plants; these findings have defined a three-way symbiosis required for an important phenotypic trait (Marquez et al., 2007; Roossinck, 2013). Several plant RNA viruses delay the symptoms of abiotic stress, such as those produced by drought and frost (dehydration, osmotic stress, and oxidative stress). Protection is mediated by increased levels of osmoprotectants and antioxidants in the infected plants (Xu et al., 2008).

Symbiotic relationships represent a state of local equilibrium between viruses and hosts, triggered by compatibility and occasionally by mutual benefits. An arms race implied by the virus-host interactions described in previous sections might have been the *modus vivendi* for viruses, only interrupted by occasional armistices. Alternatively, symbiotic and mutualistic interactions might have been the norm, only interrupted by occasional defections by killer personalities that have become the key actors of hospital wards and virology textbooks (Roossinck, 2011). It is fashionable to favor the second model, but one can find arguments in favor of either scenario. As discussed in connection with natural counterparts of the transition toward error catastrophe in viruses (Chapter 9), cells have evolved to be able to divert some physiological activities to become part of an innate immune response against viruses. This phenotypic flexibility of cellular functions to the point of being used to confront viruses attests of transient losses of an equilibrated coexistence. Some middle- and long-term equilibrium between virus and host population numbers must be continuously restored by selective events, otherwise this book would not have been written.

The evolutionary origin of defense mechanisms against viruses can be regarded as a response to an excessive number of virus-cell interactions. Superinfection exclusion is one of the mechanisms used by present-day cells to limit replication of a virus when another one is actively replicating (or incorporated) into the same cell. Exclusion has a biochemical interpretation in the competition of two viral entities for cellular resources, but it might have been boosted by early cellular adaptation to limit viral invasions (Chapter 4). Likewise, components of the intrinsic and innate immune response that prevent infection and disease might have been also endowed with activities that promoted cell variation for adaptability. Cellular editing activities such as those displayed by some of the adenosine deaminase acting on double stranded RNA and APOBEC proteins are also part of the innate immune response against some viruses. Not surprisingly, in turn, viruses have evolved multiple functions to counteract the host immune response (Chapter 4).

1.8 VIRUS AND DISEASE

The two opposite views of the activity of viruses in the biosphere (i.e., opportunistic occupation of any suitable cellular niche or intimate cooperative coevolution with host cells) would be expected to produce a different proportion of pathogenic viruses. Opportunistic invasions should lead mainly to disease-prone viruses, while long coevolutionary periods should lead to a dominance of nonpathogenic viruses. Not all viruses that have been characterized are pathogenic and, in fact, only a minority of those that exist might be. However, since only a few of the viral genomes that metagenomic surveys have detected have been characterized, it is not possible to adventure a proportion of beneficial or neutral versus harmful viruses. In the course of investigations on poliomyelitis, a search for related viruses was undertaken, and a number of new viruses later to be known as echoviruses were discovered. They were isolated because they caused cytopathology to cells in culture. The virus-containing samples were from individuals that did not show symptoms of a viral infection. The new isolates were designated as "orphans," meaning viruses without disease. The term echovirus derives from enteric cytopathogenic human orphan virus. Some of them, or their close relatives, were later associated with disease syndromes, but others not. Viruses as diverse as circoviruses, polyoma, or herpesviruses colonize a considerable proportion of animals, and only some of the virus types are the direct cause of disease (e.g., postweaning multisystemic wasting syndrome (PMWS) by porcine circovirus type 2, or cancer by some polyomaviruses, among many other examples). Disease potential is unrelated to viral genome size. PMWS is associated with the smallest mammalian DNA virus genome of only 1.7 Kb, while the almost 100-fold larger herpes virus genomes can coexist with immunocompetent humans without noticeable disease. The fact that a virus is pathogenic or not depends on intricacies of virus-host interactions that are poorly understood.

The more the knowledge of the virus-host interactions progresses, the fuzzier the border between virus being pathogenic or nonpathogenic appears to be. Viruses may not damage essential cell functions, but may affect dispensable cellular functions. Studies by M.B.A. Oldstone and his colleagues on persistent infections of lymphocytic choriomeningitis virus in neuroblastoma cells demonstrated that the resident virus altered a differentiated cell trait while preserving vital functions (Oldstone et al., 1977). This is one of many examples of virus-induced modifications of the so-called "luxury" functions of cells (Oldstone, 1984). Later work unveiled the changes in gene expression that underlie some of these alterations. Again, there is no clear cut distinction between a virus being or not pathogenic. Disease manifestations depend on multiple host and viral factors, including coinfections with viruses or other agents. With the increase of infections by HIV-1 and hepatitis C virus (HCV) experienced during the end of the twentieth century, dual infections with these two viruses are now frequent. The evolution of HCV-associated liver disease (i.e., an increasing degree of liver fibrosis) is accelerated in individuals coinfected with HIV-1. Given the number of viruses still to be discovered (Section 1.3), and the spectrum of mild to severe pathologies that can be associated with "different" variants of the "same" virus (Holland et al., 1992), we cannot anticipate whether most or only a minority (as often assumed) of the viruses that currently replicate in our biosphere are pathogenic or potentially so (Li and Delwart, 2011).

1.9 OVERVIEW AND CONCLUDING REMARKS

Contrary to usual practice in experimental virology, the contents of this chapter have forced a considerable degree of uncertainty and at times speculative argumentation. It was, however, a necessary exercise, since at least part of what we see today in viruses must have roots in the origin of life and the role of viruses in life development during epochs that humans have not witnessed. We are left with trying to reconstruct. This chapter is excitingly as close to physics and chemistry as is to biology. Many of the questions addressed are open to debate and they will probably remain open for a long time. Here are some of them. Concerning the origin of life: peptides or nucleic acids first; replication or metabolism first; sufficient time or not for life to originate on Earth; formation of protocells as an early or late event; low temperature or high temperature; and so on. Concerning the origin of viruses: early replicons or latecomers in a cellular world; first RNA viruses or DNA viruses; first cellular microbes or viruses; viruses or cells as the elements that supplied building blocks for more complex genomes; and so on. Many open questions!

Interestingly, however, the terms "heterogeneity" and "complexity" have appeared several times in this chapter in dealing with concepts coming from physics, chemistry, and biology. In fact, such key words anticipate features of present-day viruses discussed in coming chapters. Complex populations, be them of viruses, cells, protocells, lipid vesicles, primitive replicons, or peptide soups are the raw materials on which natural selection can act. Primitiveness does not imply simplicity. Complexity was likely a constant trait for early and modern life, both for the extinguished viruses we will never know anything about, and for those we strive to understand and control (see Summary Box).

SUMMARY BOX

- It is unlikely that present-day viruses are physically related to the primitive replicons that participated in early life.
- However, key features of viruses may be an inheritance of primitive replicons, notably error-prone replication and spread through membrane structures.
- At least two ancestral positive selection events might have contributed to the development of life: selection of replicating over nonreplicating polymers, and selection of splitting-prone versus nonsplitting-prone membrane vesicles.
- The geological record indicates multiple, brief, and drastic mass extinction events in Earth's history. Such events anticipate mass extinction of the resident viruses. A dynamics of emergences, re-emergences and extinctions might have operated historically at a grand-scale, with mechanisms that may have features that we observe with present-day viruses.

REFERENCES

Abrescia, N.G., Bamford, D.H., Grimes, J.M., Stuart, D.I., 2012. Structure unifies the viral universe. Annu. Rev. Biochem. 81, 795–822.

Adamala, K., Szostak, J.W., 2013. Nonenzymatic template-directed RNA synthesis inside model protocells. Science 342, 1098–1100.

Adami, C., 1998. Introduction to Artificial Life. Springer-Verlag Inc., New York.

Anthony, S.J., Epstein, J.H., Murray, K.A., Navarrete-Macias, I., Zambrana-Torrelio, C.M., et al., 2013. A strategy to estimate unknown viral diversity in mammals. mBio 4: e00598-13.

Bacarese-Hamilton, T., Mezzasoma, L., Ardizzoni, A., Bistoni, F., Crisanti, A., 2004. Serodiagnosis of infectious diseases with antigen microarrays. J. Appl. Microbiol. 96, 10–17.

Bada, J.L., Lazcano, A., 2003. Perceptions of science. Prebiotic soup—revisiting the Miller experiment. Science 300, 745–746.

Baltimore, D., 1980. Evolution of RNA viruses. Ann. N. Y. Acad. Sci. 354, 492–497.

Banfield, W.G., Woke, P.A., Mackay, C.M., Cooper, H.L., 1965. Mosquito transmission of a reticulum cell sarcoma of hamsters. Science 148, 1239–1240.

Baranowski, E., Ruiz-Jarabo, C.M., Domingo, E., 2001. Evolution of cell recognition by viruses. Science 292, 1102–1105.

Barr, J.J., Auro, R., Furlan, M., Whiteson, K.L., Erb, M.L., et al., 2013. Bacteriophage adhering to mucus provide a non-host-derived immunity. Proc. Natl. Acad. Sci. U. S. A. 110, 10771–10776.

Bengtson, S., 1994. In: Early Life on Earth. Nobel Symposium No. 84. Columbia University Press, New York.

Bernal, J.D., 1951. The Physical Basis of Life. Routledge and Kegan Paul, London.

Bernstein, A.S., 2014. Biological diversity and public health. Annu. Rev. Public Health 35, 153–167.

Berzelious, J.J., 1806. Föreläsningar i Djurkemien. vol. 1-2, Tryckte hos Carl Delén, Stockholm.

Biondi, E., Branciamore, S., Fusi, L., Gago, S., Gallori, E., 2007. Catalytic activity of hammerhead ribozymes in a clay mineral environment: implications for the RNA world. Gene 389, 10–18.

Breitbart, M., Miyake, J.H., Rohwer, F., 2004. Global distribution of nearly identical phage-encoded DNA sequences. FEMS Microbiol. Lett. 236, 249–256.

Buckling, A., Rainey, P.B., 2002. Antagonistic coevolution between a bacterium and a bacteriophage. Proc. Biol. Sci. 269, 931–936.

Burgess, S.D., Bowring, S., Shen, S.Z., 2014. High-precision timeline for Earth's most severe extinction. Proc. Natl. Acad. Sci. U. S. A. 111, 3316–3321.

Bushman, F., 2002. Lateral DNA Transfer. Mechanisms and Consequences. Cold Spring Harbor Laboratory Press, Cold Spring Harbor, New York.

Bywater, R.P., 2012. On dating stages in prebiotic chemical evolution. Naturwissenschaften 99, 167–176.

Caetano-Anolles, G., Wang, M., Caetano-Anolles, D., 2013. Structural phylogenomics retrodicts the origin of the genetic code and uncovers the evolutionary impact of protein flexibility. PLoS One 8, e72225.

Caglioti, L., Micskei, K., Palyi, G., 2011. First molecules, biological chirality, origin(s) of life. Chirality 23, 65–68.

Cairns-Smith, A.G., 1965. The origin of life and the nature of the primitive gene. J. Theor. Biol. 10, 53–88.

Canuto, V.M., Levine, J.S., Augustsson, T.R., Imhoff, C.L., 1982. UV radiation from the young Sun and oxygen and ozone levels in the prebiological palaeoatmosphere. Nature 296, 816–820.

Carrara, P., Stano, P., Luisi, P.L., 2012. Giant vesicles "colonies": a model for primitive cell communities. Chembiochem 13, 1497–1502.

Castón, J.R., Carrascosa, J.L., 2013. The basic architecture of viruses. In: Mateu, M.G. (Ed.), Structure and Physics of Viruses. Springer, Dordrecht, Heidelberg, New York, London, pp. 53–75.

Cello, J., Paul, A.V., Wimmer, E., 2002. Chemical synthesis of poliovirus cDNA: generation of infectious virus in the absence of natural template. Science 297, 1016–1018.

Chang, S., 1994. The planetary setting of prebiotic evolution. In: Bengtson, S. (Ed.), Early Life on Earth. Nobel Symposium No. 84. Columbia University Press, New York, pp. 10–23.

Chao, M., 2007. RNA recombination in hepatitis delta virus: implications regarding the abilities of mammalian RNA polymerases. Virus Res. 127, 208–215.

Ciccarelli, F.D., Doerks, T., von Mering, C., Creevey, C.J., Snel, B., et al., 2006. Toward automatic reconstruction of a highly resolved tree of life. Science 311, 1283–1287.

Cooper, G.M., Temin, R.G., Sugden, B.V., 1995. The DNA Provirus. Howard Temin's Scientific Legacy. ASM Press, Washington, DC.

Cowen, R., 2005. History of Life, fourth ed. Blackwell Publishing, Malden, USA.

Crick, F.H., 1968. The origin of the genetic code. J. Mol. Biol. 38, 367–379.

Crochu, S., Cook, S., Attoui, H., Charrel, R.N., De Chesse, R., et al., 2004. Sequences of flavivirus-related RNA viruses persist in DNA form integrated in the genome of Aedes spp. mosquitoes. J. Gen. Virol. 85, 1971–1980.

Danger, G., Plasson, R., Pascal, R., 2012. Pathways for the formation and evolution of peptides in prebiotic environments. Chem. Soc. Rev. 41, 5416–5429.

Davis, B.D., Dulbecco, R., Eisen, H.N., Ginsberg, H.S., Wood, W.B., 1968. Microbiology. Hoeber Medical Division, Harper and Row, New York.

de Duve, C., 2002. Life Evolving. Molecules, Mind and Meaning. Oxford University Press, Oxford.

Demetrius, L.A., 2013. Boltzmann, Darwin and directionality theory. Phys. Rep. 530, 1–85.

Derr, J., Manapat, M.L., Rajamani, S., Leu, K., Xulvi-Brunet, R., et al., 2011. Prebiotically plausible mechanisms increase compositional diversity of nucleic acid sequences. Nucleic Acids Res. 40, 4711–4722.

Diaz-Toledano, R., Ariza-Mateos, A., Birk, A., Martinez-Garcia, B., Gomez, J., 2009. In vitro characterization of a miR-122-sensitive double-helical switch element in the 5′ region of hepatitis C virus RNA. Nucleic Acids Res. 37, 5498–5510.

Dobson, C.M., Ellison, G.B., Tuck, A.F., Vaida, V., 2000. Atmospheric aerosols as prebiotic chemical reactors. Proc. Natl. Acad. Sci. U. S. A. 97, 11864–11868.

Domingo, E., Perales, C., 2014. Virus evolution. Encyclopedia of Life Sciences. John Wiley & Sons, Ltd, Chichester, UK. http://dx.doi.org/10.1002/9780470015902.a0000436.pub3.

Dworkin, J.P., Lazcano, A., Miller, S.L., 2003. The roads to and from the RNA world. J. Theor. Biol. 222, 127–134.

Ehrenfreund, P., Spaans, M., Holm, N.G., 2011. The evolution of organic matter in space. Philos. Transact. A Math. Phys. Eng. Sci. 369, 538–554.

Eigen, M., 1992. Steps Towards Life. Oxford University Press, Oxford.

Eigen, M., 1994. On the origin of biological information. Biophys. Chem. 50, 1.

Eigen, M., 2002. Error catastrophe and antiviral strategy. Proc. Natl. Acad. Sci. U. S. A. 99, 13374–13376.

Eigen, M., 2013. From Strange Simplicity to Complex Familiarity. Oxford University Press, Oxford.

Eigen, M., Schuster, P., 1979. The Hypercycle. A Principle of Natural Self-organization. Springer, Berlin.

Ferris, J.P., Hill Jr., A.R., Liu, R., Orgel, L.E., 1996. Synthesis of long prebiotic oligomers on mineral surfaces. Nature 381, 59–61.

Forterre, P., 2005. The two ages of the RNA world, and the transition to the DNA world: a story of viruses and cells. Biochimie 87, 793–803.

Forterre, P., 2006a. The origin of viruses and their possible roles in major evolutionary transitions. Virus Res. 117, 5–16.

Forterre, P., 2006b. Three RNA cells for ribosomal lineages and three DNA viruses to replicate their genomes: a hypothesis for the origin of cellular domain. Proc. Natl. Acad. Sci. U. S. A. 103, 3669–3674.

Fox, S.W., 1988. The Emergence of Life. Basic Books, New York.

Fox, S.W., Dose, K., 1992. Molecular Evolution and the Origins of Life. W.H. Freeman, San Francisco.

Gedulin, B., Arrhenius, G., 1994. Sources of geochemical evolution of RNA precursor molecules: the role of phosphate. In: Bengtson, S. (Ed.), Early Life on Earth. Nobel Symposium No. 84. Columbia University Press, New York, pp. 91–106.

Gesteland, R.F., Cech, T.R., Atkins, J.F., 2006. The RNA World. Cold Spring Harbor Laboratory Press, Cold Spring Harbor, New York.

Gibbs, A., Calisher, C., García-Arenal, F., 1995. Molecular Basis of Virus Evolution. Cambridge University Press, Cambridge.

Gilbert, W., 1986. The RNA world. Nature 319, 618.

Gleiser, M., Thorarinson, J., Walker, S.I., 2008. Punctuated chirality. Orig. Life Evol. Biosph. 38, 499–508.

Gorbalenya, A.E., 1995. Origin of RNA viral genomes; approaching the problem by comparative sequence analysis. In: Gibbs, A., Calisher, C.H., Garcia-Arenal, F. (Eds.), Molecular Basis of Virus Evolution. Cambridge University Press, Cambridge, pp. 49–66.

Griffith, E.C., Vaida, V., 2012. In situ observation of peptide bond formation at the water-air interface. Proc. Natl. Acad. Sci. U. S. A. 109, 15697–15701.

Hacker, J., Dobrindt, U., 2006. Pathogenomics: Genome Analysis of Pathogenic Microbes. Wiley-VCH Verlag GmbH&Co KGaA, Weinheim.

Hendrix, R.W., 2008. Evolution of dsDNA tailed phages. In: Domingo, E., Parrish, C., Holland, J.J. (Eds.), Origin and Evolution of Viruses, second ed. Elsevier, Oxford, pp. 219–228.

Holland, J.J., 2006. Transitions in understanding of RNA viruses: an historical perspective. Curr. Top. Microbiol. Immunol. 299, 371–401.

Holland, J., Domingo, E., 1998. Origin and evolution of viruses. Virus Genes 16, 13–21.

Holland, J.J., de La Torre, J.C., Steinhauer, D.A., 1992. RNA virus populations as quasispecies. Curr. Top. Microbiol. Immunol. 176, 1–20.

Horneck, G., Baumstark-Khan, C., 2002. Astrobiology, the Quest for the Conditions of Life. Springer, Berlin.

Huotari, J., Helenius, A., 2011. Endosome maturation. EMBO J. 30, 3481–3500.

Inoue, T., Orgel, L.E., 1983. A nonenzymatic RNA polymerase model. Science 219, 859–862.

Jalasvuori, M., Bamford, J.K., 2008. Structural co-evolution of viruses and cells in the primordial world. Orig. Life Evol. Biosph. 38, 165–181.

Jeffares, D.C., Poole, A.M., Penny, D., 1998. Relics from the RNA world. J. Mol. Evol. 46, 18–36.

Joklik, W.K., 1974. Evolution in viruses. In: Carlile, M.M., Skehel, J.J. (Eds.), Evolution in the Microbial World. Cambridge University Press, Cambridge, pp. 293–320.

Joshi, P.C., Aldersley, M.F., Price, J.D., Zagorevski, D.V., Ferris, J.P., 2011. Progress in studies on the RNA world. Orig. Life Evol. Biosph. 41, 575–579.

Joyce, G.F., 2004. Directed evolution of nucleic acid enzymes. Annu. Rev. Biochem. 73, 791–836.

Katzourakis, A., Tristem, M., Pybus, O.G., Gifford, R.J., 2007. Discovery and analysis of the first endogenous lentivirus. Proc. Natl. Acad. Sci. U. S. A. 104, 6261–6265.

Kauffman, S.A., 1993. The Origins of Order. Self-organization and Selection in Evolution. Oxford University Press, New York, Oxford.

Khatchikian, D., Orlich, M., Rott, R., 1989. Increased viral pathogenicity after insertion of a 28S ribosomal RNA sequence into the haemagglutinin gene of an influenza virus. Nature 340, 156–157.

Koonin, E.V., Dolja, V.V., 2013. A virocentric perspective on the evolution of life. Curr. Opin. Virol. 3, 546–557.

Krupovic, M., Bamford, D.H., 2007. Putative prophages related to lytic tailless marine dsDNA phage PM2 are widespread in the genomes of aquatic bacteria. BMC Genomics 8, 236.

La Scola, B., Desnues, C., Pagnier, I., Robert, C., Barrassi, L., et al., 2008. The virophage as a unique parasite of the giant mimivirus. Nature 455, 100–104.

Laybourn-Parry, J., Pearce, D.A., 2007. The biodiversity and ecology of Antarctic lakes: models for evolution. Philos. Trans. R. Soc. Lond. B Biol. Sci. 362, 2273–2289.

Lazcano, A., 1994a. The RNA world, its predecessors, and its descendants. In: Bengtson, S. (Ed.), Early Life on Earth. Nobel Symposium No. 84. Columbia University Press, New York, pp. 70–80.

Lazcano, A., 1994b. The transition form nonliving to living. In: Bengtson, S. (Ed.), Early Life on Earth. Nobel Symposium No. 84. Columbia University Press, New York, pp. 60–69.

Lazcano, A., 2010. Historical development of origins research. Cold Spring Harb. Perspect. Biol. 2(11): a002089.

Lazcano, A., Valverde, V., Hernandez, G., Gariglio, P., Fox, G.E., et al., 1992. On the early emergence of reverse transcription: theoretical basis and experimental evidence. J. Mol. Evol. 35, 524–536.

Li, L., Delwart, E., 2011. From orphan virus to pathogen: the path to the clinical lab. Curr. Opin. Virol. 1, 282–288.

Lopez-Bueno, A., Tamames, J., Velazquez, D., Moya, A., Quesada, A., et al., 2009. High diversity of the viral community from an Antarctic lake. Science 326, 858–861.

Luria, S.E., Darnell Jr., J.E., Baltimore, D., Campbell, A., 1978. General Virology, second ed. John Wiley & Sons, New York.

Lwoff, A., 1957. The concept of virus. J. Gen. Microbiol. 17, 239–253.

Mallet, F., Bouton, O., Prudhomme, S., Cheynet, V., Oriol, G., et al., 2004. The endogenous retroviral locus ERVWE1 is a bona fide gene involved in hominoid placental physiology. Proc. Natl. Acad. Sci. U. S. A. 101, 1731–1736.

Marquez, L.M., Redman, R.S., Rodriguez, R.J., Roossinck, M.J., 2007. A virus in a fungus in a plant: three-way symbiosis required for thermal tolerance. Science 315, 513–515.

Mateu, M.G., 2013. Structure and Physics of Viruses. Springer, Dordrecht, Heidelberg, New York, London.

Maynard Smith, J., Szathmáry, E., 1999. The Origins of Life. From the Birth of Life to the Origins of Language. Oxford University Press, Oxford.

Meyers, G., Rumenapf, T., Thiel, H.J., 1989. Ubiquitin in a togavirus. Nature 341, 491.

Miller, S.L., 1953. A production of amino acids under possible primitive Earth conditions. Science 117, 528–529.

Miller, S.L., 1987. Which organic compounds could have occurred on the prebiotic Earth? Cold Spring Harb. Symp. Quant. Biol. 52, 17–27.

Miller, S.L., Orgel, L.E., 1974. The Origins of Life on Earth. Prentice-Hall, Englewood Cliffs, New Jersey.

Miyakawa, S., Yamanashi, H., Kobayashi, K., Cleaves, H.J., Miller, S.L., 2002. Prebiotic synthesis from CO atmospheres: implications for the origins of life. Proc. Natl. Acad. Sci. U. S. A. 99, 14628–14631.

Monroe, S.S., Schlesinger, S., 1983. RNAs from two independently isolated defective interfering particles of Sindbis virus contain a cellular tRNA sequence at their 5′ ends. Proc. Natl. Acad. Sci. U. S. A. 80, 3279–3283.

Morowitz, H.J., 1992. Beginnings of Cellular Life: Metabolism Recapitulates Biogenesis. Yale University Press, New Haven and London.

Mount, D.W., 2004. Bioinformatics Sequence and Genome Analysis. Cold Spring Harbor Laboratory Press, Cold Spring Harbor, New York.

Murgia, C., Pritchard, J.K., Kim, S.Y., Fassati, A., Weiss, R.A., 2006. Clonal origin and evolution of a transmissible cancer. Cell 126, 477–487.

Nasir, A., Kim, K.M., Caetano-Anolles, G., 2012. Giant viruses coexisted with the cellular ancestors and represent a distinct supergroup along with superkingdoms Archaea. Bacteria and Eukarya. BMC Evol. Biol. 12, 156.

Oldstone, M.B.A., 1984. Virus can alter cell function without causing cell pathology: disordered function leads to imbalance of homeostasis and disease. In: Notkins, A.L., Oldstone, M.B.A. (Eds.), Concepts in Viral Pathogenesis. Springer-Verlag, New York, pp. 269–276 (chapter 38).

Oldstone, M.B., Holmstoen, J., Welsh Jr., R.M., 1977. Alterations of acetylcholine enzymes in neuroblastoma cells persistently infected with lymphocytic choriomeningitis virus. J. Cell. Physiol. 91, 459–472.

Oparin, A.I., 1938. Origin of Life. Dover Publications, New York, English version 1953.

Orgel, L.E., 2002. Is cyanoacetylene prebiotic? Orig. Life Evol. Biosph. 32, 279–281.

Orgel, L.E., 2004. Prebiotic chemistry and the origin of the RNA world. Crit. Rev. Biochem. Mol. Biol. 39, 99–123.

Parker, E.T., Cleaves, H.J., Callahan, M.P., Dworkin, J.P., Glavin, D.P., et al., 2011. Prebiotic synthesis of methionine and other sulfur-containing organic compounds on the primitive Earth: a contemporary reassessment based on an unpublished 1958 Stanley Miller experiment. Orig. Life Evol. Biosph. 41, 201–212.

Pasteur, L., 1861. Memoire sur les corpuscules organizés qui existent dans l'atmosphere. Examen de la doctrine des générations spontanés. Ann. Sci. Nat. 16, 5–98, 4esue series.

Pearse, A.M., Swift, K., 2006. Allograft theory: transmission of devil facial-tumour disease. Nature 439, 549.

Pinheiro, V.B., Taylor, A.I., Cozens, C., Abramov, M., Renders, M., et al., 2012. Synthetic genetic polymers capable of heredity and evolution. Science 336, 341–344.

Quer, J., Martell, A., Rodríguez, R., Bosch, A., Jardi, R., et al., 2008. The impact of rapid evolution of hepatitis viruses. In: Domingo, E., Parrish, C., Holland, J.J. (Eds.), Origin and Evolution of Viruses. Elsevier, Oxford, pp. 303–350.

Ravantti, J., Bamford, D., Stuart, D.I., 2013. Automatic comparison and classification of protein structures. J. Struct. Biol. 183, 47–56.

Ritson, D., Sutherland, J.D., 2012. Prebiotic synthesis of simple sugars by photoredox systems chemistry. Nat. Chem. 4, 895–899.

Robertson, H.D., 1992. Replication and evolution of viroid-like pathogens. Curr. Top. Microbiol. Immunol. 176, 213–219.

Robertson, H.D., 1996. How did replicating and coding RNAs first get together? Science 274, 66–67.

Robertson, M.P., Joyce, G.F., 2012. The origins of the RNA world. Cold Spring Harb. Perspect. Biol. 4(5): a003608.

Robertson, B.H., Jansen, R.W., Khanna, B., Totsuka, A., Nainan, O.V., et al., 1992. Genetic relatedness of hepatitis A virus strains recovered from different geographical regions. J. Gen. Virol. 73, 1365–1377.

Rodin, A.S., Szathmary, E., Rodin, S.N., 2011. On origin of genetic code and tRNA before translation. Biol. Direct 6, 14.

Rohlfing, D.L., Oparin, A.I., 1972. Molecular Evolution: Prebiological and Biological. Plenum Press, New York and London.

Roossinck, M.J., 2011. The good viruses: viral mutualistic symbioses. Nat. Rev. Microbiol. 9, 99–108.

Roossinck, M.J., 2013. Plant virus ecology. PLoS Pathog. 9, e1003304.

Ruiz-Mirazo, K., Briones, C., de la Escosura, A., 2014. Prebiotic systems chemistry: new perspectives for the origins of life. Chem. Rev. 114, 285–366.

Ryan, F.P., 2004. Human endogenous retroviruses in health and disease: a symbiotic perspective. J. R. Soc. Med. 97, 560–565.

Santoni, F.A., Guerra, J., Luban, J., 2012. HERV-H RNA is abundant in human embryonic stem cells and a precise marker for pluripotency. Retrovirology 9, 111.

Saren, A.M., Ravantti, J.J., Benson, S.D., Burnett, R.M., Paulin, L., et al., 2005. A snapshot of viral evolution from genome analysis of the tectiviridae family. J. Mol. Biol. 350, 427–440.

Schrödinger, E., 1944. What is Life? Cambridge University Press, Cambridge.

Schwartz, A.W., Orgel, L.E., 1985. Template-directed synthesis of novel, nucleic acid-like structures. Science 228, 585–587.

Sczepanski, J.T., Joyce, G.F., 2014. A cross-chiral RNA polymerase ribozyme. Nature 515, 440–442.

Selander, R.K., 1976. Genetic variation in natural populations. In: Ayala, F.J. (Ed.), Molecular Evolution. Sinauer Associates Inc., Sunderland, Massachusetts, pp. 21–45.

Shabalina, S.A., Ogurtsov, A.Y., Spiridonov, N.A., 2006. A periodic pattern of mRNA secondary structure created by the genetic code. Nucleic Acids Res. 34, 2428–2437.

Sharp, P.A., 1985. On the origin of RNA splicing and introns. Cell 42, 397–400.

Silver, L.M., 2007. Challenging Nature. HarperCollins, New York.

Stano, P., D'Aguanno, E., Bolz, J., Fahr, A., Luisi, P.L., 2013. A remarkable self-organization process as the origin of primitive functional cells. Angew. Chem. Int. Ed. Engl. 52, 13397–13400.

Suh, A., Weber, C.C., Kehlmaier, C., Braun, E.L., Green, R.E., et al., 2014. Early mesozoic coexistence of amniotes and hepadnaviridae. PLoS Genet. 10, e1004559.

Suttle, C.A., 2007. Marine viruses—major players in the global ecosystem. Nat. Rev. Microbiol. 5, 801–812.

Szathmary, E., 1999. The origin of the genetic code: amino acids as cofactors in an RNA world. Trends Genet. 15, 223–229.

Szostak, J.W., 2011. An optimal degree of physical and chemical heterogeneity for the origin of life? Philos. Trans. R. Soc. Lond. B Biol. Sci. 366, 2894–2901.

Taylor, J., Pelchat, M., 2010. Origin of hepatitis delta virus. Future Microbiol. 5, 393–402.

Tinsley, C.R., Bille, E., Nassif, X., 2006. Bacteriophages and pathogenicity: more than just providing a toxin? Microbes Infect. 8, 1365–1371.

Towe, K.M., 1994. Earth's early atmosphere: constraints and opportunities for early evolution. In: Bengton, S. (Ed.), Early Life on Earth. Nobel Symposium No. 84. Columbia University press, New York, pp. 36–47.

Trail, D., Watson, E.B., Tailby, N.D., 2011. The oxidation state of Hadean magmas and implications for early Earth's atmosphere. Nature 480, 79–82.

Trevino, S.G., Zhang, N., Elenko, M.P., Luptak, A., Szostak, J.W., 2011. Evolution of functional nucleic acids in the presence of nonheritable backbone heterogeneity. Proc. Natl. Acad. Sci. U. S. A. 108, 13492–13497.

van Rij, R.P., Andino, R., 2008. The complex interactions of viruses and the RNAi machinery: a driving force in viral evolution. In: Domingo, E., Parrish, C.R., Holland, J.J. (Eds.), Origin and Evolution of Viruses, second ed. Elsevier, Oxford, pp. 161–182.

Villarreal, L.P., 2005. Viruses and the Evolution of Life. ASM Press, Washington, DC.

Villarreal, L.P., 2008. The Widespread Evolutionary Significance of Viruses. In: Domingo, E., Parrish, C.R., Holland, J.J. (Eds.), Origin and Evolution of Viruses, second ed. Elsevier, Oxford, pp. 477–516.

Wächtershäuser, G., 1988. Before enzymes and templates: theory of surface metabolism. Microbiol. Rev. 52, 452–484.

Wächtershäuser, G., 1994. Vitalists and virulists: a theory of self-expanding reproduction. In: Bengtson, S. (Ed.), Early Life on Earth. Columbia University Press, New York, pp. 124–132.

Whon, T.W., Kim, M.S., Roh, S.W., Shin, N.R., Lee, H.W., et al., 2012. Metagenomic characterization of airborne viral DNA diversity in the near-surface atmosphere. J. Virol. 86, 8221–8231.

Wochner, A., Attwater, J., Coulson, A., Holliger, P., 2011. Ribozyme-catalyzed transcription of an active ribozyme. Science 332, 209–212.

Woese, C.R., 2002. On the evolution of cells. Proc. Natl. Acad. Sci. U. S. A. 99, 8742–8747.

Xu, P., Chen, F., Mannas, J.P., Feldman, T., Sumner, L.W., et al., 2008. Virus infection improves drought tolerance. New Phytol. 180, 911–921.

Yan, M., Wang, L.C., Hymowitz, S.G., Schilbach, S., Lee, J., et al., 2000. Two-amino acid molecular switch in an epithelial morphogen that regulates binding to two distinct receptors. Science 290, 523–527.

Yarus, M., 2010. Life from an RNA World. The Ancestor Within. Harvard University Press, Cambridge, Massachusetts and London, England.

MOLECULAR BASIS OF GENETIC VARIATION OF VIRUSES: ERROR-PRONE REPLICATION

CHAPTER CONTENTS

ABBREVIATIONS

A	adenine
ADAR	adenosine deaminase acting on double-stranded RNA
AIDS	acquired immune deficiency syndrome
AMV-RT	avian myeloblastosis virus reverse transcriptase
APOBEC	apolipoprotein B mRNA editing complex
BHK	baby hamster kidney cells
bp	base pairs
C	cytosine
cDNA	complementary DNA

Virus as Populations. http://dx.doi.org/10.1016/B978-0-12-800837-9.00002-2

CVB3	coxsackievirus B3
DI	defective interfering
DI RNAs	defective-interfering RNAs
E. coli	*Escherichia coli*
FMDV	foot-and-mouth disease virus
G	guanine
HBV	hepatitis B virus
HCV	hepatitis C virus
HIV-1	human immunodeficiency virus type 1
HIV-RT	human immunodeficiency virus reverse transcriptase
IFN	interferon
I	inosine
IRES	internal ribosome entry site
Kb	kilobase
miRNA	micro-RNA
mRNA	messenger RNA
NGS	next generation sequencing
NTP-dependent	nucleoside triphosphate-dependent
OPV	oral poliovirus vaccine
PCR	polymerase chain reaction
PV	poliovirus
RdRp	RNA-dependent RNA polymerase
rRNA	ribosomal RNA
RT	reverse transcriptase
siRNA	small interfering RNA
T	thymine
U	uracil
VSV	vesicular stomatitis virus

2.1 UNIVERSAL NEED OF GENETIC VARIATION

Genetic change was a prerequisite for the early life forms to be generated and maintained (Chapter 1), and it is also a requirement for the evolution of present-day life. We may willingly or inadvertently modify selective pressures, but genetic change is rooted in all replication machineries. Viruses use the same molecular mechanisms of genetic variation than any other form of life: mutation, hypermutation, several types of recombination, and genome segment reassortment. Mutation is observed in all viruses, with no known exceptions. Recombination is also widespread, but its role in the generation of diversity appears to vary among viruses. Its occurrence was soon accepted for DNA viruses, but it was considered uncertain for the RNA viruses. Pioneer studies by P. D. Cooper and colleagues with poliovirus (PV) and by A.M. King and colleagues with foot-and-mouth disease virus (FMDV) provided the first evidence of recombination in RNA. The present perception is that recombination is more widespread than thought only a few decades ago, but that its frequency and the types of genomic forms it generates vary among viruses. For example, as a general rule, it appears that positive strand RNA viruses recombine more easily than negative strand RNA viruses to give rise to mosaic genomes of standard length. Several negative strand RNA viruses, however, can yield defective genomes through recombination, frequently characterized by deletions in their RNA. Genome segment reassortment, the variation event in viruses closest to chromosomal exchanges in sexual reproduction, is obviously privative of segmented viral genomes. The three modes of virus genome variation are not incompatible, and reassortant-recombinant-mutant genomes have been described. The potential for genetic variation of RNA and DNA viral genomes is remarkable, and it is the ultimate molecular mechanism that lies at the origin of the virus diversity delineated in Chapter 1.

2.2 MOLECULAR BASIS OF MUTATION

Mutation is a localized alteration of a nucleotide residue in a nucleic acid. It generally refers to an inheritable modification of the genetic material. In the case of viral genomes, mutations can result from different mechanisms: (i) template miscopying (direct incorporation of an incorrect nucleotide); (ii) primer-template misalignments that include miscoding followed by realignment, and misalignment of the template relative to the growing chain (polymerase "slippage" or "stuttering"); (iii) activity of cellular enzymes (i.e., deaminases), or (iv) chemical damage to the viral nucleic acids (deamination, depurination, depyrimidination, reactions with oxygen radicals, direct and indirect effects of ionizing radiation, photochemical reactions, etc.) (Naegeli, 1997; Bloomfield et al., 2000; Friedberg et al., 2006).

The basis of nucleotide misincorporation during template copying lies mainly in the electronic structure of the bases that make up DNA or RNA. Each base in DNA (adenine, A; guanine, G; cytosine, C; thymine, T) and RNA (with uracil, U instead of T) includes potential hydrogen-bonding donor sites (amino or amino protons) and hydrogen-bonding acceptor sites (carbonyl oxygens or aromatic nitrogens) that contribute to standard Watson-Crick base pairs (Figure 2.1) as well as wobble base pairs (nonstandard Watson-Crick, but fundamental for RNA secondary structure and mRNA translation) (Figure 2.2). The conformation of the purine and pyrimidine bases is highly dynamic. Amino and methyl groups rotate about the bonds that link them to the ring structure. In dilute solution hydrogen bonds are established with water, and they can be displaced by nucleotide or amino acid residues to give rise to nucleotide-nucleotide or nucleotide-amino acid interactions. The strength difference between hydrogen bonds established in a polynucleotide chain with water, and their strength between two bases in separate polynucleotide chains determines whether a double-stranded polynucleotide will be stably formed.

Purine and pyrimidine bases can acquire different charge distributions and ionization states. As a consequence, in addition to the standard Watson-Crick and wobble, other base pairs are found in naturally occurring nucleic acids (notably cellular rRNA and tRNA) and in synthetic oligonucleotides (A-U or A-T Hoogsteen, and A-G, C-U, G-G, and U-U pairs, as well as interactions involving ionized bases). One of the types of electronic redistribution leads to tautomeric changes, such as the keto-enol and amino-imino transitions, which modify the hydrogen-bonding properties of the base; tautomeric imino and enol forms of the standard bases can produce non-Watson-Crick pairs. The proportion of the alternative tautomeric forms can be influenced by modifications in the purine and pyrimidine rings, which in turn can favor either the *syn* or *anti* conformation of a nucleoside, which is defined by the torsion angle of the bond between the 1′ carbon of the ribose and either N1 in pyrimidines or N9 in purines (Figure 2.1). The *anti* conformation is usually the more stable in standard nucleotides and polynucleotides. The transition from the *anti* to the *syn* conformation may alter the hydrogen-bonding properties of the base thereby inducing mutagenesis (Bloomfield et al., 2000; Suzuki et al., 2006). The understanding of conformational effects on the base-pairing tendencies of nucleoside analogs in the

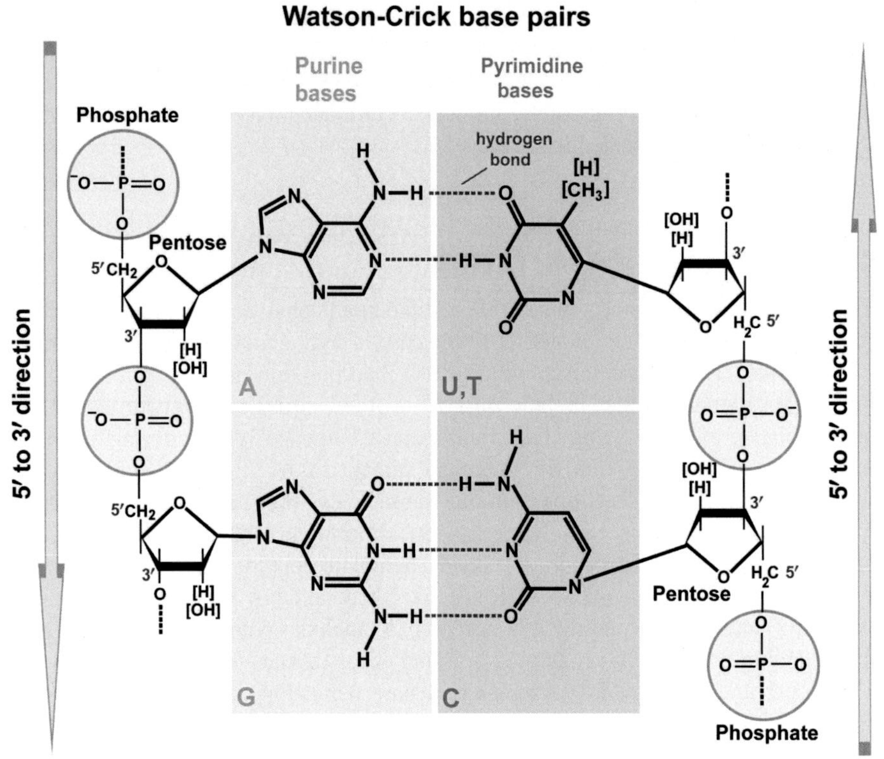

Watson-Crick base pairs

FIGURE 2.1

The standard Watson-Crick base pairs in DNA (A-T and G-C, with deoxyribose as pentose) and RNA (A-U and G-C, with ribose as pentose). Phosphodiester bonds of two potential polynucleotide chains of different polarity (outer arrows) are indicated.

context of the active site of a polymerase is very relevant to the design of specific mutagenic analogs for viral polymerases (Chapter 9).

Base-base interactions are not only responsible for part of the mutations that occur during genome replication, but also for the formation of double-stranded nucleic acids, either within the same polynucleotide chain or between two different chains. Transitions from a coil-like into an organized double-stranded (or other) structure are functionally relevant for both RNA and DNA. In the case of RNA, double-stranded regions in the adequate alternacy with single-stranded regions, determine key catalytic or macromolecule-attracting activities, as for example, the ribozyme activities mentioned in Chapter 1, the internal ribosome entry site (IRES) of several viral and cellular mRNAs, or multitudes of functional RNA-protein interactions currently being unveiled.

Stacking interactions due to electronic interactions (rather than hydrophobic bonds, as once thought) between adjacent bases in the same polynucleotide chain contribute to the stability and conformation of single-stranded nucleic acids, and also partially to duplex stability. Structural transitions due to alternative stacking conformations, particularly within polypurine or polypyrimidine tracts, can affect nucleic acid-protein interactions. In turn, replication machineries (typically including viral and host proteins

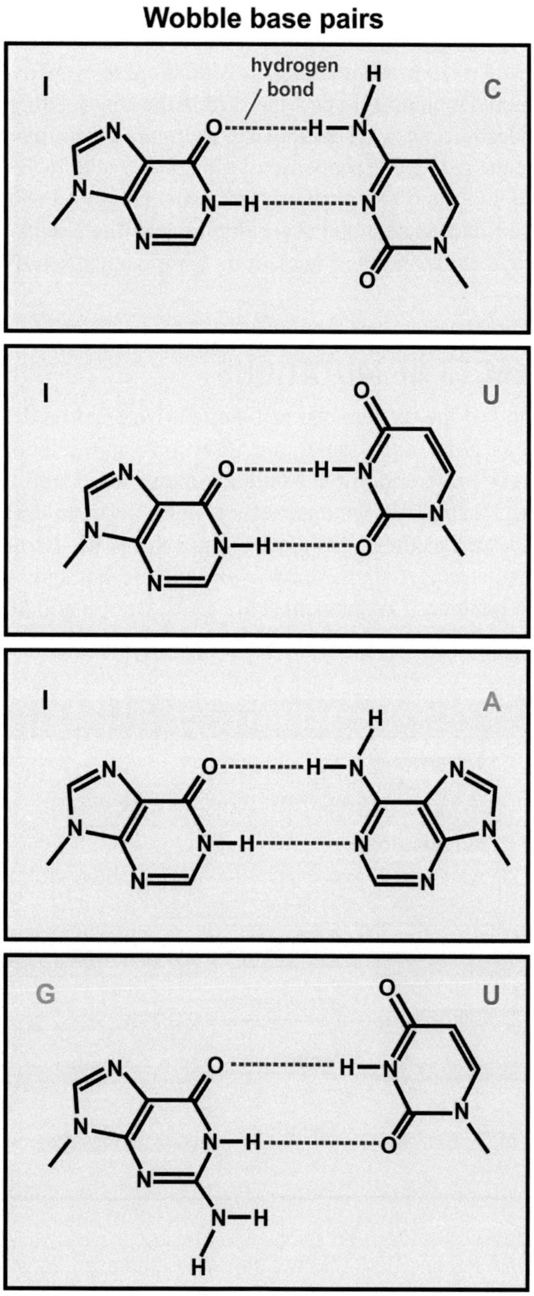

FIGURE 2.2

Examples of a class of non-Watson-Crick base pairs termed wooble base pairs. The drawing is similar to that of Figure 2.1 except that the sugar residues and phosphodiester bonds have been omitted. Hydrogen bonds (discontinuous lines in red) are shown between I (inosine) and C, U and A, and between G and U. Wooble base pairs are important for codon-anticodon interactions, as described in the text.

gathered in membrane structures) would also be expected to be affected by nucleic acid conformations; such effects are important in virology regarding consequences for mutant generation in a given template sequence context. These considerations on structural transitions are relevant to the nonneutral character of silent (also termed synonymous) mutations to be addressed in the next section. Transitions from a single-stranded into a double-stranded nucleic acid structure and the relative stability of the two forms depend on multiple factors that include the nucleotide sequence of the nucleic acid, its being a ribo- or a deoxyribo-polynucleotide, temperature and ionic composition, and ionic strength. Positively charged counterions neutralize negatively charged phosphates, and favor duplex stability [as an overview of physical and chemical properties of nucleic acids and their nucleotide components, see (Bloomfield et al., 2000)].

2.3 TYPES AND EFFECTS OF MUTATIONS

Mutations resulting from any of the mechanism just summarized can be divided in transitions, transversions (both referred to as point mutations), and insertions and deletions (referred to as *indels*) (Figure 2.3). The latter occur preferentially at homopolymeric tracts and also at short, repeated sequences which are prone to misalignment mutagenesis (Figure 2.4). An example is an editing mechanism for some viral mRNAs such as the phosphoprotein mRNA of the *Paramyxovirinae* [(Kolakofsky et al., 2005) and references therein]. Other examples *in vivo* are hot spots for variation in reiterated sequences in complex DNA genomes (Yamaguchi et al., 1998; Barrett and McFadden, 2008; McGeoch et al., 2008), or the insertion of two amino acids (often Ser-Ser, Ser-Gly, or Ser-Ala between residues

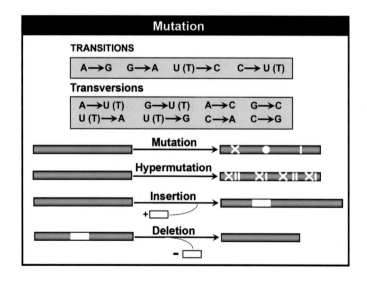

FIGURE 2.3

Major types of mutations in RNA (U) and DNA (T): four transitions and eight transversions. Below, a means to indicate point mutations, insertion or deletions (known as *indels*). A genome is depicted as an elongated rod. Symbols on the rod (cross, circle and line) represent mutations. Hypermutation is generally associated with high frequency of specific mutation types (crosses and lines). A region inserted or deleted from the genome is depicted as an empty rod.

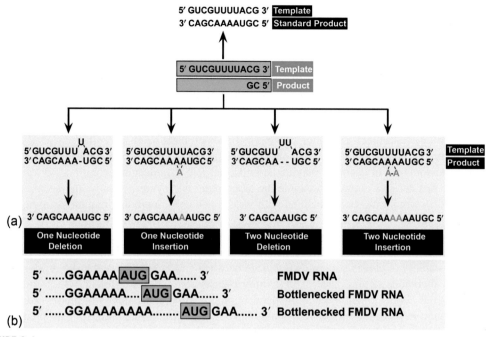

FIGURE 2.4

Misalignment mutagenesis. (a) Production of one or two nucleotide deletion or insertion at a homopolymeric tract in an RNA template. The basis of these events is a displacement of template or product residues during RNA synthesis. (b) A specific example of an internal oligoadenylated extension in the genome of FMDV subjected to plaque-to-plaque transfers. See text for biological implications and references.

69 and 70 of the HIV-1 reverse transcriptase), in concert with HIV-1 resistance to nucleoside inhibitors (Winters and Merigan, 2005) (Chapter 8). The molecular mechanisms involved in the generation of point mutations and *indels* are subject to thermodynamic and quantum-mechanical uncertainty inherent to atomic fluctuations, rendering mutagenesis a highly unpredictable event (Domingo et al., 1995; Eigen, 2013). This fact is relevant to virus evolution because it introduces an element of stochasticity (randomness) in a key motor of evolution: the generation of diversity at the molecular level.

Transition mutations occur more frequently than either transversions or *indels* during virus replication. Nucleotide discrimination at the catalytic site of viral polymerases fits this observation because of the more likely replacement of a purine or pyrimidine nucleotide by its structurally more similar nucleotide. In some cases, however, abundance of *indels* and similar numbers of transitions and transversions have been recorded (Cheynier et al., 2001; Malpica et al., 2002). The molecular bases of such unexpected behavior regarding mutational spectra are not well understood.

The effect of mutations on the structure and function of proteins is extremely relevant to penetrate into the mechanisms that drive virus evolution. Silent or synonymous mutations are those that do not give rise to an amino acid substitution despite being located in an open reading frame (protein-coding region) of a genome. The absence of amino acid substitution is due to the degeneracy of the genetic code: the same amino acid can be coded for by two or more triplets (codons), with the exception of AUG for methionine and UGG for tryptophan. Synonymous mutations are not necessarily selectively neutral, neutral meaning that they have no discernible consequence for any viral function. The assumption that synonymous mutations are selectively neutral, and the fact that the early comparison of nucleotide sequences of homologous genes showed a dominance of synonymous over nonsynonymous mutations, contributed to the foundations of the neutral theory of molecular evolution. This theory attributed the evolution of organisms at the molecular level mainly to the random drift of genomes carrying neutral or quasi (or nearly) neutral mutations (King and Jukes, 1969; Kimura, 1983, 1989). The terms quasi neutral or nearly neutral may seem ambiguous to molecular biologists. However, in the formulation of the neutral theory they had a precise meaning of the selection coefficient being lower than the inverse of the effective population size, with minor variations in the equations of some formulations (Kimura, 1983).

Despite random drift of genomes playing an important role in molecular evolution, evidence gathered over the last decades renders untenable the assumption that synonymous mutations are neutral. Evidence to the contrary has been obtained with viruses and cells, including mutations in the human genome (Novella, 2004; Novella et al., 2004; Parmley et al., 2006; Hamano et al., 2007; Resch et al., 2007; Lafforgue et al., 2011; Nevot et al., 2011). There are several mechanisms by which synonymous mutations can affect virus behavior. They may alter *cis*-acting regulatory elements in viral genomes, decrease the stability of duplex structures stabilized between viral sequences and miRNAs or siRNAs, and promote alterations of viral gene expression or escape from RNA interference (Lafforgue et al., 2011; Nevot et al., 2011). Silent mutation-dependent modification of secondary or higher order structures of genomic RNA or messenger RNAs may have several effects including splicing precision or translation fidelity through the modification of RNA-RNA or RNA-protein interactions. Synonymous codons use different tRNAs for protein synthesis, and different tRNAs do not have the same relative abundance in different host cells. Thus, the rate of protein synthesis, an extremely important phenotypic trait for cells and viruses, can be affected by the frequency of alternative synonymous codons present in a mRNA (Richmond, 1970; Akashi, 2001). Not only does codon bias affect the level of gene expression, but rare codon distribution may regulate the folding of nascent proteins during translation (Makhoul and Trifonov, 2002; Rocha, 2004; Aragones et al., 2010). As a consequence, generation of rare codons by mutation of abundant codons (or vice versa) can modify viral replicative capacity (or fitness, Chapter 5). Rare codons may also limit the fidelity of amino acid incorporation when the frequency of the required aminoacyl-tRNAS is low (Ling et al., 2009; Zaher and Green, 2009; Czech et al., 2010). The frequency of codon pairs in RNA genomes is also a fitness determinant currently under investigation to prepare attenuated viral vaccines. In Chapter 4, additional evolutionary events related to virus-host interactions that might have contributed to codon bias are discussed.

To complicate matters further, a synonymous mutation may be neutral or quasi neutral in one environment, but it may contribute to selection in a different environment, because of the phenotypic effects of RNA structure and codon usage.

Regarding the effects of mutations in viruses (Box 2.1), the following general statements are applicable to both DNA and RNA viruses:

- Truly neutral mutations (i.e., with no influence on the behavior of a virus in any environment) are probably very rare. This applies to synonymous as well as to nonsynonymous mutations.
- Of the nonsynonymous mutations, those leading to chemically conservative amino acid substitutions are more likely to be tolerated than those leading to chemically different amino acids. Tolerance must be distinguished from neutrality. A tolerated mutation may cause a reduction of fitness which is nevertheless compatible with virus replication.

BOX 2.1 THE EFFECTS OF MUTATIONS ON VIRUSES

In Noncoding Regions

Mutations may affect stem-loop or other secondary and higher order structures involved in regulatory processes through nucleic acid-nucleic acid or nucleic acid-protein interactions. The primary sequence in nonstructured, noncoding regions may be also functionally relevant.

In Coding Regions

- Synonymous or silent mutations do not affect the amino acid sequence of the encoded protein.
- Nonsynonymous mutations give rise to an amino acid substitutions in the encoded protein.
- Some mutations may generate a stop codon, leading to a truncated protein.

Regarding Functional Effects

- Neutral mutations are those that have no functional effects.
- Nonneutral mutations can have a broad range of fitness effects: from nearly complete tolerability to lethality.
- Most proteins are multifunctional. A nonsynonymous mutation can affect one but not other functions performed by the same protein.

Context Dependence

The effect of a mutation may be context dependent in two manners: it may be affected by other mutations in the same genome (epistasis) or by other genomes of the surrounding mutant spectrum.

- A conservative amino acid substitution may have important biological effects.
- The effect of any individual mutation is context dependent in two ways: it may depend on other mutations in the same genome (epistasis, see also Section 2.8 and Chapter 5) or on the mutant cloud that surrounds the genome harboring the mutation (complementing or interfering activity).
- The previous points do not deny the influence of random drift of genomes on intrahost- and interhost evolution. The currently most accepted view is that selection and drift occur continuously during virus evolution (Chapter 3).

2.4 INFERENCES ON EVOLUTION DRAWN FROM MUTATION TYPES

The proportion of transition versus transversion mutations may depend initially of the specific replication machinery of a virus that may tend to produce some mutation types preferentially over others. For a given virus, short-term evolution is often reflected in a dominance of transitions which is less apparent when distantly related sequences of the same virus are compared. The effect of evolutionary distance on the transition to transversion ratio was observed in the FMDV sequence comparison carried out in our laboratory over several decades, that ranged from analyses of mutant spectra relative to their corresponding consensus sequence to the comparison of independent viral isolates from a disease outbreak [reviews of the work on FMDV evolution in (Domingo et al., 1990, 2003)]. These two levels of sequence analyses (quasispecies vs.independent isolates) can be further understood by comparing Chapters 3 and 7.

The proportion of synonymous and nonsynonymous mutations that have mediated the diversification of viral genomic sequences that belong to the same phylogenetic lineage is often considered informative of the underlying evolutionary forces. Probably because of the rooted (albeit uncertain) notion that biological function is more likely to reside in protein than in DNA or RNA, the ratio of nonsynonymous substitutions (corrected per nonsynonymous site in the sequence under study) (termed d_n), to the number of synonymous substitutions per synonymous site (termed d_s), ($\omega = d_n/d_s$) is calculated to infer the dominant mode of evolution (Nei and Gojobori, 1986). When $\omega = 1$ the evolution is considered neutral, when $\omega < 1$ purifying (or negative) selection is dominant, and when $\omega > 1$ positive (or directional) selection prevails (Yang and Bielawski, 2000). The types of selection undergone by viruses are discussed in Section 3.4 of Chapter 3.

There are several reasons to be cautious about the significance of ω: (i) synonymous mutations need not be neutral, for reasons discussed in the previous Section 2.3. (ii) In the course of evolution, important but transient events of positive selection (termed episodic positive selection) due to one or a few amino acid substitutions may be accompanied by a larger number of synonymous, tolerated mutations. In this situation, ω will indicate purifying selection despite a critical role of positive selection triggered by one or few nonsynonymous mutations in the evolutionary outcome which are insufficient to compute as $\omega > 1$ (Crandall et al., 1999). (iii) In a striking proof of the above arguments, statistically significant mutational biases led to a value of ω indicative of positive selection in an *in vitro* evolution experiment simulating pseudogene evolution in which positive selection was not possible (Vartanian et al., 2001). (iv) A synonymous change may permit a codon to acquire a relevant nonsynonymous change through a point mutation. The term *quasisynonymous* has been used to describe codons that encode the same amino acid, but that have a different evolutionary potential regarding the accesible amino acids in the encoded protein. Alternative codons for a given amino acid approximate a replicative system to points of sequence space from which a phenotypically relevant change has a different probability (Chapters 3 and 4). (v) Finally, it has to be considered that ω was initially proposed to compare distantly related rather than closely related genomes as is often the case in short-term evolution of viruses (Kryazhimskiy and Plotkin, 2008).

For all these reasons, ω values as a diagnostic test of forces mediating DNA and RNA virus evolution must be regarded only as indirect and suggestive, not as a definitive parameter. Despite these arguments, use of ω to propose a mode of virus evolution continues being surprisingly unchallenged in the literature of virus evolution. We use ω only in a limited way in subsequent chapters, because it does not help in the interpretation of critical evolutionary events regarding viruses. Related shortcomings apply to other tests of neutrality developed to interpret the origin of DNA polymorphisms in the years following the summit of the neutralist-selectionist controversy (Fu, 1997; Achaz, 2009).

2.5 MUTATION RATES AND FREQUENCIES FOR DNA AND RNA GENOMES

Mutation rates quantify the number of misincorporations per nucleotide copied, irrespective of the fate (increase or decrease in frequency) of the error copy produced. A mutation rate for a genomic site measures a biochemical event dictated by the replication machinery and environmental parameters. In contrast, a mutant or mutation frequency describes the proportion of a mutant or a set of mutants in a genome population. The frequency of a mutant will depend on the rate at which it is generated (given by the mutation rate) and on its replication capacity relative to other genomes in the population (Drake and Holland, 1999) (Figure 2.5). A specific mutation may be produced at a modest rate, but then be found at high frequency because the mutation is advantageous for replication in that environment. The converse situation may also occur. Some mutational hot spots (in the sense of genomic sites where mutations tend to occur with high probability) may never be reflected among the repertoire of

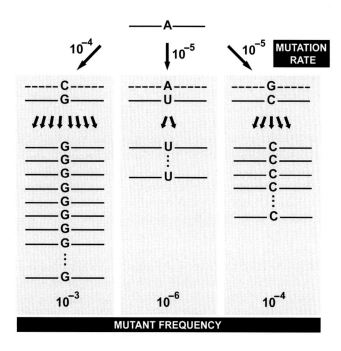

FIGURE 2.5

Scheme that illustrates the difference between mutation rate and mutant frequency. Residue A in a template residue (top) can be misread to incorporate a C, A, or G into the complementary strand (discontinuous lines), at a rate of 10^{-4}, 10^{-5}, and 10^{-5} substitutions per nucleotide, respectively. The replicative capacity of the newly generated templates (with G, U and C, continuous lines) will determine widely different mutant frequencies with $G > C > U$.

mutations found in a genome population because of the selective disadvantage they inflict upon the genome harboring them.

A very significant example is the elongation of an internal oligoadenylate tract located between the two functional AUG initiation codons in the FMDV genome. The homopolymeric tract constitutes a hot spot for variation due to polymerase slippage (Figure 2.4b). The elongation of the internal oligoadenylate was dramatic, but it was only observed when FMDV was subjected to repeated plaque-to-plaque (bottleneck) transfers, not to large population passages. In fact, this drastic genetic modification has not been recorded among natural isolates of the virus. The molecular instruction to elongate the oligoadenylate was very strong because it was observed in many independent biological clones subjected to bottleneck transfers (Escarmís et al., 1996). Despite qualifying as a hot spot for variation, the first event in fitness recovery when the clones were subjected to large population passages was the reversion of the elongated tract to its original size (Escarmís et al., 1999). The interpretation of these findings, to be further analyzed in Chapter 6, is that during plaque-to-plaque transfers the negative selection to eliminate unfit genomes is less intense than during large, highly competitive population passages. Again, a clear molecular instruction to elongate a homopolymeric track may not be reflected in a high frequency of the affected genomes. Therefore, although mutation rates and frequencies for viruses bear some relationship, rates cannot be inferred from frequencies and vice versa (Figure 2.5).

The first calculations of mutation rates for cellular organisms and for some DNA bacteriophages were carried out by J.W. Drake, who pursued comparative measurements that have generally supported a difference between mutation rates for DNA and RNA viruses. The rates estimated for bacteriophages λ and T4 were about 100 times higher than those of their host *E. coli*. An approximately constant rate of 0.003 mutations per genome per replication round was calculated for a number DNA-based microbes (Drake, 1991), an observation sometimes referred to as "Drake's rule." This rather surprising constancy suggests that different DNA organisms have accommodated the template-copying fidelity of their replication machineries to achieve a narrow window in the mutational load measured as mutations fixed per genome, a remarkable fitting of biochemistry with evolutionary needs. The basal mutation rate in mammalian cells has been estimated at about 10^{-10} substitutions per nucleotide and cell generation [reviewed in (Naegeli, 1997; Domingo et al., 2001; Friedberg et al., 2006)] (Table 2.1).

The synonymous mutation rate measured with experimental populations of bacteria has been assumed to reflect the neutral mutation rate (despite limitations explained in Section 2.3). Values for *E. coli* have ranged from 2×10^{-11} up to 5×10^{-9} substitutions per synonymous site per generation (Ochman et al., 1999) with 5×10^{-10} as the most likely estimate (Lenski et al., 2003). The latter value is in agreement with a rate of 3×10^{-10} to 4×10^{-10} substitutions per base pair and generation based on whole genome pyrosequencing of an experimentally evolved lineage of *Myxococcus xanthus* (Velicer et al., 2006). Mutation rates in cells and viruses depend on the replicative machinery (generally a multiprotein complex that includes the relevant viral polymerase with additional viral and host proteins and membrane structures) and on multiple environmental parameters. Whether bacteria are in the exponential or stationary phase of growth can affect intracellular metabolites and proton exchange rates which, in turn, may alter the proportion of tautomeric forms in nucleotides and misincorporation tendencies (Friedberg et al., 2006). The sequence context of the template nucleic acids (presence of repeated sequences that can induce misalignment mutagenesis or G-C vs. A-T rich regions in relation to relative nucleotide substrate abundances, etc.) may impel or attenuate mutability. Insertion elements may enhance mutation rates at neighboring sites in a bacterial genome (Miller and Day, 2004). Despite these influences, vesicular stomatitis virus (VSV) displayed comparable mutation rates in several host cells (Combe and Sanjuan, 2014) suggesting that there is a limited range of average error rates needed for a virus to maintain fitness (Chapters 5 and 9).

In addition to the general environmental and sequence context consequences for template-copying fidelity that may affect any genome type, mutation rates for DNA viruses will be influenced also by: (i) whether the DNA polymerase that catalyzes viral DNA synthesis includes or lacks a functional proofreading-repair activity. High copying fidelity is typical of DNA polymerases involved in cellular

Table 2.1 Mutation Rates and Frequencies for RNA and DNA Genomes	
Viriods	2×10^{-3}
RNA viruses	10^{-5} to 10^{-3}
Retroviruses	10^{-6} to 10^{-4}
DNA viruses	10^{-8} to 10^{-3}
Cellular DNA	10^{-9} to 10^{-11}

Values are expressed as substitutions per nucleotide. The range of values is the most likely according to several independent studies. No distinction is made between mutation rates and frequencies. See text for comments and references.

DNA replication, and low copying fidelity is generally a feature of DNA polymerases involved in DNA repair (Friedberg et al., 2006). Thus, repair of lesions that by themselves might not be mutagenic may lead to the introduction of mutations during the error-prone repair process. (ii) Expression of proteins active in repair encoded in the viral genome, such as uracil-DNA glycosylase, DNA repair endonucleases, etc. (iii) The mechanism of viral DNA replication, particularly the occurrence of double-stranded versus single-stranded DNA in replicative intermediates. (iv) The intracellular site of replication and the availability of postreplicative DNA repair proteins (regarding both intracellular location and concentration) to the viral replication factories. Little is known of the spatial relationships and relative affinities of cellular and viral proteins and structures that may critically affect polymerase fidelity. Comparative measurements of mutation rates at specific genome sites of DNA viruses are needed, as a first step to define the cellular and biochemical influences on the fidelity of DNA virus genome replication.

General genetic variability affecting the entire virus genome should be distinguished from localized variability at hot spots in a genome. Even the extremely complex human genome shows genetic instability at specific loci, some associated with genetic disease (Domingo et al., 2001; Alberts et al., 2002; Bushman, 2002). Genome size is a parameter pertinent to biological behavior, not only because it imposes a commensurate copying fidelity, but also because it affects the impact of genetic heterogeneity within infected organisms, and upon invasion of new hosts (Chapter 3).

Mutation frequencies measured by subjecting virus to a specific selective agent (e.g., mutants that escape the neutralizing activity of a monoclonal antibody or mutants that escape inhibition by a drug) span a broad range of values (10^{-3} to 10^{-8}) for DNA and RNA viruses (Smith and Inglis, 1987; Sarisky et al., 2000; Domingo et al., 2001) (Table 2.1). The technical details of any procedure used to calculate a mutation frequency should be carefully evaluated to translate its meaning to the genome level. Important variables are the efficacy of the antibody or drug (which will be concentration dependent) or the possibility of phenotypic hiding-mixing in the escape mutants to be quantified (Holland et al., 1989; Valcarcel and Ortin, 1989). Unexpected low levels of escape mutants (that would imply $<10^{-6}$ substitutions per nucleotide) for an RNA virus can mean either a general or site-specific high polymerase fidelity, a selective disadvantage of the genome that harbors the mutation or, when a phenotypic alteration is measured, the requirement of two or more mutations to produce the alteration. Conversely, a high mutation frequency for a DNA virus whose replication is catalyzed by a high fidelity DNA polymerase may mean that either repair activities were not functional or that the mutant displayed a selective advantage and overgrew the wild type prior to the measurement of its frequency. Mathematical treatments that take into account reversion of a low fitness mutant and its competition with wild-type virus have been used to calculate mutation rates (Batschelet et al., 1976; Coffin, 1990).

Despite difficulties and limitations in the calculations, independent genetic and biochemical methods with different viruses support mutation rates for RNA viruses in the range of 10^{-3} to 10^{-5} substitutions per nucleotide copied [as representative articles and reviews see (Batschelet et al., 1976; Domingo et al., 1978, 2001; Steinhauer and Holland, 1986; Eigen and Biebricher, 1988; Varela-Echavarria et al., 1992; Ward and Flanegan, 1992; Mansky and Temin, 1995; Preston and Dougherty, 1996; Drake and Holland, 1999; Sanjuan et al., 2010)] (Table 2.1). Historically, a few early studies indicated unusual low mutation rates or frequencies for some RNA viruses. As discussed in some of the reviews listed above, there are technical reasons that suggest that such values were probably underestimates of the true average mutations rates or frequencies. Obviously, it cannot be excluded that some genomic sites or viruses under a given environment are unusually refractory to mutations, but most evidence supports the range of values listed in Table 2.1. The near million-fold higher mutation rates for RNA viruses than cellular DNA, whose biological implications were presciently anticipated by J. Holland and colleagues (Holland et al., 1982), have been confirmed. That is, for RNA viruses of genome length between 3Kb and 32 Kb, an average of 0.1-1 mutations are introduced per template molecule copied in the replicating population. Unless most mutations impeded viral replication, a continuous input of mutant genomes is expected, as indeed found experimentally (Chapter 3).

High mutation rates for RNA genomes are also supported by measurements of template-copying fidelity by RNA polymerases, reverse transcriptases, and DNA polymerases devoid of 3′-5′ proofreading exonuclease (or under conditions in which such exonuclease is not functional) [(Steinhauer et al., 1992; Varela-Echavarría et al., 1992; Mansky and Temin, 1995; Domingo et al., 2001; Friedberg et al., 2002, 2006; Menéndez-Arias, 2002), and references therein]. *In vitro* fidelity tests may be based on genetic or biochemical assays using homopolymeric or heteropolymeric template-primers. Measurements include the kinetics of incorporation of an incorrect versus the correct nucleotide directed by a specific position of a template or the capacity of a polymerase to elongate a mismatched template-primer 3′ end, [these and other assays have been reviewed (Menéndez-Arias, 2002)] (see also Section 2.6). Differences between related enzymes (i.e., AMV RT is more accurate than HIV-1 RT), and the fact that amino acid substitutions in the polymerases affect nucleotide discrimination, demonstrate that proofreading-repair activities together with the structure of the polymerase and replication complexes are determinants of template-copying fidelity.

2.6 EVOLUTIONARY ORIGINS, EVOLVABILITY, AND CONSEQUENCES OF HIGH MUTATION RATES: FIDELITY MUTANTS

The amino acid substitutions in the core polymerase that affect fidelity can be located either close to or away from the active site of the enzyme. The change in fidelity can reach almost one order of magnitude, but virus viability is not compromised. Thus, error rates themselves can be subjected to selection, as supported by theoretical studies on evolvability (Earl and Deem, 2004). It is not clear whether mutation rates of viruses have evolved to procure a balance between adaptability and genetic stability, or whether other selective constraints have imposed the observed values. It has been suggested that because of the generally deleterious nature of most mutations, the adaptive value of the high mutation rates for RNA viruses is debatable, and that there might have been a trade-off between replication rate and copying fidelity (Elena and Sanjuan, 2005). Mutation rates would be a consequence of rapid RNA replication, and an increase in copying fidelity would come at a cost, resulting in a lower replication rate. A connection between elongation and error rate has been suggested by results with some viral and cellular polymerases [review (Kunkel and Erie, 2005)]. In an early study with the poliovirus RdRp *in vitro*, an increase in the error frequency was observed when the pH-Mg^{2+} ion conditions were modified, and the decreased fidelity correlated with increased RNA elongation rate (Ward et al., 1988). A possible connection between elongation rate and copying fidelity cannot be ruled out, but current evidence points to template-copying fidelity as being the result of multiple factors, not necessarily linked to the rate of genome replication (Vignuzzi and Andino, 2010; Campagnola et al. 2015).

Many lines of evidence support an adaptive value of high mutation rates for RNA viruses *per se*, independently of their biochemical origins. A poliovirus mutant, whose RdRp displayed a three- to five-fold higher fidelity than the wild-type enzyme, replicated at a slightly lower rate than wild-type virus in cell culture, but displayed a strong selective disadvantage regarding invasion of the brain of susceptible mice (Pfeiffer and Kirkegaard, 2005; Vignuzzi et al., 2006). The impediment to cause neuropathology was due to the limited complexity of the mutant spectrum since its broadening through mutagenesis restored the capacity to produce neuropathology. These and other studies have provided evidence that mutant spectrum complexity, by virtue of its impact on fitness, can be a virulence determinant. The work by M. Vignuzzi, J. Pfeiffer, R. Andino, C. Cameron, K. Kirkegaard, and their colleagues on poliovirus fidelity mutants opened a much needed branch of research in virus evolution and quasispecies implications. As a proof of this statement, the field of fidelity mutants is rapidly expanding, and references to the information they provide will be made in several chapters.

Theoretical models and experimental observations suggest that mechanisms for error correction had to evolve to maintain functionality of increasingly complex genomes (Swetina and Schuster, 1982; Eigen and Biebricher, 1988; Domingo et al., 2001; Eigen, 2002, 2013) (here complexity means genome size, provided no redundant information is encoded). The coronaviruses have the largest genomes among the known RNA viruses, with 30-33 kb. This is about 10-fold more genetic information than encoded in the simple RNA bacteriophages such as MS2 or Qβ. Coronaviruses are replicated by complex RNA-dependent RNA polymerases which include a domain that corresponds to a 3′-5′ exonuclease, proofreading-repair activity. The protein displays exonuclease activity *in vitro*, and its inactivation affects viral RNA synthesis (Minskaia et al., 2006), and results in increases of about 15-fold in the average mutation frequency (Eckerle et al., 2007, 2010). A coronavirus mutant devoid of this repair function is more susceptible to lethal mutagenesis (Smith et al., 2013; Smith and Denison, 2013), as expected from a connection between replication accuracy and proximity to an error threshold for the maintenance of genetic information (Chapter 9). Thus, it is likely that a proofreading activity evolved (or was captured from a cellular counterpart) in RNA genomes whose genomic complexity was in the limit compatible with the fidelity achievable by standard RNA replicases. It would be interesting to discover new RNA viruses with a single RNA molecule longer than 30 Kb as genome to analyze whether they have evolved more accurate core polymerases or exhibit a proofreading-repair function during replication. Toward the other end of the RNA size scale, viroid RNAs display a mutation rate higher than (or close to the highest) recorded for RNA viruses, consistent with the correlation between genome size and template-copying accuracy (Gago et al., 2009).

Studies with bacteria have identified some of the factors that successively increase copying fidelity. It has been estimated that during *E. coli* DNA replication the error rate would be 10^{-1} to 10^{-2} mutations per nucleotide copied if accuracy relied only upon the strength of interactions provided by base pairing (Section 2.2). The error rate would decrease to 10^{-5} to 10^{-6} with base selection and proofreading-repair, to about 10^{-7} with the contribution of additional proteins present in the replication complex, and to about 10^{-10} misincorporations per nucleotide with the participation of postreplicative mismatch correction mechanisms (Naegeli, 1997; Kunkel and Erie, 2005; Friedberg et al., 2006). The error rate of the bacteriophage φ29 DNA polymerase is about 10^{-6} without the proofreading exonuclease activity, and it decreases to 10^{-8} with the correcting activity [(de Vega et al., 2010) and references therein]. Postreplicative repair pathways act on double-stranded DNA, but not (or very inefficiently) on RNA or DNA-RNA hybrids. Therefore, the known postreplicative repair systems that operate in cellular DNA do not make a significant contribution to error correction in RNA viruses.

The importance of copying fidelity for complex genomes is reflected in the fact that more than 100 proteins are directly or indirectly involved in repair of the human genome. Elevated mutation rates in the range of those operating for RNA viruses would be lethal for mammalian genomes. Localized genetic modification occur physiologically in processes such as somatic hypermutation and class-switch recombination in B cells of the germinal centers, as mechanisms of diversification of immunoglobulin genes (Upton et al., 2011). Chromosomal instability has long been associated with cancer (Gatenby and Frieden, 2004; Stratton et al., 2009). Surveys have been (and are currently being) used to identify genes associated with chromosomal instability and their role in aging and disease (Aguilera and Garcia-Muse, 2013) (see also Chapter 10). While uncontrolled high mutability is deleterious for differentiated cellular organisms, it constitutes a *modus vivendi* for a great majority of viruses.

Despite its attractiveness, definitive proof of the hypothesis of a direct relationship between error rate and limited genome complexity will require additional functional and biochemical studies. Exceptions to the absence of repair activities in simple genetic elements have been described. A satellite RNA of the plant virus turnip crinkle carmovirus evolved a 3′-end RNA repair mechanism. It implicates the synthesis of short oligoribonucleotides by the viral replicase using the 3′-end of the viral genome as template. The mechanism consists probably of template-independent priming at the 3′-end of the damaged RNA to generate wild type, negative strand, and satellite RNA (Nagy and Simon, 1997). A reversible, NTP-dependent excision of the 3′residue of the nascent nucleic acid product has been described in some retroviruses and hepatitis C virus (Meyer et al., 1998; Jin et al., 2013). This activity is important for drug resistance and it may also modulate the overall fidelity of some polymerases. It cannot be excluded that some type of point mutation correction may operate in RNA genetic elements of less than 30 Kb. Such putative mechanisms may even diminish mutation rates that would otherwise be prohibitively deleterious, and they do not overshadow high mutation rates as a feature of RNA and some DNA genomes (Table 2.1).

Limited copying fidelity in the absence of proofreading-correction mechanisms can be regarded as an unavoidable consequence of the molecular mechanisms involved in template copying by viral polymerases. Most nucleic acid polymerases share a structure that resembles a right hand, with fingers, palm, and thumb domains (Figure 2.6a). Three-dimensional structures of viral RdRps and RTs indicate that interactions between the incoming nucleotide or residues of the template-primer with amino acids of the polymerase must permit displacement of the growing polymerase chain along the channel located at the palm domain of the polymerase (Steitz, 1999; Ferrer-Orta et al., 2006) (Figure 2.6b). If interactions around the catalytic site to ensure the correct nucleotide incorporation were so strong as to preclude misincorporations, the movement of the growing polynucleotide chain would be hampered. Again, there is a match between biochemical and evolutionary needs.

The orientation of the triphosphate moiety of the incoming nucleotide substrate is important for nucleotide incorporation (Menéndez-Arias, 2002; Graci and Cameron, 2004; Ferrer-Orta et al., 2009). One of the several steps involved in nucleotide incorporation is the formation of a ternary complex (polymerase with template-primer and the incoming nucleotide) that undergoes a conformational change (reorientation of the divalent ion-complexed triphosphate moiety of the incoming nucleotide). This conformational change activates the complex for phosphoryl transfer, to link the nucleoside-monophosphate to the 3′-terminus of the primer (or growing chain). Steps involved in the nucleotide incorporation are represented in Figure 2.6c. Both the conformational change and the relative rate of phosphoryl transfer for an incorrect nucleotide versus the correct nucleotide influence the error rate at each site of the growing chain. Critical kinetic constants in Figure 2.6c that are determined experimentally to quantify relative nucleotide incorporations and misincorporations are $K_{d,app}$ (expressed as µM), k_{pol} (expressed as s^{-1}), and the ratio $k_{pol}/K_{d,app}$ ($\mu M^{-1} s^{-1}$) termed the catalytic efficiency. The ratio of $k_{pol}/K_{d,app}$ for the incorporation of an incorrect nucleotide to $k_{pol}/K_{d,app}$ for a correct nucleotide gives

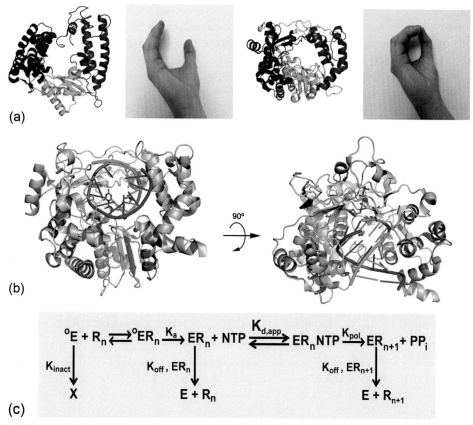

FIGURE 2.6

Polymerization by viral polymerases. (a) The structure of nucleic acid polymerases resembles that of a right hand. This was first evidenced with the structure of the Klenow fragment of the *E. coli* DNA polymerase I by T.A. Steitz and colleagues in 1985. On the left the Klenow fragment is represented with colored fingers, palm and thumb domains, next to an open right hand. The structure on the right is that of PV RdRp, next to a closed right hand. Courtesy of N. Verdaguer and L. Vives-Adrián (the hand is that of L. Vives-Adrián). (b) The structure of ternary complex between FMDV 3D, an RNA molecule and UTP as the substrate (PDB id. 2E9Z). The left panel is a front view of the complex, depicting the polymerase chain as a yellow ribbon, the RNA in dark blue (template) and cyan (primer). The incoming UTP and the pyrophosphate product are shown in atom type, and two Mg^{2+} ions as magenta balls. The right panel is the same complex in a top-down orientation. (Figure courtesy of C. Ferrer-Orta and N. Verdaguer). (c) Scheme of minimum number of steps involved in nucleotide incorporation. The first step consists of the binding of polymerase ^{0}E to the template-primer R_n (elongated up to nucleotide n) to form a complex $^{0}ER_n$. Formation of the activated complex ER_n is governed by the rate constant $k_{assembly}$ (k_a). The activated ER_n complex binds a nucleotide NTP with an apparent binding affinity given by $K_{d,app}$ to form the ER_nNTP complex. Catalysis to incorporate covalently the NTP to the growing primer chain to yield ER_{n+1} and pyrophosphate (PP_i) is governed by the rate constant k_{pol}. Other constants depicted in the scheme are the inactivation rate constant (k_{inact}) of ^{0}E, and dissociation of E from RNA (k_{off}, ER_n and k_{off}, ER_{n+1}). Based on (Arias et al., 2008), and previous studies with PV polymerase 3D by C. E. Cameron and his colleagues.

the frequency of that particular misincorporation, and an assessment of polymerase fidelity [(Castro et al., 2005) and references therein]. Modifications of polymerase residues by site-directed mutagenesis, combined with comparisons of the relevant structures, have identified critical amino acid residues involved in template-copying fidelity.

Viruses displaying higher than average template-copying fidelity have been frequently obtained by selecting viruses resistant to mutagenic nucleotide analogs such as the antiviral agent ribavirin (Beaucourt and Vignuzzi, 2014). Limited incorporation of a deleterious nucleotide can be attained either through a specific discrimination against the analog (ribavirin or other) or through a general decrease of all types of misincorporations, that is, a high fidelity phenotype. Structural modifications of viral polymerases that lead to high fidelity have inspired the design and testing of additional substitutions. In this manner, mutant viral polymerases showing either an increase or decrease of copying fidelity achieved through a single amino acid substitution have been characterized (Wainberg et al., 1996; Menéndez-Arias, 2002; Mansky et al., 2003; Pfeiffer and Kirkegaard, 2003; Arnold et al., 2005; Domingo, 2005; Vignuzzi et al., 2006; Coffey et al., 2011; Gnadig et al., 2012; Meng and Kwang, 2014; Rozen-Gagnon et al., 2014; Borderia et al., 2016). These results imply that the capacity of virus to evolve at higher or lower rates than their ancestors is easily achievable through modest numbers of mutations (limited movements in sequence space, Chapter 3), again emphasizing the evolvability of mutation rates.

2.7 HYPERMUTAGENESIS AND ITS APPLICATION TO GENERATE VARIATION: APOBEC AND ADAR ACTIVITIES

Some viral genomes either isolated from biological samples or evolved in cell culture show biased mutation types (e.g., monotonous $G \rightarrow A$ or $C \rightarrow U$ substitutions in the same genome), generally at frequencies of around 10^{-2} substitutions per nucleotide (10- to 1000-fold higher than standard mutation rates and frequencies) (Table 2.1). Biased hypermutation was first observed in some defective-interfering (DI) RNAs of vesicular stomatitis virus (VSV) (Holland et al., 1982), and in variant forms of measles virus, associated with postmeasles neurological disease such as subacute sclerosing panencephalitis (Cattaneo and Billeter, 1992). Hypermutation is mainly due to the activity of cellular deaminases such as, the apolipoprotein B mRNA and the editing complex (APOBEC), or the adenosine deaminase acting on double-stranded RNA (ADAR) families, that are involved in cellular editing and regulatory functions (Sheehy et al., 2002; Santiago and Greene, 2008; Nishikura, 2010; Stavrou et al., 2014). In the event of a viral infection, such cellular functions can become part of an innate defense mechanism against the invading virus. Viral proteins (i.e., Vif in HIV-1) bind some APOBEC proteins thus inhibiting mutagenesis and permitting virus survival (Sheehy et al., 2002). In oncoretroviruses, retroviruses, and hepatitis B virus (HBV), the APOBEC-3 cytidine deaminase acts on single-stranded DNA and results mainly in $G \rightarrow A$ and $C \rightarrow U$ hypermutation, that may affect 40% to 100% of the G residues. The preferred sequence context for G hypermutation in HIV-1 observed *in vivo* is $GpA > GpG > GpT \approx GpC$. The specific dinucleotide context of the hypermutated sites provides a means to distinguish genomes that have undergone hypermutation by cellular activities from those that are heavily mutated by other mechanism such as the action of mutagenic agents (Chapter 9).

The ADAR-associated hypermutation was identified in negative strand RNA riboviruses, and results mainly in $A \rightarrow G$ and $U \rightarrow C$ hypermutation. It is originated by $A \rightarrow I$ (inosine) modification in double-stranded viral RNA catalyzed by ADAR-1 L, one of more than one hundred proteins inducible by type I IFN (Maas et al., 2003). Inosine can be recognized as G by the replication machinery (Valente and Nishikura, 2005), although it can form wobble base pairs also with A and U (Figure 2.2).

There are additional mechanisms of hypermutagenesis. Higher than average mutation frequencies can occur as a result of replication in the presence of biased concentrations of the standard nucleotide substrates; this has been applied to the *in vitro* generation of genes mutated at frequencies of 10^{-1} to 10^{-2} (mutagenic PCR), as a powerful tool to study sequence-function relationships and functional robustness of nucleic acids and proteins (Meyerhans and Vartanian, 1999). Error-prone PCR has been used in experiments of *in vitro* evolution of nucleic acid enzymes to generate heterogeneous collections of nucleic acid sequences to select for molecules capable of catalyzing specific reactions (Joyce, 2004) (Chapter 1).

2.8 ERROR-PRONE REPLICATION AND MAINTENANCE OF GENETIC INFORMATION: INSTABILITY OF LABORATORY VIRAL CONSTRUCTS

High mutation rates have practical implications in laboratory studies on the behavior of virus mutants obtained by molecular cloning of a biological sample, or constructed by site-directed mutagenesis. A transition mutation that causes a strong fitness decrease but that still allows residual RNA genome replication, will most likely revert following infection or transfection of cells with the mutant construct. Double or triple mutants (preferentially including transversions) should be engineered (when possible according to the genetic code) to study the behavior of a viral mutant with an amino acid replacement of interest that may produce a fitness decrease. As an example, a $C \rightarrow U$ transition found in an open-reading frame of an RNA virus may convert a Pro into a Ser (CCG $\rightarrow$ UCG). Since Ser will revert to Pro through a $U \rightarrow C$ transition in the triplet (a common type of misincorporation by most polymerases), Ser should be engineered to be encoded by AGU; in the course of replication, reversion to Pro would require at least two transversions since the codons for Pro are CCU, CCC, CCA, or CCG. Thus, if effects derived from the difference in primary sequence of the RNA or codon bias do not intervene in the behavior of the viral genome, codons with a high genetic barrier to reversion should be engineered for studies involving viral replication.

In general, deletions revert at a much lower frequency than point mutations, and, when appropriate for the question under study, a deletion should be introduced within the gene of interest to probe gene function in reverse genetic studies. High mutation rates also imply that infection or transfection with debilitated mutant viruses may result in progeny with sequences that differ from the input. V. I. Agol and colleagues have coined the term *quasi-infectious* to refer to mutant viruses that are capable of yielding progeny, but the progeny differs from the initial genome (pseudorevertants) (Gmyl et al., 1993). The difference between the input mutant and the rescued progeny virus will depend on the type of genetic lesion in the input virus and its consequences for replication. A single point mutation that decreases replication is likely to lead to a true revertant (return to the original sequence). If the same reversion depends on two or more mutations, a true revertant will be generated only after extended replication to achieve the required exploration of sequence space (Chapter 3). In this case, selection of compensatory mutations elsewhere in the genome (sometimes referred to as second site revertants) is likely. The term compensatory is often used for mutations that compensate the deleterious effects of other mutations. A typical example is a mutation that decreases the stability of a stem in an RNA stem-loop that functions as a *cis*-acting element. Frequently, a compensatory mutation occurs that restores a stable stem. Thus, an engineered virus may produce progeny which differs from the parent. There is an uncertainty regarding the representativeness of a virus mutant or construct when the objective is to infer the behavior of genetic lesions in the natural context. If a substantial loss of replicative capacity is produced by a drastic genetic change (an *indel*, loss of a stem-loop structure, etc.) selection of a true revertant becomes extremely unlikely. The compensatory generation of alternative structures (or constellations of point mutations) that restore replication (partially or completely) becomes an interesting and informative possibility.

Procedures to copy an entire viral RNA genome into a cDNA for reverse genetics studies are now available (Fan and Di Bisceglie, 2010). If for technical reasons an infectious cDNA clone is constructed from several molecules which were copied from different genomes present in the mutant spectrum, the ligation product may be transcribed into an RNA which is not infectious. This is because while some constellations of mutations may be compatible with infectivity, others may not, or may allow limited, suboptimal replication, thus favoring the selection of additional mutations or reversions. The effect of different mutations in the same genome is often referred to as epistasis. Mutations that reinforce each other with regard to a viral function are said to produce positive epistasis and those that interfere with each other produce negative epistasis. Thus, a mutation may have an effect on its own, and an effect that depends on accompanying mutations in the same genome (also mentioned regarding the context-dependent effect of mutations in Section 2.3). Epistasis in RNA viruses may be blurred by the considerable weight of mutant spectra in determining viral behavior through intergenome interactions (Chapter 3).

An interesting contrast that recapitulates concepts given in Sections 2.5 and 2.6 is the effect of an active proofreading-repair activity in maintaining the infectivity of a viral genome upon its extended replication *in vitro* (in a test tube, in the absence of cellular extracts). The 19,285 bp bacteriophage φ29 DNA can be amplified at least 4000-fold without detectable loss of infectivity due to the fidelity of φ29 DNA polymerase conferred by a 3′-5′ proofreading-repair exonuclease activity (Bernad et al., 1989). Engineered φ29 DNA polymerases provide a powerful amplification tool in genomics (de Vega et al., 2010). In contrast, the 4,220 nucleotides long Qβ RNA rapidly loses its infectivity when replicated by Qβ replicase *in vitro* due to accumulation of mutations and deletions in the viral RNA (Mills et al., 1967; Sabo et al., 1977). The error-prone Qβ replicase is not adequate to amplify infectious viral RNA, but it was at the origin of the quasispecies concept to be discussed in Chapter 3.

2.9 RECOMBINATION IN DNA AND RNA VIRUSES

Recombination is the formation of a new genome by covalent linkage of genetic material from two or more different parental genomes (Figure 2.7). Recombination can also involve different sites of the same genome to yield insertions or deletions, such as in the formation of defective interfering (DI) genomes. It is a widespread mechanism of genetic variation in all biological systems, and in cells it underlies critical physiological and developmental processes (splicing, generation of diversity in immunoglobulin genes and T cell receptors, transposition events, phase variation in bacteria, repair pathways that promote postreplicative error correction, etc.). Cellular DNA recombination relates to replication, repair, and completion of DNA replication, operations that involve multiple proteins displaying a variety of activities (Smith and Jones, 1999; Alberts et al., 2002; Nimonkar and Boehmer, 2003; Friedberg et al., 2006).

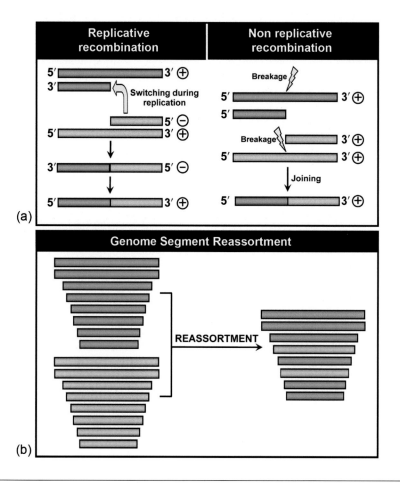

FIGURE 2.7

RNA recombination and segment reassorment. (a) Scheme of replicative and nonreplicative RNA recombination. RNA polarity is indicated by +, − symbols. Replicative recombination is displayed as the result of template switching during minus strand RNA synthesis. Nonreplicative recombination is depicted as the outcome of breakage and ligation (joining) of fragments of plus strand RNA. (b) An example of genome segment reassortment in which a new segment constellation is produced in which six genomic segments originate from one parent (blue) and two from the other (gold). Influenza virus is the best known example (see text).

Recombination occurs both with DNA and RNA viruses, often with participation of the virus replication machinery. Several types of recombination have been distinguished in viruses: homologous versus nonhomologous recombination, according to the extent of nucleotide sequence identity around the recombination (crossover) site, and replicative versus nonreplicative recombination, according to the requirement of viral genome replication for recombination to occur (Kirkegaard and Baltimore, 1986; King, 1988; Lai, 1992; Nagy and Simon, 1997; Plyusnin et al., 2002; Boehmer and Nimonkar, 2003; Gmyl et al., 2003; Chetverin et al., 2005; Agol, 2010; Simmonds, 2010; Bujarski, 2013).

As in the case of cells, homologous recombination in double-stranded DNA viruses is intimately connected with DNA replication and repair. It implicates multiple viral gene products (DNA polymerase, single-stranded DNA-binding proteins, processivity factors, helicase-primase, eukaryotic topoisomerase I, etc.), and a succession of protein-catalyzed steps. In the copy choice (or template switching) mechanism, the nascent DNA switches from one template molecule to another, resulting in synthesis of recombinant, daughter DNAs. In its basic form, recombination by breakage and rejoining starts with the introduction of a nick at one of the strands of each parental DNA, strand invasion of one parental DNA by the other, branch migration, ligation at the nicks (linking DNA strands from the two parents), and further isomerization and cleavage reactions. DNA recombination is responsible for the endonuclease-mediated isomerization of some herpesvirus genomes [four isomers defined by the orientation of the long (L) and short (S) regions of the viral genome]. During the late phase of herpes simplex virus-1 replication, the frequency of recombination has been estimated at 0.6% per Kb of genome (Boehmer and Nimonkar, 2003). Integration or excision of proviral DNA or temperate bacteriophage DNA, are examples of site-specific recombination that involves specific enzyme activities (i.e., retroviral integrases), and requires a short stretch of nucleotide sequence identity.

The copy choice mechanism of homologous RNA recombination is also associated with genome replication. An RNA polymerase molecule with its nascent RNA product jumps into the corresponding position of another template molecule, to complete synthesis of the RNA product (Figure 2.7). Given the large numbers of viral genomes often present in replication complexes in each infected cell, it is not surprising that this mechanism may give rise to frequent recombinant progeny genomes. Formation of mosaic genomes has long been recognized as an essential feature of the genetics of some retroviruses and plant RNA viruses. For HIV-1 and some plant RNA viruses recombination frequencies have been estimated at 2% to 10% of progeny per 100 nucleotides; for picornaviruses and coronaviruses the number of recombinants amount to 10% to 20% of the progeny (King, 1988; Lai, 1992; Nagy and Simon, 1997; Levy et al., 2004; Urbanowicz et al., 2005; Sztuba-Solinska et al., 2011). A phylogenetic approach estimated an average recombination rate of HIV-1 *in vivo* of 1.4×10^{-4} recombination events/site/generation, which is about fivefold greater than the average point mutation rate (Shriner et al., 2004). Recombination may be inefficient or absent in some negative strand RNA viruses. However, some of them undergo homologous recombination (Plyusnin et al., 2002), and certainly high frequency recombination in the generation of DI RNAs (Roux et al., 1991).

Recombination frequency may be altered by environmental factors that affect viral replication. A decrease of intracellular nucleotide levels as a result of treatment of cells with hydroxyurea may favor template switching reflected in an increase of intra- and intermolecular recombination (Pfeiffer et al., 1999; Svarovskaia et al., 2000).

Nonreplicative recombination does not require replication of the viral genome, and has been described upon cotransfection of cells with viral RNA fragments that could not replicate by themselves (Gmyl et al., 2003; Gallei et al., 2004; Agol, 2010). It appears to be a promiscuous event with a required 3′-phosphate in the 5′ partner RNA and a 5′-hydroxyl residue in the 3′ partner RNA. The cellular activities involved in nonreplicative RNA recombination have not been characterized.

2.9.1 MOLECULAR OCCURRENCE VERSUS OBSERVED RECOMBINATION

The emerging picture is that the frequency of recombination varies among viruses, and that as new tools for genome analyses have become available, recombination has been detected in an increasing number of viruses. Recognition of recombination in a viral system is facilitated when a cell culture system is available. Controlled infection of cells with genetically marked parental viruses has been essential to estimate recombination frequencies, and to distinguish true recombination from mutation-reversion events that may mimic the formation of recombinants. As with the concept of high genetic variation in RNA viruses, recombination has often gone from being considered marginal to prominent and relevant; HCV is a typical example (Galli and Bukh, 2014).

Viral replicative machineries may be endowed with features that influence the occurrence of recombination. One such feature is processivity of the viral polymerase (capacity of continued copying of the same template molecule). Genome detachment of the polymerase complex from one genome to bind either to a different genome or to a distant site of the same genome is part of the standard replicative cycle of viruses such as retroviruses and coronaviruses. Reverse transcriptase participates in strand transfer during DNA synthesis, and coronavirus polymerase switches from one template site to another during discontinuous RNA synthesis. It may be significant that they belong to viral families displaying high recombination frequencies (Makino et al., 1986). Thus, here again we encounter a "molecular instruction" that evolved as an essential feature of viral genome replication and that can be exploited to generate variation, and permit new genomic forms to undergo the scrutiny of selection (Chapter 3).

The occurrence of recombination is diagnosed by discordant positions of different genes or genomic regions in phylogenetic trees, as a result of the transfer of part of a viral genome from one lineage to another (Chapter 7). A commonly used procedure measures similarity values between sequences using a sliding-window scanning method. The recombination crossover point (where the two parental sequences meet) is identified by the point (or region) where the similarity plot crosses from one sequence into another (Salemi and Vandamme, 2004; Martin et al., 2005; Kosakovsky Pond et al., 2006). Crossover points along a viral genome are not distributed at random, either because polymerase detachment from the template is sequence dependent or because many of the recombination events do not lead to viable progeny. Absence of recombinant viability introduces a parallel with the distinction we made in Section 2.5 between mutation rate and frequency; that is, a distinction between what does occur and what is observed. Many newly arising recombinants may be subjected to negative selection, and only a viable subset of all recombinants produced might be detectable in progeny virus (King, 1988; Lai, 1992). Thus, as in the case of mutations, many recombinant genomes are not detected because of their elimination by negative selection. An elegant study by D.J. Evans and colleagues documented "imprecise" enterovirus recombinant intermediates that were lost upon serial virus passage (Lowry et al., 2014). Recombination is viewed as a biphasic process consisting of initial imprecise events followed by a stage of resolution in favor of fit recombinants.

The distinction between generation and resolution events that applies both to mutants and recombinants has yet another implication for RNA virus genetics. Some mutants or recombinants that in isolation do not display sufficient replicative fitness to acquire dominance in a population may nevertheless persist as minority genomes. They may display low-level replication or be maintained by complementation by partner genomes (as in the case of two FMDV genome segments that are described in Section 2.11). As minority genomes, they may engage in modulatory activities such as those described in Chapter 3. When technically feasible, application of next generation sequencing (NGS) to characterize recombination intermediates should provide important information.

Homologous RNA recombination can be influenced also by amino acid substitutions in the polymerase, the primary sequence in the RNA (i.e., high frequency of template switching in AU-rich regions), the sequence identity between the nascent strand and acceptor template, and secondary structures at or around the crossover sites, among other influences [(Nagy and Simon, 1997; Alejska et al., 2005; Agol, 2010) and references therein]. At a cellular level, since recombination necessitates coinfections of the same cell by at least two parental genomes, persistence of a viral genome in a cell increases the likelihood of sequential coinfections, unless some reinfection or superinfection exclusion mechanism operates [(Webster et al., 2013) and references therein]. Without such restrictions, persistently infected cells may be an environment with higher probability of recombination than transiently infected cells, assuming comparable genome loads at the sites of replication.

2.10 GENOME SEGMENT REASSORTMENT

In viruses whose genomes are composed of two or more RNA or DNA segments, genome segment reassortment consists in the formation of new constellations of viral genomic segments from two or more parental genomes (Figure 2.7). Reassortment can produce new phenotypic traits. It is the main mechanism of antigenic shift of influenza A viruses—often associated with new influenza pandemics (Webster et al., 1992; Morse, 1994; Gibbs et al., 1995; Domingo et al., 2001)—as opposed to antigenic drift which is mediated by amino acid substitutions in the surface proteins hemagglutinin and neuraminidase. Reassortments occur among the 9-12 double-stranded RNA segments of the widespread *Reoviridae* family (Tanaka et al., 2012). Fitness differences among all possible segment combinations (2^n, for two types of coinfecting particles with n genome segments) determine the types of genome segment groupings that dominate subsequent rounds of infection. In the laboratory, analysis of reassortant viruses has been applied to map a viral function into one segment or a combination of segments.

Genomic segments can be encapsidated either into a single virus particle (as in Orthomyxoviruses or Arenaviruses) or into separate particles (as in multipartite plant viruses). Multipartite virus can have either RNA or DNA as genetic material. The plant Nanoviruses have 6-8 molecules of single-stranded circular DNA of about 900 to 1000 nucleotides, and each segment encodes a single protein. In the case of the nanovirus Faba bean necrotic stunt virus, its 8 segments vary in frequency in a host-dependent manner (Sicard et al., 2013). This observation led the authors to propose a "setpoint genome formula" which may reflect a control of segment (gene) copy number that may provide some still unrecognized benefit to the multipartite phenotype (see next section). In principle, replication of multipartite viruses requires that each cell be coinfected by at least one of each type or particle harboring a different genome type, which in fact represents a remarkable cost for replicative efficiency. The fact that unsegmented and segmented RNA viruses are well represented in our biosphere suggests that neither of the two organizations confers a definitive advantage for long-term survival.

2.11 TRANSITION TOWARD VIRAL GENOME SEGMENTATION: IMPLICATIONS FOR GENERAL EVOLUTION

The origin of viral genome segmentation is a debated issue, although there is general agreement that it may confer adaptive flexibility to viruses. Most proposals have been based on theoretical studies. Segmentation has been viewed as a form of sex that facilitates genomic exchanges, to counteract the effect of deleterious mutations (Chao, 1988; Szathmary, 1992). An alternative, not mutually exclusive model, is that genome segmentation may have an advantage because replication of shorter RNA molecules is completed earlier than the unsegmented counterpart (Nee, 1987). Yet another possibility is that the life style of a virus (in particular, the particle yield in connection with the number of surrounding susceptible cells), established over long evolutionary periods, may favor segmentation over intactness of a genetic message or vice versa.

An experimental system of genome segmentation has become available with the picornavirus foot-and-mouth disease virus (FMDV). Its single-stranded RNA genome underwent a modification akin to genome segmentation when the standard virus was subjected to more than 200 passages in BHK-21 cells at high multiplicity of infection. The experiment was originally intended to investigate the limits of fitness gain during virus optimization in a defined environment, in this case BHK-21 cells in culture. The FMDV used was a biological clone obtained in BHK-21 cells; the virus had not adapted to the BHK-21 cell environment since it derived from a diseased swine during a disease outbreak. Upon extensive replication in BHK-21 cells, the virus evolved toward a bipartite genome (García-Arriaza et al., 2004) thus adding reassortment to mutation and recombination as potential mechanisms of genetic variation of this laboratory-adapted picornavirus.

Each of the two pieces of RNA that composed the bipartite (or segmented) genome version contained in-frame deletions affecting *trans*-acting proteins (Figure 2.8). Each segment in isolation could not infect cells productively, but, when present together, they were infectious by complementation, and killed cells in the absence of standard FMDV. Low multiplicity of infection rapidly selected the full length genome as a result of recombination of the two parental, defective segments (García-Arriaza et al., 2004). The particles containing the shortened RNA were thermically more stable than the standard particles (Ojosnegros et al., 2011), but this difference did not explain the initial trigger of the segmentation event. The solution to this question came with the demonstration that the transition toward genome segmentation was possible because of an extensive exploration of sequence space by the standard virus. Indeed, the mutations accumulated during serial passages enhanced fitness of the segmented genome version to a much higher extent than fitness of the standard genome, thus conferring a selective advantage to the segmented form over its unsegmented (monopartite) counterpart (Moreno et al., 2014) (Figure 2.8). Thus, in this segmentation event, gradual evolution (drift in sequence space) was a requirement for the major transition toward segmentation. This experimental result suggests that in evolution there is no unsurmountable barrier that allows the conversion between intact and split forms of the same genome, reflecting a remarkable genome flexibility that will be emphasized in Chapter 7 in the context of the relevance of virus variation in the emergence of viral pathogens.

2.12 MUTATION, RECOMBINATION, AND REASSORTMENT AS INDIVIDUAL AND COMBINED EVOLUTIONARY FORCES

Mutation, recombination, and segment reassortment contribute to the evolution of most DNA and RNA viruses. Sometimes one form of genetic change appears to be more prominent than another, and sometimes the concerted action of recombination or reassortment with mutation is apparent [i.e., antigenic drift in influenza virus, following the origin of a new antigenic type through reassortment (Ghedin et al., 2005)]. Mutation is a universal form of genetic change. It underlies numerous adaptive responses and critical biological transitions in viruses, and it is a prerequisite for recombination and reassortment to have a biological impact. If mutations were not present in different template molecules during replication, recombinants with the crossover point at equivalent positions

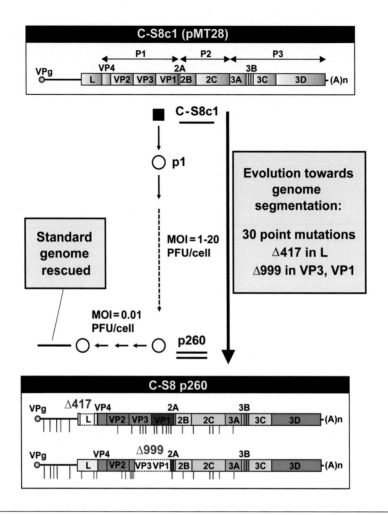

FIGURE 2.8

Evolution toward RNA genome segmentation in the laboratory. The monopartite, standard FMDV genome (clone C-S8c1 or pMT28, top) was subjected to 260 passages in BHK-21 cells. The resulting population p260 lacked detectable standard genome that could be rescued by low MOI passages. The evolved C-S8p260 accumulated 30 point mutations (depicted as vertical lines on the genome at the bottom) and consisted in two segments that were infections by complementation: Δ417, that lacked most of the L protease-coding region, and Δ999 that lacked most of the capsid proteins VP3, VP1-coding region. See text for further details and references.

of the parental genomes would be "silent," and display the same behavior as the parental genomes. Apparently "silent" recombination events may take place within replicative units; even if some mutations distinguished individual genomes of the same quasispecies swarm, a recombinant would not be distinguished from a mutant genome. The frequency of recombination in HIV-1 was noticed only when the acquired immune deficiency syndrome (AIDS) pandemic had advanced and the virus had diversified through accumulation of mutations. Similar arguments apply to segment reassortment.

Genomes necessitate mutation-driven diversification for reassortment to provide a biological advantage (see also Chapter 10).

The evolutionary significance of recombination has been viewed in two opposite ways: as a means to rescue fit genomes from less fit parents (a conservative force that eliminates deleterious mutations), or as a means to explore new genomic forms for adaptive potential (a vast substrate for the exploration of sequence space; Chapter 3) [reviewed in (Zimmern, 1988; Lai, 1992; Worobey and Holmes, 1999; Simmonds, 2010)]. Recombination has been probably at the origin of new viruses that presently occupy a well-established niche, and it is also at play today to expand diversity during the spread of viruses. As a historical event, the coronavirus mouse hepatitis virus appears to have acquired its hemagglutinin-esterase gene by recombination with an influenza C virus. The alphavirus Western equine encephalitis virus originated probably by recombination between Sindbis-like and Eastern equine encephalitis-like viruses [reviewed in different chapters of (Domingo et al., 2008)].

Several recent poliomyelitis outbreaks have been associated with recombinants between oral poliovirus vaccine (OPV) viruses and other circulating enteroviruses (Gavrilin et al., 2000; Kew et al., 2002; Oberste et al., 2004). Intersubtype HIV-1 recombinants play a key role in current HIV-1 diversification, with an increasing number of circulating recombinant forms displaying complex mosaic structures (multiple crossover sites) (Thomson et al., 2002; Gerhardt et al., 2005). Positive selection of HIV-1 recombinants that unite different drug-resistant mutations in the same genome offers an example of the conservative force of recombination to rescue fit viruses in the face of a strong selective constraint (Menéndez-Arias, 2002). Recombination is expected to play an increasing role in the spread of drug resistance among viruses for which new antiviral agents are in use, such as HBV and HCV.

Some defective DNA and RNA genomes that include *indels*, notably DI RNAs, originate by recombination. DI RNAs can play an important role in the establishment and maintenance of persistent infections in cell culture, and can modulate viral infections *in vivo* (Holland and Villarreal, 1974; Roux et al., 1991). Detailed genetic and biochemical analyses by A. Huang, J.J. Holland and their colleagues on the generation of VSV DI's and their interplay with the standard, infectious VSV contributed to unveil a continuous dynamics of genetic variation, competition, and selection, observable within short time intervals, a hallmark of RNA genetics (Palma and Huang, 1974; Holland et al., 1982). Despite the continuous dynamics of escape of infectious virus to the interfering activities of DIs, some authors consider DIs as potential antiviral agents (Dimmock and Easton, 2014). DI RNAs can be regarded as the tip of the iceberg of many classes of defective genomes with a range of interfering or potentiating capacities that may coexist with standard animal, plant, insect, and bacterial viruses, and that may facilitate persistence and modulate disease symptoms (Holland et al., 1982; Vogt and Jackson, 1999; López-Ferber et al., 2003; Rosario et al., 2005; Sachs and Bull, 2005; Villarreal, 2005; Aaskov et al., 2006). Noncytopathic coxsackievirus B3 (CVB3) variants with deletions at the untranslated 5'-genomic region were isolated from hearts of mice inoculated with CVB3. The variants replicated *in vivo* and were associated with long-term viral persistence (Kim et al., 2005).

The presence of defective viruses may influence the behavior of viral populations by different means. If defective genomes are competent in RNA (or DNA) synthesis, or are complemented to replicate, they can act as dominant-negative swarms, provided they reach a sufficient load. In this manner, they may underlie suppressive effects of mutant spectra of viral quasispecies. Intramutant spectrum decrease of replicative capacity due to the presence of defective genomes is one of the mechanisms of virus extinction evoked by enhanced mutagenesis (Chapters 3 and 9).

Recombination events must have been the last step in ancestral processes of horizontal gene transfer that mediated the incorporation of host genes (or gene segments) into viral genomes, and vice versa. Host genes related to immune responses were probably captured by complex DNA viruses at early stages of their evolution (Alcami, 2003; McFadden, 2005). Mosaicism associated with nonhomologous recombination events is the norm among tailed bacteriophages (Canchaya et al., 2003). Nonhomologous recombination can give rise to genomic sequences with a viral and a nonviral moiety. They include DI RNAs of Sindbis virus containing cellular RNA sequences at their 5′ ends, some cytopathic forms of bovine viral diarrhea virus, RNA of potato leaf roll virus containing tobacco chloroplast RNA, or an influenza virus with an insertion of ribosomal RNA into the hemagglutinin gene, mentioned in Chapter 5 regarding transient, high fitness levels [reviewed in (Domingo et al., 2001)].

Phylogenetic analyses have suggested that recombination between RNA and DNA viruses might have occurred to give rise to some present-day single-stranded DNA viruses (Stedman, 2013). However, the evidence for this attractive possibility is indirect and, to my knowledge, no experimental evidence of viral RNA-viral DNA recombination in cell culture or *in vivo* has been reported. Viability of mutant and recombinant viral genomes is severely constrained by the evolutionary history of the virus that has shaped viral genomes as coordinated sets of modules (Botstein, 1980, 1981; Zimmern, 1988; Koonin and Dolja, 2014). Experimental studies with engineered recombinant viruses have shown that modularity can restrict recombination (Martin et al., 2005).

It would be truly remarkable if a viral system could be proven to be totally devoid of one of the mechanisms of genetic variation to which it has access. It would imply that there are powerful molecular reasons to dispense with an effective adaptive mechanism. Absence of a mechanism is extremely difficult to demonstrate but, if we could, its basis would open a new chapter of molecular virology.

2.13 OVERVIEW AND CONCLUDING REMARKS

All forms of genetic variation of viruses must be viewed essentially as blind processes despite preferences of nucleotide sequences or structures for mutation and recombination events: *hot spots* with higher than average rate values, and cold spots with lower than average rate values. Mutation originates largely in fluctuations of electronic structure that modify base-pairing properties, and from features of polymerase-template interactions, not subjected to regulation. Recombination and reassortment are probably not regulated either. It is selection acting at the very center of replication and recombinant complexes that preserves some mutant and recombinant forms in detriment of others. Subsequent levels of selection occur when variant forms expand in multiple rounds of infection first within cells, then within an organism and then at the epidemiological level. The very nature of life in our planet has been built upon an inherent tendency to instruct variation in an incessant fashion, as necessary and unavoidable as the physical principles that dictate the behavior of our universe.

The net result of all mechanisms of genetic variation available to a virus is the generation of repertoires of variant genomes for random drift and selective forces to act upon. In other terms, genetic variation sets the scene for the actors of evolution to play their roles, and secure a continuous input of new forms despite subtle or catastrophic environmental perturbations. The same forces that drive general evolution have produced the dominant virus forms we see in nature, with all their nuances in the interaction with cell components. The adaptation of viruses to participate in intracellular processes with cells dictates that genetic variation of viruses has its limits to prevent deleteriousness. This is currently exemplified by the effects of amino acid substitutions in viral polymerases that either increase or decrease template-copying fidelity. Viruses have reached a compromise between stability of a core information and flexibility for adaptability. Although not yet treated in this chapter, viral population numbers are a key parameter in the evolutionary events. Next chapters address some of these questions, not only in general conceptual terms but also in the way evolution affects our daily confrontation with viral disease (see Summary Box).

SUMMARY BOX

- Mutation, recombination and genome segment reassortment are the mechanisms of genetic variation used by DNA and RNA viruses. Mutations are due mainly to changes in the electronic distributions of the standard nucleotides, to damage of nucleotides by external influences, and by alignment alterations of template relative to product polynucleotide chains. The effect of a mutation can range from being well tolerated to highly detrimental or lethal.
- Mutation frequencies are only an indirect consequence of mutation rates. Their values for viruses whose replication is catalyzed by polymerases devoid of proofreading-repair activity are 10^5 to 10^6-fold higher than those displayed by replicative cellular DNA polymerases. Error-prone replication is a hallmark of RNA viruses and some DNA viruses. The larger the amount of genetic information encoded in a viral genome, the lower the mutation rate must be to maintain the genetic message.
- Several mechanisms of genetic recombination have been described for DNA and RNA viruses. The best characterized is homologous recombination whose frequency of occurrence is dependent on the replicative machinery, in particular polymerase processivity. Genome segment reassortment is operative in segmented genomes and it gives rise to biologically relevant changes such as antigenic shift in the influenza type A viruses.
- Studies with foot-and-mouth disease virus have shown that extensive evolution of an unsegmented RNA genome has the potential to undergo a recombination-mediated transition akin to genome segmentation. Therefore, segmented and unsegmented forms of RNA viruses need not be considered as completely unrelated classes of genome organization.
- Recombination and genome segment reassortment have been viewed as conservative forces to rescue viable genomes from a damaged pool, and also as means to explore new genomic compositions that deviate from their parents. All forms of genetic variation give rise to repertoires of variant genomes on which selection and random drift act to produce the viral forms that we isolate and study.

REFERENCES

Aaskov, J., Buzacott, K., Thu, H.M., Lowry, K., Holmes, E.C., 2006. Long-term transmission of defective RNA viruses in humans and Aedes mosquitoes. Science 311, 236–238.

Achaz, G., 2009. Frequency spectrum neutrality tests: one for all and all for one. Genetics 183, 249–258.

Agol, V.I., 2010. Picornaviruses as a model for studying the nature of RNA recombination. In: Ehrenfeld, E., Domingo, E., Roos, R.P. (Eds.), The Picornaviruses. ASM Press, Whashington, DC, pp. 239–252.

Aguilera, A., Garcia-Muse, T., 2013. Causes of genome instability. Annu. Rev. Genet. 47, 1–32.

Akashi, H., 2001. Gene expression and molecular evolution. Curr. Opin. Genet. Dev. 11, 660–666.

Alberts, B., Johnson, A., Lewis, J., Raff, M., Roberts, K., et al., 2002. Molecular Biology of the Cell. Garland Science, New York, N.Y.

Alcami, A., 2003. Viral mimicry of cytokines, chemokines and their receptors. Nat. Rev. Immunol. 3, 36–50.

Alejska, M., Figlerowicz, M., Malinowska, N., Urbanowicz, A., Figlerowicz, M., 2005. A universal BMV-based RNA recombination system—how to search for general rules in RNA recombination. Nucleic Acids Res. 33, e105.

Aragones, L., Guix, S., Ribes, E., Bosch, A., Pinto, R.M., 2010. Fine-tuning translation kinetics selection as the driving force of codon usage bias in the hepatitis a virus capsid. PLoS Pathog. 6, e1000797.

Arias, A., Arnold, J.J., Sierra, M., Smidansky, E.D., Domingo, E., et al., 2008. Determinants of RNA-dependent RNA polymerase (in)fidelity revealed by kinetic analysis of the polymerase encoded by a foot-and-mouth disease virus mutant with reduced sensitivity to ribavirin. J. Virol. 82, 12346–12355.

Arnold, J.J., Vignuzzi, M., Stone, J.K., Andino, R., Cameron, C.E., 2005. Remote site control of an active site fidelity checkpoint in a viral RNA-dependent RNA polymerase. J. Biol. Chem. 280, 25706–25716.

Barrett, J.W., McFadden, G., 2008. Origin and evolution of poxviruses. In: Domingo, E., Parrish, C.R., Holland, J.J. (Eds.), Origin and Evolution of Viruses, second ed. Elsevier, Oxford, pp. 431–446.

Batschelet, E., Domingo, E., Weissmann, C., 1976. The proportion of revertant and mutant phage in a growing population, as a function of mutation and growth rate. Gene 1, 27–32.

Beaucourt, S., Vignuzzi, M., 2014. Ribavirin: a drug active against many viruses with multiple effects on virus replication and propagation. Molecular basis of ribavirin resistance. Curr. Opin. Virol. 8C, 10–15.

Bernad, A., Blanco, L., Lazaro, J.M., Martin, G., Salas, M., 1989. A conserved 3' → 5' exonuclease active site in prokaryotic and eukaryotic DNA polymerases. Cell 59, 219–228.

Bloomfield, V.A., Crothers, D.M., Tinoco Jr., I., 2000. Nucleic acids. Structures, properties, and functions. University Science Books, Sausalito, CA.

Boehmer, P.E., Nimonkar, A.V., 2003. Herpes virus replication. IUBMB Life. 55, 13–22.

Borderia, A.V., Rozen-Gagnon, K., Vignuzzi, M., 2016. Fidelity variants and RNA quasispecies. Curr. Top. Microbiol. Immunol.

Botstein, D., 1980. A theory of modular evolution for bacteriophages. Ann. NY Acad. Sci. 354, 484–491.

Botstein, D., 1981. A modular theory of virus evolution. In: Fields, B.N., Jaenisch, R., Fox, C.F. (Eds.), Animal Virus Genetics. Academic Press, New York, pp. 363–384.

Bujarski, J.J., 2013. Genetic recombination in plant-infecting messenger-sense RNA viruses: overview and research perspectives. Front Plant Sci. 4, 68.

Bushman, F., 2002. Lateral DNA Transfer. Mechanisms and Consequences. Cold Spring Harbor Laboratory Press, Cold Spring Harbor, New York.

Campagnola, G., McDonald, S., Beaucourt, S., Vignuzzi, M., Peersen, O.B., 2015. Structure-function relationships underlying the replication fidelity of viral RNA-dependent RNA polymerases. J. Virol. 89, 275–286.

Canchaya, C., Fournous, G., Chibani-Chennoufi, S., Dillmann, M.L., Brussow, H., 2003. Phage as agents of lateral gene transfer. Curr. Opin. Microbiol. 6, 417–424.

Castro, C., Arnold, J.J., Cameron, C.E., 2005. Incorporation fidelity of the viral RNA-dependent RNA polymerase: a kinetic, thermodynamic and structural perspective. Virus Res. 107, 141–149.

Cattaneo, R., Billeter, M.A., 1992. Mutations and A/I hypermutations in measles virus persistent infections. Curr. Top. Microbiol. Immunol. 176, 63–74.

Chao, L., 1988. Evolution of sex in RNA viruses. J. Theor. Biol. 133, 99–112.

Chetverin, A.B., Kopein, D.S., Chetverina, H.V., Demidenko, A.A., Ugarov, V.I., 2005. Viral RNA-directed RNA polymerases use diverse mechanisms to promote recombination between RNA molecules. J. Biol. Chem. 280, 8748–8755.

Cheynier, R., Kils-Hutten, L., Meyerhans, A., Wain-Hobson, S., 2001. Insertion/deletion frequencies match those of point mutations in the hypervariable regions of the simian immunodeficiency virus surface envelope gene. J. Gen. Virol. 82, 1613–1619.

Coffey, L.L., Beeharry, Y., Borderia, A.V., Blanc, H., Vignuzzi, M., 2011. Arbovirus high fidelity variant loses fitness in mosquitoes and mice. Proc. Natl. Acad. Sci. U. S. A. 108, 16038–16043.

Coffin, J.M., 1990. Genetic variation in retroviruses. Appl. Virol. Res. 2, 11–33.

Combe, M., Sanjuan, R., 2014. Variation in RNA virus mutation rates across host cells. PLoS Pathog. 10, e1003855.

Crandall, K.A., Kelsey, C.R., Imamichi, H., Lane, H.C., Salzman, N.P., 1999. Parallel evolution of drug resistance in HIV: failure of nonsynonymous/synonymous substitution rate ratio to detect selection. Mol. Biol. Evol. 16, 372–382.

Czech, A., Fedyunin, I., Zhang, G., Ignatova, Z., 2010. Silent mutations in sight: co-variations in tRNA abundance as a key to unravel consequences of silent mutations. Mol. Biosyst. 6, 1767–1772.

de Vega, M., Lazaro, J.M., Mencia, M., Blanco, L., Salas, M., 2010. Improvement of phi29 DNA polymerase amplification performance by fusion of DNA binding motifs. Proc. Natl. Acad. Sci. U. S. A. 107, 16506–16511.

Dimmock, N.J., Easton, A.J., 2014. Defective interfering influenza virus RNAs: time to reevaluate their clinical potential as broad-spectrum antivirals? J. Virol. 88, 5217–5227.

Domingo, E., 2005. Virus entry into error catastrophe as a new antiviral strategy. Virus Res. 107 (Special Issue), 115–228.

Domingo, E., Sabo, D., Taniguchi, T., Weissmann, C., 1978. Nucleotide sequence heterogeneity of an RNA phage population. Cell 13, 735–744.

Domingo, E., Mateu, M.G., Martínez, M.A., Dopazo, J., Moya, A., et al., 1990. Genetic variability and antigenic diversity of foot-and-mouth disease virus. In: Kurkstak, E., Marusyk, R.G., Murphy, S.A., Van-Regenmortel, M.H.V. (Eds.), Applied Virology Research. Plenum Publishing Co., New York, pp. 233–266.

Domingo, E., Holland, J.J., Biebricher, C., Eigen, M., 1995. Quasispecies: the concept and the word. In: Gibbs, A., Calisher, C., García-Arenal, F. (Eds.), Molecular Evolution of the Viruses. Cambridge University Press, Cambridge, pp. 171–180.

Domingo, E., Biebricher, C., Eigen, M., Holland, J.J., 2001. Quasispecies and RNA Virus Evolution: Principles and Consequences. Landes Bioscience, Austin.

Domingo, E., Escarmis, C., Baranowski, E., Ruiz-Jarabo, C.M., Carrillo, E., et al., 2003. Evolution of foot-and-mouth disease virus. Virus Res. 91, 47–63.

Domingo, E., Escarmís, C., Menéndez-Arias, L., Perales, C., Herrera, M., et al., 2008. Viral quasispecies: dynamics, interactions and pathogenesis. In: Domingo, E., Parrish, C., Holland, J.J. (Eds.), Origin and Evolution of Viruses. Elsevier, Oxford, pp. 87–118.

Drake, J.W., 1991. A constant rate of spontaneous mutation in DNA-based microbes. Proc. Natl. Acad. Sci. U. S. A. 88, 7160–7164.

Drake, J.W., Holland, J.J., 1999. Mutation rates among RNA viruses. Proc. Natl. Acad. Sci. U. S. A. 96, 13910–13913.

Earl, D.J., Deem, M.W., 2004. Evolvability is a selectable trait. Proc. Natl. Acad. Sci. U. S. A. 101, 11531–11536.

Eckerle, L.D., Lu, X., Sperry, S.M., Choi, L., Denison, M.R., 2007. High fidelity of murine hepatitis virus replication is decreased in nsp14 exoribonuclease mutants. J. Virol. 81, 12135–12144.

Eckerle, L.D., Becker, M.M., Halpin, R.A., Li, K., Venter, E., et al., 2010. Infidelity of SARS-CoV Nsp14-exonuclease mutant virus replication is revealed by complete genome sequencing. PLoS Pathog. 6, e1000896.

Eigen, M., 2002. Error catastrophe and antiviral strategy. Proc. Natl. Acad. Sci. U. S. A. 99, 13374–13376.

Eigen, M., 2013. From strange simplicity to complex familiarity. Oxford University Press, Oxford.

Eigen, M., Biebricher, C.K., 1988. Sequence space and quasispecies distribution. In: Domingo, E., Ahlquist, P., Holland, J.J. (Eds.), RNA Genetics. CRC Press Inc, Boca Raton, FL, pp. 211–245.

Elena, S.F., Sanjuan, R., 2005. Adaptive value of high mutation rates of RNA viruses: separating causes from consequences. J. Virol. 79, 11555–11558.

Escarmís, C., Dávila, M., Charpentier, N., Bracho, A., Moya, A., et al., 1996. Genetic lesions associated with Muller's ratchet in an RNA virus. J. Mol. Biol. 264, 255–267.

Escarmís, C., Dávila, M., Domingo, E., 1999. Multiple molecular pathways for fitness recovery of an RNA virus debilitated by operation of Muller's ratchet. J. Mol. Biol. 285, 495–505.

Fan, X., Di Bisceglie, A.M., 2010. RT-PCR amplification and cloning of large viral sequences. Methods Mol. Biol. 630, 139–149.

Ferrer-Orta, C., Arias, A., Escarmis, C., Verdaguer, N., 2006. A comparison of viral RNA-dependent RNA polymerases. Curr. Opin. Struct. Biol. 16, 27–34.

Ferrer-Orta, C., Agudo, R., Domingo, E., Verdaguer, N., 2009. Structural insights into replication initiation and elongation processes by the FMDV RNA-dependent RNA polymerase. Curr. Opin. Struct. Biol. 19, 752–758.

Friedberg, E.C., Wagner, R., Radman, M., 2002. Specialized DNA polymerases, cellular survival, and the genesis of mutations. Science 296, 1627–1630.

Friedberg, E.C., Walker, G.C., Siede, W., Wood, R.D., Schultz, R.A., et al., 2006. DNA repair and mutagenesis. American Society for Microbiology, Washington, DC.

Fu, Y.X., 1997. Statistical tests of neutrality of mutations against population growth, hitchhiking and background selection. Genetics 147, 915–925.

Gago, S., Elena, S.F., Flores, R., Sanjuan, R., 2009. Extremely high mutation rate of a hammerhead viroid. Science 323, 1308.

Gallei, A., Pankraz, A., Thiel, H.J., Becher, P., 2004. RNA recombination in vivo in the absence of viral replication. J. Virol. 78, 6271–6281.

Galli, A., Bukh, J., 2014. Comparative analysis of the molecular mechanisms of recombination in hepatitis C virus. Trends Microbiol. 22 (6), 354–364.

García-Arriaza, J., Manrubia, S.C., Toja, M., Domingo, E., Escarmís, C., 2004. Evolutionary transition toward defective RNAs that are infectious by complementation. J. Virol. 78, 11678–11685.

Gatenby, R.A., Frieden, B.R., 2004. Information dynamics in carcinogenesis and tumor growth. Mutat. Res. 568, 259–273.

Gavrilin, G.V., Cherkasova, E.A., Lipskaya, G.Y., Kew, O.M., Agol, V.I., 2000. Evolution of circulating wild poliovirus and of vaccine-derived poliovirus in an immunodeficient patient: a unifying model. J. Virol. 74, 7381–7390.

Gerhardt, M., Mloka, D., Tovanabutra, S., Sanders-Buell, E., Hoffmann, O., et al., 2005. In-depth, longitudinal analysis of viral quasispecies from an individual triply infected with late-stage human immunodeficiency virus type 1, using a multiple PCR primer approach. J. Virol. 79, 8249–8261.

Ghedin, E., Sengamalay, N.A., Shumway, M., Zaborsky, J., Feldblyum, T., et al., 2005. Large-scale sequencing of human influenza reveals the dynamic nature of viral genome evolution. Nature 437, 1162–1166.

Gibbs, A.J., Calisher, C.H., García-Arenal, F. (Eds.), 1995. Molecular Basis of Virus Evolution. Cambridge University Press, Cambridge.

Gmyl, A.P., Pilipenko, E.V., Maslova, S.V., Belov, G.A., Agol, V.I., 1993. Functional and genetic plasticities of the poliovirus genome: quasi- infectious RNAs modified in the 5'-untranslated region yield a variety of pseudorevertants. J. Virol. 67, 6309–6316.

Gmyl, A.P., Korshenko, S.A., Belousov, E.V., Khitrina, E.V., Agol, V.I., 2003. Nonreplicative homologous RNA recombination: promiscuous joining of RNA pieces? RNA 9, 1221–1231.

Gnadig, N.F., Beaucourt, S., Campagnola, G., Borderia, A.V., Sanz-Ramos, M., et al., 2012. Coxsackievirus B3 mutator strains are attenuated in vivo. Proc. Natl. Acad. Sci. U. S. A. 109, E2294–2303.

Graci, J.D., Cameron, C.E., 2004. Challenges for the development of ribonucleoside analogues as inducers of error catastrophe. Antivir. Chem. Chemother. 15, 1–13.

Hamano, T., Matsuo, K., Hibi, Y., Victoriano, A.F., Takahashi, N., et al., 2007. A single-nucleotide synonymous mutation in the gag gene controlling human immunodeficiency virus type 1 virion production. J. Virol. 81, 1528–1533.

Holland, J.J., Villarreal, L.P., 1974. Persistent noncytocidal vesicular stomatitis virus infections mediated by defective T particles that suppress virion transcriptase. Proc. Natl. Acad. Sci. U. S. A. 71, 2956–2960.

Holland, J.J., Spindler, K., Horodyski, F., Grabau, E., Nichol, S., et al., 1982. Rapid evolution of RNA genomes. Science 215, 1577–1585.

Holland, J.J., de la Torre, J.C., Steinhauer, D.A., Clarke, D., Duarte, E., et al., 1989. Virus mutation frequencies can be greatly underestimated by monoclonal antibody neutralization of virions. J. Virol. 63, 5030–5036.

Jin, Z., Leveque, V., Ma, H., Johnson, K.A., Klumpp, K., 2013. NTP-mediated nucleotide excision activity of hepatitis C virus RNA-dependent RNA polymerase. Proc. Natl. Acad. Sci. U. S. A. 110, E348–357.

Joyce, G.F., 2004. Directed evolution of nucleic acid enzymes. Annu. Rev. Biochem. 73, 791–836.

Kew, O., Morris-Glasgow, V., Landaverde, M., Burns, C., Shaw, J., et al., 2002. Outbreak of poliomyelitis in Hispaniola associated with circulating type 1 vaccine-derived poliovirus. Science 296, 356–359.

Kim, K.S., Tracy, S., Tapprich, W., Bailey, J., Lee, C.K., et al., 2005. 5'-Terminal deletions occur in coxsackievirus B3 during replication in murine hearts and cardiac myocyte cultures and correlate with encapsidation of negative-strand viral RNA. J. Virol. 79, 7024–7041.

Kimura, M., 1983. The Neutral Theory of Molecular Evolution. Cambridge University Press, Cambridge, UK.

Kimura, M., 1989. The neutral theory of molecular evolution and the world view of the neutralists. Genome 31, 24–31.

King, A.M.Q., 1988. Genetic recombination in positive strand RNA viruses. In: Domingo, E., Holland, J.J., Ahlquist, P. (Eds.), RNA Genetics. CRC Press Inc, Boca Raton, FL, pp. 149–165.

King, J.L., Jukes, T.H., 1969. Non-Darwinian evolution. Science 164, 788–798.

Kirkegaard, K., Baltimore, D., 1986. The mechanism of RNA recombination in poliovirus. Cell 47, 433–443.

Kolakofsky, D., Roux, L., Garcin, D., Ruigrok, R.W., 2005. Paramyxovirus mRNA editing, the "rule of six" and error catastrophe: a hypothesis. J. Gen. Virol. 86, 1869–1877.

Koonin, E.V., Dolja, V.V., 2014. Virus world as an evolutionary network of viruses and capsidless selfish elements. Microbiol. Mol. Biol. Rev. 78, 278–303.

Kosakovsky Pond, S.L., Posada, D., Gravenor, M.B., Woelk, C.H., Frost, S.D., 2006. GARD: a genetic algorithm for recombination detection. Bioinformatics 22, 3096–3098.

Kryazhimskiy, S., Plotkin, J.B., 2008. The population genetics of dN/dS. PLoS Genet. 4, e1000304.

Kunkel, T.A., Erie, D.A., 2005. DNA mismatch repair. Annu. Rev. Biochem. 74, 681–710.

Lafforgue, G., Martinez, F., Sardanyes, J., de la Iglesia, F., Niu, Q.W., et al., 2011. Tempo and mode of plant RNA virus escape from RNA interference-mediated resistance. J. Virol. 85, 9686–9695.

Lai, M.M.C., 1992. Genetic recombination in RNA viruses. Curr. Top. Microbiol. Immunol. 176, 21–32.

Lenski, R.E., Winkworth, C.L., Riley, M.A., 2003. Rates of DNA sequence evolution in experimental populations of Escherichia coli during 20,000 generations. J. Mol. Evol. 56, 498–508.

Levy, D.N., Aldrovandi, G.M., Kutsch, O., Shaw, G.M., 2004. Dynamics of HIV-1 recombination in its natural target cells. Proc. Natl. Acad. Sci. U. S. A. 101, 4204–4209.

Ling, J., Reynolds, N., Ibba, M., 2009. Aminoacyl-tRNA synthesis and translational quality control. Annu. Rev. Microbiol. 63, 61–78.

López-Ferber, M., Simon, O., Williams, T., Caballero, P., 2003. Defective or effective? Mutualistic interactions between virus genotypes. Proc. R. Soc. Lond. B Biol. Sci. 270, 2249–2255.

Lowry, K., Woodman, A., Cook, J., Evans, D.J., 2014. Recombination in Enteroviruses Is a Biphasic Replicative Process Involving the Generation of Greater-than Genome Length 'Imprecise' Intermediates. PLoS Pathog. 10, e1004191.

Maas, S., Rich, A., Nishikura, K., 2003. A-to-I RNA editing: recent news and residual mysteries. J. Biol. Chem. 278, 1391–1394.

Makhoul, C.H., Trifonov, E.N., 2002. Distribution of rare triplets along mRNA and their relation to protein folding. J. Biomol. Struct. Dyn. 20, 413–420.

Makino, S., Keck, J.G., Stohlman, S.A., Lai, M.M., 1986. High-frequency RNA recombination of murine coronaviruses. J. Virol. 57, 729–737.

Malpica, J.M., Fraile, A., Moreno, I., Obies, C.I., Drake, J.W., et al., 2002. The rate and character of spontaneous mutation in an RNA virus. Genetics 162, 1505–1511.

Mansky, L.M., Temin, H.M., 1995. Lower *in vivo* mutation rate of human immunodeficiency virus type 1 than that predicted from the fidelity of purified reverse transcriptase. J. Virol. 69, 5087–5094.

Mansky, L.M., Le Rouzic, E., Benichou, S., Gajary, L.C., 2003. Influence of reverse transcriptase variants, drugs, and Vpr on human immunodeficiency virus type 1 mutant frequencies. J. Virol. 77, 2071–2080.

Martin, D.P., Williamson, C., Posada, D., 2005. RDP2: recombination detection and analysis from sequence alignments. Bioinformatics 21, 260–262.

McFadden, G., 2005. Poxvirus tropism. Nat. Rev. Microbiol. 3, 201–213.

McGeoch, D.J., Davison, A.J., Dolan, A., Gatherer, D., Sevilla-Reyes, E.E., 2008. Molecular evolution of the *Herpesvirales*. In: Domingo, E., Parrish, C.R., Holland, J.J. (Eds.), Origin and Evolution of Viruses, second ed. Elsevier, Oxford, pp. 447–476.

Menéndez-Arias, L., 2002. Molecular basis of fidelity of DNA synthesis and nucleotide specificity of retroviral reverse transcriptases. Prog. Nucleic Acid Res. Mol. Biol. 71, 91–147.

Meng, T., Kwang, J., 2014. Attenuation of human enterovirus 71 high-replication-fidelity variants in AG129 mice. J. Virol. 88, 5803–5815.

Meyer, P.R., Matsuura, S.E., So, A.G., Scott, W.A., 1998. Unblocking of chain-terminated primer by HIV-1 reverse transcriptase through a nucleotide-dependent mechanism. Proc. Natl. Acad. Sci. U. S. A. 95, 13471–13476.

Meyerhans, A., Vartanian, J.-P., 1999. The fidelity of cellular and viral polymerases and its manipulation for hypermutagenesis. In: Domingo, E., Webster, R.G., Holland, J.J. (Eds.), Origin and Evolution of Viruses. Academic Press, San Diego, pp. 87–114.

Miller, R.V., Day, M.J., 2004. Microbial Evolution. Gene Establishment, Survival and Exchange. ASM Press, Washington, D.C.

Mills, D.R., Peterson, R.L., Spiegelman, S., 1967. An extracellular Darwinian experiment with a self-duplicating nucleic acid molecule. Proc. Natl. Acad. Sci. U. S. A. 58, 217–224.

Minskaia, E., Hertzig, T., Gorbalenya, A.E., Campanacci, V., Cambillau, C., et al., 2006. Discovery of an RNA virus 3' → 5' exoribonuclease that is critically involved in coronavirus RNA synthesis. Proc. Natl. Acad. Sci. U. S. A. 103, 5108–5113.

Moreno, E., Ojosnegros, S., Garcia-Arriaza, J., Escarmis, C., Domingo, E., et al., 2014. Exploration of sequence space as the basis of viral RNA genome segmentation. Proc. Natl. Acad. Sci. U. S. A. 111, 6678–6683.

Morse, S.S. (Ed.), 1994. The Evolutionary Biology of Viruses. Raven Press, New York.

Naegeli, H., 1997. Mechanisms of DNA damage recognition in mammalian cells. Landes Bioscience, Austin, Texas.

Nagy, P.D., Simon, A.E., 1997. New insights into the mechanisms of RNA recombination. Virology 235, 1–9.

Nee, S., 1987. The evolution of multicompartmental genomes in viruses. J. Mol. Evol. 25, 277–281.

Nei, M., Gojobori, T., 1986. Simple methods for estimating the numbers of synonymous and nonsynonymous nucleotide substitutions. Mol. Biol. Evol. 3, 418–426.

Nevot, M., Martrus, G., Clotet, B., Martinez, M.A., 2011. RNA interference as a tool for exploring HIV-1 robustness. J. Mol. Biol. 413, 84–96.

Nimonkar, A.V., Boehmer, P.E., 2003. Reconstitution of recombination-dependent DNA synthesis in herpes simplex virus 1. Proc. Natl. Acad. Sci. U. S. A. 100, 10201–10206.

Nishikura, K., 2010. Functions and regulation of RNA editing by ADAR deaminases. Annu. Rev. Biochem. 79, 321–349.

Novella, I.S., 2004. Negative effect of genetic bottlenecks on the adaptability of vesicular stomatitis virus. J. Mol. Biol. 336, 61–67.

Novella, I.S., Zarate, S., Metzgar, D., Ebendick-Corpus, B.E., 2004. Positive selection of synonymous mutations in vesicular stomatitis virus. J. Mol. Biol. 342, 1415–1421.

Oberste, M.S., Maher, K., Pallansch, M.A., 2004. Evidence for frequent recombination within species human enterovirus B based on complete genomic sequences of all thirty-seven serotypes. J. Virol. 78, 855–867.

Ochman, H., Elwyn, S., Moran, N.A., 1999. Calibrating bacterial evolution. Proc. Natl. Acad. Sci. U. S. A. 96, 12638–12643.

Ojosnegros, S., Garcia-Arriaza, J., Escarmis, C., Manrubia, S.C., Perales, C., et al., 2011. Viral genome segmentation can result from a trade-off between genetic content and particle stability. PLoS Genet. 7, e1001344.

Palma, E.L., Huang, A.S., 1974. Cyclic production of vesicular stomatitis virus caused by defective interfering particles. J. Infect. Dis. 126, 402–410.

Parmley, J.L., Chamary, J.V., Hurst, L.D., 2006. Evidence for purifying selection against synonymous mutations in mammalian exonic splicing enhancers. Mol. Biol. Evol. 23, 301–309.

Pfeiffer, J.K., Kirkegaard, K., 2003. A single mutation in poliovirus RNA-dependent RNA polymerase confers resistance to mutagenic nucleotide analogs via increased fidelity. Proc. Natl. Acad. Sci. U. S. A. 100, 7289–7294.

Pfeiffer, J.K., Kirkegaard, K., 2005. Ribavirin resistance in hepatitis C virus replicon-containing cell lines conferred by changes in the cell line or mutations in the replicon RNA. J. Virol. 79, 2346–2355.

Pfeiffer, J.K., Topping, R.S., Shin, N.H., Telesnitsky, A., 1999. Altering the intracellular environment increases the frequency of tandem repeat deletion during Moloney murine leukemia virus reverse transcription. J. Virol. 73, 8441–8447.

Plyusnin, A., Kukkonen, S.K., Plyusnina, A., Vapalahti, O., Vaheri, A., 2002. Transfection-mediated generation of functionally competent Tula hantavirus with recombinant S RNA segment. EMBO J. 21, 1497–1503.

Preston, B.D., Dougherty, J.P., 1996. Mechanisms of retroviral mutation. Trends Microbiol. 4, 16–21.

Resch, A.M., Carmel, L., Marino-Ramirez, L., Ogurtsov, A.Y., Shabalina, S.A., et al., 2007. Widespread positive selection in synonymous sites of mammalian genes. Mol. Biol. Evol. 24, 1821–1831.

Richmond, R.C., 1970. Non-Darwinian evolution: a critique. Nature 225, 1025–1028.

Rocha, E.P., 2004. Codon usage bias from tRNA's point of view: redundancy, specialization, and efficient decoding for translation optimization. Genome Res. 14, 2279–2286.

Rosario, D., Perez, M., de la Torre, J.C., 2005. Functional characterization of the genomic promoter of borna disease virus (BDV): implications of 3'-terminal sequence heterogeneity for BDV persistence. J. Virol. 79, 6544–6550.

Roux, L., Simon, A.E., Holland, J.J., 1991. Effects of defective interfering viruses on virus replication and pathogenesis in vitro and in vivo. Adv. Virus Res. 40, 181–211.

Rozen-Gagnon, K., Stapleford, K.A., Mongelli, V., Blanc, H., Failloux, A.B., et al., 2014. Alphavirus mutator variants present host-specific defects and attenuation in mammalian and insect models. PLoS Pathog. 10, e1003877.

Sabo, D.L., Domingo, E., Bandle, E.F., Flavell, R.A., Weissmann, C., 1977. A guanosine to adenosine transition in the 3' terminal extracistronic region of bacteriophage Qβ RNA leading to loss of infectivity. J. Mol. Biol. 112, 235–252.

Sachs, J.L., Bull, J.J., 2005. Experimental evolution of conflict mediation between genomes. Proc. Natl. Acad. Sci. U. S. A. 102, 390–395.

Salemi, M., Vandamme, A.M., 2004. The phylogenetic Handbook. A practical approach to DNA and Protein Phylogeny. Cambridge University Press, Cambridge.

Sanjuan, R., Nebot, M.R., Chirico, N., Mansky, L.M., Belshaw, R., 2010. Viral mutation rates. J. Virol. 84, 9733–9748.

Santiago, M.L., Greene, W.C., 2008. The Role of the APOBEC3 Family of Cytidine Deaminase in Innate Immunity, G-to-A Hypermutation, and Evolution of Retroviruses. In: Domingo, E., Webster, R.G., Holland, J.J. (Eds.), Origin and Evolution of Viruses, second ed. Elsevier, Oxford, pp. 183–206.

Sarisky, R.T., Nguyen, T.T., Duffy, K.E., Wittrock, R.J., Leary, J.J., 2000. Difference in incidence of spontaneous mutations between Herpes simplex virus types 1 and 2. Antimicrob. Agents Chemother. 44, 1524–1529.

Sheehy, A.M., Gaddis, N.C., Choi, J.D., Malim, M.H., 2002. Isolation of a human gene that inhibits HIV-1 infection and is suppressed by the viral Vif protein. Nature 418, 646–650.

Shriner, D., Rodrigo, A.G., Nickle, D.C., Mullins, J.I., 2004. Pervasive genomic recombination of HIV-1 in vivo. Genetics 167, 1573–1583.

Sicard, A., Yvon, M., Timchenko, T., Gronenborn, B., Michalakis, Y., et al., 2013. Gene copy number is differentially regulated in a multipartite virus. Nat. Commun. 4, 2248.

Simmonds, P., 2010. Recombination in the evolution of picornaviruses. In: Ehrenfeld, E., Domingo, E., Roos, R.P. (Eds.), The Picornaviruses. ASM Press, Whashington, DC, pp. 229–237.

Smith, E.C., Denison, M.R., 2013. Coronaviruses as DNA wannabes: a new model for the regulation of RNA virus replication fidelity. PLoS Pathog. 9, e1003760.

Smith, D.B., Inglis, S.C., 1987. The mutation rate and variability of eukaryotic viruses: an analytical review. J. Gen. Virol. 68, 2729–2740.

Smith, P.J., Jones, C.J., 1999. DNA Recombination and Repair. Oxfird University Press, Oxford.

Smith, E.C., Blanc, H., Vignuzzi, M., Denison, M.R., 2013. Coronaviruses lacking exoribonuclease activity are susceptible to lethal mutagenesis: evidence for proofreading and potential therapeutics. PLoS Pathog. 9, e1003565.

Stavrou, S., Crawford, D., Blouch, K., Browne, E.P., Kohli, R.M., et al., 2014. Different modes of retrovirus restriction by human APOBEC3A and APOBEC3G in vivo. PLoS Pathog. 10, e1004145.

Stedman, K., 2013. Mechanisms for RNA capture by ssDNA viruses: grand theft RNA. J. Mol. Evol. 76, 359–364.

Steinhauer, D.A., Holland, J.J., 1986. Direct method for quantitation of extreme polymerase error frequencies at selected single base sites in viral RNA. J. Virol. 57, 219–228.

Steinhauer, D.A., Domingo, E., Holland, J.J., 1992. Lack of evidence for proofreading mechanisms associated with an RNA virus polymerase. Gene 122, 281–288.

Steitz, T.A., 1999. DNA polymerases: structural diversity and common mechanisms. J. Biol. Chem. 274, 17395–17398.

Stratton, M.R., Campbell, P.J., Futreal, P.A., 2009. The cancer genome. Nature 458, 719–724.

Suzuki, T., Moriyama, K., Otsuka, C., Loakes, D., Negishi, K., 2006. Template properties of mutagenic cytosine analogues in reverse transcription. Nucleic Acids Res. 34, 6438–6449.

Svarovskaia, E.S., Delviks, K.A., Hwang, C.K., Pathak, V.K., 2000. Structural determinants of murine leukemia virus reverse transcriptase that affect the frequency of template switching. J. Virol. 74, 7171–7178.

Swetina, J., Schuster, P., 1982. Self-replication with errors. A model for polynucleotide replication. Biophys. Chem. 16, 329–345.

Szathmary, E., 1992. Viral sex, levels of selection, and the origin of life. J. Theor. Biol. 159, 99–109.

Sztuba-Solinska, J., Urbanowicz, A., Figlerowicz, M., Bujarski, J.J., 2011. RNA-RNA recombination in plant virus replication and evolution. Annu. Rev. Phytopathol. 49, 415–443.

Tanaka, T., Eusebio-Cope, A., Sun, L., Suzuki, N., 2012. Mycoreovirus genome alterations: similarities to and differences from rearrangements reported for other reoviruses. Front. Microbiol. 3, 186.

Thomson, M.M., Perez-Alvarez, L., Najera, R., 2002. Molecular epidemiology of HIV-1 genetic forms and its significance for vaccine development and therapy. Lancet Infect. Dis. 2, 461–471.

Upton, D.C., Gregory, B.L., Arya, R., Unniraman, S., 2011. AID: a riddle wrapped in a mystery inside an enigma. Immunol. Res. 49, 14–24.

Urbanowicz, A., Alejska, M., Formanowicz, P., Blazewicz, J., Figlerowicz, M., et al., 2005. Homologous crossovers among molecules of brome mosaic bromovirus RNA1 or RNA2 segments in vivo. J. Virol. 79, 5732–5742.

Valcarcel, J., Ortin, J., 1989. Phenotypic hiding: the carryover of mutations in RNA viruses as shown by detection of *mar* mutants in influenza virus. J. Virol. 63, 4107–4109.

Valente, L., Nishikura, K., 2005. ADAR gene family and A-to-I RNA editing: diverse roles in posttranscriptional gene regulation. Prog. Nucleic Acid Res. Mol. Biol. 79, 299–338.

Varela-Echavarria, A., Garvey, N., Preston, B.D., Dougherty, J.P., 1992. Comparison of Moloney murine leukemia virus mutation rate with the fidelity of its reverse transcriptase in vitro. J. Biol. Chem. 267, 24681–24688.

Vartanian, J.P., Henry, M., Wain-Hobson, S., 2001. Simulating pseudogene evolution in vitro: determining the true number of mutations in a lineage. Proc. Natl. Acad. Sci. U. S. A. 98, 13172–13176.

Velicer, G.J., Raddatz, G., Keller, H., Deiss, S., Lanz, C., et al., 2006. Comprehensive mutation identification in an evolved bacterial cooperator and its cheating ancestor. Proc. Natl. Acad. Sci. U. S. A. 103, 8107–8112.

Vignuzzi, M., Andino, R., 2010. Biological implications of picornavirus fidelity mutants. In: Ehrenfeld, E., Domingo, E., Roos, R.P. (Eds.), The Picornaviruses. ASM Press, Washington DC, pp. 213–228.

Vignuzzi, M., Stone, J.K., Arnold, J.J., Cameron, C.E., Andino, R., 2006. Quasispecies diversity determines pathogenesis through cooperative interactions in a viral population. Nature 439, 344–348.

Villarreal, L.P., 2005. Viruses and the Evolution of Life. ASM Press, Washington, D.C.

Vogt, P.K., Jackson, A.O. (Eds.), 1999. Satellites and defective viral RNAs. Current Topics in Microbiol. and Immunol. 239. Springer, Berlin.

Wainberg, M.A., Drosopoulos, W.C., Salomon, H., Hsu, M., Borkow, G., et al., 1996. Enhanced fidelity of 3TC-selected mutant HIV-1 reverse transcriptase. Science 271, 1282–1285.

Ward, C.D., Flanegan, J.B., 1992. Determination of the poliovirus RNA polymerase error frequency at eight sites in the viral genome. J. Virol. 66, 3784–3793.

Ward, C.D., Stokes, M.A., Flanegan, J.B., 1988. Direct measurement of the poliovirus RNA polymerase error frequency in vitro. J. Virol. 62, 558–562.

Webster, R.G., Bean, W.J., Gorman, O.T., Chambers, T.M., Kawaoka, Y., 1992. Evolution and ecology of influenza A viruses. Microbiol. Rev. 56, 152–179.

Webster, B., Ott, M., Greene, W.C., 2013. Evasion of superinfection exclusion and elimination of primary viral RNA by an adapted strain of hepatitis C virus. J. Virol. 87, 13354–13369.

Winters, M.A., Merigan, T.C., 2005. Insertions in the human immunodeficiency virus type 1 protease and reverse transcriptase genes: clinical impact and molecular mechanisms. Antimicrob. Agents Chemother. 49, 2575–2582.

Worobey, M., Holmes, E.C., 1999. Evolutionary aspects of recombination in RNA viruses. J. Gen. Virol. 80 (Pt 10), 2535–2543.

Yamaguchi, T., Yamashita, Y., Kasamo, K., Inoue, M., Sakaoka, H., et al., 1998. Genomic heterogeneity maps to tandem repeat sequences in the herpes simplex virus type 2 UL region. Virus Res. 55, 221–231.

Yang, Z., Bielawski, J.P., 2000. Statistical methods for detecting molecular adaptation. Trends Ecol. Evol. 15, 496–503.

Zaher, H.S., Green, R., 2009. Fidelity at the molecular level: lessons from protein synthesis. Cell 136, 746–762.

Zimmern, D., 1988. Evolution of RNA viruses. In: Domingo, E., Holland, J.J., Ahlquist, P. (Eds.), RNA Genetics. CRC Press Inc, Boca Raton, FL, pp. 211–240.

DARWINIAN PRINCIPLES ACTING ON HIGHLY MUTABLE VIRUSES

3

CHAPTER CONTENTS

ABBREVIATIONS

A   adenine
AD   average distance
AZC   azacytidine

Virus as Populations. http://dx.doi.org/10.1016/B978-0-12-800837-9.00003-4

cDNA	complementary deoxyribonucleic acid
C	cytosine
CTL	cytotoxic T lymphocyte
DI	defective interfering
E. coli	*Escherichia coli*
FMDV	foot-and-mouth disease virus
FU	5-fluorouracil
G	guanine
HBV	hepatitis B virus
HIV-1	human immunodeficiency virus type 1
HSV-1	virus herpes simplex type 1
HTLV-1	human T-lymphotropic virus type 1
HTLV-2	human T-lymphotropic virus type 2
Kb	kilobase
LCMV	lymphocytic choriomeningitis virus
MARLS	monoclonal antibody resistant FMDV mutant
MOI	multiplicity of infection
MV	measles virus
NGS	next generation sequencing
PAQ	partition analysis of quasispecies
PCR	polymerase chain reaction
PV	poliovirus
R	ribavirin
RdRp	RNA-dependent RNA polymerase
RNase	ribonuclease
RT-PCR	reverse transcription followed by polymerase chain reaction
s/nt	substitutions per nucleotide
s/s/y	substitutions per site and year
U	uracil
VSV	vesicular stomatitis virus

3.1 THEORETICAL FRAMEWORKS TO APPROACH VIRUS EVOLUTION

Population thinking is the term to refer to a view of biology that started at the beginning of the nineteenth century, and that grew as the result of observations on differences among individuals of the same biological group that were made by collectors of natural specimens and by animal and plant breeders. The accumulated evidence prepared a scientific environment that was ready to react in 1859 to Darwin's Origin of Species. As described by E. Mayr, the contribution of Darwin was threefold: (i) He provided evidence that evolution indeed occurred, (ii) He proposed natural selection as a mechanism for evolutionary change, and (iii) He replaced "typological" thinking by "population" thinking (Mayr, 1959). Concerning the third point, the traditional Platonic view of reality conferred to fixed "ideas" or "types" an essential value in the description of nature. Variations of the "types" were considered secondary, when not irrelevant, and not worthy of study. Plato had an impact on subsequent philosophical currents and permeated scientific thinking. The lesser value of "variations" of the standard types also impacted microbiology and virology (Domingo, 1999). As an extension of what E. Mayr explains for general evolution, virology is now confronted with the evidence that a virus "type" (we may call it the "wild type") is an abstraction, while the variations are real.

A major breakthrough in clarifying evolutionary thinking was the formulation known as "the modern synthesis" through work led by R.A. Fisher, J.B.S. Haldane, and S. Wright. The "synthesis" recognized the role of Darwinian natural selection acting on phenotypic variation, itself the result of genetic variation. It represented a unification of natural selection with Mendelian genetics. The "modern synthesis" provided a very clarifying framework that recognized the major forces of evolution, mainly genetic change and natural selection. The concepts contributed by the "synthesis" were in agreement with the molecular mechanisms of heredity whose details were unveiled in subsequent decades, and buried other appealing but incorrect views such as Lamarckism (the inheritance of modifications introduced in the body of organisms) (Mayr and Provine, 1980).

Since the first nucleotide and amino acid sequences of biological systems were obtained, procedures to establish evolutionary relationships among them were developed. The extent of sequence similarity indicated how closely homologous genes (those with recognized common ancestry) were. Phylogenetic trees were invented by W.M. Fitch and E. Margoliash in the middle of the twentieth century (Fitch and Margoliash, 1967). They calculate ancestry relationships and quantify the relatedness among multiple traits, mainly nucleotide or amino acid sequences. Phylogenetic procedures are currently abundantly used to approach virus evolution, and to estimate the time at which the most common ancestor of a set of genomes under comparison existed. Relatively conserved viral genes (typically those encoding nonstructural proteins) are useful to establish connections—order of ancestry and degree of genetic diversification—among relatively distant viruses, while variable genes (typically those encoding capsid and surface proteins) can define short-term relationships among epidemiologically related isolates (Chapter 7). There are excellent accounts of phylogenetic procedures that expand on the concepts and procedures sketched here (Salemi and Vandamme, 2004; Holmes, 2008, 2009). In particular, Chapter 1 of the M. Salemi and A.M. Vandamme book, written by A.M. Vandamme, is a concise and insightful introduction to key concepts which are relevant to virus evolution.

The extremely rapid formation of mutants during replication of viruses, in particular RNA viruses and also many DNA viruses (Chapter 2), introduces a new element to be considered in the studies of evolution. A relevant and debated question is whether biological features which are dependent on high mutability of viruses are amenable to description by classic population genetics or some new paradigm is needed. One such paradigm that has permeated virology is quasispecies theory, one of the mathematical formulations of molecular evolution that has exerted great impact in the understanding of RNA viruses (Section 3.6). A unanimous view is, however, that no matter which theoretical framework is adopted, the fundamental mechanisms involved in viral evolution are presided by Darwinian principles. We refer to extended Darwinian mechanisms which can be succinctly expressed as four major successive events: genetic variation, competition among variants, selection of the most fit, and random events.

Darwinian principles have been verified experimentally with many rapidly replicating cells and organisms, not only viruses. However, due to their mutability and rapid replication, viruses have become accepted systems to address experimentally basic questions of general evolution (Morse, 1994; Gibbs et al., 1995; Moya et al., 2000; Domingo et al., 2001). In fact, viruses have gone from being totally ignored by the most influential population geneticists in the middle of the twentieth century to being recognized as ideal systems to address problems in evolution that are very difficult to approach with differentiated organisms, including *Drosophila* (the anecdote goes that when last century a prominent population geneticist (whose name I omit) was asked to comment on microbial evolution, the response was: "Next question please"). One of the reasons to ignore viruses as systems to explore evolution could have been a lack of understanding of the real contribution of viruses to the biological world. For many decades it was not clear what the influence of virus replication was on their host cells and organisms, except being disease agents. There was no overview on what viruses might have represented either historically (in the early phases of evolution of life, Chapter 1) or currently (in present-day biology). Once the replicative mechanisms displayed by viruses were put into the perspective of general evolution, it became clear that at least part of the fundamental questions that have been addressed in basic general genetics can be studied experimentally with viruses.

Not a single theory has been able to capture the multiple features of virus evolution, and not a single formulation of population dynamics is totally independent and unrelated to others that approach variation of biological entities. K.M. Page and M.A. Nowak explained the relationships among several mathematical equations that describe evolutionary and ecological dynamics (i.e., replicator-mutator, Price, Lotka-Volterra, game dynamical, and quasispecies) (Page and Nowak, 2002) (Figure 3.1). The Price equation (from the American geneticist G. Price) is considered key to describe the behavior of evolving populations in general (for review of theoretical aspects, see Okasha, 2013). However, because of the conspicuous participation of mutation in virus evolution and the extreme heterogeneity

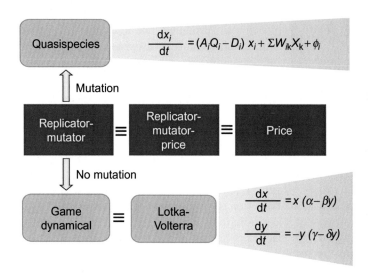

FIGURE 3.1

Correspondence among different equations that describe evolutionary dynamics, with or without participation of mutation. The middle row indicates the equivalence of three important equations for general evolutionary dynamics: the replicator-mutator, Price, and replicator-mutator-Price equations. The mathematical equivalences were described by K.M. Page and M.A. Nowak. From the replicator-mutator equation, the quasispecies equation is obtained when mutation is present, while the game-dynamical and Lotka-Volterra equations do not consider mutation. In the quasispecies equation dx_i/dt describes the concentration of mutant i as a function of time; x_k describes the concentration of mutant k; the replication, degradation and the synthesis of i from k are described by A_i, D_i and W_{ik}, respectively. Q_i expresses the fraction of correct copies of i produced as a result of copying i. Finally, ϕ_i corresponds to the flux of molecule i that in the most simple case it will be proportional to x_i. Concerning the Lotka-Volterra equation that describes the predator-prey relationship, x is the number of prey, y is the number of predator, dx/dt and dy/dt their variation as a function of time, and α, β, γ, and δ are parameters that describe the predator-prey interactions.

Figure based in data published in Page and Nowak (2002), Domingo et al. (2012), Domingo and Schuster (2016).

(or "polyploidy" using a classical term) of RNA viruses within infected cells, quasispecies theory is suitable for RNA viruses that display elevated mutation rates (Domingo and Schuster, 2016). Other equations of evolutionary dynamics may suffice to describe the population dynamics of DNA-based organisms, viruses displaying lower average mutation rates than RNA viruses, or competition dynamics when the mutation rate is not a relevant parameter (prey-predator relationship expressed by the Lotka-Volterra equation in Figure 3.1). The theoretical concepts contributed by classic genetics and the new quasispecies-based formulations have inspired experimental designs that are currently providing a deeper understanding of general evolutionary mechanisms, not only for viruses (Chapter 10).

3.1.1 THEORY AND EXPERIMENT

The connections that can be established between experimental studies and theoretical formulations (models of population dynamics with virtual replicons, *in silico* simulations that mimic virus replication, phase transitions in complex adaptive systems, bioinformatic approaches for the analysis of massive numbers of sequences, metagenomic investigations, etc.) are increasingly important in virology. The input of theory in virology has been greatly aided by the availability of powerful computers, together with developments in bioinformatics, including extensive information repositories in data banks (Chapter 7). Theory has inspired experiments and vice versa. For theoretical studies to help understanding virus behavior, it is important that the relevant algorithms include parameters based on solid experimental determinations. Realistic values are needed for mutation rates, rounds of replication, population size, frequency of advantageous and deleterious mutations, and distributions of fitness values in mutant spectra. Very often, predictions of virus behavior based upon simulations with virtual replicons cannot take into consideration the array of selective pressures that viruses encounter during their infectious cycles, because they are difficult to identify and thus to mimic. Keeping such limitations in mind, complementation between experiment and theory (using experimental data-driven models, in combination with exploratory models and predictive algorithms) has produced remarkable new insights on virus behavior as described in several chapters of this book.

Theoretical predictions based on unrealistic parameters are a continuous temptation to theoretical biologists eager to anticipate the behavior of real objects, in this case viruses. This is legitimate, provided predictions do not contradict experimental findings. An excessive use of theory in detriment of experimentation has created skepticism among experimentalists regarding the value of theoretical models. Excesses have to be avoided since, as M. Eigen puts it, at least in biology "pure theory" may be "poor theory" (Eigen, 2013). In particular, and to give a specific example, when a theoretical formulation includes the term "infinite" (∞), either referring to size or time, strange conclusions may be reached. Extreme numbers (zero or ∞) are justified as a tool, but predictions derived from their use in theoretical models should be subjected to the scrutiny of experimentation.

As clarifying as it was, the new theoretical framework provided by the "modern synthesis" did not consider a major concept that, having originated in physics, has pervaded modern biology: complexity (Solé and Goodwin, 2000) (Section 3.9). Important issues pertinent to the study of viruses at the population level, what we may term "population virology," relate to complexity. It is not obvious how to establish a connection between the theoretical treatment of complexity and experimental findings on virus evolution, but such connection would provide a fresh approach to important aspects of virus evolution and viral pathogenesis (see Sections 3.8 and 3.9).

3.2 GENETIC VARIATION, COMPETITION, AND SELECTION

Current evidence supports the view that virus evolution, both within infected host organisms, and as a result of extended rounds of replication and transmission between hosts, is guided by general Darwinian principles in their broad sense, including random events in addition to natural selection. Viruses undergo genetic variation through the mechanisms (mutation, recombination, and genome segment reassortment) summarized in Chapter 2. Variant forms of the same virus are continuously generated and compete with each other in changing environments. As a result of such competitions, and also as a consequence of random events, many (probably most) viral variants that are generated during replication are extinguished, while others continue to multiply in their host organisms with different replication efficiencies. The long-term reiteration of such processes has conformed the multiple pathogenic and nonpathogenic viruses we isolate and study.

Replication with the production of error copies in natural hosts has been extensively documented by infecting or transfecting cells or organisms with either biological or molecular clones (DNA or RNA transcripts from molecular cDNA clones that encode information to express an infectious virus). Transfection is the penetration of viral genetic material into cells rendered transiently permeable by physical or chemical means. Because mutations are produced during preparation of a biological clone from a single genome, and transcription has also limited fidelity (Fong et al., 2013; Xu et al., 2014), infection or transfection are necessarily performed with multiple variant genomes present at low frequencies. Infection and transfection are intrinsically inaccurate processes regarding which precise genomic sequences enter the cells. New repertoires of variants that differ from the input are unavoidably generated in the infected cells or organisms (examples and reviews in Domingo et al., 1978, 2001, 2008; Martell et al., 1992; Pfeiffer and Kirkegaard, 2005; Vignuzzi et al., 2006). Consensus sequences may give the appearance of invariance between an infecting and a newly replicating population, but mutant spectra must be analyzed to properly describe the events at the molecular level (Section 3.6.3). The second phase, that is, competition among variants, has been profusely documented *in vivo* and in cell culture. Competition is a continuous process due to limitation of resources. The latter include, in its broadest sense, host cell numbers, space confinements in the compartments of differentiated organisms, intracellular structures, biomolecules and metabolites, and others.

Following competition, some variants become dominant over others due to positive selection, unless random events intervene (Figure 3.2). While genetic variation takes place in genomes and affects genotypes, selection acts on phenotypes, the traits that result from the expression of genomes that have acquired the relevant mutations. The current picture is that different forms of selection (Section 3.4) together with random sampling are continuously and transiently shaping viral populations. The experiments designed to test relative viral fitness (an important parameter that captures the overall replicative capacity of a virus in a given environment) are based on competition between variants under controlled environmental conditions (Chapter 5).

Populations of viruses, and also microbes and unicellular organisms that can grow in culture, have the tremendous advantage that experiments to test Darwinian principles can be designed in the laboratory and the results recorded in days or even hours. A round of viral replication can be completed in minutes, and changes in viral genomes can be recorded in real time. Viruses provide scientists with an accelerated version of those basic molecular events that have presided long-term evolution of life on Earth. Although it is rather ambiguous to refer to numbers of "generations" of a virus (because viral genome replication is not merely a regular process of genome division), simple laboratory experiments can involve hundreds of viral passages, each one comprising several rounds of genome copying in thousands or millions of cells. For comparison, our human ancestors living 300 generations ago were in the preagriculture era!

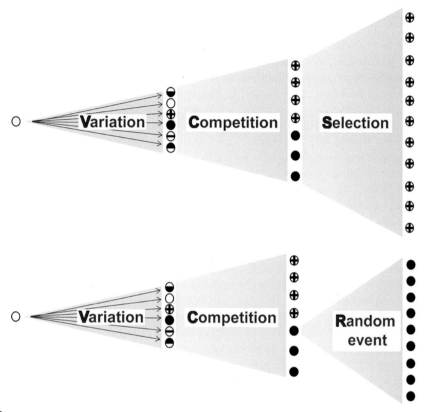

FIGURE 3.2

Random events may prevent a potentially advantageous mutant swarm from reaching dominance. Top: variation of an initial population, followed by competition and selection results in dominance of genomes represented by circles with a cross. Bottom: a random event may modify the outcome of selection and result in dominance of genomes represented by a filled circle.

3.3 MUTANT DISTRIBUTIONS DURING DNA AND RNA VIRUS INFECTIONS

The continuous presence of mutant distributions during RNA virus infections *in vivo* defines a key feature of viral populations which is relevant not only to interpret viral evolution but also viral pathogenesis. This statement is based on overwhelming evidence that has been gathered over the last four decades, and that will be further analyzed in this and subsequent chapters of this book. The main points that in my view constitute the foundation of what might be called the relevance of viral variation and population complexity in virus behavior have resulted from contributions of many authors, and several aspects can be found in specific articles and reviews (Holland et al., 1982; Kimata et al., 1999; Farci and Purcell, 2000; Domingo et al., 2001, 2008; Figlerowicz et al., 2003; Pfeiffer and Kirkegaard, 2005; Domingo, 2006; Vignuzzi et al., 2006; Lauring and Andino, 2010; Perales et al., 2010; Domingo and Perales, 2012; Domingo and Schuster, 2016).

Despite emphasis having been placed on the extreme variability and population heterogeneity of RNA viruses, studies on drug- and monoclonal antibody-escape mutants, as well as nucleotide sequence surveys of molecular and biological clones from virus isolates, indicate the presence of heterogeneous populations during infections by DNA viruses, including those with complex genomes that encode many proteins (Smith and Inglis, 1987; Sarisky et al., 2000; Domingo et al., 2001; López-Bueno et al., 2003; Quinlivan et al., 2004; Domingo, 2006). How heterogeneous are the populations of DNA viruses with large genomes, as compared to RNA viruses, is not known because the experimental evidence is limited. Comparative experiments of multiplication of viral clones of the simplest and the most complex DNA viruses in their host cells, as well as examination of their population heterogeneity after passage, are needed to respond to this question. It is generally accepted that simple DNA viruses whose replication is carried out by low-fidelity cellular DNA polymerases display population heterogeneities comparable to RNA viruses. Well-documented cases are those of parvoviruses, and the plant geminiviruses with the presence of multiple variants in one individual host (Isnard et al., 1998; Battilani et al., 2006; Lopez-Bueno et al., 2006; Ge et al., 2007). However, characterization of the DNA polymerases that replicate variable DNA viruses requires further research.

DNA viruses carrying a large amount of information in their genomes may have limitations to exploit generalized high mutation rates due to the lethal character of accumulation of mutations in a large genome, because of the inverse correlation between genome size and mutation tolerability (Chapter 2). However, variations at specific *loci* (such as at recombination or mutation hot spots, gain or loss of tandem repeats, copy number expansions, etc.) may occur frequently and may have an adaptive value (Elde et al., 2012). In an experiment intended to examine the genetic and phenotypic consequences of repeated bottleneck events, the complex DNA virus herpes simplex type 1 (HSV-1) behaved very similarly to RNA viruses (Jaramillo et al., 2012; further discussed in Chapter 6). Thus, as our capacity to penetrate into the composition of viral populations improves, nucleotide sequence heterogeneity of genomes during replication is increasingly seen as a shared feature of all viruses.

The average genetic distance among genomes of a population is expected to increase with the number of rounds of replication, following an infection initiated by one or few founder genomes (see also Section 4.1.2 in Chapter 4 on gain of mutant spectrum complexity following a population bottleneck). It is often difficult to know whether *in vivo* an initial founder infection (an infection started by one or a limited number of infectious particles) has occurred close to or far from the time at which heterogeneity of the viral population is examined. This is one of several reasons that limit our capacity to interpret comparative heterogeneity levels among natural viral isolates. In the case of retroviruses, another factor that influences heterogeneity is the time that the virus spends as a proviral element integrated into host DNA versus the time during which viral sequences experience transcription from DNA (proviral DNA copied into RNA) and reverse transcription (viral RNA copied into DNA). During viral DNA integration without expression (silent integration which is a form of latency), the provirus adheres to the evolutionary stasis typical of cellular DNA, with million-fold lower mutation rates than during reverse transcription. The two Deltaretroviruses human T-lymphotropic virus types 1 and 2 (HTLV-1 and HTLV-2), appear to be less variable than other retrovirus genera because of their extended latency periods as integrated proviruses. Their population dynamics is characterized by clonal expansions of the immune cells they infect (Chapter 7). Looking now at a Gammaretrovirus, it was calculated that the *v-mos* gene of Moloney murine sarcoma virus evolved at a rate of 1.3×10^{-3} substitutions per site and year (s/s/y) while its cellular homologue *c-mos* evolved at the million-fold lower rate of 1.7×10^{-9} s/s/y (Gojobori and Yokoyama, 1985). Several early studies confirmed a very different evolutionary rate of RNA viruses versus cells, an extremely important point first emphasized by J.J. Holland and his colleagues (Holland et al., 1982; Steinhauer and Holland, 1987). The biological consequences of divergent evolutionary tempos among components of our biosphere are probably still underappreciated.

The concept of population heterogeneity, shared by the great majority of viruses that display extensive replication as autonomous elements, was reached by comparing limited numbers of sequences from molecular and biological clones, due to technical limitations. The capacities of deep sequencing (also termed massive or new generation sequencing (NGS), with several, rapidly changing platforms (Beerenwinkel et al., 2012; Marz et al., 2014)) are expanding while this book is being written, and have confirmed the extreme heterogeneity of viral populations (see Section 3.6.4).

3.4 POSITIVE VERSUS NEGATIVE SELECTION: TWO SIDES OF THE SAME COIN

Two types of selection act on viruses: positive (also termed Darwinian, linear, dynamic, or directional) and negative (also termed purifying, normalizing, or stabilizing selection). Positive selection results in survival of subsets of variants because their replication and capacity to persist are superior to other variants of the same entourage. In terms of M. Eigen, "selection is the fixation of the genotype by the continual evaluation of the phenotype" (Eigen, 1992). A drug-resistant viral mutant is selected because its capacity to replicate in the presence of the drug is higher than that of other components of the viral population; the fixed genetic change is the nucleotide that encodes the amino acid that determines the resistance phenotype. Important examples of medically relevant, positive selection events are provided by the frequent occurrence in viral quasispecies of mutants with amino acid substitutions that mediate escape from antibodies, cytotoxic T lymphocytes (CTLs), or antiviral inhibitors (reviewed in Domingo et al., 2001; Figlerowicz et al., 2003; Gerrish and Garcia-Lerma, 2003; Domingo, 2006; relate to the concept of fitness treated in Chapter 5).

When a selective constraint that favors one such trait occurs naturally or is applied externally, the subset of genomes expressing the relevant trait will contribute the following generations of genomes. Thus, the level of heterogeneity of the parental viral population prior to the application of selection will be generally higher than that of the selected population because the latter has been founded by a minority of genomes that will participate in the ensuing rounds of replication in the presence of the selective agent. These types of selection events that lead to a decrease in the heterogeneity of viral populations or to a decrease in the polymorphism in host organisms have been termed "sweeping" selection events. A selection sweep can produce the hitchhiking of those mutations that accompany the selected mutation in the same genome. If hitchhiked mutations are independent of the selected mutation, their chance occurrence in the selected genomes can influence the future course of virus evolution. This is one of the mechanisms of antigenic variation in the absence of immune selection described in Chapter 4. Sweeping events may have an effect similar to a population bottleneck regarding a reduction of population complexity (Section 3.8.3).

Positive selection can reduce the diversity of a viral population by focusing the population toward one or a few genome types, but it may also result in a repertoire of variants harboring different genetic changes compatible with replication in the face of the selective constraint (Ciurea et al., 2000; Perales et al., 2005; Martin and Domingo, 2008). During evolution of parallel viral lineages, identical adaptive mutations may be found (de la Torre et al., 1992; Borrego et al., 1993; Couderc et al., 1994; Chumakov et al., 1994; Martín Hernández et al., 1994; Mateu et al., 1994; Lu et al., 1996; Brown et al., 2001; Ruiz-Jarabo et al., 2003; Herrera et al., 2008) while, in other cases, multiple molecular pathways for fitness gain can be observed (Escarmís et al., 1999; Wichman et al., 1999; Domingo et al., 2001; Tsibris et al., 2009).

Application of NGS to the quantification of mutant viral genomes when the population is responding to a selection event often reveals multiple, transient episodes of selection until a particularly well-suited (in terms of fitness) subpopulation overtakes the population. It would be interesting to explore whether an increasing complexity (genetic information content) of the virus genome may limit the number of alternative molecular pathways to respond to a selective constraint. Positive selection can contribute to biological novelty, and it is an integral part of the establishment of an emergent or reemergent virus in a host population (Chapter 7).

Negative selection acts to eliminate or to diminish the frequency of subsets of variants. Mutations that abolish regulatory elements or result in inactivation of an essential viral enzyme will generally be lethal. Lethality can be regarded as the extreme manifestation of negative selection. When genomic RNA acquires mutations that modify the secondary structure of an RNA motif involved in replication, the mutant genomes will be subjected to negative selection. Not all mutations inflict a disadvantage specifically to the genomes that harbor them. When an amino acid substitution diminishes by a factor of two the catalytic efficiency of a viral polymerase (or polymerization complex) the RNAs whose replication is catalyzed by the altered polymerase will not compete favorably with other genomes of the same replicative ensemble whose replication is catalyzed by a fully functional polymerase. Deleterious mutations in *cis*-acting functional elements—those that affect the same genome that harbors the element but not other genomes of the population—will result in negative selection acting on the genomes that include the deleterious mutations, but not others. In contrast, deleterious mutations that alter *trans*-acting functions—those that are expressed by one genome but that influence the replication of other genomes of the same population—may impair replication of many genomes, not only those that include the mutation. The negative effect of *trans*-acting gene products is difficult to quantify since it depends on the dose of the product and its relative affinity for the sites whose recognition determines the relevant biological activity. The proportion of *cis*- versus *trans*-acting functions in viruses, together with their capacity to accept nonlethal deleterious mutations, is relevant to the ease of generation and maintenance of defective interfering (DI) particles in a virus-host system. That proportion is also pertinent to the understanding of internal interactions within mutant spectra (Section 3.8), and to the interpretation of virus extinction by lethal mutagenesis through the mechanism of lethal defection by a class of defective genomes (Chapter 9). Deleterious mutations affecting *trans*-acting functions may be maintained in the viral population by genetic complementation, rendering negative selection dependent on the population context. A mutation in a genome that expresses a phenotype which is selected against when the genome replicates in isolation, may be perpetuated because the corresponding function is expressed by other genomes of the same replicative ensemble.

Despite general agreement in the distinction between positive and negative selection, there might be instances during virus evolution in which difference between the two modes of selection is not obvious. Suppression by negative selection of all but a limited subset of variant viruses in a population will have the same effect as positive selection acting on the spared subset (Domingo and Holland, 1994). Thus, we accept that a monoclonal antibody-resistant mutant of a virus has been positively selected because replication of all other viruses that did not include the required escape mutation(s) was suppressed (subjected to negative selection).

Positive and negative selection can be viewed as two facets of the same phenomenon, and in some theoretical treatments no distinction is made between these two modes of selection. The intensity of selection, expressed as a selection coefficient, can be measured through competition between two populations in the absence and presence of a selective constraint, often in the form of a fitness assay (Chapter 5).

Positive selection can be regarded both as a response for survival and as an exploratory process. Virus variants selected for a specific trait may, unpredictably, manifest an unexpected attribute, therefore, contributing to biological novelty (Baranowski et al., 2003) (Chapter 7). In contrast, negative selection is basically a conservative force that maintains the biological identity of viruses despite a dissipative mutational pressure, particularly accentuated in the case of RNA and DNA viruses which display error-prone replication.

3.5 SELECTION AND RANDOM DRIFT

Viral populations are of finite size, and the size undergoes modifications in the course of the natural history of any virus. It is very often reduced during transmission from infected hosts into uninfected recipients. The viral population size may decrease in an infected individual as a result of the immune response even though the virus may not be cleared. In this case, a subset of the initial viral population may be perpetuated by positive selection, for example, because it included key amino acid substitutions in B-cell or T-cell epitopes that conferred an advantage in the face of the immune response. Subsets of genomes need not continue replicating only because the others have been selected against, but merely by their chance survival during a random event of population size reduction such as when a virus abundant in an organ invades another organ through a limited access pathway. Therefore, the viral subsets that go on to reproduce and populate new niches are not necessarily those dictated purely by selection. In an infected organism a virus with high replicative capacity may multiply mainly in an internal organ such as the liver, causing liver injury. Yet, a particle that is not necessarily among the best adapted to multiply in the internal organ, may surface the infected host by chance and be transmitted to a susceptible individual through contact with secretions, excretions, or blood. Successive transmissions of this sort introduce an additional element of randomness in virus evolution (see also Section 3.2 and Chapter 4).

As a brief digression from viral populations, and to emphasize the role of contingency in natural events, we are all aware that our own personal existence is due to a succession of chance events that need not be detailed here. As science has progressed, we have learned that not only our existence but also that of the universe we inhabit is probably the result of chance episodes such as the generation of a particular pocket universe (one of the "megaverse") in which matter could be made and life thrive (according to recent cosmological theories). Even if this cosmological argument is not accepted, one should consider that mammals might not have had a chance to evolve toward hominids if the mass extinction 65 million years ago—with more than 50% of existing species, including dinosaurs, gone extinct—had not occurred. That particular mass extinction was a key event for the future of our biosphere, and it was likely produced by the chance collision of a large asteroid with Earth. Thus, random events are rooted in the way nature works, virus evolution included.

A point of interest and controversy in evolutionary virology as well as in general population genetics is the relative contribution of random events versus positive selection in shaping the genetic composition of viral populations. Random drift of sequences harboring neutral or quasi-neutral mutations (defined in Chapter 2), independent of positive selection, was proposed as the main mechanism of molecular evolution in the neutral theory championed by M. Kimura and his school (Kimura, 1983). In the second half of the twentieth century, the term "non-Darwinian evolution" was coined to refer to evolutionary changes due to random processes rather than to natural selection; the term has lost much of its initial impact.

Not only chance events are not incompatible with the ensemble of processes that we call "Darwinian," but it is also apparent that random evolutionary events act on biological objects that at some point must have been selected. The nature of the viral replication cycle, involving overtly competitive steps, suggests that before reaching a privileged survival position through random drift, a virus (or virus variant) had to face the scrutiny of selection. Viruses that survive by chance (i.e., the transmitted subset vs. the majority present in a donor infected host) will meet a situation in which they will compete with other variants. Extending the example of a virus replicating in the liver mentioned in the previous paragraph, if the liver is the major site of replication, the viral particles that circulate in the blood—and therefore can be transmitted through blood-to-blood contact—are likely to reflect those particles that replicated best in the liver. M. Eigen in his recent treatise (Eigen, 2013) discloses a conversation in which M. Kimura proposed to reformulate the Darwinian principle as "survival of the luckiest" (also published in Kimura, 1989). Eigen agreed but with the addition that the luckiest "always has to be a member of the elite club of the fittest" (Eigen, 2013). J. Holland and I used similar terms: "In RNA virus evolution it is not exactly the luckiest that perpetuates itself, it is the selected luckiest that does" (Domingo and Holland, 1994). I do not know if M. Eigen read our publication, but I take this coincidence to mean that J.J. Holland and I were on the right track in this issue.

Despite these arguments, the neutralist school of molecular evolution exerted a great influence in the understanding of the mechanisms that preside biological evolution, and enriched the way we think about the operation of selection, and the multiple factors that determine evolutionary outcomes. As we learn more about virus evolution, it appears that both, positive selection and random drift of viral genomes occur regularly during viral multiplication and spread. Viral populations continuously face episodes of selection and random drift, acting either conjointly or successively.

3.6 VIRAL QUASISPECIES

RNA viruses and some DNA viruses have high mutation rates as one of their distinctive features (Chapter 2), with a continuous production of variant genomes that provide a rich substrate for selection and random drift to act upon. Quasispecies became the most adequate theoretical framework for RNA virus dynamics because it had limited copying fidelity as a key parameter in its mathematical formulation (Figure 3.1), and emphasized the presence of mutant distributions during replication. No other models of population biology dealt with mutant spectra. Mutation is parallel to correct replication (Domingo and Schuster, 2016).

3.6.1 THE ORIGINS OF QUASISPECIES THEORY

The term quasispecies was first proposed by M. Eigen and P. Schuster to describe error-prone replication and the self-organization potential of primitive replicons postulated to have populated the Earth at the onset of life, some 4×10^9 years ago (Eigen and Schuster, 1979) (Chapter 1). Quasispecies theory was built on a first theoretical treatment by M. Eigen that described a system characterized by the regular generation of error copies during replication with limited template-copying fidelity (Eigen, 1971) (Figure 3.1). As explained by M. Eigen in his pioneer publication (this seminal paper was not accepted for publication in several journals because of its length (personal communication from C. Biebricher)), his interest was stimulated by S. Spiegelman's experiments on *in vitro* replication of Qβ RNA—which evidenced production and selection of mutant RNAs—as well as by a "breakfast

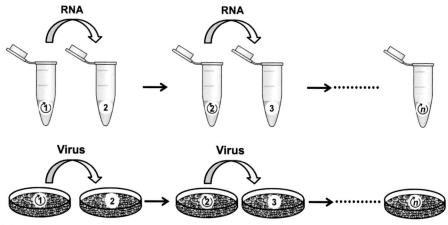

FIGURE 3.3

Serial passages with viral nucleic acid or with virus. Schematic depiction of serial passages as performed in experimental evolution. Top: The Spiegelman-Weissmann passages of Qβ RNA replicated by the enzyme Qβ replicase. Tube 1 contains the complete reaction mixture (Qβ RNA, Qβ replicase, host factor, nucleoside-triphosphates, and buffer with Mg^{2+}); after a chosen time of reaction, an aliquot from tube 1 is transferred to tube 2 which contains the complete reaction mixture except Qβ RNA. The process is repeated the desired number of times (n). Analogous designs have been implemented for the amplification of prions. Bottom: Serial passages of virus in cell monolayers or cells in suspension. Once an initial infection is completed in plate 1, the virus is titrated and the desired amount is used to infect the cells in plate 2. The process can be repeated n times. Several examples with cytolytic and persistent infections are described in connection with experimental evolution designs in Chapter 6.

discussion" with F. Crick, in which Crick expressed the interest of developing a mathematical description of replication with production of mutants. The so-called Spiegelman experiments referred to by Eigen involved serial amplification passages of bacteriophage Qβ RNA in the test tube. The basic design preceded many subsequent experiments on RNA evolution *in vitro*, the "serial passage" experiments of viruses and bacteria that today are familiar to virologists and bacteriologists (Figure 3.3), and even the protein misfolding cyclic amplification in the case of prions. The origin of serial passages may well go back to the attenuating passages carried out by L. Pasteur to develop the first human vaccines. Serial passages of viruses in cell culture or in live animal and plant hosts constitute one of the basic designs for experimental studies on virus evolution discussed in Chapter 6.

The experiments by S. Spiegelman and colleagues did not involve infection of living cells, but amplification of bacteriophage Qβ RNA (or its low molecular mass RNA derivatives) by purified Qβ replicase—the RNA-dependent RNA polymerase (RdRp) of the virus—in the test tube (Figure 3.3). Bacteriophage Qβ belongs to a group of RNA-containing bacterial viruses that infect *E. coli*, and that were discovered by N. Zinder and his colleagues (Zinder, 1975). RNA bacteriophages were instrumental in the development of molecular biology, as a follow-up of the fundamental work of M. Delbrück and his school with DNA bacteriophages. Their RNA genomes furnished the first heteropolymeric RNAs for studies on *in vitro* protein synthesis. Moreover, genomic Qβ RNA can be copied into a full length complementary (minus) strand by the enzyme Qβ replicase in the test tube, and continuous production of plus and minus strands can be achieved in the presence of the replicase, a host factor, nucleotide substrates, and a suitable ionic environment (Weissmann, 1974). To avoid the nucleotide substrates being exhausted, an aliquot of the reaction mixture is transferred into a fresh mixture containing the necessary ingredients for synthesis to proceed. Many transfers can be performed, attaining remarkable amplifications of RNA in the test tube so that the initial template molecules are no longer present at the end of the transfer series (representative publications of Spiegelman's group on *in vitro* RNA evolution are Haruna et al., 1963; Mills et al., 1967; Saffhill et al., 1970). The Darwinian mechanisms acting on Qβ RNA molecules were one of the inspirations for M. Eigen to develop quasispecies theory.

The origin of self-organization of primitive replicons, one of the main aims of M. Eigen's treatment, represented a link between the principles of Darwinian evolution and classical information theory (Eigen, 1971). "Self-instructive" behavior was distinguished from "general autocatalytic" behavior, and it was pointed out that "self-instruction" is required for "template" activity. A quality factor was defined that determines the fraction of copying processes that leads to an exact copy of the template (the relevant equation is given in Figure 3.1). The "master copy" will produce "error copies" which will occur with a certain probability distribution. The error copy distribution is what today we term "mutant spectrum" or "mutant cloud," but "comet tail" of error copies was the term used by Eigen (1971).

In subsequent elaborations of the theory by M. Eigen, P. Schuster, and their colleagues, quasispecies typified infinite, steady-state distributions of related genomes in equilibrium (Eigen and Schuster, 1979; Eigen and Biebricher, 1988; Eigen, 1992; Schuster and Stadler, 1999). The equilibrium is usually metastable: it will be replaced by a new one represented by another mutant distribution when an advantageous mutant appears in the first distribution. A treatment that considers a mutant spectrum of infinite size in equilibrium implies evolution as a deterministic process. This means that whenever a mutant with a selective advantage is produced, it will inevitably outgrow the former distribution. A deterministic model has a mathematical solution, and it is a frequent first formulation, to be followed by more realistic extensions that include stochastic (chance) components. This is a common way to proceed in theoretical biology, not only in the progression of quasispecies theory.

3.6.2 DETERMINISTIC VERSUS STOCHASTIC QUASISPECIES

In considering an application of quasispecies theory to the dynamics of viral populations, it was obvious that viruses (as opposed to theoretical mutant spectra) are not steady-state populations of infinite size in equilibrium when they replicate in their natural hosts. Neither were in equilibrium the primitive replicons at the onset of life whose behavior was addressed by quasispecies theory (unless the Earth 4 billion years ago included oceans filled with million-trillions of replicons under steady-state conditions that defy our imagination). A deterministic model is not realistic for RNA viruses or for any other natural replicative entity, a limitation that has been amply recognized by theoreticians and virologists. The nondeterministic nature or mutant generation has deep roots in how mutant nucleic acids are generated, and how a mutant may reach dominance in a viral population. First, mutagenesis results from atomic fluctuations subject to quantum-mechanical uncertainty; for example, those fluctuations that determine the presence of one tautomeric form or another in a template nucleotide, precisely at the time at which an incoming nucleotide reaches the catalytic site of the polymerase elongation complex (Chapter 2). The precise protein conformation at the catalytic site and its surrounding residues is also subjected to fluctuations and can contribute to incorporation uncertainties. Second, any growth-competition process is also subjected to statistical fluctuations, which may alter the outcome anticipated by a deterministic theory. Third, we know now that a new ingredient has to be introduced in the way selection operates: a newly arising mutant that shows a selective advantage need not outgrow the population. Its dominance is dependent on the mutant spectrum context, a key issue in viral quasispecies behavior addressed in Section 3.8.

A deterministic behavior can be approached when the viral population size is large, and the physical and biological environments remain constant during virus multiplication. In relation to viral pathogenesis, the alternations of episodes of replication of large populations and reductions to small populations, including severe bottleneck events, can give rise to successions of nearly deterministic steps followed by highly nondeterministic episodes (compare with selection and random drift discussed in Section 3.5). Alternations among population regimes are a key factor of the inherent unpredictability of virus evolution.

An important consideration is whether the generation of mutant distributions adheres to the quasi-equilibrium model as defined in theoretical biology (Domingo and Schuster, 2016). That is, whether the formation of mutants and their equilibration occurs faster than the environmental changes that the mutant distributions must confront. This is likely to be fulfilled by the RNA replicons displaying high mutation rates in the microenvironment of a single replicative unit inside a cell. Thus, in keeping with the original deterministic formulation of quasispecies theory, it would be legitimate to view the modification of mutant spectra as a succession of short equilibrium steps, each of them following the requirements of the deterministic quasispecies theory. In fact, such an abstraction provides an intuitively useful way to view virus evolution as a function of time (Figure 3.4), and it is not an unusual notion in biological studies. In a discussion of virulence models for host-parasite relationships, successive equilibrium steps were proposed in relation to numbers of uninfected hosts, parasite virulence levels, and variation in the optimal transmission rate (Lenski and May, 1994). Even more fundamental, the Hardy-Weinberg law of population genetics states the constancy of the frequency of alleles in a randomly mating, large population (also termed "panmictic" population), as a direct consequence of Mendelian heredity. It has attached to it a formula that predicts the frequency of alternative alleles in a randomly mating diploid population. Yet, as emphasized by E.D. Wilson, there are "some large conditions attached to the use of the Hardy-Weinberg formula in the real world" (Wilson, 1998). The same limitations have been expressed by other authors. A number of simplifications—including finite population sizes—must be conceded to approximate the behavior of real populations to the predictions of the Hardy-Weinberg law. This does not diminish its value as a theoretical framework to understand the behavior of mating diploid populations.

The criticism that a theoretical model does not reflect real situations in a precise manner could be made virtually of any model, including those of population genetics. Also significant in this regard is a statement by M. Eigen when referring to specific conditions in the experimental implementation of the theoretical treatment of selection: "Important as these specifications of defined conditions are for an understanding of the principles of evolution and for a quantitative evaluation of experimental data, it is not in the least necessary that any real evolutionary process in nature should have taken place under these special conditions—just as no steam engine ever had to work under the exactly specified thermodynamic equilibrium conditions of the Carnot cycle" (Eigen, 1971). Thus, based on these several scientific precedents, as well as many others one could draw from connections between theoretical principles and their experimental implementation in physics, chemistry, and biology, it is not justified to raise the non-equilibrium argument against viral quasispecies.

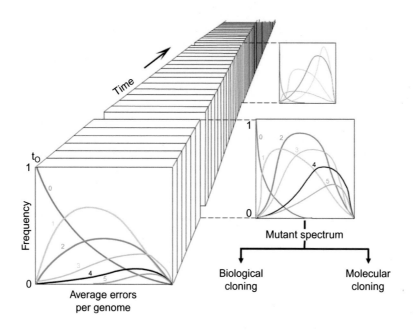

FIGURE 3.4

A representation of temporal evolution of a viral queasispecies as a succession of very short equilibrium steps. At any time point in the evolution of a viral quasispecies (here represented by a "slice" of time) a mutant distribution can be considered a "frozen" distribution of genomes with 0, 1, 2, 3, 4, and 5, etc. mutations at the frequencies given in ordinate. As time elapses the different genome categories vary in frequency, and the process of change can be considered as consisting of a succession of equilibrium steps ("slices" of time). Molecular and biological cloning provide a snapshot of mutant distributions that in reality are changing continuously.

The issue has been also brought up whether quasispecies is a necessary concept to understand virus evolution or not. C.O. Wilke emphasized that there is no conflict between quasispecies and classical Fisher-Wright models (Wilke, 2005). A reason why virologists did not adhere to classical population genetics models of evolution might be that in the middle of the twentieth century microbes were not considered satisfactory systems to approach evolution. In addition, most virologists did not find in the formulations of population genetics a means to grasp the key observations related to the presence of mutant spectra and the potential of rapid evolution of RNA viruses. Population geneticists ignored viruses, and virologists did not establish the connections between observations on viral populations and the body of conceptual and mathematical tools available from population genetics. In contrast, since the very beginning, in the 1970s, actors involved in the development of quasispecies theory, and those that contributed the first experimental observations on RNA virus heterogeneity, made a point of connecting the two areas of scientific activity (Domingo and Wain-Hobson, 2009). Perhaps the key issue can be summarized with the following statement which was addressed to authors who object to the term "quasispecies": "Whether they are termed quasispecies or not, we are dealing with a very unique population structure that has defied antiviral interventions despite huge economic investments, and that, presently, is helping to interpret virus behavior and to design new means to control viral disease. As Shakespeare said: *A rose by any other name would smell as sweet*" (quoted from Perales et al., 2010).

3.6.3 MUTANT SPECTRA, MASTER GENOMES, AND CONSENSUS SEQUENCES

Extensions of quasispecies theory to finite populations and changing environments (variable fitness landscapes (Chapter 5)) have been developed (Eigen, 2000; Wilke et al., 2001; Nowak, 2006; Saakian and Hu, 2006; Saakian et al., 2006, 2009; Park et al., 2010; Schuster, 2010; Domingo et al., 2012); reviewed in Domingo, 2006; Domingo and Schuster, 2016). In the way virologists view quasispecies, a mutant distribution may be dominated by one or several master sequences whose preponderance may be transient or extended, dependent partly on environmental stability. The master genome dominates because it displays the highest fitness value in the mutant spectrum. A dominant master and the consensus sequence may or may not be identical (Figure 3.5). Box 3.1 describes some terms used in quasispecies descriptions.

Recombination (Chapter 2) is an added element of stochasticity and a factor of population disequilibrium that has been also integrated into quasispecies theory (Boerlijst et al., 1996; Park and Deem, 2007; Muñoz et al., 2008). Thus, quasispecies theory has incorporated environmental and molecular

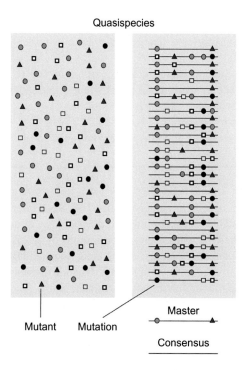

FIGURE 3.5

Two ways of representing a viral quasispecies. The symbols in the population depicted on the left mean different mutants expressing a relevant phenotypic trait (drug-resistance or other); identical symbols express the same phenotype but they may differ in genomic nucleotide sequence. The representation on the right illustrates individual genomes as horizontal lines and mutations as symbols on the lines. Note that the master (dominant) sequence does not coincide with the consensus, as explained in the text. The two types of representation are used in this book, depending on the concept to be conveyed.

BOX 3.1 DEFINITIONS AND TERMS RELATED TO VIRAL QUASISPECIES

- Physical definition of quasispecies
  Cloud in sequence space.
- Chemical definition of quasispecies
  Rated distribution of related but non-identical nucleic acid sequences.
- Biological definition of quasispecies
  Target of selection.
- Medical significance of quasispecies
  The many drug resistant mutants present or arising in infected patients that may result in treatment failure.
- Usual definition by virologists
  An ensemble of viral genomes subjected to genetic variation, competition, and selection, and that can act as a unit of selection.
- Master sequence in a viral quasispecies
  The genomic nucleotide sequence that is present at the highest frequency in a mutant distribution because of its higher fitness than the other genomes. The master sequence may still be a minority relative to the ensemble of low frequency variants.
- Consensus sequence
  The sequence that has at each position (nucleotide or amino acid) the residue which is most abundant at the corresponding position in a set of aligned sequences. The consensus sequence need not coincide with the master sequence.
- Mutant spectrum of a viral quasispecies
  The ensemble of mutant genomes that make up a viral quasispecies. Alternative terms are mutant distributions, clouds or swarms.

---

events that perturb equilibrium, as well as epigenetic effects or RNA interference in the process of template copying. Overall, these additions have reinforced the adequacy of quasispecies as a framework to understand virus evolution. There have been several definitions of quasispecies depending on whether a perspective from physics, chemistry, biology, or medicine is taken (Box 3.1). The key concept for virology is that viral populations consist of complex and dynamic distributions of related genomes, and that the ensemble often plays a role in determining virus behavior.

The first experimental evidence that an RNA virus population consisted of a spectrum of mutants was obtained in C. Weissmann's laboratory in Zürich, working with bacteriophage Qβ replicating in its natural host *E. coli*. The studies included the calculation of a mutation rate for a specific nucleotide transition at the 3′-extracistronic region of Qβ RNA (Batschelet et al., 1976), and the demonstration that biological clones (the progeny of a single viral genome) evolved rapidly to form complex spectra of mutants. A key sentence that summarized the results is: "A Qβ phage population is in a dynamic equilibrium, with viable mutants arising at a high rate on the one hand, and being strongly selected against on the other. The genome of Qβ phage cannot be described as a defined unique structure, but rather as a weighted average of a large number of different individual sequences" (Domingo et al., 1978). Subsequent studies with other RNA viruses, some DNA viruses, and subviral replicons revealed that the same statement applies to their population structure in cell culture and *in vivo* (Holland et al., 1982; Eigen and Biebricher, 1988; Martell et al., 1992; Duarte et al., 1994; Morse, 1994; Coffin, 1995; Gibbs et al., 1995; Isnard et al., 1998; Vogt and Jackson, 1999; Gavrilin et al., 2000; Sarisky et al., 2000; Domingo et al., 2001; López-Bueno et al., 2003; Domingo, 2005, 2006); for review see Domingo et al. (2012), and for historical accounts of the early studies on theoretical and viral quasispecies and their impact, see Domingo and Wain-Hobson (2009) and Domingo and Schuster (2016). The experimental results with viruses fit the theoretical quasispecies introduced in previous sections, and provided the biological definition of quasispecies (Box 3.1) (Eigen and Biebricher, 1988; Domingo et al., 2001; Domingo and Schuster, 2016). The quasispecies concept has extended its applicability to nonviral, cellular and macromolecular assemblies (Domingo and Schuster, 2016), and such interesting ramifications are treated in Chapter 10.

The mutant spectrum nature of viral populations is full of implications. Modifications of host cell tropism, failures of synthetic vaccines to produce solid and durable protection, and the very frequent selection of virus mutants resistant to antiviral agents, are some of the important events now readily understood on the basis of the dynamic mutant spectra on which selection acts. Indeed, swarms of mutant genomes are continuously being placed under the test of selection, and mutants resistant to antibodies or inhibitors can arise and be maintained as minority components in the population until they increase in frequency in the presence of the relevant antibodies or inhibitors. Moreover, the application of quasispecies theory to viruses has opened new avenues of antiviral research, including one termed virus entry into error catastrophe or lethal mutagenesis. The problem that quasispecies dynamics poses for the control of viral disease and the different strategies to overcome the adaptive capacity of mutant spectra are treated in Chapters 8 and 9. Lethal mutagenesis is based on one of the corollaries of quasispecies theory that asserts that in any replication system there is a maximum information content which can be maintained for a given average copying fidelity (termed by M. Eigen the "recognition parameter" q). This is the origin of the error threshold concept, and of virus entry into error catastrophe or lethal mutagenesis, the basis of new strategies to combat viral disease (Chapter 9). Here again, the concept of error threshold has been extended to finite populations and variable fitness landscapes, that is, under nonequilibrium conditions (Nowak and Schuster, 1989; Alves and Fontanari, 1998; Hogeweg and Takeuchi, 2003; Ochoa, 2006; Domingo and Schuster, 2016).

3.6.4 MEASUREMENT OF QUASISPECIES COMPLEXITY: INSIGHTS FROM NEW GENERATION SEQUENCING

At any given time during replication a viral quasispecies consists of a complex mutant spectrum, from which a consensus sequence can be derived. (Figure 3.5 and Box 3.1). Contrary to the term "consensus" applied to molecular biology (consensus promoter sequence, TATA box, polyadenylation signal, nuclear localization signal, etc.) in the case of viral quasispecies the use of consensus does not imply universality. Independent or even sequential populations of the same virus can be characterized by different consensus sequences. An identical consensus may characterize different mutant spectra displaying disparate behaviors. The same consensus may hide mutant spectra of very different complexity.

Complexity of the mutant spectrum can be quantitated by measuring the mutation frequency (minimum and maximum), Shannon entropy, and average genetic distances (or Hamming distances),

---

BOX 3.2 PARAMETERS USED TO ESTIMATE THE COMPLEXITY OF MUTANT SPECTRA

- Minimum mutation frequency
  The number of different mutations scored in a set of sequences divided by the total number of nucleotides sequenced (expressed as substitutions/nucleotide (s/nt)).
- Maximum mutation frequency
  The total number of mutations (be they repeated or not) in a set of sequences divided by the total number of nucleotides sequenced (expressed as substitutions/nucleotide (s/nt)).
- When mutation frequency refers to a specific mutation it is called mutant frequency.
- Shannon entropy
  The proportion of different sequences in a set of sequences. A value of 0 means that all compared sequences are identical; a value of 1 means that each compared sequence is different from the others.
- Average genetic (or Hamming) distance
  The average number of mutations that distinguishes any two sequences in the set under study.
- Parameters also obtained from NGS: number of different mutations, number of polymorphic sites, number of haplotypes.

---

and additional parameters derived from NGS (Gregori et al., 2014; Seifert and Beerenwinkel, 2016) (Box 3.2). To obtain Sanger nucleotide sequences of the individual components of a mutant spectrum, either RNA or DNA from biological clones (the progeny of a single viable genome), or DNA or cDNA from molecular clones (the progeny of a genome copied and amplified *in vitro*) have been used (Figure 3.6). Each of these procedures has some advantages and limitations. When biological clones (virus from individual viral plaques formed on a cell monolayer, or virus from end-point dilution series) are the starting materials, the genomic nucleic acid from each clone or clonal population is extracted. Then, the viral DNA or RNA is subjected to amplification by PCR (polymerase chain reaction) or RT-PCR (reverse transcription followed by polymerase chain reaction), respectively, and the amplified material is sequenced. The result is an average or consensus sequence of the sampled population of genomes that are part of each biological clone. As expected from viral mutation rates and from the rounds of viral amplification needed for plaque development, a viral biological clone is generally not homogeneous (Yin, 1993). The nucleotide sequence obtained cannot be affected by possible misincorporations in individual copies introduced as a consequence of the limited copying fidelity of the enzymes used for the *in vitro* amplification, unless the amount of template used for the amplification reaction is very small (amounting to a few molecules). Misincorporations can alter a genomic position in one of the amplified genomes, but not the average or consensus sequence representative of the clone (Figure 3.5). Obviously, this method biases the sequence information in favor of clones that are more infectious in the cells (that is, produce visible plaques or infection foci) used for the biological cloning. Genomes from noninfectious viral particles are excluded from the analysis unless complemented within a growing plaque, a possibility that cannot be excluded.

The alternative procedure, also using Sanger sequencing, is to obtain molecular clones from total viral DNA or RNA (or an end-point dilution series) followed by PCR or RT-PCR amplification and cloning of individual DNA or cDNA copies in a vector (generally an *E. coli* plasmid) (Figure 3.6).

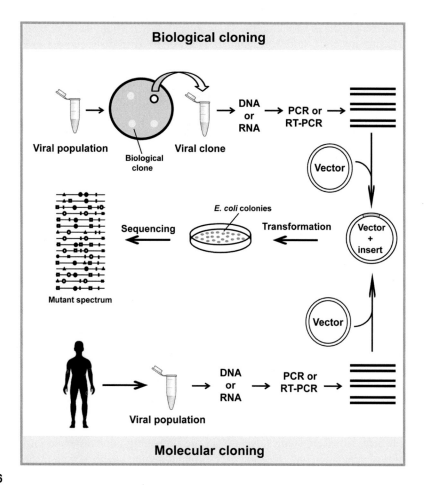

FIGURE 3.6

A schematic representation of steps involved in biological and molecular cloning for the analysis of viral quasispecies. Top: Biological cloning. A viral population is plated on a cell monolayer to obtain virus from an individual plaque (biological clone). DNA or RNA is extracted and the viral sequences amplified by PCR or RT-PCR, respectively. The amplified DNA is ligated to a vector and the product used to transform *E. coli*. Viral DNA inserted into individual bacterial colonies is further amplified and sequenced to provide a sample of genomic sequences present in the initial viral population. Bottom: Molecular cloning. DNA or RNA is extracted from a biological sample and the viral sequences amplified by PCR or RT-PCR, respectively. The products are ligated to vector DNA and the procedure completed as indicated for biological cloning. See text for controls and possible biases in mutant representation.

This method does not depend on infectivity of viral particles or genomes and, therefore, can analyze defective, noninfectious genomes, or probe the composition of viral preparations independently of their infectivity in an infected organism or cell culture. A bias may be produced by the limited copying fidelity of enzymes used in the *in vitro* amplification. The problem can be largely solved using high-fidelity polymerases and correct amplification conditions, together with control experiments to determine the basal error rate inherent to the *in vitro* amplification protocol. When a population of viral DNA or RNA is used as the starting material, sufficient template DNA or RNA must be present to ensure that the sequence repertoire obtained approximates that of the viral genomes present in the population under study (Arias et al., 2001; Airaksinen et al., 2003; Polyak et al., 2005; Domingo, 2006).

Despite their value in revealing the complexity of viral populations, cloning methods are limited by the number of sequences that are obtained, often less than 100 from a viral sample. The advent of NGS in its several platforms has represented an important progress in genetics (Mardis, 2008; Fox et al., 2009). Some of the NGS procedures can provide ten thousands to several hundred thousands sequences from the same sample, be it a biological specimen (tissue, secretion, excretion, tumor, entire animal, or plant) or an environmental sample (sea or lake water, soil, or toothbrush). Putative new viruses and others related to established viral pathogens have been identified by NGS, attesting of the ubiquity of viruses in our biosphere (Delwart, 2013; Drexler et al., 2013) (Chapter 1). NGS is gradually superseding cloning-Sanger-based analyses of mutant spectra.

The increased penetration into the composition of mutant spectra has not only confirmed, but even reinforced a generalized extreme population heterogeneity of viral populations, with myriads of low-frequency variants that could not be detected by classic clonal surveys. Reliable frequencies of minority mutations reach 1% level in viral populations, and some analyses have achieved 10- and even 100-fold lower cut-off frequency values. Some studies have compared NGS with molecular cloning-Sanger sequencing to characterize virus populations (Eriksson et al., 2008; Kuroda et al., 2010; Zagordi et al., 2010; Liang et al., 2011; Vandenbroucke et al., 2011; Wright et al., 2011; Gregori et al., 2014). Not only the well-confirmed cases of HBV, HCV, and HIV-1, but also pathogens that produce sporadic but severely lethal outbreaks such as Ebola virus replicate as complex quasispecies as revealed by NGS (Gire et al., 2014). Despite their discriminatory potential, NGS techniques have not unveiled any fundamentally new concept that had not been known from application of cloning-Sanger methodology. This may be due to the established fact in sampling theory that the accuracy of sampling to represent a collection of objects (in this case genomic sequences) depends more on ensuring randomness than in increased sample size. However, studies using NGS are furnishing unprecedented information about the fine dynamics of adaptation and selection events, including the kinetics of dominance of drug- or immune-escape mutant subpopulations, as discussed in coming Chapters.

For the characterization of mutant spectra, NGS techniques have the limitation of a high frequency of technical errors (*indels* and point mutations) that may artifactually bias mutation frequency values. Molecular amplification and bioinformatic procedures have been developed that allow reliable assignment of mutations present at frequencies in the range of 0.01-1% (depending on the genome copying and amplification protocols, as well as the correcting operations and controls performed) (Zagordi et al., 2010; Jabara et al., 2011; Vandenbroucke et al., 2011; Chen-Harris et al., 2013; Poh et al., 2013; Acevedo et al., 2014; Gregori et al., 2014). These procedures have established the number of different mutations, number of haplotypes and nucleotide diversity as parameters to reflect the complexity of a viral population using NGS. The parameters calculated from the NGS data can be complemented by the measurement of mutation frequency using standard Sanger sequencing of biological or molecular clones. To limit the number of artifactual mutations, R. Andino and colleagues have devised a method that consists in fragmenting and circularizing the RNA genomes to be sequenced, and synthesizing tandem cDNA repeats (redundant copies) by rolling circle replication. Only those mutations present in each of the repeats of the same original templates are *bona fide* mutations existing in the RNA population (Acevedo et al., 2014). Applied to poliovirus (PV), this CirSeq procedure using the Illumina platform has confirmed the presence of multitudes of PV variants at frequencies of 10^{-3} to 10^{-5}, giving an average of 1 to 2 mutations per genome. Curiously, but perhaps not coincidental, this estimate of heterogeneity is exactly the same obtained for bacteriophage Qβ 40 years ago using T1-oligonucleotide fingerprinting of genomic RNA (Domingo et al., 1978). Comparative massive sequencing allows a molecular dissection of quasispecies dynamics during cross-species transmissions, and the identification of minority mutations that become dominant at later stages of evolution (Borucki et al., 2013) (see concept of "harbinger" mutations in Chapter 5). Moreover, NGS reveals that there are "underground" swarms of very low frequency genomes whose fitness, despite being very low, is not zero. NGS is like a magnifying glass that visualizes genomes that did not exist according to prior techniques. The once considered "lethal class" may actually include poorly replicating genomes; NGS is acting as an underground police that watches events occurring at a previously hidden level.

The unprecedented penetration in the genomic composition of mutant spectra has important clinical implications for patient management. It allows the quantification of drug-resistance mutations in virus from treated or untreated infected patients, and of patients prior to treatment (Wang et al., 2007; Mitsuya et al., 2008; Solmone et al., 2009; Tsibris et al., 2009; Hedskog et al., 2010; Svicher et al., 2011). The methodology permits detection of mixed infections with different genotypes of the same virus that would go unnoticed by standard cloning analyses (Margeridon-Thermet et al., 2009; Wu et al., 2013; Quer et al., 2014). The capacity to interrogate dynamics of minority quasispecies components has also revealed multiple antibody- or CTL-escape routes, some of them only transient, others successful, during episodes of immune selection (Fischer et al., 2010; Cale et al., 2011) (Figure 3.7). These important implications for the control of viral disease are further treated in Chapters 8 and 9.

The extremely complex and dynamic mutant "mixtures" render the current data banks a minimal and insufficient representation of viral identity (Domingo et al., 2006). The real composition of viral populations demands also new classification approaches to group components of mutant spectra (either from one isolate, from sequential isolates from one infected host, or from different hosts). A nonhierarchical, clustering methods termed partition analysis of quasispecies (PAQ) was developed prior to the routine use of deep sequencing (Baccam et al., 2001). PAQ groups those viral sequences that are separated by the shortest genetic distances. In a pictorial representation, a center genotype nucleates a group of sequences within a circle with a previously set radius. In this fashion, multiple, overlapping or nonoverlapping groups of sequences can be defined within mutant spectra of viral quasispecies. PAQ has been applied for the analysis of subpopulations of equine infectious anemia virus (Baccam et al., 2003), regulatory sequences of bovine viral diarrhea virus (Jones et al., 2002), sequential hepatitis A virus isolates (Costa-Mattioli et al., 2006), and mutant spectra of foot-and-mouth disease virus (FMDV) subjected to lethal mutagenesis (Ojosnegros et al., 2008) (Figure 3.8). Other algorithms for quasispecies reconstruction are based on a sliding window approach to compare reads derived from massive sequencing with a reference sequence. Computational methods to organize and interpret the increasing numbers of minority variants being discovered in viral quasispecies have been developed (Prosperi et al., 2011; Poh et al., 2013; Gregori et al., 2014; Mangul et al., 2014; Topfer et al., 2014; for review see Marz et al., 2014). Elucidating which information is biologically and medically relevant, and worthy of inclusion in data banks is an important challenge.

Although generally less informative than sequencing, alternative, rapid sampling procedures have been successfully used to compare relative genetic heterogeneities of viral populations: restriction enzyme site polymorphisms, RNase A mismatch cleavage (Sánchez-Palomino et al., 1993), DNA heteroduplex mobility assay, oligonucleotide mapping, walking within mutant spectra by successive, specific

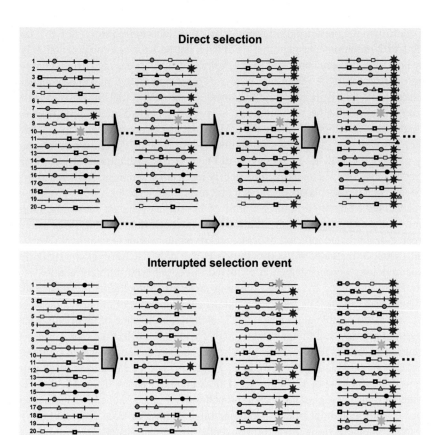

FIGURE 3.7

Transient selection events. In the upper part, genomes with a mutation represented by a red asterisk are directly selected because it confers an adequate phenotype for the existing environment. In the bottom part, genomes with a mutation represented by a green asterisk are transiently selected, until they are overgrown by genomes with the alternative mutation represented by the red asterisk. Application of NGS to viral populations undergoing selection is revealing large numbers of transient and unsuccessful selection pathways that were previously unnoticed due to insufficient penetration into the mutant spectrum. See text for references and operating mechanisms.

PCR amplifications (Garcia-Arriaza et al., 2007), among others. DNA microarrays have become a promising technology for diagnostic virology and to characterize viral populations, with prospects of an application to quantify the complexity of mutant spectra (Amexis et al., 2001; Wang et al., 2002; Mount, 2004; Briones and Domingo, 2008; Lauring and Andino, 2011). These methods have the advantage of comparing very large numbers of samples by automated procedures. If not coupled with sequence determinations, however, they lack the precision of information conveyed by sets of nucleotide sequences.

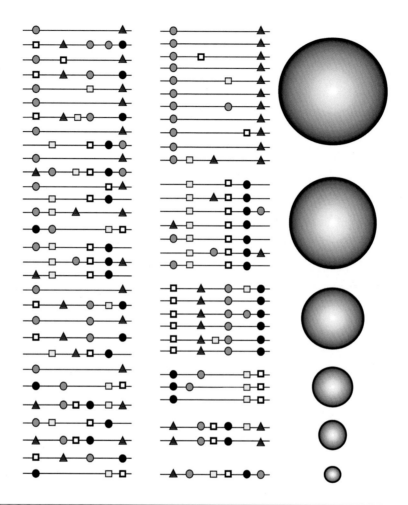

FIGURE 3.8

Representation of viral quasispecies as a number of subpopulations, inspired in partition analysis of quasispecies developed by P. Baccam and colleagues. A seemingly random distribution of genomes (left) may actually include distinct groups according to mutation types (second distribution). On the right the mutant classes are represented as spheres whose size is proportional to the number of genomes in the class they represent.

The figure is modified from Domingo et al. (2012), with permission from de American Society for Microbiology, Washington DC, USA.

3.6.5 SOME KEY POINTS ON THE IMPACT OF QUASISPECIES IN VIROLOGY

Quasispecies insights have been instrumental for the understanding of the nature of viral populations, and to interpret intrahost evolution and pathogenesis of RNA viruses for the following reasons:

• Quasispecies provides a new definition of wild type

One of the most inspiring departures from classic concepts contributed by quasispecies theory is that the wild type is no longer a defined genome with a precise nucleotide sequence, but a collection of related genomes. This means that forces that promote evolution (selection acting on phenotypes and random drift acting on genotypes (Sections 3.5 and 3.6)) can operate on collections of genomes. Using viral quasispecies reconstructed with minority genomes expressing a similar phenotype (resistance to a neutralizing antibody) it was shown that what is selected is indeed a cloud of mutants (Perales et al., 2005) (Section 3.8.3). The new structure of the wild-type influences several aspects of virus biology, as summarized in the bullet statements that follow, and in subsequent chapters.

- Mutant spectra increase the adaptive potential of viruses because phenotypic variants are present in the population or can be rapidly generated

The adaptedness conferred by mutant spectra was elegantly proven by a decreased pathogenic potential for mice of a poliovirus mutant whose polymerase displayed higher copying fidelity than the wild-type enzyme. The virus that harbored such mutant polymerase formed a narrower mutant spectrum (a lower average number of mutations per genome) than the virus with the wild-type polymerase (Pfeiffer and Kirkegaard, 2005; Vignuzzi et al., 2006) (Chapter 2). Several fidelity mutants of other viruses have been subsequently characterized and they offer a clear (but by no means the only) example of selective advantage conferred by high mutation rates, and of the involvement of quasispecies in viral pathogenesis and persistence (Borderia et al., 2016). Mutant spectrum complexity may influence the outcome of a viral infection, and the response to antiviral treatments (Chapters 8 and 9).

- The information provided by consensus sequences may be insufficient to understand virus behavior

Consensus sequences are commonly determined for virus isolates, and they provide adequate information for purposes such as virus identification and classification, structural determinations, or studies on molecular epidemiology. However, for an interpretation of viral evolution, pathogenesis, and in general what we may call the genotype-phenotype relationship, it is essential to consider that a consensus sequence is an average of many different sequences, and that a genome identical with the consensus may not exist in the population (Figure 3.5). It is sometimes wrongly believed that the consensus sequence is the sequence of the major component of the population. This is not necessarily the case. Following a reflection by M. Eigen, the consensus of the numbers of a telephone directory is no telephone number (Eigen, 2013). Furthermore, the most abundant genomes at one time point need not be the ancestors of the genomes that dominate the population at a subsequent time point. The latter may derive from a hidden minority at an earlier phase of the evolution of the same lineage (Briones and Domingo, 2008). One or a few point mutations may not affect the phylogenetic position of a virus, and yet they may modify some of its relevant biological traits. This comprises even a single synonymous mutation within a protein-coding region, which may control virion production (Novella and Ebendick-Corpus, 2004; Hamano et al., 2007). The recent literature includes many additional cases, too numerous to describe here, of single nucleotide or amino acid substitutions that produce salient phenotypic modifications. One of the most serious mistakes of viral evolutionary genetics has been to ignore mutant spectra.

- The individual genomes that compose a mutant spectrum may behave in dependence of other genomes

Either positive interactions through complementation, or negative (suppressive or interfering) interactions, can modulate the behavior of the ensemble, and this important aspect is detailed in Section 3.8. Profound fitness differences can follow from alterations in the composition of the mutant spectrum, without modification of the consensus nucleotide sequence (González-López et al., 2004, 2005;

Novella and Ebendick-Corpus, 2004; Grande-Pérez et al., 2005), including very drastic events such as the transition toward extinction by lethal mutagenesis (Chapter 9).

As the complexity of mutant spectra of RNA viruses is unveiled, aided by the zooming-in power of NGS, additional implications of quasispecies become apparent, and the reader will find several of them in the following chapters.

3.7 SEQUENCE SPACE AND STATE TRANSITIONS

"Sequence space" applied to genomes refers to a theoretical representation of all possible variants of any nucleotide sequence (Figure 3.9). In the language of physics, it is a "hypercube" (a geometrical representation that gathers multiple cubic forms) or a multidimensional set of points, analogous to a cube in a three-dimensional space (Eigen, 1992) (not easy to visualize). "Space" is a concept that originated in information theory (Hamming, 1980), and has found broad application to indicate the range of positions or values for a given parameter or alternative states for physical or biological objects. Experts refer to the "structural space" of viral particles, or the "conformation space" of a given protein. Coexistence of minority conformers, that is, the presence of multiple points in the space of conformations within a protein population has been identified as a factor of prion adaptability to different host species (Chapter 10). In pharmacology, exploration of the "chemical space" of drugs with some desired biological activity is often used to describe the synthesis of new molecules by computer-assisted design. The following modest digression in psychology is justified because it relates to "quasispecies memory" of viral populations treated in Chapter 5. "Psychological space" was proposed in an attempt to apply physical principles to psychological science (Shepard, 1987). An individual's psychological space was mapped to a physical space that defined metric distances between brain stimuli able to elicit a response. The probability that a response to a stimulus extends to another stimulus will depend on the distance between them. In this view, upon removal of an external stimulus, its memory trace may undergo "diffusion" in psychological space (Staddon, 2001). The term "diffusion" will later be used as a metaphor to refer to delocalizations in sequence space undergone by viruses (Chapter 9). Psychological space has been previously discussed in the literature regarding an analogy between quasispecies molecular memory and neurological memory, as well as quasispecies memory loss viewed as a "diffusion" in sequence space (Chapter 5). For biological macromolecules, protein sequence space (Maynard Smith, 1970) and nucleic acid sequence space (Rechenberg, 1973; Eigen and Biebricher, 1988) are commonly used.

3.7.1 VIRUS EVOLUTION AS A MOVEMENT IN SEQUENCE SPACE

At any given time a virus occupies a cloud (multiple points, each corresponding to a genome with a nucleotide sequence) in a region of sequence space. A selective pressure or a bottleneck event acting on the replicating cloud reorganizes and relocalizes the cloud. Mutagenesis expands at least transiently the occupation of sequence space by a virus, sometimes with irreversible loss of genetic information when many nonfunctional points of the space dominate. Lethal mutagenesis can be regarded as a drift in sequence space until nonviable areas for the virus are reached; this view reconciles some theoretical models of lethal mutagenesis, and the theoretical concept of error catastrophe with experimental observations (Chapter 9). A key feature of sequence space for virus adaptation or

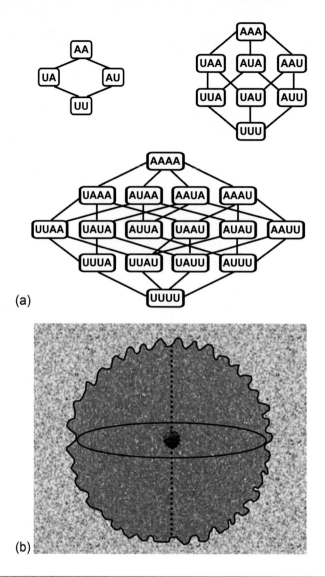

(a)

(b)

FIGURE 3.9

Representations of sequence space. (a) Connections between hypothetical genomes consisting of only two nucleotides (A,U) of length (ν) two (top left), three (top right), and four (middle). The diagrams are intended to illustrate the structure of simple sequence spaces, the connectivity among points of the space, and the increase of complexity with genome length. The sequence space for a real virus is impossible to depict in the terms used in these three diagrams (Figure inspired and modified from several articles by M. Eigen and colleagues that can be found in references quoted in this and other chapters of the book). (b) A useful surrogate representation of the viable (replication competent) sequence space of a virus (four nucleotides with ν=3000-32,000 for RNA genomes) is in the form of spheres of size proportional to sequence space occupation. The outer square in grey represents a huge theoretical multidimensional space that includes all possible viable and non-viable genomes. The outer blue and internal red spheres have been drawn with a 10-fold difference in radius, that results in 100-fold difference in surface area. The population represented by the large sphere has access to many more contacts with other genomes present in the sphere surface than the population represented by the small sphere. See text for the relevance of movements in sequence space in virus evolution.

(Part (b) of the figure is reproduced from Domingo et al. (2012),
with permission from the American Society for Microbiology, Washington DC, USA).

de-adaptation is its high connectivity (Eigen, 1992). Points separated by one mutation are linked by a line (Figure 3.9). In the gigantic hypercube that represents the entire sequence space for a genome of ν nucleotides, the minimal distance between any two points is never larger than ν. Fitness differences guide movements of mutant clouds through sequence space and allow distant points to be reached, with relevant consequences for virus adaptation (Chapter 5). Once the virus is in a distant point, the cloud produced by mutational pressure has a different domain of exploration than the previous one, and reiteration of this process allows the virus to increase its fitness in each specific environment. What we are doing here is using physics notions to visualize important events that take place during increased or decreased adaptation of a virus to an environment. The mathematics behind this and other abstractions should not deter us from the usage of physics concepts since mathematics is a formal representation of a concept, as we also saw with the equations that describe Darwinian evolution (Figure 3.1).

Another, more every day type of image for a virus reaching its optimal local fitness through successions of exploratory mutations is to think how you reach a specific word in a classic printed dictionary: you open the dictionary at random and then you approach the target page by sometimes going a few times to pages that precede or that follow the target, until you reach and stay in the correct page to read what you wanted. The major difference between a virus searching an ideal position in sequence space and your searching a world is that while neighbor positions in sequence space often have related fitness (with exceptions though), words next to each other do not need to have related meanings (again, with exceptions). Thus, for highly mutable entities, adaptation is a gradual approximation to the desired end point; to reach the objective, fitness is the guiding force and mutation is the major tool box.

At any time in an evolutionary process multiple points in sequence space are present in "one" viral population. Even if many such points are incompatible with genome replication, those that maintain some replicative ability can still guide the genome ensemble toward regions of sequence space which are compatible with replication and survival. Striking virus "resurrections" are sometimes observed, with viruses that have nearly lost their infectivity due to the action of a mutagenic agent, and that can nevertheless regain fitness when the mutagen is removed (Chapter 9). This "Lazarus effect" in virology is reminiscent of the "Lazarus phenomenon" in physiology (spontaneous return of blood circulation). In patients considered "cured" of an infection, on occasions the virus may reappear after days or months.

Replication-competent viruses occupy only a tiny fraction of the immense theoretical sequence space: $4^{10,000}$ for a viral genome of 10,000 nucleotides! a palpably unimaginable number. Obviously, only very restricted subsets of genomic sequences are compatible with virus viability no matter which the biological environment might be (Eigen and Biebricher, 1988; Domingo et al., 2001). Importantly, the functional sequence space can be subdivided broadly into fully and conditionally functional subsets of genomes. Such subdivision is of relevance to interpret internal interactions among components of a mutant spectrum, one of the events that determines that a viral quasispecies acts as a "unit of selection," meaning that what is selected and evolves may be an ensemble of genomes, not individual genomes (Section 3.8).

3.7.2 EXPLORATION OF SEQUENCE SPACE AND THE SAMPLING PROBLEM: VIRAL POPULATION SIZE AS A KEY PARAMETER

It may be thought that genomic diversity being a general feature of all replicating entities, be they viruses or cells, there are no distinctive features in RNA viruses (or the highly variable DNA viruses) that can justify the emphasis placed on their genetic heterogeneity. This is incorrect on several grounds, and here we consider one that relates sequence space occupation with adaptability.

The capacity to explore a large proportion of the theoretically available sequence space is a determinant of adaptability. In this respect, the impact of existing as quasispecies to permit adaptation is far greater for viruses than for cells. This is best understood when four parameters are considered: genome size, population size, average number of mutations per genome, and fecundity and number of mutations needed for a phenotypic change (explained in Box 3.3). The connection between genome and virus population size can be illustrated with the following arguments. A genome of 10,000 nucleotides can produce a total of 3×10^4 possible single mutants, even though many will not survive or contribute to the next generation due to low fitness. The potential number of single mutants is below the size of most viral populations that multiply in differentiated host organisms, and often even in a single cell from an organism! For a mammalian genome the potential number of single mutants is of the order of 10^{10}, which is well above the population size of mammalian species (some values are discussed in Chapter 4). This comparison illustrates that the capacity to explore sequence space (and to find adaptive pathways) is vastly superior for simple replicons displaying high mutation rates than for complex, cellular genomes (Eigen and Biebricher, 1988; Eigen, 1992; Domingo et al., 2001; Domingo, 2006).

Expressed in another way: a large mammalian genome has many nucleotide positions that are doomed to remain invariant through mutation, no matter what the environmental demands are. It can explore solutions through multiple recombination events, but not by mutation. In contrast, a small

---

BOX 3.3 SOME PARAMETERS THAT INFLUENCE THE ADAPTABILITY OF VIRAL QUASISPECIES

- Genome size
  For RNA viruses, typically 3-33 Kb (see Chapter 1 for comparative genome size of viruses and cellular organisms). The limited genome size render RNA viruses more effective explorers of sequence space than cells.
- Population size
  It is highly variable. An acutely infected organism may harbor up to 10^{12} viral particles at a given time, each with a certain probability of productive infection. A viral plaque (the progeny of a single genome) on a cell monolayer can include 10^3 to 10^{10} viral particles. (See Chapter 4 for comparisons of population size of viruses and host organisms.)
- Average number of mutations per genome calculated for the components of a mutant spectrum.
  It is a wide range (that can be placed at 1-100 mutations per genome) that depends on error rate and proximity to a clonal origin. Application of deep sequencing has generally increased the estimates first obtained by cloning-Sanger sequencing.
- Fecundity or average yield of viral particles per infected cell
  Again, a broad range of values (<1-10^6 particles produced per cell) have been reported for infections in nature. High fecundity promotes exploration of sequence space.
- Number of mutations needed for a phenotypic change
  Critical to the pertinence of quasispecies to understand viruses is that one or few mutations are often sufficient to modify a relevant phenotypic trait.
  See text for illustration of the interplay among these parameters.

viral genome has the potential to explore the effect of changes in each of its nucleotide positions. Exploration of sequence space that leads to adaptation is facilitated by two additional parameters listed in Box 3.3 that relate to the complexity of mutant spectra: the great fecundity of most viruses, mainly those associated with lytic infections or with acute disease, and the number of mutations needed for a phenotypic change (see also Section 3.7). It is worth adding that if the number of mutations needed for an adaptive change of a virus were in the order of hundreds, the biological relevance of quasispecies with the present replicative and mutational parameters would be minimal.

A numerical example of how the nonmutated class of genomes, and single and multiple mutants may be distributed in a viral population is given in Table 3.1. Several conclusions can be drawn from examining the table. One is that the genomes without mutations ($K=0$) are 25-fold less abundant than genomes with 4 or 5 mutations. Another is that genomes with 20 mutations are a million-fold less abundant than genomes with 3-6 mutations. Finally, it can be observed that an increase in population size (from 10^8 to 2×10^{11} individuals in the example of Table 3.1) entails a correspondingly larger presence of both nonmutated and mutated genomes. From what we know of the negative impact of mutations on fitness, real mutant distributions will be skewed toward a lower proportion of genomes with the highest numbers of mutations. Yet, current estimates either from the sequencing of molecular or biological clones or from NGS reveal profound mutant spectra with many single and multiple mutants in viral populations. The concept that mutants are a minority and that the nonmutated class dominates the population is obsolete in RNA virology. Viruses "are" collections of mutants.

Regarding measurements of population complexity, a sampling limitation must be commented. The available methodology (including NGS) only allows for the analysis of an extremely tiny proportion of all genomic sequences present in many of the viral quasispecies of biological interest. Typically an infected organism, or a laboratory dish with a monolayer of infected cells, can include or produce 10^{10} to 10^{12} infectious particles, measured at a given time point. These numbers exclude noninfectious viruses (or viral genomes) that may outnumber their infectious counterparts and that may play important

Table 3.1 Mutant Distributions in Viral Quasispecies

Number of Mutations in the Genome	Number of Virus Particles with Each Number of Mutations	
	Total Particles in Population	
	1×10^8	2×10^{11}
0	6.7×10^5	6.7×10^8
1	3.3×10^6	3.3×10^9
2	8.4×10^6	8.4×10^9
3	1.4×10^7	1.4×10^{10}
4	1.7×10^7	1.7×10^{10}
5	1.7×10^7	1.7×10^{10}
6	1.4×10^7	1.4×10^{10}
–	–	–
20	2.6×10^1	2.6×10^3

This calculation is based on a Poisson distribution of mutations among genomes, ignoring fitness effects of mutations. The formula is: $P(K) = m^k e^{-m}/K!$ where $P(K)$ is the frequency of a genome with K mutations, and m is the average number of mutations per genome. In the present example $m = 5$.

biological roles (Section 3.8). All together, we may be talking about 10^{11} to 10^{16} viral genomes in many infected hosts. The standard biological or molecular cloning procedures, followed by Sanger sequencing can realistically be used to provide sequences of about 100 (at most 1000) out of the total 10^{11} to 10^{16} genomes present in a host (i.e., 0.000001% to 0.00000000001% of the genomes!). Currently, deep sequencing can increase such proportion to 0.0001% to 0.000000001%, still a tiny minority of the population. Both procedures offer complementary sampling methods to represent a far more complex reality (Ortega-Prieto et al., 2013; Gregori et al., 2014) (see also Section 3.6.4 and Figure 3.6). There might even be additional "underground" levels in mutant spectra awaiting discovery when methodology will allow.

We could assume that if the sampling of 100 genomes is performed adequately (i.e., the 100 genomes are an unbiased representation of the ranking of genomes in the entire genome population in the sense that no minority genomes are overrepresented and no dominant genomes are underrepresented), a 100 genomes sample may be sufficient to characterize the population. Indeed, the great majority of analyses of mutant spectra carried out until 2010 were based on molecular and biological cloning and Sanger sequencing of not more than 20 genomes per sample. To the extent that this procedure unveiled the complexity of viral populations it can be regarded as extremely successful. Its conclusions and predictions have been largely confirmed by NGS. However, in addition to the limitations of biological and molecular cloning discussed in Section 3.6.4, it is often assumed that the PCR amplification procedures using standard protocols (at annealing and extension temperatures suitable to the primers available) will produce an unbiased representation of sequences. This assumption is not justified, as documented by studies that have allowed accession to new regions of sequence space in viral populations using PCR at low denaturing temperatures, a technique termed 3D PCR, developed by J.P. Vartanian, S. Wain-Hobson, and their colleagues at the Pasteur Institute in Paris (Suspène et al., 2005, 2008; Perales et al., 2011). The results obtained with 3D PCR cannot be regarded as a mere extension of the results using standard RT-PCR amplification. When applied to parallel FMDV populations passaged in the absence and presence of the mutagenic nucleoside analog ribavirin, AU-rich sequences were about as complex in the ribavirin-treated and untreated populations. However, the analysis unveiled that the major effect of ribavirin was to accelerate the occupation of the AU-rich portion of sequence space by the virus in its way toward extinction (Perales et al., 2011) (Chapter 9). The sampling representativeness is unavoidably conditioned by methodological filters.

PCR methodologies to explore either AU- or GC-enriched viral subpopulations, together with application of NGS, including whole individual genome sequencing now under development, combined with microarray-based methodologies, should permit increasingly detailed and accurate characterization of mutant spectra and their dynamics. In particular, they are expected to define in quantitative terms changes in population structure that accompany selection and random drift events, or transitions toward increased or decreased fitness for viral survival or extinction. Transitions in population structure have a number of implications for medical practice that are discussed in Chapter 9.

3.8 MODULATING EFFECTS OF MUTANT SPECTRA: COMPLEMENTATION AND INTERFERENCE: AN ENSEMBLE AS THE UNIT OF SELECTION

A major consequence of virus replicating as mutant spectra is that interactions among genomes are established, mainly through their expression products acting in *trans*. The result is a modulation of phenotypic traits displayed by an entire population, that may not be attributable to specific single genomes that are part of the population. Observations both *in vivo* and in cell culture have substantiated that what could have been a dominant trait expressed by a genome in isolation (i.e., the same genome type filling an entire population) may be suppressed by the surrounding mutant spectrum. This important feature of viral quasispecies was predicted by computer simulation studies with short-chain replicons subjected to error-prone replication. The simulations suggested that a master sequence could dominate a fitter one by virtue of being surrounded by a better (more closely related) mutant spectrum (Swetina and Schuster, 1982; Eigen and Biebricher, 1988). Applied to virus behavior, the mutant spectrum of a virus can suppress replication of a mutant of superior fitness, as shown experimentally for the first time by J.C. de la Torre and J. Holland working with vesicular stomatitis virus (VSV) quasispecies (de la Torre and Holland, 1990), and then documented with different viruses in cell culture and *in vivo* (Box 3.4).

Of particular significance is that suppressive or interfering effects are accentuated when viral populations are mutagenized and approach extinction (González-López et al., 2004; Grande-Pérez et al., 2005). Interference has been experimentally verified with individual mutants of RNA viruses in the

---

BOX 3.4 COLLECTIVE BEHAVIOR, INTERFERENCE, AND COMPLEMENTATION AMONG COMPONENTS OF A MUTANT SPECTRUM

- Biological clones of Qβ and VSV display higher fitness than the populations from which they were isolated (Domingo et al., 1978; Duarte et al., 1994).
- High fitness VSV clone was suppressed by a VSV mutant spectrum (de la Torre and Holland, 1990).
- Live-attenuated PV vaccines can suppress minority virulent PV (Chumakov et al., 1991).
- Antibody-resistant FMDV can suppress high fitness FMDV subpopulations (Borrego et al., 1993).
- LCMV that produces a growth hormone deficiency syndrome can be suppressed by non-pathogenic LCMV (Teng et al., 1996).
- Quasispecies memory affects the response to selective pressures (Ruiz-Jarabo et al., 2000; Briones and Domingo, 2008).
- Mutagenized FMDV populations interfere with replication of standard FMDV (González-López et al., 2004).
- Defector LCMV genomes interfere with replication of standard LCMV (Grande-Pérez et al., 2005).
- PV mutant spectra can suppress inhibitor-resistant PV mutants, and compounds directed to a dominant drug target can suppress selection of drug-resistant variants (Crowder and Kirkegaard, 2005; Tanner et al., 2014).
- A PV quasispecies was needed to mediate penetration of a PV mutant into the central nervous system (Pfeiffer and Kirkegaard, 2005; Vignuzzi et al., 2006).
- Two measles virus mutants were needed to mediate membrane fusion (Shirogane et al., 2012).
- Coexistence of HBV variants enhanced viral replication and immune response in a mouse model (Cao et al., 2014).
  Quasispecies memory is discussed in Chapter 5, and its implications for antiviral treatments in Chapter 9.

absence of mutagenic treatments (Crowder and Kirkegaard, 2005; Perales et al., 2007). In studies with FMDV, the mutants that exerted an interference were those whose RNA replicated in the infected cells, irrespective of their ability to yield infectious particles. A single amino acid substitution in an interfering mutant abolished its interfering activity, implying that minimal movements in sequence space can generate or abrogate potentially modulating effects on mutant ensembles. A synergistic interference that slowed down replication of the wild-type virus was produced by mixtures of FMDV capsid and polymerase mutants (Perales et al., 2007). Inhibition by several defective genomes acting conjointly probably mimics what is encountered with a mutagenized quasispecies in which a dispersed, abundant, and dynamic mutated swarm exerts a collective interfering activity (González-López et al., 2004). We may call it a dominant-negative effect of a mutagenized viral swarm, and it is one of the molecular mechanisms that underlies lethal mutagenesis as an antiviral strategy (Chapter 9). The collective behavior of mutant spectra is a major departure from the classic mutation-equilibrium built on mere generation, competition and coexistence of mutants. Components of a mutant spectrum are connected in such a manner that an ensemble can command behavior.

Inhibition of standard virus replication by DI particles can be viewed as a specific case of interference by heavily mutated genomes (with point mutations and deletions that render DI's noninfectious). Studies with VSV showed that the presence of DI resulted in a cyclic production of infectious virus (Palma and Huang, 1974). The observation that standard VSV and its DIs engaged in a continuous dynamics of mutation and competition led J. Holland and his colleagues to recognize the value of quasispecies dynamics to understand virus behavior (Holland et al., 1982; Holland, 1984). More generally, interference may be exerted by viable genomes that replicate suboptimally because they have acquired mutations by the action of mutagenic agents, as proposed for the first stages of a transition toward the error threshold (Chapter 9). When interference is prevalent (under increased mutagenesis) the average fitness of rescued viable biological clones tends to be higher than the fitness of the average parental population (Borrego et al., 1993; González-López et al., 2004) (Figure 3.10). As a consequence, bottleneck events may serve to rescue higher fitness subpopulations (Bergstrom et al., 1999; Grande-Pérez et al., 2005; Domingo, 2006). Genomes displaying various degrees of defectiveness are likely to play relevant roles as modulators of viral replication, and there is direct evidence that defective viral genomes are produced *in vivo*. In patients acutely infected with HIV-1, two-thirds of the infected cells included defective genomes, and early diversification from the founder virus resulted in populations of HIV-1 with about 30% of genomes with mutations that conferred defectiveness (Salazar-Gonzalez et al., 2009).

In contrast to suppression or interference, components of mutant spectra can complement each other, thereby producing ensembles that show higher fitness than many of its individual components (Domingo et al., 1978; Duarte et al., 1994; Moreno et al., 1997; Novella, 2003; Wilke et al., 2004; Perales et al., 2007) (Figure 3.10). Among other examples in cell culture and *in vivo*, it is worth underlining the case of a marked, non-neurotropic poliovirus mutant that could not make its way into the brain of mice when inoculated alone, but reached the brain when coinoculated with neuropathogenic poliovirus populations (Vignuzzi et al., 2006) (discussed in Chapter 2 regarding PV fidelity mutants). A mutant measles virus (MV) possessed two types of genomes within the same virus particle: one of the genomes encoded unsubstituted fusion protein while the other encoded the fusion protein with an amino acid substitution. Both proteins together displayed enhanced fusion activity through hetero-oligomer formation (Shirogane et al., 2012). This intra-quasispecies cooperation through oligomeric proteins is the reverse of the interference in polymeric complexes depicted in Figure 3.10b. In a mouse

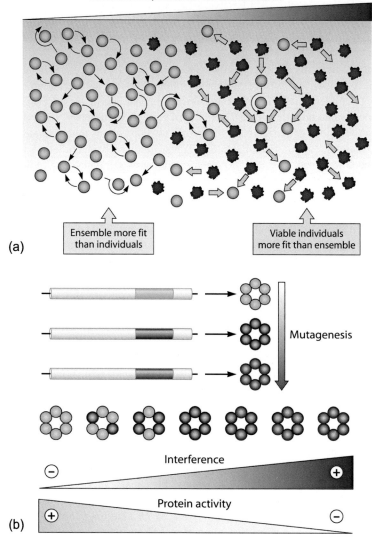

Mutation rate, transition toward extinction

(a) Ensemble more fit than individuals

Viable individuals more fit than ensemble

(b) Mutagenesis

Interference

Protein activity

FIGURE 3.10

Representation of complementation and interference within the mutant spectrum of viral populations. (a) Representation of the transition from a dominance of complementation (left) toward dominance of interference (right). Standard genomes are drawn as blue circles and complementing interactions as thin arrows. Interfering genomes are depicted as rough red symbols and interference by large yellow arrows. Interference increases with the mutation rate operating during quasispecies replication. When complementation dominates, individual clones display lower relative fitness than the entire population; when interference dominates, individual viable genomes display higher fitness than the entire population. (b) A model of complementation and interference mediated by a *trans*-acting protein. The standard gene and encoded protein is represented in tan, and the increasingly mutagenized counterparts in blue and red. It is assumed that the protein is active as an hexameric complex, with a fully functional protein composed of six tan subunits, and decreasingly functional proteins composed of blue and red subunits. It should be noted that the blue protein can either decrease the activity of a functional (tan) ensemble or rescue partially the activity of a non-functional (red) ensemble, when present as a subunit component. Progressive decrease of protein activity is accompanied by increased interference. See text for references that justify this model.

Parts (a) and (b) of the figure are reproduced from Domingo et al. (2012), with permission from the American Society from Microbiology, Washington DC, USA.

model, coexistence of hepatitis B virus (HBV) variants enhanced viral replication and evoked stronger intrahepatic CTL responses and surface protein-specific antibody responses (Cao et al., 2014). Thus, intra-quasispecies complementation may be not only a mechanism of enhanced viral replication but also of virus-induced immunopathogenesis. Analysis of viral populations by NGS has documented high frequencies of noninfectious genomes (e.g., genomes with stop codons that truncate essential proteins) that are probably maintained by complementation. Transient maintenance of lethal or deleterious mutations by complementation has a parallel in deleterious recessive mutants of diploid organisms which avoid negative selection when carried in a heterozygote with the dominant normal allele. The prevalence of complementation or interference depends on the quasispecies mutational status (Figure 3.10).

Some theoretical models have introduced cooperation and complementation in quasispecies dynamics and the error threshold for the maintenance of genetic information (Arbiza et al., 2010; Sardanyés and Elena, 2010; Moreno et al., 2012). Here again, the interplay between theory and experiment may provide new insights into quasispecies behavior.

3.8.1 MOLECULAR MECHANISMS OF COMPLEMENTATION AND INTERFERENCE

Complementation and interference in viral populations can be viewed as two sides of the same coin, with a related molecular basis. A *trans*-acting functional protein expressed by a subset of genomes from the mutant spectrum can rescue a nonfunctional version of the same protein expressed by mutated genomes. The result is complementation. Yet, expression above a certain level of nonfunctional forms of the same protein encoded by mutated genomes in the quasispecies may impede the protein fulfilling its functions. The result is interference. Malfunctioning due to incapacity to establish adequate protein-to-protein contacts can be easily visualized with proteins that act as homopolymeric and heteropolymeric ensembles during the virus life cycle (i.e., viral capsids whose stability is based on intersubunit contacts among capsid proteins, or the case of replication complexes which include viral and host proteins). A defective subunit in these and other multisubunit complexes may compromise function of the molecular assembly, leading to an unstable capsid, suboptimal replication complex, or other (Perales et al., 2007) (Figure 3.10).

Complementation and suppression can be viewed as an emerging (collective) property that results from interactions among quasispecies components. Experimental results (Box 3.4) are coherent with theoretical predictions on emerging traits (Solé and Goodwin, 2000). The behavior of an ensemble of related mutants is not necessarily predicted from the behavior of the individual mutants (see Section 3.9). The influence that a mutant cloud exerts on an individual mutant establishes a conceptual link with interference between different viruses, and with the effect of dominant-negative mutants, both well-established concepts in viral genetics.

Several biochemical mechanisms explain the inhibitory activities among distantly related viruses or between mutant viruses on their wild-type counterparts (Youngner and Whitaker-Dowling, 1999). They include interferon induction, production of a defective polypeptide that enters a multimeric complex (termed the "rotten apple" hypothesis), a defective function that may sequester a factor necessary for the replication of the two viruses (the "road-block" hypothesis), imbalances in gene dosage (the "all things in moderation" hypothesis), or competition for host factors (direct competition hypothesis). Some of these mechanisms may underlie also the suppressive effect exerted by viral quasispecies which can act as a "collective dominant-negative mutant." To establish the operation of such interference mechanisms, it is important to ascertain the specificity of interfering interactions. For example, in experiments with FMDV, altered capsid, and polymerase mutants of FMDV inhibited replication of FMDV but not of the other picornavirus encephalomyocarditis virus (Perales et al., 2007), thereby excluding some general inhibitory effect evoked by the presence of the mutants. Likewise, production of interferon type I or III (or other general antiviral activity associated with innate immunity) was excluded in the interference exerted by mutagenized FMDV on standard FMDV (González-López et al., 2004). Thus, despite interference being sometimes operational between distantly related viruses the mechanisms (e.g., interferon production or competition for a cellular resource) are not the same as those operating in the interference that requires intimate protein-protein interaction during the viral life cycle (Figure 3.10).

The negative effects of mutant spectra acting synergistically as a dominant-negative population aggregate are expected to be reinforced by the multifunctional nature of most viral proteins. A malfunction due to a defective protein will jeopardize any step in the virus life cycle in which the protein participates, unless such participation involves distinct protein domains, and the protein lesion affects only one of the domains. Both in virology and cell biology a current trend is to unveil the participation of one protein in multiple functions. This fact, together with the presence of several isoforms of a protein, the discovery of the relevance of small proteins (expressed from open reading frames once thought to be meaningless or secondary), along with the realization that different domains of the same protein may be involved in distinct functions, complicates interpretation of protein-protein interaction maps. These maps, that once were thought to provide a solution to relevant problems of cell biology, succumb also to an aspect of biological complexity, arising from multifunctionality. If clarification tools are developed, protein-protein interaction maps might provide insights into the mechanisms of intra-quasispecies interactions.

For a mutant cloud to interfere with a specific class of genomes, the cloud has either to be generated inside the same cell in which the specific class of genomes replicate or the cloud has to infect the same cell than the specific genome class. Intracellular generation of a mutant cloud occurs during mutagenesis by nucleotide analogs; because the cloud is generated intracellularly, it is one of the mechanisms of the transition of viruses toward extinction that does not require infection of the same cell by multiple viral particles (Chapter 9). In addition, infection of a cell with a mutant cloud may be simultaneous with infection by a specific mutant, or the two infections may be successive, provided no super-infection exclusion mechanisms operate. Although double or multiple infections of the same cells in differentiated organisms may appear as unlikely, they may be more frequent than expected from dual hits occurring at random (Cicin-Sain et al., 2005; Chohan et al., 2005). This may underlie frequent natural recombination events, and facilitate modulating effects among variant viruses observed *in vivo* (Box 3.4). Complementing interactions between mouse cytomegalovirus variants improved their dissemination and replication in the host, which agrees with poor prognosis in the case of patients infected with multiple variants of this virus (Cicin-Sain et al., 2005). In addition to intracellular complementation, altered expression of a diffusible regulatory factor, triggered by one class of virus variant, altered the behavior of another variant (Cicin-Sain et al., 2005; Sala et al., 2006). Thus, direct and indirect interactions among components of mutant spectra may result in an emergent behavior, with potentially important clinical implications. Complex viral quasispecies constitute inviting systems to explore experimentally molecular events behind properties that emerge from interacting individuals. Remarkably, mutual influences among mutants and their gene products render viral quasispecies units of selection, one of the essential features of the theoretical quasispecies.

3.8.2 INDIVIDUAL VERSUS GROUP SELECTION

Viral population numbers are very relevant on several grounds, and several implications have been discussed in the previous Section 3.7.2, and others are treated in Chapters 4 and 8, the latter in connection with virus escape from inhibition by antiviral treatment. One of the influences of viral population numbers results from the number of infectious particles that penetrate a susceptible cell, a parameter known as the multiplicity of infection (MOI). The importance of the MOI from the point of view of virus evolution was documented for the first time by N. Sevilla and colleagues by showing that FMDV adapted to the MOI: a population with a passage history at low MOI overgrew a related population when the competition passages were carried out at low MOI but not at high MOI (Sevilla et al., 1998). Thus, the evolutionary history in terms of how massive have been prior infections may influence an evolutionary outcome should a competition with other variants arise.

I.S. Novella and colleagues obtained evidence of density-dependent selection in VSV, also using the MOI as the key variable. In competition assays, the fitness of a deleterious mutant relative to the wild type increased as a function of the MOI. The results were interpreted in terms of an increased complementation at high MOI, and were supported by a theoretical model that assumed that the two viruses shared their gene products (generalized *trans*-acting functions) during coinfections (Novella et al., 2004).

The evidence of interactions within quasispecies raises the issue of the target of selection being either an individual or a group of individuals. This has been a debated issue in general genetics with ramifications in social behavior (Wright, 1945; Maynard Smith, 1964; Lewontin, 1970; Williams, 1992; Okasha, 2013). In the case of viruses this debate has not reached prominence yet, probably because the mechanisms that may favor group selection have been unveiled only recently.

The intra-mutant spectrum interactions (previous Sections 3.8 and 3.8.1) confer selective forces with the potential to act at the level of a single viral genome (liberated from possible links with its surrounding genome ensemble) or a group of genomes (tied by molecular dependences). Here I suggest a model according to which the operation of individual or group selection in viral quasispecies may depend on the mutant spectrum context (i.e., how firmly have the internal interactions been established) as well as on the intensity and specificity of the selective constraint. A specific antiviral inhibitor (directed to a viral protein) or a neutralizing monoclonal antibody (defined as being directed to a single epitope within an antigenic site) will tent to select resistant "individuals." The selection process will be little affected by the surrounding community of viral genomes, unless the latter is strongly suppressive, for example due to low fitness (several examples were described in Section 3.8). In contrast, adaptation to a less focused multicomponent environment (e.g., the penetration of a virus into a new tissue or organ) may be prone to modulation by the collectivity. In this view, still to be largely confirmed by experiments, the duality of individual versus group selection would in itself confer a selective advantage to viruses because of an inherent flexibility in the means to respond to selective constraints.

3.8.3 STOCHASTIC EFFECTS IN SELECTED COLLECTIVITIES

The evidence put forward in previous sections suggests that selection may operate either at the level of a collectivity of related mutants or at the level of an individual (or a very specific set of individuals). In the latter instance, the outcome (success or not) of the selected individual may depend on the progeny mutant spectrum generated, with the indeterminations inherent to the atomic fluctuations that determine mutagenesis (the atomic basis for the stochastic nature of the generation of mutations)

(Chapter 2). The statement of G.C. Williams (Williams, 1992) "… a particular mutation in a particular individual at a particular time could have important effects on all descendants of the mutant individual and all others that interact with these descendants…" is very relevant to the development of mutant spectra as collections of interacting individuals. In the initial stages of generation of a mutant spectrum, the stochastic nature of the mutations that are generated may alter the types of the successive mutations that will conform the dynamic mutant spectrum (see Chapter 5 for the contrast between mutant generation and establishment of high fitness genomes facilitated by epistatic interactions).

Positive selection may lead to a decrease of heterogeneity in a viral population. A complex mutant spectrum may include variants that cover a range of phenotypic traits (cell tropism, antibody-escape, CTL-escape, drug-resistance, among others). When selection for one such trait occurs naturally or is applied externally, the subset of genomes expressing the relevant trait will contribute to the following generations of genomes. The level of heterogeneity of the parental viral population prior to the application of selection will be generally higher than that of the selected population because the latter was founded by a minority of genomes that will participate in the ensuing rounds of replication in the presence of the selective agent. These types of "sweeping" selection events (Section 3.4) may have an effect similar to a population bottleneck regarding a reduction of population complexity of the population. Fitness may also decrease if measured in the absence of the selective agent that was responsible of the selective sweep (Chapter 5).

When selection acts on a highly heterogeneous viral population, multiple, nonidentical genomes can possess different genomic sequences all of them amenable to selection. In a study with FMDV, a quasispecies was reconstructed with multiple antigenic variants resistant to a specific monoclonal antibody, each at a frequency that was very similar to their expected frequency in viral populations. When the reconstructed quasispecies was subjected to selection by the relevant antibody, a swarm of antibody-escape mutants was selected (Perales et al., 2005). This observation indicates that what could be regarded as a sweeping selection event does not necessarily lead to loss of heterogeneity since, depending on the mutant spectrum composition, multiple, nonidentical genomes may share a selectable trait and they will soon regenerate a complex mutant spectrum. The result also provided evidence that the target of selection in viral quasispecies can be a collection of related individuals, irrespective of their being linked by complementation.

3.9 VIRAL POPULATIONS IN CONNECTION WITH BIOLOGICAL COMPLEXITY

Complexity has at least two meanings in virology. It may denote the amount of nonredundant information contents (in terms of regulatory regions and protein coding, open reading frames) embodied in a viral genome, or the level of heterogeneity within a population. According to the first meaning of complexity, poxviruses—with up to 375,000 base pairs of a terminally covalently linked DNA encoding 200 to 300 proteins—are more complex than the RNA picornaviruses which typically contain between 7500 and 8000 nucleotides encoding about a dozen fully processed proteins. The difference in complexity is related to the strategies of virus-host interactions, in particular the molecular mechanisms evolved to cope with the host immune response (Chapter 4). A compact RNA genome anticipates that evolution has selected a maximum number of functions associated with a small number of nucleotide residues, reflected in multifunctional proteins or protein precursors in the case of genome expression through polyproteins, a quite widespread strategy. The expectation, still to be confirmed, is that for the more compact (i.e., less complex from the point of view of the genetic information being carried) a more extensive multifunctionality of virus-coded proteins will be found.

A second meaning of complexity, of relevance to the understanding of viruses as populations, relates to the variation in nucleotide sequence among individual genomes that compose a virus population, and recapitulates features of viral quasispecies delineated in Section 3.6. All viruses possess a core information required to complete their life cycle: penetration into the cell, uncoating and replication of the RNA or DNA genome, viral protein expression, particle assembly, and release from cells. The features embodied in this minimal central information largely determine virus identity; it is what allows a new viral isolate to be unambiguously assigned to a group (influenza virus, papilloma virus, or other). Once the core information is secured, variations in nucleotide sequence (and in the corresponding amino acid sequence of the encoded proteins) are allowed among different isolates or the individual genomes within an isolate. Despite not altering the core information such variations are essential because they can affect interactions of viruses with their physical and biological environment that determine virus survival. Suffice to mention as an example the B-cell and T-cell epitopes in their interaction with components of the immune response. Variations in a few specific viral residues may have as the most extreme outcomes either virus evasion and survival or viral inactivation and extinction. From the point of view of viral pathogenesis, despite sharing the core information, only some influenza viruses are highly lethal, and only some papilloma viruses are oncogenic. Core information embodies minimal requirements for short-term survival, but variations of the core are essential for host interactions and long-term survival.

We refer to "population complexity" to portray the numbers of different genomes that harbor mutations in viral populations. This meaning of complexity bears on the concept of complexity that originated in physics, and that has ramifications in disparate fields of science such as neurobiology, astrophysics or even economics (reviews in Gell-Mann, 1994a,b; Simon, 1996). No single parameter captures what we would like to convey to indicate that a viral population is more complex than another. Mutation frequencies, number of polymorphic sites, or average genetic distances used to describe quantitatively mutant spectra of a viral quasispecies are only an approximation to the complexity of the population (Gregori et al., 2014) (Section 3.6.4).

Keeping in mind that we need to apprehend what may be more crucial in virus population complexity, it is relevant to examine the effective algorithmic complexity described by M. Gell-Mann. When his view is applied to viruses, a homogeneous collection of genomes (all identical) has the lowest complexity possible, as compared with any collection of genomes in which some of its components differ by some mutations. In the other end of the scale of homogeneity, a population consisting of random sequences, devoid of any information content in the considered environment, would not qualify as complex regarding the amount of information or the effective algorithmic complexity (Gell-Mann, 1994a), simply because it does not contain meaningful information. A degree of complexity must be defined relative to an information content since without information, complexity looses its meaning. According to the algorithmic complexity, population size (the number of genomes in the distribution) is not a significant variable. However, we know that the total viral population size is highly relevant to virus biological behavior, fundamentally because it determines the capacity to explore sequence space with its consequences (Section 3.7). The justification of such statement can be illustrated with the following example: if the frequency of mutations that can contribute to antibody-escape is 10^{-4} per nucleotide (not considering fitness effects), for a viral genome of 10^4 nucleotides, a population of at least 10^4 is required to find one mutation relevant to antibody-escape. With the same mutation frequency, multiple antibody-escape mutations will be present in a 1.6×10^5 genomes population, and none in a 16 genomes population. Yet, the 16, 16,000, and 160,000 viral populations may share identical mutation frequency. Thus, in the case of mutant spectra of viral quasispecies the statement of P. Anderson "more is different" (quoted in Levin, 2006) acquires its full validity. Following again physics language, properties of a viral population that do not depend on population size (e.g., the mutation frequency) have been termed "intrinsic" while properties that depend on the population size (e.g., the presence or not of a mutant present at a frequency of 10^{-4}) have been termed "extrinsic" properties (Domingo and Perales, 2012).

Physics has contributed yet another meaning of complexity that relates to emerging properties, applicable also to viral quasispecies. Some biological phenomena are the result of multiple influences, and cannot be explained by the mere addition of the individual influences (Solé and Goodwin, 2000). The term "emergence" describes the inability to explain a whole as the sum of its parts, a concept not too appreciated by some schools of biochemistry and classical genetics. (The rejection of "emergence" arises probably from the halo of mystery that it conveys, and that on occasions has been exploited to cast doubts on the capacity of science to explain the origin of some biological systems or mechanisms, and the proposal of "intelligent design" as a better alternative). Obviously, the term "emergence" does not imply any mysterious (scientifically unexplainable) intervention; it manifests that there may be transitional behaviors not directly explained by the parts involved, considered in isolation. Mental activity, as an outcome of neuronal connections, is an example. This meaning of complexity bears on the understanding of virus evolution, pathogenesis, and disease emergence (Chapter 7) which are an outcome of sets of host-virus-environmental interactions. Only very recently we have begun to understand part of the molecular basis of some "emergent" properties of viral populations. Molecular virology is probably immersed in complexities of the types summarized here with us hardly noticing it.

3.10 OVERVIEW AND CONCLUDING REMARKS

Experimental observations and theoretical treatments have contributed to our understanding of how Darwinian principles guide virus evolution. There are several mathematics equations that have been used to describe evolutionary dynamics in general. The equations are interconnected, and emphasize different aspects of the Darwinian processes. One of the equations corresponds to quasispecies theory which includes mutation as a prominent event in the evolutionary behavior. The attention to error-prone replication renders quasispecies an adequate theoretical framework to describe the evolution of RNA viruses, and generally any highly variable virus. Quasispecies has impacted virology in several key aspects that range from a new concept of "wild type" as a collection of genomes to the evidence of a molecular mechanisms of adaptation to changing environments that was not obvious before quasispecies came into play.

An individual viral genome is characterized by a defined nucleotide sequence that has the information to generate infectious progeny in an infected cell. However, unavoidably the replication of such a genome generates a distribution of mutants that consists of multitudes of closely related, but different genomes. The "sequence" of the genome population is the average of a multitude of different sequences. Populations are often treated as if they had "a genomic sequence," but they have not. To assign "a sequence" to a viral population is a gross simplification perpetuated in textbooks and data banks. This seemingly inconsequential practice masks fundamental properties of viruses that cannot be anticipated by examining a consensus sequence. NGS methodologies have enhanced the potential to penetrate into the composition of mutant spectra, and to dissect the molecular details of quasispecies dynamics. Virus evolution consists fundamentally in the replacement of some viral subpopulations by others in response to selective constraints and random fluctuations in population size.

Mutant spectra are not mere aggregate of mutants which act independently one from the other. Rather, interactions of complementation or interference can be established within quasispecies that may dictate their behavior. The concept that the entire quasispecies can act as a unit of selection has been supported by many experimental observations in cell culture and *in vivo*, and it has found endorsement in theoretical models. The intramutant spectrum interactions raise the issue of selection acting at the level of the individual or ensembles of individuals. Current evidence suggests that viral quasispecies may be endowed with the dual potential of individual versus group selection, a feature that, if further substantiated by experimentation, would add to their repertoire of adaptive strategies.

Finally, this chapter has concluded with the concepts of complexity and emerging properties applied to viral population. Despite being in its infancy, research on properties that have a multicomponent origin starts playing an important role in the understanding of viruses as populations. Emergent behavior will be encountered in coming chapters (see Summary Box).

SUMMARY BOX

- Extended Darwinian principles consisting on generation of variant genomes, competition among them, selection of the most fit in the considered environment, and random drift of genomes preside virus evolution.
- The several mathematical equations of Darwinian formulations are related. Quasispecies emphasizes the effect of high mutation rates and, therefore, is adequate as the theoretical framework for highly variable viruses, in particular the RNA viruses.
- Viral quasispecies are complex and dynamic mutant distributions that have explained the adaptive potential of RNA viruses. Virus evolution can be regarded as a movement in sequence space, in which some genome subpopulations are replaced by others.
- Interactions among components of a mutant spectrum can influence the behavior of the ensemble. Selection can act either on individuals or group of individuals in viral quasispecies.
- Viruses are adequate experimental systems to approach complexity, in the sense of probing why the behavior of a genome collectivity may not be explained by the sum of contributions of the individual members of the collectivity.

REFERENCES

Acevedo, A., Brodsky, L., Andino, R., 2014. Mutational and fitness landscapes of an RNA virus revealed through population sequencing. Nature 505, 686–690.

Airaksinen, A., Pariente, N., Menendez-Arias, L., Domingo, E., 2003. Curing of foot-and-mouth disease virus from persistently infected cells by ribavirin involves enhanced mutagenesis. Virology 311, 339–349.

Alves, D., Fontanari, J.F., 1998. Error threshold in finite populations. Phys. Rev. E 57, 7008–7013.

Amexis, G., Oeth, P., Abel, K., Ivshina, A., Pelloquin, F., et al., 2001. Quantitative mutant analysis of viral quasispecies by chip-based matrix- assisted laser desorption/ionization time-of-flight mass spectrometry. Proc. Natl. Acad. Sci. U. S. A. 98, 12097–12102.

Arbiza, J., Mirazo, S., Fort, H., 2010. Viral quasispecies profiles as the result of the interplay of competition and cooperation. BMC Evol. Biol. 10, 137.

Arias, A., Lázaro, E., Escarmís, C., Domingo, E., 2001. Molecular intermediates of fitness gain of an RNA virus: characterization of a mutant spectrum by biological and molecular cloning. J. Gen. Virol. 82, 1049–1060.

Baccam, P., Thompson, R.J., Fedrigo, O., Carpenter, S., Cornette, J.L., 2001. PAQ: partition analysis of quasispecies. Bioinformatics 17, 16–22.

Baccam, P., Thompson, R.J., Li, Y., Sparks, W.O., Belshan, M., et al., 2003. Subpopulations of equine infectious anemia virus Rev coexist in vivo and differ in phenotype. J. Virol. 77, 12122–12131.

Baranowski, E., Ruiz-Jarabo, C.M., Pariente, N., Verdaguer, N., Domingo, E., 2003. Evolution of cell recognition by viruses: a source of biological novelty with medical implications. Adv. Virus Res. 62, 19–111.

Batschelet, E., Domingo, E., Weissmann, C., 1976. The proportion of revertant and mutant phage in a growing population, as a function of mutation and growth rate. Gene 1, 27–32.

Battilani, M., Scagliarini, A., Ciulli, S., Morganti, L., Prosperi, S., 2006. High genetic diversity of the VP2 gene of a canine parvovirus strain detected in a domestic cat. Virology 352, 22–26.

Beerenwinkel, N., Gunthard, H.F., Roth, V., Metzner, K.J., 2012. Challenges and opportunities in estimating viral genetic diversity from next-generation sequencing data. Front. Microbiol. 3, 329.

Bergstrom, C.T., McElhany, P., Real, L.A., 1999. Transmission bottlenecks as determinants of virulence in rapidly evolving pathogens. Proc. Natl. Acad. Sci. U. S. A. 96, 5095–5100.

Boerlijst, M.C., Boenhoefer, S., Nowak, M.A., 1996. Viral quasispecies and recombination. Proc. R. Soc. Lond. B 263, 1577–1584.

Borderia, A.V., Rozen-Gagnon, K., Vignuzzi, M., 2016. Fidelity variants and RNA quasispecies. Curr. Top. Microbiol. Immunol.

Borrego, B., Novella, I.S., Giralt, E., Andreu, D., Domingo, E., 1993. Distinct repertoire of antigenic variants of foot-and-mouth disease virus in the presence or absence of immune selection. J. Virol. 67, 6071–6079.

Borucki, M.K., Chen-Harris, H., Lao, V., Vanier, G., Wadford, D.A., et al., 2013. Ultra-deep sequencing of intra-host rabies virus populations during cross-species transmission. PLoS Negl. Trop. Dis. 7, e2555.

Briones, C., Domingo, E., 2008. Minority report: hidden memory genomes in HIV-1 quasispecies and possible clinical implications. AIDS Rev. 10, 93–109.

Brown, E.G., Liu, H., Kit, L.C., Baird, S., Nesrallah, M., 2001. Pattern of mutation in the genome of influenza A virus on adaptation to increased virulence in the mouse lung: identification of functional themes. Proc. Natl. Acad. Sci. U. S. A. 98, 6883–6888.

Cale, E.M., Hraber, P., Giorgi, E.E., Fischer, W., Bhattacharya, T., et al., 2011. Epitope-specific CD8+ T lymphocytes cross-recognize mutant simian immunodeficiency virus (SIV) sequences but fail to contain very early evolution and eventual fixation of epitope escape mutations during SIV infection. J. Virol. 85, 3746–3757.

Cao, L., Wu, C., Shi, H., Gong, Z., Zhang, E., et al., 2014. Coexistence of hepatitis B virus quasispecies enhances viral replication and the ability to induce host antibody and cellular immune responses. J. Virol. 88, 8656–8666.

Chen-Harris, H., Borucki, M.K., Torres, C., Slezak, T.R., Allen, J.E., 2013. Ultra-deep mutant spectrum profiling: improving sequencing accuracy using overlapping read pairs. BMC Genomics 14, 96.

Chohan, B., Lavreys, L., Rainwater, S.M., Overbaugh, J., 2005. Evidence for frequent reinfection with human immunodeficiency virus type 1 of a different subtype. J. Virol. 79, 10701–10708.

Chumakov, K.M., Powers, L.B., Noonan, K.E., Roninson, I.B., Levenbook, I.S., 1991. Correlation between amount of virus with altered nucleotide sequence and the monkey test for acceptability of oral poliovirus vaccine. Proc. Natl. Acad. Sci. U. S. A. 88, 199–203.

Chumakov, K.M., Dragunsky, E.M., Norwood, L.P., Douthitt, M.P., Ran, Y., et al., 1994. Consistent selection of mutations in the 5'-untranslated region of oral poliovirus vaccine upon passaging in vitro. J. Med. Virol. 42, 79–85.

Cicin-Sain, L., Podlech, J., Messerle, M., Reddehase, M.J., Koszinowski, U.H., 2005. Frequent coinfection of cells explains functional in vivo complementation between cytomegalovirus variants in the multiply infected host. J. Virol. 79, 9492–9502.

Ciurea, A., Klenerman, P., Hunziker, L., Horvath, E., Senn, B.M., et al., 2000. Viral persistence *in vivo* through selection of neutralizing antibody- escape variants. Proc. Natl. Acad. Sci. U. S. A. 97, 2749–2754.

Coffin, J.M., 1995. HIV population dynamics in vivo: implications for genetic variation, pathogenesis, and therapy. Science 267, 483–489.

Costa-Mattioli, M., Domingo, E., Cristina, J., 2006. Analysis of sequential hepatitis A virus strains reveals coexistence of distinct viral subpopulations. J. Gen. Virol. 87, 115–118.

Couderc, T., Guedo, N., Calvez, V., Pelletier, I., Hogle, J., et al., 1994. Substitutions in the capsids of poliovirus mutants selected in human neuroblastoma cells confer on the Mahoney type 1 strain a phenotype neurovirulent in mice. J. Virol. 68, 8386–8391.

Crowder, S., Kirkegaard, K., 2005. Trans-dominant inhibition of RNA viral replication can slow growth of drug-resistant viruses. Nat. Genet. 37, 701–709.

de la Torre, J.C., Holland, J.J., 1990. RNA virus quasispecies populations can suppress vastly superior mutant progeny. J. Virol. 64, 6278–6281.

de la Torre, J.C., Giachetti, C., Semler, B.L., Holland, J.J., 1992. High frequency of single-base transitions and extreme frequency of precise multiple-base reversion mutations in poliovirus. Proc. Natl. Acad. Sci. U. S. A. 89, 2531–2535.

Delwart, E., 2013. A roadmap to the human virome. PLoS Pathog. 9, e1003146.

Domingo, E., 1999. Vers une compréhension des virus comme systèmes complexes et dynamiques. Ann. Med. Vet. 143, 225–235.

Domingo, E., 2005. Virus entry into error catastrophe as a new antiviral strategy. Virus Res. 107, 115–228.

Domingo, E., 2006. Quasispecies: concepts and implications for virology. Curr. Top. Microbiol. Immunol. 299.

Domingo, E., Holland, J.J., 1994. Mutation rates and rapid evolution of RNA viruses. In: Morse, S.S. (Ed.), Evolutionary Biology of Viruses. Raven Press, New York, pp. 161–184.

Domingo, E., Perales, C., 2012. From quasispecies theory to viral quasispecies: how complexity has permeated virology. Math. Model. Nat. Phenom. 7, 32–49.

Domingo, E., Schuster, P., 2016. Quasispecies: from theory to experimental systems. Curr. Top. Microbiol. Immunol.

Domingo, E., Wain-Hobson, S., 2009. The 30th anniversary of quasispecies. Meeting on 'Quasispecies: past, present and future'. EMBO Rep. 10, 444–448.

Domingo, E., Sabo, D., Taniguchi, T., Weissmann, C., 1978. Nucleotide sequence heterogeneity of an RNA phage population. Cell 13, 735–744.

Domingo, E., Biebricher, C., Eigen, M., Holland, J.J., 2001. Quasispecies and RNA Virus Evolution: Principles and Consequences. Landes Bioscience, Austin.

Domingo, E., Brun, A., Núñez, J.I., Cristina, J., Briones, C., et al., 2006. Genomics of viruses. In: Hacker, J., Dobrindt, U. (Eds.), Pathogenomics: Genome Analysis of Pathogenic Microbes. Wiley-VCH Verlag GmbH & Co. KGaA, Weinheim, pp. 369–388.

Domingo, E., Parrish, C., Holland, J.J.E., 2008. Origin and Evolution of Viruses, second ed. Elsevier, Oxford.

Domingo, E., Sheldon, J., Perales, C., 2012. Viral quasispecies evolution. Microbiol. Mol. Biol. Rev. 76, 159–216.

Drexler, J.F., Geipel, A., Konig, A., Corman, V.M., van Riel, D., et al., 2013. Bats carry pathogenic hepadnaviruses antigenically related to hepatitis B virus and capable of infecting human hepatocytes. Proc. Natl. Acad. Sci. U. S. A. 110, 16151–16156.

Duarte, E.A., Novella, I.S., Ledesma, S., Clarke, D.K., Moya, A., et al., 1994. Subclonal components of consensus fitness in an RNA virus clone. J. Virol. 68, 4295–4301.

Eigen, M., 1971. Self-organization of matter and the evolution of biological macromolecules. Die Naturwissenschaften. 58, 465–523.

Eigen, M., 1992. Steps Towards Life. Oxford University Press, Oxford.

Eigen, M., 2000. Natural selection: a phase transition? Biophys. Chem. 85, 101–123.

Eigen, M., 2013. From Strange Simplicity to Complex Familiarity. Oxford University Press, Oxford.

Eigen, M., Biebricher, C.K., 1988. Sequence space and quasispecies distribution. In: Domingo, E., Ahlquist, P., Holland, J.J. (Eds.), RNA Genetics. CRC Press, Boca Raton, FL, pp. 211–245.

Eigen, M., Schuster, P., 1979. The Hypercycle. A Principle of Natural Self-Organization. Springer, Berlin.

Elde, N.C., Child, S.J., Eickbush, M.T., Kitzman, J.O., Rogers, K.S., et al., 2012. Poxviruses deploy genomic accordions to adapt rapidly against host antiviral defenses. Cell 150, 831–841.

Eriksson, N., Pachter, L., Mitsuya, Y., Rhee, S.Y., Wang, C., et al., 2008. Viral population estimation using pyrosequencing. PLoS Comput. Biol. 4, e1000074.

Escarmís, C., Dávila, M., Domingo, E., 1999. Multiple molecular pathways for fitness recovery of an RNA virus debilitated by operation of Muller's ratchet. J. Mol. Biol. 285, 495–505.

Farci, P., Purcell, R.H., 2000. Clinical significance of hepatitis C virus genotypes and quasispecies. Semin. Liver Dis. 20, 103–126.

Figlerowicz, M., Alejska, M., Kurzynska-Kokorniak, A., Figlerowicz, M., 2003. Genetic variability: the key problem in the prevention and therapy of RNA-based virus infections. Med. Res. Rev. 23, 488–518.

Fischer, W., Ganusov, V.V., Giorgi, E.E., Hraber, P.T., Keele, B.F., et al., 2010. Transmission of single HIV-1 genomes and dynamics of early immune escape revealed by ultra-deep sequencing. PLoS One 5, e12303.

Fitch, W.M., Margoliash, E., 1967. Construction of phylogenetic trees. Science 155, 279–284.

Fong, Y.W., Cattoglio, C., Tjian, R., 2013. The intertwined roles of transcription and repair proteins. Mol. Cell 52, 291–302.

Fox, S., Filichkin, S., Mockler, T.C., 2009. Applications of ultra-high-throughput sequencing. Methods Mol. Biol. 553, 79–108.

Garcia-Arriaza, J., Domingo, E., Briones, C., 2007. Characterization of minority subpopulations in the mutant spectrum of HIV-1 quasispecies by successive specific amplifications. Virus Res. 129, 123–134.

Gavrilin, G.V., Cherkasova, E.A., Lipskaya, G.Y., Kew, O.M., Agol, V.I., 2000. Evolution of circulating wild poliovirus and of vaccine-derived poliovirus in an immunodeficient patient: a unifying model. J. Virol. 74, 7381–7390.

Ge, L., Zhang, J., Zhou, X., Li, H., 2007. Genetic structure and population variability of tomato yellow leaf curl china virus. J. Virol. 81, 5902–5907.

Gell-Mann, M., 1994a. Complex adaptive systems. In: Cowan, G.A., Pines, D., Meltzer, D. (Eds.), Complexity. Metaphors, Models and Reality. Wesley Publishing Co., Reading, MA, pp. 17–45.

Gell-Mann, M., 1994b. The Quark and the Jaguar. Freeman, New York.

Gerrish, P.J., Garcia-Lerma, J.G., 2003. Mutation rate and the efficacy of antimicrobial drug treatment. Lancet Infect. Dis. 3, 28–32.

Gibbs, A., Calisher, C., García-Arenal, F., 1995. Molecular Basis of Virus Evolution. Cambridge University Press, Cambridge.

Gire, S.K., Goba, A., Andersen, K.G., Sealfon, R.S., Park, D.J., et al., 2014. Genomic surveillance elucidates Ebola virus origin and transmission during the 2014 outbreak. Science 345, 1369–1372.

Gojobori, T., Yokoyama, S., 1985. Rates of evolution of the retroviral oncogene of Moloney murine sarcoma virus and of its cellular homologues. Proc. Natl. Acad. Sci. U. S. A. 82, 4198–4201.

González-López, C., Arias, A., Pariente, N., Gómez-Mariano, G., Domingo, E., 2004. Preextinction viral RNA can interfere with infectivity. J. Virol. 78, 3319–3324.

González-López, C., Gómez-Mariano, G., Escarmís, C., Domingo, E., 2005. Invariant aphthovirus consensus nucleotide sequence in the transition to error catastrophe. Infect. Genet. Evol. 5, 366–374.

Grande-Pérez, A., Lázaro, E., Lowenstein, P., Domingo, E., Manrubia, S.C., 2005. Suppression of viral infectivity through lethal defection. Proc. Natl. Acad. Sci. U. S. A. 102, 4448–4452.

Gregori, J., Salicru, M., Domingo, E., Sanchez, A., Esteban, J.I., et al., 2014. Inference with viral quasispecies diversity indices: clonal and NGS approaches. Bioinformatics 30, 1104–1111.

Hamano, T., Matsuo, K., Hibi, Y., Victoriano, A.F., Takahashi, N., et al., 2007. A single-nucleotide synonymous mutation in the gag gene controlling human immunodeficiency virus type 1 virion production. J. Virol. 81, 1528–1533.

Hamming, R.W., 1980. Coding and Information Theory. Prentice-Hall, Englewood Cliffs.

Haruna, I., Nozu, K., Ohtaka, Y., Spiegelman, S., 1963. An RNA "Replicase" induced by and selective for a viral RNA: isolation and properties. Proc. Natl. Acad. Sci. U. S. A. 50, 905–911.

Hedskog, C., Mild, M., Jernberg, J., Sherwood, E., Bratt, G., et al., 2010. Dynamics of HIV-1 quasispecies during antiviral treatment dissected using ultra-deep pyrosequencing. PLoS One 5, e11345.

Herrera, M., Grande-Perez, A., Perales, C., Domingo, E., 2008. Persistence of foot-and-mouth disease virus in cell culture revisited: implications for contingency in evolution. J. Gen. Virol. 89, 232–244.

Hogeweg, P., Takeuchi, N., 2003. Multilevel selection in models of prebiotic evolution: compartments and spatial self-organization. Orig. Life Evol. Biosph. 33, 375–403.

Holland, J.J., 1984. Continuum of change in RNA virus genomes. In: Notkins, A.L., Oldstone, M.B.A. (Eds.), Concepts in Viral Pathogenesis. Springer-Verlag, New York.

Holland, J.J., Spindler, K., Horodyski, F., Grabau, E., Nichol, S., et al., 1982. Rapid evolution of RNA genomes. Science 215, 1577–1585.

Holmes, E.C., 2008. Comparative studies of RNA virus evolution. In: Domingo, E., Parrish, C.R., Holland, J.J. (Eds.), Origin and Evolution of Viruses, second ed. J.J. Elsevier, Oxford, pp. 119–134.

Holmes, E.C., 2009. The Evolution and Emergence of RNA Viruses. Oxford Series in Ecology and Evolution. Oxford University Press, New York.

Isnard, M., Granier, M., Frutos, R., Reynaud, B., Peterschmitt, M., 1998. Quasispecies nature of three maize streak virus isolates obtained through different modes of selection from a population used to assess response to infection of maize cultivars. J. Gen. Virol. 79, 3091–3099.

Jabara, C.B., Jones, C.D., Roach, J., Anderson, J.A., Swanstrom, R., 2011. Accurate sampling and deep sequencing of the HIV-1 protease gene using a Primer ID. Proc. Natl. Acad. Sci. U. S. A. 108, 20166–20171.

Jaramillo, N., Domingo, E., Muñoz, M.C., Tabarés, E., Gadea, I., 2012. Evidence of Muller's ratchet in herpes simplex virus type 1. J. Gen. Virol. 94, 366–375.

Jones, L.R., Zandomeni, R., Weber, E.L., 2002. Quasispecies in the 5' untranslated genomic region of bovine viral diarrhoea virus from a single individual. J. Gen. Virol. 83, 2161–2168.

Kimata, J.T., Kuller, L., Anderson, D.B., Dailey, P., Overbaugh, J., 1999. Emerging cytopathic and antigenic simian immunodeficiency virus variants influence AIDS progression. Nat. Med. 5, 535–541.

Kimura, M., 1983. The Neutral Theory of Molecular Evolution. Cambridge University Press, Cambridge, UK.

Kimura, M., 1989. The neutral theory of molecular evolution and the world view of the neutralists. Genome 31, 24–31.

Kuroda, M., Katano, H., Nakajima, N., Tobiume, M., Ainai, A., et al., 2010. Characterization of quasispecies of pandemic 2009 influenza A virus (A/H1N1/2009) by de novo sequencing using a next-generation DNA sequencer. PLoS One 5, e10256.

Lauring, A.S., Andino, R., 2010. Quasispecies theory and the behavior of RNA viruses. PLoS Pathog. 6, e1001005.

Lauring, A.S., Andino, R., 2011. Exploring the fitness landscape of an RNA virus by using a universal barcode microarray. J. Virol. 85, 3780–3791.

Lenski, R.E., May, R.M., 1994. The evolution of virulence in parasites and pathogens: reconciliation between two competing hypotheses. J. Theor. Biol. 169, 253–265.

Levin, S.A., 2006. Fundamental questions in biology. PLoS Biol. 4, e300.

Lewontin, R.C., 1970. The units of selection. Annu. Rev. Ecol. Syst. 1, 1–18.

Liang, B., Luo, M., Scott-Herridge, J., Semeniuk, C., Mendoza, M., et al., 2011. A comparison of parallel pyrosequencing and sanger clone-based sequencing and its impact on the characterization of the genetic diversity of HIV-1. PLoS One 6, e26745.

López-Bueno, A., Mateu, M.G., Almendral, J.M., 2003. High mutant frequency in populations of a DNA virus allows evasion from antibody therapy in an immunodeficient host. J. Virol. 77, 2701–2708.

Lopez-Bueno, A., Rubio, M.P., Bryant, N., McKenna, R., Agbandje-McKenna, M., et al., 2006. Host-selected amino acid changes at the sialic acid binding pocket of the parvovirus capsid modulate cell binding affinity and determine virulence. J. Virol. 80, 1563–1573.

Lu, Z., Rezapkin, G.V., Douthitt, M.P., Ran, Y., Asher, D.M., et al., 1996. Limited genetic changes in the Sabin 1 strain of poliovirus occurring in the central nervous system of monkeys. J. Gen. Virol. 77 (Pt 2), 273–280.

Mangul, S., Wu, N.C., Mancuso, N., Zelikovsky, A., Sun, R., et al., 2014. Accurate viral population assembly from ultra-deep sequencing data. Bioinformatics 30, i329–i337.

Mardis, E.R., 2008. The impact of next-generation sequencing technology on genetics. Trends Genet. 24, 133–141.

Margeridon-Thermet, S., Shulman, N.S., Ahmed, A., Shahriar, R., Liu, T., et al., 2009. Ultra-deep pyrosequencing of hepatitis B virus quasispecies from nucleoside and nucleotide reverse-transcriptase inhibitor (NRTI)-treated patients and NRTI-naive patients. J. Infect. Dis. 199, 1275–1285.

Martell, M., Esteban, J.I., Quer, J., Genesca, J., Weiner, A., et al., 1992. Hepatitis C virus (HCV) circulates as a population of different but closely related genomes: quasispecies nature of HCV genome distribution. J. Virol. 66, 3225–3229.

Martín Hernández, A.M., Carrillo, E.C., Sevilla, N., Domingo, E., 1994. Rapid cell variation can determine the establishment of a persistent viral infection. Proc. Natl. Acad. Sci. U. S. A. 91, 3705–3709.

Martin, V., Domingo, E., 2008. Influence of the mutant spectrum in viral evolution: focused selection of antigenic variants in a reconstructed viral quasispecies. Mol. Biol. Evol. 25, 1544–1554.

Marz, M., Beerenwinkel, N., Drosten, C., Fricke, M., Frishman, D., et al., 2014. Challenges in RNA virus bioinformatics. Bioinformatics 30, 1793–1799.

Mateu, M.G., Hernández, J., Martínez, M.A., Feigelstock, D., Lea, S., et al., 1994. Antigenic heterogeneity of a foot-and-mouth disease virus serotype in the field is mediated by very limited sequence variation at several antigenic sites. J. Virol. 68, 1407–1417.

Maynard Smith, J., 1964. Group selection and kin selection. Nature 201, 1145–1147.

Maynard Smith, J.M., 1970. Natural selection and the concept of a protein space. Nature 225, 563–564.

Mayr, E., 1959. Evolution and the Diversity of Life. Harvard University Press, Cambridge, Massachussets.

Mayr, E., Provine, W.B., 1980. The Evolutionary Synthesis. University Press, Cambridge, MA.

Mills, D.R., Peterson, R.L., Spiegelman, S., 1967. An extracellular Darwinian experiment with a self-duplicating nucleic acid molecule. Proc. Natl. Acad. Sci. U. S. A. 58, 217–224.

Mitsuya, Y., Varghese, V., Wang, C., Liu, T.F., Holmes, S.P., et al., 2008. Minority human immunodeficiency virus type 1 variants in antiretroviral-naive persons with reverse transcriptase codon 215 revertant mutations. J. Virol. 82, 10747–10755.

Moreno, I.M., Malpica, J.M., Rodriguez-Cerezo, E., Garcia-Arenal, F., 1997. A mutation in tomato aspermy cucumovirus that abolishes cell-to-cell movement is maintained to high levels in the viral RNA population by complementation. J. Virol. 71, 9157–9162.

Moreno, H., Tejero, H., de la Torre, J.C., Domingo, E., Martin, V., 2012. Mutagenesis-mediated virus extinction: virus-dependent effect of viral load on sensitivity to lethal defection. PLoS One 7, e32550.

Morse, S.S., 1994. The Evolutionary Biology of Viruses. Raven Press, New York.

Mount, D.W., 2004. Bioinformatics. Sequence and Genome Analysis. Cold Spring Harbor Laboratory Press, Cold Spring Harbor, New York.

Moya, A., Elena, S.F., Bracho, A., Miralles, R., Barrio, E., 2000. The evolution of RNA viruses: a population genetics view. Proc. Natl. Acad. Sci. U. S. A. 97, 6967–6973.

Muñoz, E., Park, J.M., Deem, M.W., 2008. Quasispecies theory for horizontal gene transfer and recombination. Phys. Rev. 78, 061921.

Novella, I.S., 2003. Contributions of vesicular stomatitis virus to the understanding of RNA virus evolution. Curr. Opin. Microbiol. 6, 399–405.

Novella, I.S., Ebendick-Corpus, B.E., 2004. Molecular basis of fitness loss and fitness recovery in vesicular stomatitis virus. J. Mol. Biol. 342, 1423–1430.

Novella, I.S., Reissig, D.D., Wilke, C.O., 2004. Density-dependent selection in vesicular stomatitis virus. J. Virol. 78, 5799–5804.

Nowak, M.A., 2006. Evolutionary Dynamics. The Belknap Press of Harvard University Press, Cambridge, Massachusetts and London, England.

Nowak, M., Schuster, P., 1989. Error thresholds of replication in finite populations mutation frequencies and the onset of Muller's ratchet. J. Theor. Biol. 137, 375–395.

Ochoa, G., 2006. Error thresholds in genetic algorithms. Evol. Comput. 14, 157–182.

Ojosnegros, S., Agudo, R., Sierra, M., Briones, C., Sierra, S., et al., 2008. Topology of evolving, mutagenized viral populations: quasispecies expansion, compression, and operation of negative selection. BMC Evol. Biol. 8, 207.

Okasha, S., 2013. Evolution and the Levels of Selection. Clarendon Press, Oxford.

Ortega-Prieto, A.M., Sheldon, J., Grande-Pérez, A., Tejero, H., Gregori, J., et al., 2013. Extinction of hepatitis C virus by ribavirin in hepatoma cells involves lethal mutagenesis. PLoS One 8, e71039.

Page, K.M., Nowak, M.A., 2002. Unifying evolutionary dynamics. J. Theor. Biol. 219, 93–98.

Palma, E.L., Huang, A.S., 1974. Cyclic production of vesicular stomatitis virus caused by defective interfering particles. J. Infect. Dis. 126, 402–410.

Park, J.M., Deem, M.W., 2007. Phase diagrams of quasispecies theory with recombination and horizontal gene transfer. Phys. Rev. Lett. 98, 058101.

Park, J.M., Munoz, E., Deem, M.W., 2010. Quasispecies theory for finite populations. Phys. Rev. 81, 011902.

Perales, C., Martin, V., Ruiz-Jarabo, C.M., Domingo, E., 2005. Monitoring sequence space as a test for the target of selection in viruses. J. Mol. Biol. 345, 451–459.

Perales, C., Mateo, R., Mateu, M.G., Domingo, E., 2007. Insights into RNA virus mutant spectrum and lethal mutagenesis events: replicative interference and complementation by multiple point mutants. J. Mol. Biol. 369, 985–1000.

Perales, C., Lorenzo-Redondo, R., López-Galíndez, C., Martínez, M.A., Domingo, E., 2010. Mutant spectra in virus behavior. Futur. Virol. 5, 679–698.

Perales, C., Henry, M., Domingo, E., Wain-Hobson, S., Vartanian, J.P., 2011. Lethal mutagenesis of foot-and-mouth disease virus involves shifts in sequence space. J. Virol. 85, 12227–12240.

Pfeiffer, J.K., Kirkegaard, K., 2005. Increased fidelity reduces poliovirus fitness under selective pressure in mice. PLoS Pathog. 1, 102–110.

Poh, W.T., Xia, E., Chin-Inmanu, K., Wong, L.P., Cheng, A.Y., et al., 2013. Viral quasispecies inference from 454 pyrosequencing. BMC Bioinform. 14, 355.

Polyak, S.J., Sullivan, D.G., Austin, M.A., Dai, J.Y., Shuhart, M.C., et al., 2005. Comparison of amplification enzymes for Hepatitis C Virus quasispecies analysis. J. Virol. 2, 41.

Prosperi, M.C., Prosperi, L., Bruselles, A., Abbate, I., Rozera, G., et al., 2011. Combinatorial analysis and algorithms for quasispecies reconstruction using next-generation sequencing. BMC Bioinform. 12, 5.

Quer, J., Gregori, J., Rodriguez-Frias, F., Buti, M., Madejon, A., et al., 2014. High-resolution hepatitis C virus subtyping using NS5B deep sequencing and phylogeny, an alternative to current methods. J. Clin. Microbiol. 53, 219–226.

Quinlivan, M.L., Gershon, A.A., Steinberg, S.P., Breuer, J., 2004. Rashes occurring after immunization with a mixture of viruses in the Oka vaccine are derived from single clones of virus. J. Infect. Dis. 190, 793–796.

Rechenberg, I., 1973. Evolutionsstrategie. Problemata frommann-holzboog, Stuttgart-Bad Cannstatt.

Ruiz-Jarabo, C.M., Arias, A., Baranowski, E., Escarmís, C., Domingo, E., 2000. Memory in viral quasispecies. J. Virol. 74, 3543–3547.

Ruiz-Jarabo, C.M., Miller, E., Gómez-Mariano, G., Domingo, E., 2003. Synchronous loss of quasispecies memory in parallel viral lineages: a deterministic feature of viral quasispecies. J. Mol. Biol. 333, 553–563.

Saakian, D.B., Hu, C.K., 2006. Exact solution of the Eigen model with general fitness functions and degradation rates. Proc. Natl. Acad. Sci. U. S. A. 103, 4935–4939.

Saakian, D.B., Munoz, E., Hu, C.K., Deem, M.W., 2006. Quasispecies theory for multiple-peak fitness landscapes. Phys. Rev. E 73, 041913.

Saakian, D.B., Biebricher, C.K., Hu, C.K., 2009. Phase diagram for the Eigen quasispecies theory with a truncated fitness landscape. Phys. Rev. 79, 041905.

Saffhill, R., Schneider-Bernloehr, H., Orgel, L.E., Spiegelman, S., 1970. *In vitro* selection of bacteriophage Q-beta ribonucleic acid variants resistant to ethidium bromide. J. Mol. Biol. 51, 531–539.

Sala, M., Centlivre, M., Wain-Hobson, S., 2006. Clade-specific differences in active viral replication and compartmentalization. Curr. Opin. HIV AIDS 1, 108–114.

Salazar-Gonzalez, J.F., Salazar, M.G., Keele, B.F., Learn, G.H., Giorgi, E.E., et al., 2009. Genetic identity, biological phenotype, and evolutionary pathways of transmitted/founder viruses in acute and early HIV-1 infection. J. Exp. Med. 206, 1273–1289.

Salemi, M., Vandamme, A.M. (Eds.), 2004. The Phylogeny Handbook. A Practical Approach to DNA and Protein Phylogeny. Cambridge University Press, Cambridge.

Sánchez-Palomino, S., Rojas, J.M., Martínez, M.A., Fenyo, E.M., Najera, R., et al., 1993. Dilute passage promotes expression of genetic and phenotypic variants of human immunodeficiency virus type 1 in cell culture. J. Virol. 67, 2938–2943.

Sardanyés, J., Elena, S.F., 2010. Error threshold in RNA quasispecies models with complementation. J. Theor. Biol. 265, 278–286.

Sarisky, R.T., Nguyen, T.T., Duffy, K.E., Wittrock, R.J., Leary, J.J., 2000. Difference in incidence of spontaneous mutations between Herpes simplex virus types 1 and 2. Antimicrob. Agents Chemother. 44, 1524–1529.

Schuster, P., 2010. Genotypes and phenotypes in the evolution of molecules. In: Caetono-Anolles, G. (Ed.), Evolutionary Genomics and Systems Biology. Wiley-Blackwell, Hoboken, New Jersey, pp. 123–152.

Schuster, P., Stadler, P.F., 1999. Nature and evolution of early replicons. In: Domingo, E., Webster, R.G., Holland, J.J. (Eds.), Origin and Evolution of Viruses. Academic Press, San Diego, pp. 1–24.

Seifert, D., Beerenwinkel, N., 2016. Estimating fitness of viral quasispecies from next-generation sequencing data. Curr. Top. Microbiol, Immunol.

Sevilla, N., Ruiz-Jarabo, C.M., Gómez-Mariano, G., Baranowski, E., Domingo, E., 1998. An RNA virus can adapt to the multiplicity of infection. J. Gen. Virol. 79, 2971–2980.

Shepard, R.N., 1987. Toward a universal law of generalization for psychological science. Science 237, 1317–1323.

Shirogane, Y., Watanabe, S., Yanagi, Y., 2012. Cooperation between different RNA virus genomes produces a new phenotype. Nat. Commun. 3, 1235.

Simon, H.A., 1996. The Sciences of the Artificial, third ed. The MIT Press, Cambridge, Massachusetts.

Smith, D.B., Inglis, S.C., 1987. The mutation rate and variability of eukaryotic viruses: an analytical review. J. Gen. Virol. 68, 2729–2740.

Solé, R., Goodwin, B., 2000. Signs of Life. How Complexity Pervades Biology. Basic Books, New York.

Solmone, M., Vincenti, D., Prosperi, M.C., Bruselles, A., Ippolito, G., et al., 2009. Use of massively parallel ultradeep pyrosequencing to characterize the genetic diversity of hepatitis B virus in drug-resistant and drug-naive patients and to detect minor variants in reverse transcriptase and hepatitis B S antigen. J. Virol. 83, 1718–1726.

Staddon, J.E.R., 2001. Adaptive Dynamics. The Theoretical Analysis of Behavior. The MIT Press, Cambridge, Massachusetts.

Steinhauer, D.A., Holland, J.J., 1987. Rapid evolution of RNA viruses. Annu. Rev. Microbiol. 41, 409–433.

Suspène, R., Henry, M., Guillot, S., Wain-Hobson, S., Vartanian, J.P., 2005. Recovery of APOBEC3-edited human immunodeficiency virus G->A hypermutants by differential DNA denaturation PCR. J. Gen. Virol. 86, 125–129.

Suspène, R., Renard, M., Henry, M., Guetard, D., Puyraimond-Zemmour, D., et al., 2008. Inversing the natural hydrogen bonding rule to selectively amplify GC-rich ADAR-edited RNAs. Nucleic Acids Res. 36, e72.

Svicher, V., Balestra, E., Cento, V., Sarmati, L., Dori, L., et al., 2011. HIV-1 dual/mixed tropic isolates show different genetic and phenotypic characteristics and response to maraviroc in vitro. Antiviral Res. 90, 42–53.

Swetina, J., Schuster, P., 1982. Self-replication with errors. A model for polynucleotide replication. Biophys. Chem. 16, 329–345.

Tanner, E.J., Liu, H.-M., Oberste, M.S., Pallansch, M., Collett, M.S., Kirkegaard, K., 2014. Dominant drug targets suppress the emergence of antiviral resistance. eLife 3:e03830.

Teng, M.N., Oldstone, M.B., de la Torre, J.C., 1996. Suppression of lymphocytic choriomeningitis virus-induced growth hormone deficiency syndrome by disease-negative virus variants. Virology 223, 113–119.

Topfer, A., Marschall, T., Bull, R.A., Luciani, F., Schonhuth, A., et al., 2014. Viral quasispecies assembly via maximal clique enumeration. PLoS Comput. Biol. 10, e1003515.

Tsibris, A.M., Korber, B., Arnaout, R., Russ, C., Lo, C.C., et al., 2009. Quantitative deep sequencing reveals dynamic HIV-1 escape and large population shifts during CCR5 antagonist therapy in vivo. PLoS One 4, e5683.

Vandenbroucke, I., Van Marck, H., Verhasselt, P., Thys, K., Mostmans, W., et al., 2011. Minor variant detection in amplicons using 454 massive parallel pyrosequencing: experiences and considerations for successful applications. BioTechniques 51, 167–177.

Vignuzzi, M., Stone, J.K., Arnold, J.J., Cameron, C.E., Andino, R., 2006. Quasispecies diversity determines pathogenesis through cooperative interactions in a viral population. Nature 439, 344–348.

Vogt, P.K., Jackson, A.O., 1999. Satellites and Defective Viral RNAs. Current Topics in Microbiology and Immunology. Springer-Verlag, Berlin.

Wang, D., Coscoy, L., Zylberberg, M., Avila, P.C., Boushey, H.A., et al., 2002. Microarray-based detection and genotyping of viral pathogens. Proc. Natl. Acad. Sci. U. S. A. 99, 15687–15692.

Wang, C., Mitsuya, Y., Gharizadeh, B., Ronaghi, M., Shafer, R.W., 2007. Characterization of mutation spectra with ultra-deep pyrosequencing: application to HIV-1 drug resistance. Genome Res. 17, 1195–1201.

Weissmann, C., 1974. The making of a phage. FEBS Lett. 40 (suppl), S10–S18.

Wichman, H.A., Badgett, M.R., Scott, L.A., Boulianne, C.M., Bull, J.J., 1999. Different trajectories of parallel evolution during viral adaptation. Science 285, 422–424.

Wilke, C.O., 2005. Quasispecies theory in the context of population genetics. BMC Evol. Biol. 5, 44.

Wilke, C.O., Ronnewinkel, C., Martinetz, T., 2001. Dynamic fitness landscapes in molecular evolution. Phys. Rep. 349, 395–446.

Wilke, C.O., Reissig, D.D., Novella, I.S., 2004. Replication at periodically changing multiplicity of infection promotes stable coexistence of competing viral populations. Evolution 58, 900–905.

Williams, G.C., 1992. Natural Selection. Domains, Levels and Challenges. Oxford University Press, New York, Oxford.

Wilson, E.D., 1998. Consilience. Abacus, Little, Brown and Co., U.K.

Wright, S., 1945. Tempo and mode of evolution. A critical review. Ecology 26, 415–419.

Wright, C.F., Morelli, M.J., Thebaud, G., Knowles, N.J., Herzyk, P., et al., 2011. Beyond the consensus: dissecting within-host viral population diversity of foot-and-mouth disease virus by using next-generation genome sequencing. J. Virol. 85, 2266–2275.

Wu, S., Kanda, T., Nakamoto, S., Jiang, X., Miyamura, T., et al., 2013. Prevalence of hepatitis C virus subgenotypes 1a and 1b in Japanese patients: ultra-deep sequencing analysis of HCV NS5B genotype-specific region. PLoS One 8, e73615.

Xu, L., Da, L., Plouffe, S.W., Chong, J., Kool, E., et al., 2014. Molecular basis of transcriptional fidelity and DNA lesion-induced transcriptional mutagenesis. DNA Repair 19, 71–83.

Yin, J., 1993. Evolution of bacteriophage T7 in a growing plaque. J. Bacteriol. 175, 1272–1277.

Youngner, J.S., Whitaker-Dowling, P., 1999. Interference. In: Granoff, A., Webster, R.G. (Eds.), Encyclopedia of Virology. Academic Press, San Diego, California, pp. 850–854.

Zagordi, O., Klein, R., Daumer, M., Beerenwinkel, N., 2010. Error correction of next-generation sequencing data and reliable estimation of HIV quasispecies. Nucleic Acids Res. 38, 7400–7409.

Zinder, N.D., 1975. RNA Phages. Cold Spring Harbor Monograph Series. Cold Spring Harbor Laboratory New York.

INTERACTION OF VIRUS POPULATIONS WITH THEIR HOSTS

CHAPTER CONTENTS

ABBREVIATIONS

A	adenine
ABV	avian bornavirus
ACE 2	angiotensin-converting enzyme 2
AIDS	acquired immune deficiency syndrome
Arg (R)	arginine
BDV	Borna disease virus
C	cytosine

CAR	coxsackievirus and adenovirus receptor
CTL	cytotoxic T lymphocyte
CVB	group B coxsackieviruses
DAF	decay-accelerating factor
EIAV	equine infectious anemia virus
FMDV	foot-and-mouth disease virus
G	guanine
HA	hemagglutinin
HAV	hepatitis A virus
HBV	hepatitis B virus
HCV	hepatitis C virus
HIV-1	human immunodeficiency virus type 1
IFN	interferon
IV	influenza virus
LCMV	lymphocytic choriomeningitis virus
LTR	long terminal repeat
MHC	major histocompatibility complex
MIC	mutual information criterion
MOI	multiplicity of infection
mRNA	messenger RNA
N	population size
Ne	effective population size
NGS	next generation sequencing
NK cells	natural killer cells
NTPase	nucleoside-triphosphatase
PV	poliovirus
PVR	poliovirus receptor
RGD	Arginine-Glycine-Aspartic acid
RNAi	RNA interference
SARS CoV	severe acute respiratory syndrome coronavirus
SIV	simian immunodeficiency virus
T	thymine
TCID$_{50}$	tissue culture infectious dose 50 (the amount of virus needed to infect 50% of a cell population)
tRNA	transfer RNA
Trp (W)	tryptophan
U	uracil
VPg	viral protein genome linked
VSV	vesicular stomatitis virus

4.1 CONTRASTING VIRAL AND HOST POPULATION NUMBERS

Although we often refer to viruses as "autonomous" genetic elements, their replication is dependent on host functions. This dependence occurs at two levels: at the stage of replication within individual cells and at the stage of dissemination among populations of susceptible cells, animals, or plants. Host cells and organisms have mediated survival of the viruses that we can isolate and study, but they have conditioned their persistence to the capacity to overcome selective constraints imposed by the cellular world. The paradox that represents that host functions are sometimes recruited by the virus to ensure its replication, and other times they become part of the innate immune response can be interpreted as a result of a long-term coevolution between viruses and cells, and the basic mechanisms by which biological evolution is working in our biosphere. The concept of evolutionary tinkering proposed by F. Jacob in the last century (Jacob, 1977) is most adequate to interpret the paradoxical interplay of viral and cellular functions (Domingo, 2011). In this view, viruses and cells, in their coevolutionary race for survival, must have taken advantage, as needed, of what existed at any given stage of the evolutionary process. Here, again, viral population numbers and molecular instructions to ensure genome diversity have played essential roles.

The infection of a host by a virus can be viewed as a specific example of a predator (virus)-pray (host) relationship, with a very distinctive attribute: the disproportionate difference in population size between viruses and their hosts. The difference is dramatically in favor of viruses. The large population size of many viral populations (both, as replicative ensembles inside individual cells and as particles available for new rounds of infection) favors their adaptability, as a consequence of the ease of exploration of sequence space (Section 3.7 in Chapter 3). It is well established in ecology that, except in the case of being a pathogenic agent, the predator cannot be more numerous than the prey (Remmert, 1980). A virus in an infected organism can reach 10^9 to 10^{12} potentially infectious particles at a given point in time. Early calculations were made in cattle infected with foot-and-mouth disease virus (FMDV) (Sellers, 1971), and during togavirus infections (Halstead, 1980). More recently, similar estimates have been obtained for human immunodeficiency virus type 1 (HIV-1), hepatitis B virus (HBV), and hepatitis C virus (HCV) (Wei et al., 1995; Nowak et al., 1996; Neumann et al., 1998) or for a tobacco leaf infected with tobacco mosaic virus (review in Gutierrez et al., 2012). Large amounts of virus can also be present in excretions and secretions from infected individuals, where the total amount of virus and its concentration is a factor of virus spread. In adult volunteers infected with influenza virus (IV) type A, maximum titers of 10^3 to 10^7 tissue culture infectious dose 50 per ml ($TCID_{50}$/ml) were determined in nasopharyngeal fluids 1 day after infection (Murphy and Webster, 1985). In children with type B IV, titers of 10^4 $TCID_{50}$/ml were present in nasal washings also 1 day after infection (Hall et al., 1979).

In the laboratory, for viruses that infect cells in culture, a visible viral plaque on a cell monolayer (or a focus formed in a cluster of infected cells) contains a variable number of infectious units, but that generally exceeds 10^3. A range of 10^3 to 10^9 infectious units per plaque is quite frequent for cytopathic viruses such as FMDV and VSV used in studies of experimental evolution (Chapter 6). The lytic plaque and infected cell focus size, and the number of viral particles in them depends on the replication rate of the viral clone, the way of transmission from cell to cell (either exit into the intercellular medium prior to penetration into a neighbor cell, or direct cell-to-cell transmission), virus stability in the extracellular environment (agar overlay or other medium), and the host cell viability during the time of plaque or focus development. Culture medium added to confluent bacteriophage plaques on a bacterial lawn is an effective method to achieve high bacteriophage titers for viral purification and physical studies. For animal viruses that infect cells in culture, infections of cell monolayers or cells in suspension can be scaled up to produce viral populations of 10^{10} to 10^{11} particles in the case of the most fecund viruses.

The viral population numbers have been given to underline the sharp contrast with the population numbers of the hosts that viruses infect. Considering mammals, each of many primate species include a total of hundreds to a few thousand individuals. Among the most abundant, the gibbon *Hylobates muelleri* is represented by 3×10^5 to 4×10^5 individuals, and humans by 7×10^9 individuals, with a projection of 9×10^9 individuals for 2040, values which are orders of magnitude lower than the total number of HCV particles in a liver acutely infected with HCV. Mammals are, however, modest in representation as compared with other types of organisms. The number of insect species is uncertain, with estimates broadly ranging from 1×10^6 to 20×10^6, with about 10^{18}-10^{19} individual insects alive in our planet at a given time. Despite these impressive numbers (imagine how many viruses might be hosted by insects that have never been analyzed!), the estimated number of individual insects is still 10^{13}-fold lower than the total number of viral particles on Earth (compare with figures given in Chapter 1). Insects are only exceeded by zooplankton (about 10^{21} individuals) and nematodes (about 10^{22} individuals). To give some additional comparative figures that will become pertinent when dealing with zoonotic transmissions and viral disease emergence (Chapter 7), the total number of livestock is 2.4×10^{10}, and the total number of birds, mammals, reptiles, amphibians, or fish is 10^{10} to 10^{13}. Estimates of the number of biological species and of individuals within species are regularly published, and the reader will find numbers that are all extremely modest compared with the VIROME (Viral Informatics Resource for Metagenome Exploration) regarding virus diversity and anticipated number of individual viral particles per group (Wommack et al., 2012; Virgin, 2014).

Long-term virus survival has been based not only in multiple strategies to cope with the host immune response, but also in their life cycles having generally evolved to produce vast numbers of progeny. Using terminology of ecology, viruses as *r* strategists in the sense that they base their success in rapid reproduction to confront multiple habitats (intrahost compartmentalization and multiple selective constraints, as discussed in Section 4.2). In contrast, large animals are *K* strategists that produce limited progeny, have a long life span and inhabit relatively stable environments (Remmert, 1980).

4.1.1 PRODUCTIVE POWER OF SOME VIRAL INFECTIONS

The exploration of sequence space is commensurate with the number of newly synthesized viral genomes per unit time in infected organisms. Only for a few virus-host systems the velocity of genome replication (number of nucleotides incorporated into a growing viral RNA or DNA genome per unit time) has been calculated. Early studies indicated that the average time needed to synthesize an entire plus strand of bacteriophage Qβ RNA (4220 nucleotides) *in vivo* was about 90 s (Robertson, 1975). For poliovirus (PV) it has been estimated that it takes about 1 min to synthesize a full length genomic RNA (7440 nucleotides), and that when PV RNA synthesis reaches its maximum, 2000-3000 RNA molecules are produced per cell and minute (Richards and Ehrenfeld, 1990; Paul, 2002). HCV polymerase incorporates 5 to 20 nucleotides per second (reviewed in Fung et al., 2014) These values imply that, with the mutation rates and frequencies typical of RNA viruses (Chapter 2) mutant distributions of 10^5 to 10^7 genomes can be produced in infected cell cultures or host organisms in minutes.

In the course of infections by HIV-1, it has been estimated that 10^{10} to 10^{11} new virions are produced each day (Coffin, 1995; Ho et al., 1995). The average life span of cells productively infected with HIV-1 has been estimated in 1-2.2 days, with a half-life ($t\frac{1}{2}$) of about 1.5 days. The average life span of HIV-1 virions in plasma is about 6 h, with a $t\frac{1}{2}$ of 2-4 h (Ho, 1995; Wei et al., 1995; Perelson et al., 1996; Markowitz et al., 2003). A rapid turnover of virions occurs also during HCV and HBV infections. The half-life of HCV particles circulating in infected individuals is about 2.7 h, and about

10^{12} particles are produced and cleared every day (Neumann et al., 1998; Ramratnam et al., 1999). A typical active HBV infection can produce 10^{13} viral particles per day; with an average mutation rate of 10^{-4} mutations per nucleotide, 10^9 new mutations can be tested every day in the 3200 bp HBV genome (Whalley et al., 2001) (review in Quer et al., 2008). Thus, replication of some important viral pathogens is extremely rapid, viruses undergo continuous genetic change, and are constantly replaced by new variants (rapid turnover). Genetic and phenotypic diversification is fast and observable. This has been directly noticed in the case of HIV-1 upon reconstruction of the genomic nucleotide sequence of the transmitted (or founder) virus in a number of patients (Keele et al., 2008; Salazar-Gonzalez et al., 2009). These studies revealed a rapid diversification of the founder, biologically active HIV-1 into multiple replication-competent and defective progeny. There is little question that quasispecies dynamics, as defined in Chapter 3, is operating *in vivo*, and it implies an effective exploration of the permissive area of sequence space. Rapid, error-prone replication is the basis of virus behavior as *r* strategists, an adaptation to their long-term survival in heterogeneous environments.

4.1.2 POPULATION SIZE LIMITATIONS AND THE EFFECT OF BOTTLENECKS: THE EFFECTIVE POPULATION SIZE

High viral yields are not universal during viral infections. Viral production can be very high in acute infections *in vivo* and in cytopathic infections in cell culture. However, viruses can establish also latent infections with intermittent periods of virus production and intervals without detection of infectious virus. In latent infections by DNA viruses or retroviruses, the virus can be undetectable or present in minimal quantities until recurrence of the infection by activation of the latent reservoir takes place. Latency can occur with or without integration of viral DNA into the host DNA. Chronic infections involve continuous but variable production of infectious virus, with or without disease manifestations that may become apparent only after prolonged chronicity. An acute infection can be followed by a persistent stage, sometimes producing highly mutated forms of the acute virus that give rise to new pathologies. This is the case of subacute sclerosing panencephalitis, a rare brain disease associated with hypermutated variants of measles virus (Chapter 2). Persistent infections in cell culture have been divided into steady-state and carrier cell infections. The maintenance of a steady-state system is not dependent on reinfections by virus particles produced from cells. In contrast, in carrier cell cultures there is a continuous supply of a small number of uninfected cells that engage in a sustained, low level viral production. Persistent infections in cell culture have been instrumental to learn about the consequences of virus-host cell interactions, and they are studied in Chapter 6.

In steady-state persistent infections, cells often produce and release a limited amount of virus, while cells divide with little metabolic affectation. Persistent infections by Borna disease virus (BDV) are particularly illustrative because as a little as 0.01-0.05 infectious units are present per infected cell (Pauli and Ludwig, 1985). This behavior may relate to the frequent occurrence of asymptomatic infection of several animal species by this unique viral pathogen (de la Torre, 2002). Limited Borna virus replication may also explain its relative evolutionary stasis in some hosts, while its rate of evolution and extent of diversification appear to be larger among newly described avian bornaviruses (ABVs) (Philadelpho et al., 2014), with 17-fold higher substitutions rates for ABV than the mammal-infecting BDV (He et al., 2014).

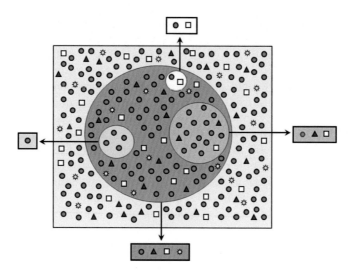

FIGURE 4.1

A depiction of population bottlenecks of different intensity. The big rectangle represents a viral population with multiple variants distinguished by various symbols. Internal circles denote reductions in population size that result in the isolation of one, two, three or four variant types (different external rectangles) to initiate a next round of infection. The evolutionary outcome will depend on the number of founder variants that participate in the infection.

The figure is modified from Domingo et al. (2012), with permission from the American society for Microbiology, Washington DC, USA.

The maintenance of a large population size of viruses in cytolytic or acute infections *in vivo* is conditioned to the absence of population bottlenecks (strong reductions in population size that can alter the course of selective events, as discussed in Chapter 3) (Figure 4.1). Bottlenecks have been characterized during viral transmission from an infected into a susceptible host, and also within infected hosts (reviewed in Domingo et al., 2012; Gutierrez et al., 2012; Forrester et al., 2014). During host-to-host transmission, bottlenecks are favored by aerosol spread of viruses in the case of respiratory viruses, but also when infection is initiated through contact with small volumes of secretions or excretions. The microdroplets in aerosols should contain a very small proportion of the total number of viruses present in an infected individual (estimates were reported in Artenstein and Miller, 1966; Gerone et al., 1966; see also Clarke et al., 1994 and Chapter 6 for the effect of serial bottlenecks, as assessed in laboratory experiments). Different evolutionary outcomes can be expected when a virus is transmitted through a small or large amount of infected fluids, for example, sharing a syringe versus a transfusion with contaminated blood in the case of HIV-1 or HCV when blood screening had not been implemented (Shepard et al., 2005; Sharma and Sherker, 2009; De Cock et al., 2012). Not only the probability of infection is higher when a susceptible individual is exposed to a large amount of a contaminated fluid, but a massive amount of initial virus facilitates adaptation of the quasispecies to the recipient host.

Severe population bottlenecks occur during plant leave inoculation, and seed or aphid transmission of plant viruses. Studies have included the transmission of cucumber mosaic virus, tobacco mosaic virus, pea seed-borne mosaic virus, and potato virus Y variants (Li and Roossinck, 2004; Ali et al., 2006; Moury et al., 2007; Betancourt et al., 2008; Sacristan et al., 2011; Fabre et al., 2014;

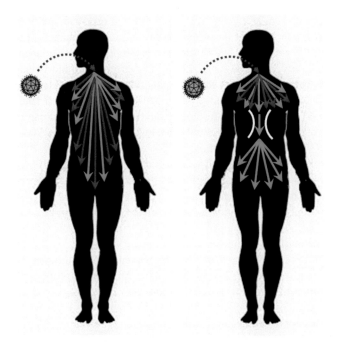

FIGURE 4.2

A simplified scheme of the consequences of a bottleneck event *in vivo*. On the left, a virus spreads in an infected individual without intervening bottlenecks. Any variant produced (here only two are represented by red and blue arrows) can reach any target organ. On the right, the occurrence of a bottleneck restricts the types of variants (here represented by a blue arrow) that can reach some target organs.

reviewed in Roossinck, 2008; Gutierrez et al., 2012). Aphids transmit an average of 0.5-5 virus particles into the recipient plant host, a range of values that is very similar to that estimated for HIV-1, with evidence that in about 75% of patients studied a single founder genome initiated the infection while in the others a minimum of two to five viruses were involved (Keele et al., 2008). Bottlenecks have been identified also during transmission and early stages of HCV infections (Quer et al., 2005; Bull et al., 2011).

Bottlenecks happen within infected host once the infection is already ongoing (Foy et al., 2004; Scholle et al., 2004; Quer et al., 2005; Ali et al., 2006; Kuss et al., 2008; Smith et al., 2008; Haaland et al., 2009; Forrester et al., 2014). In other cases, no severe transmission bottlenecks or major barriers for dissemination seem to operate during intrahost virus expansion (Murcia et al., 2010; Erickson and Pfeiffer, 2013) (Figure 4.2). As emphasized by S. Gutierrez, S. Blanc and colleagues a variety of population bottleneck sizes is probably encountered during virus-host interactions (Gutierrez et al., 2012). Additional studies are needed to elucidate how variations in population size can affect evolutionary outcomes in viruses. Application of next generation sequencing (NGS) should provide new insights in the consequence of bottlenecks for the genetic heterogeneity of viral populations (see Section 4.10 on bottlenecks during the arbovirus life cycle).

Not all viral particles present at a given time in an infected organism are productively replicating. The term effective population size (N_e) has been used to indicate the fraction of the total viral population (N) that contributes to progeny. N_e in virology was adopted from a concept initially introduced by S. Wright in population genetics to mean the number of breeding individuals and, therefore, the number of the total N that contributes to the next generation. It is considered important to evaluate the variability in a population and the relative participation of selection and random drift in evolution (Charlesworth, 2009). N_e is also relevant for conservation biology since it can predict the rate of inbreeding and the level of genetic diversity in wildlife (Palstra and Ruzzante, 2008).

The application of N_e to viral populations is not easy and has not been without controversy. Measurements of N_e for HIV-1 have yielded values that range from 10^3 to 10^6 (Rouzine et al., 2014). There are at least two reasons to account for the fact that not all viral particles present in an infected individual are replicating to produce infectious progeny at any given time: (i) despite all infectious virus being potentially replication competent, for stochastic reasons not all of them make their way into a susceptible cell. (ii) A proportion of viral particles that are counted in virus censuses based on quantification of genetic material, are intrinsically noninfectious (they could not infect despite having available permissive host cells). The reasons for defectiveness are multiple, encompassing genetic lesions, and virion assembly defects. It is very difficult to estimate the proportion of defective viruses in the different host compartments in which replication takes place. Regarding transmission, it means that when one to four particles are transmitted (external rectangles in Figure 4.1) some particles may be defective or unfit thus contributing to possible unproductive infections. Viral N and N_e values may vary locally in different tissues and organs, rendering the interest of the actual N_e value highly dependent on the purpose of the determination (i.e., during viremia to interpret the neutralizing activity of circulating antibodies, or in a specific host compartment to evaluate pathogenic consequences of the infection, or the probability of generation of mutants resistant to an inhibitor). The distinction between N and N_e has been attempted with only a few virus-host systems (Gutierrez et al., 2012; Rouzine et al., 2014, and references therein). In the cases in which estimates of N_e have been obtained, the values are such that both selection and random drift can influence the evolutionary outcomes. For all these reasons, in the present book we refer only to virus population size, without attempting a distinction between N and N_e. The interpretation of several observations related to differences in population size that will be discussed in Chapters 5 and 6 is not critically dependent on such a distinction, further justifying our simplification, despite acknowledging the relevance of N_e in some particular cases or for other biological systems.

Differences in population size and the severity and frequency of bottleneck events determine the contributions of random drift versus selection as evolutionary influences (Chapter 3). In addition, a bottleneck event, irrespective of its perturbing effect in a process of selection, will reduce the diversity of the postbottleneck population. A few rounds of replication are needed to restore the mutant spectrum amplitude of the initial populations (Figure 4.3). It has been proposed that one of the driving forces for RNA viruses to maintain as a universal trait high mutation rates is to favor a rapid repertoire of mutants to ensure adaptability following bottleneck events that they have to undergo as part of their life cycles (Vignuzzi and Andino, 2010).

Despite having population sizes far smaller than viruses, differentiated organisms can also undergo bottleneck events, for example, through geographical isolation of a subset of individuals of an animal species. Such founder events by which progeny with a subset of alleles has a chance to proliferate, can contribute to geographical differentiation of host species, and it is one of the models proposed for the generation of new species (speciation).

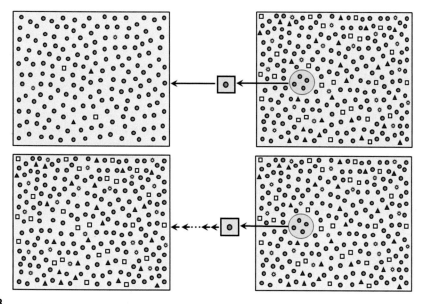

FIGURE 4.3

A population bottleneck results in reduction of population complexity, and complexity restoration may require multiple rounds of infection. A severe bottleneck is represented by a single type of virus (blue sphere in the small square). In the upper drawing, a single infection leads to limited relevant phenotypic diversification (although virtually any progeny virus can be a mutant). In the bottom drawing, multiple infective cycles have restored a phenotypic complexity similar to that of the parental population.

4.2 TYPES OF CONSTRAINTS AND EVOLUTIONARY *TRADE-OFFS* IN VIRUS-HOST INTERACTIONS

Infection of host organisms by a vast amount of viruses must be necessarily limited to ensure long-term survival of both. The term constraint is used to refer to multiple mechanisms that prevent generalized and devastating infections by viruses, notably the host immune response. In Chapter 3 positive and negative selection acting on viruses were discussed in conceptual terms, using some specific examples, but without detailing types of constraints that viruses must face in their hosts. At the cellular level, superinfection exclusion provides a set of mechanisms for an infected cell to prevent the infection by other related viruses. However, virus mutants can be selected to partly overcome the exclusion mechanism (specific examples and some introductory references can be found in Webster et al., 2013; Campbell et al., 2014).

In addition to immune responses and exclusion mechanisms, viruses are continuously and unavoidably subjected to functional and structural constraints, even when replication occurs in relatively constant environments such as those provided by established cell lines in culture. Several classes of selective constraints can be distinguished (Morse, 1994; Gibbs et al., 1995; Domingo et al., 2001; Simmonds et al., 2004):

- Basal constraints inherent to the virus, and to the requirements for viral genome replication, that operate independently of the environment in which a virus is immersed. Basal constraints prevent the deterioration of the "core information" that determines viral identity.

- Selective constraints intrinsic to individual host cells. They include the availability of membrane structures, nucleotide pool levels, tRNA abundances, ionic composition, accessibility of proteins that can act as receptors for viruses, host factors that have to be incorporated into replication complexes or used for viral protein synthesis, presence of RNA or protein chaperones, etc. A balance of many host cell-dependent influences is necessary for a virus to complete its replication cycle.
- Selective constraints internal to the host as an organism understood as an integrated aggregate of cells, tissues, and organs. These constraints include immune responses, metabolic alterations that may result in fever or abnormal concentrations of metabolites, increased levels of reactive oxygen radicals, or other perturbations, phenotypic heterogeneity among cells of the tissues where virus replicates, among others.
- Selective constraints due to external influences exerted on the host. They relate to human intervention and include administration of antiviral drugs or therapeutic antibodies, vaccination, immunosuppressant treatments, and others.

4.2.1 LONG-TERM HISTORY DICTATES BASAL CONSTRAINTS

According to our current understanding of cellular and viral genomics, a present day virus is the result of a long evolutionary history. Using the molecular mechanisms and the building blocks available at the different evolutionary periods, viruses have attained a viable solution as autonomous, cell-dependent, genetic elements. As in the developmental program of an organism, a present day virus can be viewed as a result of a historical process (Maynard Smith et al., 1985). A viral genome was shaped by natural selection that involved transfers of functional modules, duplication of genomic regions, constellations of mutations acting in concert, and other molecular events. It is not easy to find pathways to partially dismantle the coordinated set of elements without destroying the replicative capacity of the construct. The requirement of a finely tuned assemblage of elements introduces what we refer to as basal constraints. Their existence at multiple structural and functional levels has become more and more evident as knowledge of the molecular biology of viruses has progressed. This is one of the reasons why synonymous mutations are not necessarily neutral, as justified in Chapter 2.

High-order structures include the internal ribosome entry site (IRES) for the initiation of protein synthesis of some viruses, and pseudoknot structures involved in protein recognition and ribosomal frameshifting. These structures generally tolerate limited numbers of mutations, and very often compensatory mutations occur to maintain their folding (Olsthoorn et al., 1994; Pleij, 1994; Escarmís et al., 1995; Arora et al., 1996; Belsham and Sonenberg, 1996; Martinez-Salas et al., 1996). Open-reading frames may include regulatory signals that are independent of the protein-coding function of the same RNA region. Restrictions to the fixation of silent substitutions within open-reading frames may be imposed also by codon usage and translation efficiency (Britten, 1993; Eyre-Walker, 1996) (Section 4.3). The long evolutionary history of viruses (Chapter 1) has derived in the array of basal constraints that we are beginning to characterize at the molecular level.

4.2.2 CELL-DEPENDENT CONSTRAINTS: NO FREE LUNCH

Other constraints for DNA and RNA viruses are a consequence of the complete dependence of virus replication on the cell, and the need for the virus to be competent for transmission from a donor cell into a recipient cell or from a donor organism into a recipient organism. Constraints to variation of surface residues in the viral particle is one of the mechanisms proposed to explain antigenic stability of some viruses despite their displaying high mutation rates and frequencies (e.g., measles or rabies virus), or the presence of widely different numbers of serotypes among picornavirus genera (Chapter 7). Constraints introduce conflicts regarding evolutionary change. A viral capsid must confer stability to virions outside the cell, but be labile enough (or capable of a transition toward a labile form) to permit uncoating and release of the viral genome as a result of some intracellular environmental change.

The integrin-recognition site of FMDV illustrates constraints acting on this rapidly evolving virus. Contrary to other picornaviruses, the capsid of FMDV lacks a canyon or pit where residues involved in cell receptor recognition lie. FMDV has a smooth surface with a protruding, mobile loop that has the dual function of interacting with integrin receptors and with neutralizing antibodies (Acharya et al., 1989; Verdaguer et al., 1995). Thus, antibody escape through amino acid substitutions in this loop must be compatible with receptor recognition. This explains the limited repertoire of amino acid substitutions at this multifunctional loop among field isolates and laboratory populations of FMDV (Martínez et al., 1992; Borrego et al., 1993; Mateu, 1995) (Section 4.4).

Other types of conflicting requirements are becoming evident as we learn about the multiple molecular interactions between viruses and cells. They force viruses to evolutionary *trade-offs* by which some nucleotide or amino acid substitutions are introduced to confront a constraint, despite rendering the genome suboptimal for another trait. Suboptimality, however, has a limit, and any *trade-off* must fulfill a balance regarding how a genetic change is positive for a trait and negative for other traits. Most viral functions (replication rate, viral protein synthesis, protein processing, particle assembly, etc.) are unlikely to have been optimized for maximum rate; rather, they are the result of *trade-offs* to fulfill multiple requirements to complete the virus life cycle, and each individual function has a performance level which is conditioned by the requirements of other viral functions.

In the context of *trade-offs*, the theorem of "no free lunch," amply used in economy and algorithm solution searching, applies. Stated in simple terms, a person or a society cannot obtain "something for nothing." There is always a cost, even if that cost is hidden. A virus will endure a fitness cost as a result of having successfully confronted a selective constraint, as illustrated with specific examples in Chapters 5 and 8, but a universal optimization is impossible. Application of the "no free lunch" theorem to complexity theory is currently under investigation regarding its applicability to biological processes such as genetic optimization algorithms (Whitley and Watson, 2005; Manning et al., 2013; Buenno et al., 2015). Data for virus evolution suggest that due to the extremely compact information packed into a physically small genome, each nucleotide is exploited for multiple functions, even those nucleotides that do not belong to overlapping genes (alternative open-reading frames with two different proteins encoded by the same nucleotide sequence). As a consequence, possible neutral sites ("neutral" meaning that their modification does not entail any functional difference) are probably very rare, as explained in Chapters 2 and 3. Some sites may be referred to as neutral only in the sense that a modification in them still allows the virus to survive. Few nucleotides of an RNA virus genome will conform to the definition of neutral alleles as that "one could be substituted for the other.... without affecting the altered individual's prospect of survival and reproduction under any environmental circumstance" (Reeve et al., 1990). The scarcity of neutral mutations is one of the major reasons for fitness landscapes being extremely rugged for viruses (Chapter 5). Finding *trade-offs* is a way of life for viruses. No free lunch for individuals, societies, or viruses.

The constraints often cited as inherent to the host cells (supply of cellular membranes, components of the translation apparatus exploited by viruses for their own replication machinery, etc.) constitute only a minority of the cellular functions that are, one way or another, involved in virus replication.

Application of microarray hybridization analysis to quantitate the alteration of host gene expression during viral infections typically shows upregulation or downregulation of hundreds of host genes. Testing the influence of cellular functions by RNAi (interference) screens often reveal unsuspected effects of host gene products in virus replication, even when appropriate controls to exclude off-target effects of the interfering RNAs are used. Part of the host modifications in gene expression observed in infected cells may be just the indirect consequence of other cellular perturbations, and, thus, they may not be essential to the progression of the infection. However, the massive response of collectivities of genes illustrates the multiple connections that a replicating virus establishes with cells, and suggests that disarrangements of some cellular functions lead to other (perhaps compensatory) changes in the cells. Viruses are structurally and functionally deeply integrated into the cells in which they replicate (compare with the theories of virus origins that favor long-term coexistence of viral elements with precellular and primitive cell organizations discussed in Chapter 1).

4.2.3 CONSTRAINTS IN HOST ORGANISMS: CONTRAST WITH MAN-MADE ANTIVIRAL INTERVENTIONS

Concerning constrains internal to organisms, the latter have evolved at least three lines of defense against pathogens: intrinsic (preexisting factors that restrict virus replication), innate (activated when a virus enters the organism; i.e., several interferons, apoptosis, NK cells, etc.), and adaptive (immune interferon, T cells, B cells, that expand and evoke a specific cellular and antibody response against the invading virus, among other activities). Long-term coevolution of host organisms and their viruses must have contributed to the survival of both (Woolhouse et al., 2002; Switzer et al., 2005; Villarreal, 2005). Viruses may respond to short-term perturbations by the dominance of subsets of variants that, in addition to increasing fitness under the new environment, must not compromise viral survival. Many examples can be cited: an amino acid substitution permitting antibody escape should be compatible with virion stability and receptor recognition; a substitution that decreases virus affinity for a soluble cellular receptor should still allow recognition and binding to the cell-anchored receptor for cell entry, or allow an alternative entry pathway, etc. They are changes that must conform to the *trade-off* concept described in Section 4.2.2.

The external influences exerted on host organisms intended to limit their replication, typically during antiviral treatments, can be divided in two major groups: those that consist of inducing or mimicking an immune response (vaccination or passive immune therapy, respectively), and those that consist of challenging the virus with antiviral inhibitors which are not among the metabolites of their hosts. Viruses do not have an evolutionary history of confrontation with man-made antiviral agents, and nevertheless viruses can overcome the effect of this class of inhibitors. This reflects that viruses have evolved survival mechanisms of such a general nature (basically the different genetic variation strategies described in Chapter 2) that they constitute a flexible "tool-box" ready for contingencies. One of the major problems in antiviral therapy has been the realization that viruses are capable of selecting mutants resistant virtually to any drug used in therapy. In fact, the capacity of a drug to select resistant mutants has been traditionally considered a proof of the selectivity of the drug (Herrmann and Herrmann, 1977). Although many studies on inhibitor-resistant mutants have involved HIV-1, HBV, and HCV infections, all viruses respond to specific antiviral agents by selecting escape mutants, as described in Chapter 8. Remarkably, different mutations in the same viral genes may determine resistance to, or dependence on, an antiviral agent (de la Torre et al., 1990; Baldwin et al., 2004; Baldwin and Berkhout, 2007, and references therein). This adaptive flexibility is obviously of great practical relevance, and it has encouraged the exploration of new antiviral strategies that aim at avoiding selection of virus mutants resistant to antiviral agents (Chapter 9).

There are antiviral inhibitors that establish a connection between external and internal constraints. They target a cellular metabolic pathway whose inhibition results in a stimulation of innate immune response genes. Inhibitors of pyrimidine biosynthesis (A3, DD264, and brequinar) may affect specific replicative steps of some viruses, and they trigger induction of components of the innate immune response. They confront the virus with a broad (multifactorial) antiviral response (Lucas-Hourani et al., 2013; Ortiz-Riano et al., 2014, and references therein). For this reason, this class of inhibitors are broad-spectrum antiviral agents, and the hope is that they may be less prone that standard inhibitors to select resistant mutants. If this expectation is confirmed, this class of inhibitors could be incorporated into new antiviral strategies to control viral quasispecies (Chapter 9).

4.3 CODON USAGE AS A SELECTIVE CONSTRAINT: VIRUS ATTENUATION THROUGH CODON AND CODON-PAIR DEOPTIMIZATION

Codon use bias in relation to cellular tRNA abundances was listed as one of the mechanisms by which synonymous mutations can affect virus behavior (Section 2.3 in Chapter 2), and the possibility of codon choice is part of the relevant information harbored by the genetic code (Maraia and Iben, 2014). Here we expand the discussion of codon usage as a constraint for viral infections and its implications for long-term virus-host interactions. Several computations have established that the choice among synonymous codons is not random in many biological systems, not only in viruses. Evolutionary events must have resulted in the preferential use of some codons over others, and several possibilities have been proposed. One is that misincorporation tendencies of nucleic acid polymerases may have led to selection of codons rich in the bases preferentially introduced during genome replication. Long-term priority for some mutation types (because of the catalytic and fidelity properties of cellular or viral polymerases) may decant synonymous codons in favor of those containing the nucleotides that arise more frequently as a consequence of the mutational bias. An alternative, but not mutually exclusive possibility, is that codon bias might have been an evolutionary outcome for optimal RNA secondary structure or RNA-RNA interactions. RNA is rarely a linear unstructured polynucleotide (as usually drawn for simplicity), but rather a complex molecule with a number of high-order structures (stem-loops, pseudoknots, kissing loops, etc.) that play functional roles, and contribute to RNA stability. The appearance of an RNA molecule is certainly closer to that of a protein than to double-stranded DNA (as examples, see Cantara et al., 2014; de Borba et al., 2015 and references therein). Regulatory signals in viral RNA may be located either in untranslated regions or within open-reading frames. In the latter case, preservation of the higher order structure may restrict the possibility of some triplets to mutate to synonymous ones. This occurrence would constitute an example of negative selection acting on synonymous mutations amply documented during RNA virus evolution (Chapter 2).

tRNAs are among the oldest biological molecules dating back to the time in which the genetic code was developed (Eigen, 1992). Given their ancient nature and the critical role they play in the transmission of information, tRNAs are extremely conserved among cellular organisms, and their sequences have served to date the origin of the genetic code (discussed in Chapter 1). tRNAs must have been extremely restricted with regard to nucleotide sequence changes, due to folding requirements and role as "adaptor" molecules. Regulation of translation could not be achieved (at least in an effective manner) through tRNA sequence modifications. Instead, regulation could be attained through differences in the abundances of tRNAs that recognize different synonymous codons.

The accommodation of synonymous codon usage to the cellular tRNA pool is known as translational selection. Other mechanisms related to selection of base and dinucleotide frequencies, composition of enhancers of splicing, or translation kinetics have also been proposed as underlying the variation of codon usage (dos Reis et al., 2004; Chamary and Hurst, 2005; Lavner and Kotlar, 2005; Shackelton et al., 2006; Yang and Nielsen, 2008; Aragones et al., 2010). Picornaviruses have been studied regarding synonymous codon usage, and the response of a virus when codon frequencies are artificially altered. Introduction of unpreferred synonymous codons in the capsid-coding region of PV resulted in fitness decrease, attributed to alteration of an early step in the virus replication cycle (Burns et al., 2006). The relative fitness of the modified virus, measured in HeLa cells, decreased in proportion to the number of replaced codons. Codon deoptimization resulted in reduced viral RNA yields, and decreased specific infectivities of purified virus. The specific infectivity is the ratio between infectious and physical particles, an important parameter that is further discussed in Chapter 9. Not only codon usage, but also codon-pair frequencies can affect PV fitness. It has been suggested that viruses deoptimized for condon pairs may open the way to a new generation of antiviral vaccines (Coleman et al., 2008), and this approach is currently investigated with other viral systems [(Martrus et al., 2013; Le Nouen et al., 2014; Nogales et al., 2014; Cheng et al., 2015), among others]. These observations reflect once again the multiple ways in which a viral RNA *per se* can be part of the viral phenotype, independently of its protein-coding function.

In contrast to PV, hepatitis A virus (HAV) uses rare synonymous codons (those that correspond to tRNAs that are present at low concentrations) to control the rate of translation. The adequate combination of common and rare codons allows HAV to regulate ribosome traffic and to slow down the synthesis of capsid proteins to facilitate their proper folding and capsid stability (Sánchez et al., 2003; Aragones et al., 2010; Costafreda et al., 2014). In an elegant study, HAV was replicated in cells treated with actinomycin D (a specific inhibitor of DNA-dependent RNA polymerases), which provided an altered cellular environment in which the tRNA pool available for translation of viral RNA was increased. HAV adaptation to this environment resulted in a new deoptimization of codon usage in the capsid-coding region, again supporting translation kinetics selection as the basis for biased codon usage by HAV. Proper protein folding may be essential for nonenveloped viruses that are transmitted via the fecal-oral route to survive for prolonged time periods in the external environment (Aragones et al., 2010). For viruses whose infectivity is maintained in the external environment, the time-dependent difference between intra- and interhost evolutionary rate might be accentuated (discussed in Chapter 7). The differences between PV and HAV regarding fitness effects of codon usage modification were reviewed by Bosch et al. (2010).

In the course of the studies with HAV, evidence of quasispecies memory (explained in Chapter 5) and of selection for fine-tuning translation kinetics acting on the mutant spectrum as a whole were obtained (Aragones et al., 2010). Also, the continuous re-deoptimization of HAV to the new environment to maintain fitness constitutes further support of the "Red Queen" hypothesis (Van Valen, 1973; Krakauer and Jansen, 2002), one of the concepts of population genetics shown to operate with RNA viruses. (Concepts first proposed in general population genetics and then shown to operate in viruses are treated in Chapter 6.)

Codon usage has also biotechnological implications. When the codon frequencies in an expression system do not match the codon usage of the viral genomes to be expressed, viral yields may be diminished. This is a relevant factor to be taken into consideration for the choice of expression systems. When elevated expression levels are not obtained because of codon usage-related effects on translation efficiency, the problem may be circumvented by engineering genetic forms of the virus with a modified codon composition that matches the requirements for expression in the desired host (Lanza et al., 2014 and references therein).

4.3.1 **THE SYNONYMOUS CODON SPACE CAN AFFECT AN EVOLUTIONARY OUTCOME**

It is worth expanding on the concept outlined in Chapter 2 that synonymous codons may lie at a different distance from a nonsynonymous codon in sequence space, and this may modify the extent of genetic change needed for adaptation to a new environment (e.g., to reach an amino acid substitution to confer resistance to an antiviral inhibitor). In fact, some mutations have been termed quasisynonymous because, despite not leading to an amino acid replacement, they can affect the evolutionary course (Salemi and Vandamme, 2004). As an example, there are six triplets that encode the amino acid R (Arg), and they all have a G in the middle position (AGG, AGA, CGG, CGA, CGC, CGU); out of these, only two (AGG and CGG) are within a one nucleotide distance from W (Trp) (UGG); transversion A → U is required to change the triplet AGG (R) to UGG (W), and transition C → U is required to change CGG (R) to UGG (W). The only triplet encoding W (UGG) is within a single nucleotide distance from two termination codons (UGA and UAG) (Figure 4.4). Thus, a difference in the R codons that a virus uses is not a neutral trait since two of them are within a one nucleotide distance of the codon for W, and the latter codon is at one nucleotide distance of two termination (Stop) codons. The latter may be reached as a result of increased mutational pressure, either due to a decrease of polymerase fidelity or to antiviral lethal mutagenesis treatments (Chapter 9). Furthermore, in several viral populations that have been examined by deep sequencing, a remarkably high frequency of termination codons is observed that presumably denotes the presence of defective genomes that represent either dead-end evolutionary pathways or are maintained by complementation during quasispecies replication (Rodriguez-Frías

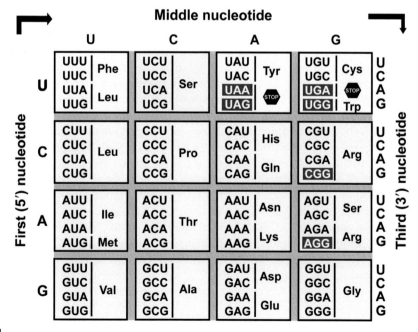

FIGURE 4.4

The genetic code with indication of triplets and encoded amino acids. The triplets in red are those discussed in the text in relation of the phenotypic nonequivalence of synonymous codons, regarding the proximity to STOP (protein synthesis terminating) codons.

et al., 2012). The frequency of termination codons may increase as a result of treatment with antiviral agents and may decrease fitness of the quasispecies ensemble.

The studies and concepts summarized in this section bear on the increasing realization of the potential biological impact of synonymous codon modifications (Hunt et al., 2014) and how movements in a neutral sequences space can actually affect the phenotypic space (Schuster, 2011). Position in sequence space is pertinent to the viral dynamics in the infected hosts.

4.4 MODIFICATIONS OF HOST CELL TROPISM AND HOST RANGE

Any step in the replication cycle can determine the permissivity of a cell type to sustain the replication of a virus. However, recognition by viruses of cellular receptors is a key determinant of cell tropism. Macromolecules that act as viral receptors are diverse, and they include cell adhesion and cell-to-cell contact proteins, extracellular matrix components, sugar and lipid derivatives, chemokine and G-protein-coupled receptors, growth factor receptors, complement control protein superfamily, low- and high-density lipoprotein receptor, tumor necrosis factor-related proteins, and transporter proteins, among others (as review, see Baranowski et al., 2003; Bhella, 2015). The expression of a virus receptor is a necessary but not a sufficient condition for the virus to infect a cell. The PV receptor (PVR or CD155) belongs to the immunoglobulin-like superfamily of cell adhesion molecules that constitute an important group of cellular receptor for viruses (Mendelsohn et al., 1989; Bhella, 2015). Many tissues that express PVR are not infected by PV *in vivo*. Likewise, the sialic acids that act as IV receptors are common in cell surfaces and the virus attaches to them, but productive infection is highly restricted to the epithelial cells of the respiratory tract (Bergelson, 2010).

Early and more recent measurements indicate that for several viral receptors about 10^3 to 10^4 molecules are present per cell (Lonberg-Holm et al., 1976; Consigli et al., 1986; Thulke et al., 2006). Although the MOI *in vivo* is difficult to estimate, it is unlikely that the number of receptor molecules represents a limitation to an infection. Bottlenecks are more probably due to restrictions of the number of infectious virus that reaches a target tissue or organ than to insufficient number of receptor molecules on the cell surface. The interaction of a virus with one or several receptors (or a receptor and a coreceptor) will generally allow virus entry which is a multistep process that involves changes in virion structure, a succession of low- and high-affinity binding to one or more cellular proteins, and membrane fusion in the case of enveloped viruses (Verdaguer et al., 2014; Strauss et al., 2015).

Here we describe cell tropism changes of viruses that document the relevance of viral population size in the process. Features of virus-receptor interactions relevant to viral evolution are summarized in Box 4.1, based on concepts reviewed in Baranowski et al. (2003). Of note is that differentiated organisms do no express the same set of cell surface macromolecules in different tissues and organs. This implies not only compartmentalization of viral infections but also the possibility of selection of viral subpopulations in specific host compartments, a translation *in vivo* of one of the tenets of quasispecies dynamics. Also, members of the same virus family, or viruses associated with related disease manifestations, may use different receptors. Among other examples, consider the different viruses that cause brain or liver disease. Since tropism changes are one of the biologically most relevant consequences of viruses existing as mutant spectra, several studies have been listed in Table 4.1 to emphasize that tropism changes are not exceptional; the list is by no means exhaustive.

BOX 4.1 FACTS RELATED TO RECEPTOR USAGE BY VIRUSES

- Different compartments within an organism do not express the same surface macromolecules that can act as viral receptors.
- A virus may use different receptors and coreceptors.
- A receptor type can be shared by different viruses and other microbial pathogens.
- A phylogenetic position or biological features of a virus do not predict the use of some receptor types. Members of the same virus family or viruses that are associated with similar disease manifestations may use different receptors.
- One or a few amino acid substitutions in capsid or surface proteins may modify receptor recognition, with consequences for viral pathogenesis.

Table 4.1 Examples of One or Few Amino Acid Substitutions That Can Modify Virus Cell Tropism

Observation	References
Host restriction of avian polyomavirus budgerigar fledgling disease virus	Stoll et al. (1994)
Colonic tropism and persistence of murine norovirus	Nice et al. (2013)
N versus B tropism of murine leukemia virus	Jung and Kozak (2000)
Tissue tropism of adeno-associated viruses	Wu et al. (2006)
Cellular tropism of feline immunodeficiency virus	Verschoor et al. (1995), Vahlenkamp et al. (1997), Lerner and Elder (2000)
Loss of enteric tropism of transmissible gastroenteritis coronavirus by two amino acid substitutions in the spike protein	Ballesteros et al. (1997)
HIV-1 tropism by a single amino acid substitution in gp120	Takeuchi et al. (1991), Boyd et al. (1993)
SARS coronavirus recognition of ACE2 receptor (see also text)	Li et al. (2005)
Receptor preferences of herpes simplex virus	Spear et al. (2000)
Hemagglutin residues in influenza virus tropism (see also text)	Rogers et al. (1983), Connor et al. (1994)
Substitutions in poliovirus capsid expand receptor recognition	Colston and Racaniello (1995)
Conversion of encephalomyocarditis D into a diabetogenic variant through altered cell tropism	Bae and Yoon (1993)
High-affinity binding of measles virus to CD46	Hsu et al. (1998)
Decreased neurovirulence of Sindbis virus through impaired receptor recognition in neural cells	Tucker and Griffin (1991), Lee et al. (2002)
Parvovirus host range (see also text and Figure 4.5)	Hueffer et al. (2003)
Change in receptor recognition by foot-and-mouth disease virus *in vivo* and in cell culture (see also text and Figure 4.6)	Baranowski et al. (2000, 2001), Ruiz-Jarabo et al. (2004)

Next we comment on a few specific cases with the aim of deriving general concepts. Some residues of the IV hemagglutinin (HA) dictate the specificity for sialic acid linked to galactose by either an α-2, 3 or an α-2, 6 linkage, and linkage preference is a determinant of host specificity (a preference for α-2, 6 linkage by human IVs, and for α-2, 3 linkage by avian IVs) (Skehel and Wiley, 2000; Parrish and Kawaoka, 2005; Yang et al., 2015). In lymphocytic choriomeningitis virus (LCMV), high-affinity binding to its receptor α-dystroglycan is associated with immunosuppression and viral persistence in mice, whereas low-affinity binding results in clearance of infection (Sevilla et al., 2000, 2002; Smelt et al., 2001).

One or few amino acid substitutions have been associated with changes in cell tropism and host range of parvoviruses (Hueffer and Parrish, 2003; Parrish and Kawaoka, 2005; Shackelton et al., 2005; Lopez-Bueno et al., 2006) (Figure 4.5), adenoviruses (Huang et al., 1999), herpes simplex viruses (Spear et al., 2000), and lentiviruses. HIV-1 coreceptor usage varies in the course of infection in humans. Most primary HIV-1 isolates belong to the R5 receptor specificity (they use coreceptor CCR5) and, as infection progresses, dualtropic (R5X4) and X4 variants (that use coreceptor CXCR4) arise and often become dominant. The expansion by HIV-1 of coreceptor usage to include CXCR4 is associated with loss of CD4$^+$T cells, and to progression to AIDS (Connor et al., 1997). The R5 to X4 transition in coreceptor usage constitutes an example of tropism change with a direct impact in viral pathogenesis, a change that is observed in about half of HIV-1-infected patients. Multiple, additional coreceptors can be used by HIV and simian immunodeficiency virus (SIV) variants, and transitions in receptor usage vary with virus types and subtypes. Two amino acids in the spike (S) protein of the SARS coronavirus (SARS CoV) modulate the binding to either human or palm civet angiotensin-converting enzyme 2 (ACE 2), a functional receptor for the virus (Li et al., 2005).

It must be clarified that the examples in Table 4.1 and those commented in the text are centered on the effects of a few amino acid substitutions on cell tropism. Some of these substitutions and additional ones may affect virus pathogenesis which is often (but not always) related to tropism alterations. Moreover, other genetic lesions such as insertions or deletions (not only point mutations) may also affect receptor recognition and pathogenic potential.

The selective forces that trigger a modification of cell tropism are not easy to identify although several observations suggest that the availability of a receptor type may select a variant subpopulation out of a mutant distribution. HIV-1 variants enter CD8$^+$ cells at late disease stages (Saha et al., 2001). In a mouse model, a modified form of RANTES, a natural ligand for CCR5, selected HIV-1 mutants that used CXCR4 as a coreceptor (Mosier et al., 1999) (discussed also in Chapter 9 in connection with antiviral agents directed to cellular targets). The bicyclam AMD3100, a selective antagonist of CXCR4, led to suppression of X4 variants in cell culture, and prevented the switch from R5 HIV-1 to X4 HIV-1 (Este et al., 1999). Group B coxsackieviruses (CVB) use the coxsackievirus and adenovirus receptor to infect cells (Bergelson, 2010). When a CVB was passaged in a cell line expressing a limited amount of CAR, the virus expanded its cell receptor specificity to bind multiple molecules including CAR and decay-accelerating factor (DAF) The modification involved a limited number of amino acid substitutions in the viral capsid (Carson et al., 2011).

Multiplication of a virus in a given cell line (or primary culture) may select virus subpopulations present as a minority in a biological sample from a naturally infected host. Passage of biological clone of FMDV in BHK-21 cells resulted in the dispensability of the RGD (integrin recognition sequence) and expansion of cell tropism (Ruiz-Jarabo et al., 2004) (Figure 4.6). Thus, paradoxically, repeated replication in a cell line may lead to relaxation of virus specialization for that cell line. Despite this example of tropism expansion, one of the tenets of quasispecies dynamics, the generation of a mutant spectrum as a source of minority variants that can become dominant in response to environmental

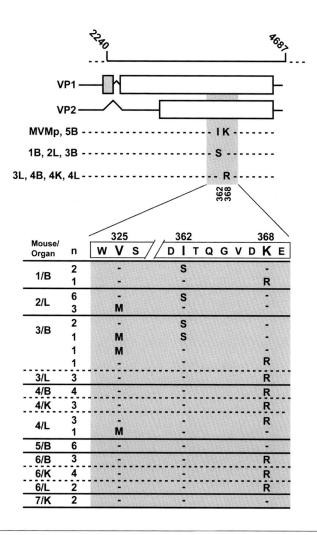

FIGURE 4.5

Amino acid substitutions found in the capsid protein of the prototypic strain of the parvovirus minute virus of mice (MVMp) upon passage in immunodeficient mice. At the top genomic residue numbers and amino acid substitutions (single letter code) found in virus from several organs are listed. At the bottom a more detailed list of substitutions found in many individual clones is given. The numbers (*n*) indicate the number of clones analyzed, and the letters the organ from which the clones were isolated (B, brain; K, kidney; L, liver).

The figure is modified from Lopez-Bueno et al. (2006), with permission from the American Society for Microbiology, Washington DC, USA.

demands, probably underlies several cases of tropism changes in RNA and DNA viruses. Application of NGS to probe into the minority levels of viral quasispecies *in vivo* should help establishing whether low-frequency mutants with altered receptor recognition sites are frequently present in evolving quasispecies or they remain below the level of detection. Their basal frequency will depend on their fitness relative to other components of the mutant spectrum. Here again the relevance of virus population

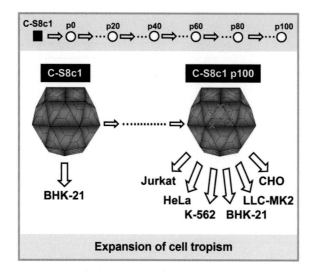

FIGURE 4.6

Expansion of host cell tropism. FMDV biological clone C-S8c1 (filled square at the top) was passaged in BHK-21 cell at a MOI of 1 to 5 PFU per cell (open circles at the top). The resulting population included amino acid substitutions in the viral capsid and acquired the capacity to infect additional cell lines (large arrows below the capsid structures).

Scheme based on the results reported in Ruiz-Jarabo et al. (2004).

numbers is evident: the larger the population size, the higher the probability that mutants with altered cell recognition are present. The fitness cost associated with amino acid substitutions in viral proteins is discussed in Chapter 5.

4.4.1 NONSTRUCTURAL VIRAL PROTEINS AND RNA IN CELL TROPISM AND HOST RANGE OF VIRUSES

The event of being able to penetrate a new cell type has consequences for the host range of viruses, and the potential emergence of new viral pathogens (Chapter 7). In the present section, we document that nonstructural proteins (those that are not present in the viral particles) and regulatory regions in viral genomes may also mediate cellular tropism and host range changes in viruses. Deletions and point mutations in the nonstructural protein 3A of FMDV have been associated with attenuation for cattle (reviewed in Baranowski et al., 2003). A single amino acid substitution in 3A was critical for the adaptation of a swine FMDV to guinea pig (Núñez et al., 2001) (Figure 4.7). Despite the conservation of the LTR of a macrophage-tropic strain of the lentivirus equine infectious anemia virus (EIAV) *in vivo*, the stepwise generation of a new transcription factor-binding motif within the enhancer element was associated with the expansion of tropism to endothelial cells and fibroblasts (Maury et al., 2005). A comparison of the minimal promoter-enhancer element of HIV-1 of clades B, C, and E, engineered into the same SIV genetic background, indicated that this element can modulate viral replication in different cell subsets *in vivo* (Centlivre et al., 2005; Sala et al., 2006). Polymorphism in the binding sites of transcription factors is probably involved in differential expression of HIV-1 clades in different tissues. The polymerase genes of

FMDV	Lesion score	Amino acid VP1 (147)	2C (248)	3A (44)
C-S8c1		L	I	Q
V 2.1		L	I/T(10%)	Q
V 2.2	+	L	T	Q/R (40%)
V 2.3	+	L	T	R
V 2.4a	+	L/P(40%)	T	R
V 2.4b	++	L/P(20%)	T	R
V 2.5a	+	P	T	R
V 2.5b	+	L/P(50%)	T	R
V 2.6a	++	P	T	R
V 2.6a	+	P	T	R
V 2.6b	+	P	T	R
V 2.7b	++	P	T	R
V 2.7b	++	P	T	R
V 2.10b	++	P	T	R

FIGURE 4.7

Amino acid substitutions in proteins VP1, 2C and 3A in the course of adaptation of swine FMDV C-S8c1 to the guinea pig. Sequential virus isolates from different animals are represented in the first column. The lesion score indicates the presence of a vesicle at the inoculation site (+) or the presence of vesicles at the inoculation site and other sites (secondary lesions) (++). The amino acid substitutions (single letter code) in VP1, 2C and 3A (amino acid position given for each protein) are indicated in the three columns on the right. Results are based on consensus sequences; the presence of amino acid mixtures (percentage in parenthesis) is based on the peaks of the sequencing experiments.

The figure has been adapted from Núñez et al. (2001), with permission from the American Society for Microbiology, Washington DC, USA.

IV may influence host range, and, probably, gene constellations affects the relative replication capacity of viruses in different hosts (Parrish and Kawaoka, 2005).

Dominance of viral subpopulations *in vivo* can be due to cells offering higher permissivity to variant forms of a virus, or to depletion of cells due to cytolytic infection, and survival of other cell types permissive to other viral variants (Centlivre et al., 2005; Sala et al., 2006). Cell dynamics can exert an important influence on short-term virus evolution and pathogenesis. (See Chapter 6 for a description of the dependence on host cell variation for the initiation of a persistent infection of FMDV in cell culture.) Not only nonstructural proteins, modifications in viral RNA may also contribute to host-specific viral fitness. RNA structural elements and mutations within the elements may influence the adaptation of arboviruses to alternative hosts (Ventoso, 2012; Villordo et al., 2015).

From an evolutionary perspective, the capacity of viruses to use alternative receptors and to modify receptor specificity or intracellular preferences by modest genetic change—involving short distances in the genotypic sequence space (Chapter 3)—may manifest a necessity of viruses to parasitize increasingly differentiated organisms. Any virus whose capacity to expand cell tropism was limited by genetic constraints (e.g., the need to move to long distances in sequence space, incompatible with viral population numbers) would have a lower probability of long-term survival in an increasingly differentiated cellular world. A large virus population size is the key to provide a sufficient number of variants with new potential cell recognition specificities. That is, the application of the Darwinian principles to long-term evolution of viruses is expected to have produced a flexible virus-host cell interaction in the sense that it may be modified by limited genetic change in the virus.

4.5 TRAIT COEVOLUTION: MUTUAL INFLUENCES BETWEEN ANTIGENIC VARIATION AND TROPISM CHANGE

Coevolution between organisms and their pathogenic agents means that reciprocal and adaptive genetic modifications have occurred in them because of their interaction as biological systems that have shared space over prolonged time periods. Coevolution is a general concept that applies not only to host-parasite relationships, but also to other interacting biological species, that have shaped biological systems up to the present epoch (Futuyma and Slatkin, 1983; Woolhouse et al., 2002; Gomez et al., 2015). Long-term coevolution has probably led to increasingly subtle molecular mechanisms to deal with the host immune response, based on virus-coded proteins that interact with host proteins. In addition, mutual evolutionary influences may be exerted among different sites within a virus. Such influences are favored by overlaps between domains involved in distinct functions in viral nucleic acids or proteins, that may also introduce additional evolutionary constraints. Coevolution of amino acids at or around important functional domains may contribute to functional stability (Gloor et al., 2005). Intramolecular coevolution can be measured for amino acid pairs by the probability of their occurrence at some defined positions, also termed mutual information criterion (MIC). A MIC value of zero means independent evolution of the two amino acids tested while increasingly positive values denote enhanced covariation (Korber et al., 1993).

Structural studies with several viruses have documented that frequently there is an overlap between antibody and receptor recognition sites (reviewed in Baranowski et al., 2003) (Tables 4.2 and 4.3). As mentioned in section 4.2.2, the FMDV capsid includes an Arg-Gly-Asp (RGD) triplet at an exposed mobile loop in protein VP1 (Acharya et al., 1989; Fry and Stuart, 2010) which is involved in the binding of neutralizing antibodies and in recognition of integrin receptors (Verdaguer et al., 1995). Functional alterations in FMDV can be interpreted as a consequence of the overlap between a major antigenic site and the integrin recognition domain: variants selected with neutralizing monoclonal antibodies displayed altered integrin recognition (Martinez et al., 1997; Baranowski et al., 2000; Ruiz-Jarabo et al., 2004). Adaptation of the virus to cell culture may result in antigenic variation and in the use of heparan sulfate as a molecule that facilities virus entry; some of the residues involved in heparin binding map at antigenic sites (Curry et al., 1996; Sa-Carvalho et al., 1997; Fry et al., 1999; Baranowski et al., 2000). Cattle that were partially immunized with synthetic peptides representing the VP1 loop sequence, selected FMDV mutants with substitutions within the RGD or at neighboring sites. The mutants showed altered host cell tropism (Taboga et al., 1997; Tami et al., 2003) (see also Chapter 8).

Studies with IVs have shown several consequences of the close connection between antigenic and receptor interaction sites. HA antigenic variants were selected upon egg adaptation of the virus (Robertson et al., 1987). Treatment with antibodies resulted in selection of variants with altered receptor binding (Laeeq et al., 1997). The hemagglutinating activity of type C IV can be modulated by amino

Table 4.2 Examples of Overlap Between Antigenic Sites and Receptor Recognition Sites in DNA Viruses

Virus	Observation	References
Adenovirus-3	Fiber knob includes receptor-binding and antigenic sites	Liebermann et al. (1998)
Adeno-associated virus (serotypes 1 and 5)	Virus-antibody complex structures show epitopes at receptor recognition sites	Tseng et al. (2015)
Bovine herpesvirus-1	Anti-idiotypic antibodies bind to cellular receptors	Thaker et al. (1994), Varthakavi and Minocha (1996)
Human cytomegalovirus	Anti-idiotypic antibodies bind to cellular receptors	Keay et al. (1989), Keay and Baldwin (1991)
Herpes simplex virus	Anti-idiotypic antibodies bind to cellular receptors	Huang and Campadelli-Fiume (1996)
	Overlap between a receptor binding domain in gD and an antigenic site	Whitbeck et al. (1999)
Hepatitis B virus	Anti-idiotypic antibodies that mimic cellular structures bind to small HBV surface antigen. Synthetic peptide analog is recognized by anti-HBV antibodies and cell receptors	Neurath et al. (1986)
	Anti-idiotypic antibodies bind to cellular receptors	Petit et al. (1992), Hertogs et al. (1994), Budkowska et al. (1995)
Duck hepatitis B virus	Residues involved in the interaction with cells are also critical for virus neutralization	Tong et al. 1995, Li et al. (1996), Sunyach et al. (1999)

acid residues involved in antibody binding (Matsuzaki et al., 1992). Passage of this virus in MHV-II cells resulted in antigenic variants that displayed an advantage in receptor binding (Umetsu et al., 1992). The receptor binding specificity of IV can modify the antigenic profile of the virus as analyzed by reactivity of monoclonal antibodies with the HA (Yamada et al., 1984). The mutual influence between antigenic variation and receptor site modifications has been extended to recent IV isolates (Koel et al., 2013, 2014). These summarized accounts for FMDV and IV, in addition to the studies listed in Tables 4.2 and 4.3 provide incontestable evidence that antigenic changes that are extremely frequent because of the continuous confrontation of viruses with antibodies, may contribute to modification of cell-binding preferences.

Two facts facilitate antigenicity-cell tropism coevolution: (i) the limited number of amino acid substitutions, and hence of mutations, that are needed for an antigenic change (one of the parameters that render the quasispecies nature of viruses biologically relevant; see Box 3.3 and text in Chapter 3). (ii) Surface residues tend to be less constrained structurally than internal residues in viral capsids or envelopes. Their tolerance to accept amino acid substitutions is one of the mechanisms of antigenic variation in the absence of immune selection (Section 4.7). Surface residues appear as the most variable when all the historically recorded substitutions are depicted on the three-dimensional structures of viral particles. In the case of FMDV high variability of surface amino acids has been corroborated with several virion structures determined by D. Stuart, E. Fry and their colleagues (one of the comparisons between two isolates of a subtype and an antigenic variant is described in Lea et al., 1995). Conformational

Table 4.3 Examples of Overlap Between Antigenic Sites and Receptor Recognition Sites in RNA Viruses

Virus	Observation	References
Poliovirus	Receptor recognition influenced by residues of antigenic sites	Murray et al. (1988), Harber et al. (1995)
	Critical role of VP1 BC loop in receptor interaction	Yeates et al. (1991)
Human rhinovirus	Neutralizing antibody to HRV14 penetrates the receptor-binding canyon	Smith et al. (1996)
	Exposed VP1 BC- and HI-loops covered by footprint of very low-density lipoprotein receptor	Hewat et al. (2000)
Theiler's encephalomyelitis virus	Neutralizing antibodies map close to putative receptor-binding site	Sato et al. (1996)
	Substitutions of adaptation to some cells map in antigenic sites	Jnaoui and Michiels (1998)
Foot-and-mouth disease virus	Overlap of integrin- and antibody-binding sites (Additional studies presented in the text)	Verdaguer et al. (1995)
Human influenza virus	Amino acid residues of the sialic acid-binding pocket are accessible to neutralizing antibodies (Additional studies presented in the text)	Stewart and Nemerow (1997)
Newcastle disease virus	Monoclonal antibodies to HN glycoprotein prevent virus attachment	Iorio et al. (1989)
Rabies virus	Anti-idiotypic antibodies bind to cellular receptors	Hanham et al. (1993)
	Residues critical for neurotropism are involved in antibody binding	Coulon et al. (1998)
Bovine viral diarrhea virus	Anti-idiotypic antibodies bind to cellular receptors	Xue and Minocha (1993), Minocha et al. (1997)
Dengue virus	Residues critical for mouse neurovirulence are involved in antibody binding	Hiramatsu et al. (1996)
Yellow fever virus	Residues critical for neurotropism are involved in antibody binding	Jennings et al. (1994)
Murine coronavirus	Overlap between epitopes and receptor-binding sites	Kubo et al. (1993, 1994)
Middle East respiratory syndrome (MERS) coronavirus	Antibodies bind to receptor recognition site	Ying et al. (2014)
Sindbis virus	Anti-idiotypic antibodies bind to cellular receptors	Ubol and Griffin (1991), Wang et al. (1991), Strauss et al. (1994)
Ross River virus	Binding of antibodies to cell-receptor recognition regions	Smith et al. (1995)
Reovirus	Anti-idiotypic antibodies bind to cellular receptors	Co et al. (1985), Gaulton et al. (1985), Williams et al. (1988, 1989, 1991)
Bluetongue virus	Anti-idiotypic antibodies bind to cellular receptors	Xu et al. (1997)

antigenic sites, similarly to internal capsid residues, are involved in interactions which may be necessary for virion stability and, therefore, their tolerance of amino acid substitutions is more restricted than in linear, continuous epitopes. This is the case with discontinuous epitopes that have been characterized within antigenic site D_2 of FMDV. The conformation of this site in FMDV O_1BFS and C_1 is conserved despite different primary sequences. The substitutions in antibody-escape mutants of D_2 map in amino acids which are not involved in interactions with surrounding residues. Significantly, the only substitution found in a residue involved in hydrogen bonding, led to an amino acid that maintained the hydrogen bond with the same neighbor amino acid, according to the modeling of the change based on crystallographic data (Lea et al., 1994). It is expected that disordered protein regions free of structural constraints can tolerate amino acid substitutions. In some cases, surprisingly, even a viral polymerase may contain domains that can be extensively changed and remain functional (Gitlin et al., 2014).

4.6 ESCAPE FROM ANTIBODY AND CYTOTOXIC T CELL RESPONSES IN VIRAL PERSISTENCE: FITNESS COST

Viruses use two major strategies to cope with the host response: modulation and escape. By modulation we mean the expression of viral gene products that by any mechanism can alter components of the immune response. The consequence is to facilitate virus survival to increase the probability of transmission, or virus persistence in the infected organism. By escape we mean mutations in the viral genome that render the virus resistant to inactivation by components of the immune response, typically neutralization of viral particles by antibodies or elimination of infected cells by specific cytotoxic T cells. Both modulation and escape strategies are exploited by DNA and RNA viruses, although complex DNA viruses encode several proteins whose primary function is to interfere with the host defense mechanisms. They include homologues of cytokines, chemokines, viral proteins that act as a decoy for antiviral antibodies, proteins that block complement activation, that suppress MHC class I and II molecules, that interfere with ubiquitin-dependent proteolysis, that induce or inhibit apoptosis, among other proteins and activities (Alcami, 2003; Seet et al., 2003; Rustagi and Gale, 2014). Viral proteins block interferon induction (i.e., influenza virus NS1, Ebola virus P35, and others) (Basler and Garcia-Sastre, 2002; Katze et al., 2002; Weber et al., 2004). Some of the proteins involved provide remarkable examples of protein multifunctionality. The leader L proteinase of FMDV catalyzes its own cleavage from the polyprotein, cleaves the host cell translation factor eIF4G—leading to the shutoff of host cell translation dependent on capped mRNAs—and inhibits IFN induction in the infected cells (de Los Santos et al., 2006). HIV-1 Nef contributes to HIV-1 evasion of immune surveillance through interaction with membrane traffic regulators (Pawlak and Dikeakos, 2015).

Despite expressing proteins that interfere with the immune response, RNA viruses exploit evasion through genetic variation as a major means to cope with host defenses. This is probably an evolutionary coadaptation of high mutability and genome compactness. Genomic compression is evidenced by overlapping reading frames, ambisense RNA genomes, RNA editing, partial read-through of termination codons, overlap between regulatory and protein-coding regions, leaky ribosome scanning with initiation of protein synthesis at two in-frame AUGs, ribosome frameshifting, hopping, shunting and bypassing, synthesis of polyproteins whose partial or complete processing leads to several functional proteins, etc. (Domingo et al., 2001; Alberts et al., 2002). Evolution has offered high mutation rates, small genome size, and escape pathways as an alternative to the modulation strategy.

Mutations that mediate escape from neutralizing antibodies and from CTLs can be readily observed *in vivo* (Weiner et al., 1995; Borrow et al., 1997; McMichael and Phillips, 1997; Ciurea et al., 2000; Sevilla et al., 2002). Rather than being a secondary phenomenon in the course of viral infections, the generation of antibody- and CTL escape may indeed contribute to viral persistence (Gebauer et al., 1988; Ciurea et al., 2000; Richman et al., 2003; Domingo, 2006). Evasion of an immune response, as the outcome of a selection event in viruses, may entail a fitness cost. Such a cost may bring about reversion to the initial sequences when selective forces (antibodies or CTLs) are no longer present (Borrow et al., 1997). Viruses often display multiple antibody-escape routes, and the preferred pathway may be imposed by the number and concentration of the antibodies (Borrego et al., 1993; Keck et al., 2014).

There is an important difference between the selective constraint imposed by antiviral drugs and by the host immune response. While drugs inhibit a specific step of the virus replication cycle (or two or more steps if two or multiple drugs are administered simultaneously), the immune response gives rise to multiple constraints that act upon the virus. As an example, 100 HCV passages in cell culture in the presence of IFN-α were necessary to select HCV mutants displaying resistance to IFN-α (Perales et al., 2013). The resistant HCV displayed higher fitness than the populations passaged in the absence of IFN-α when fitness was measured in the presence of IFN-α, but not in its absence. Sequence analysis documented that amino acid substitutions that contributed to resistance were present in most viral proteins and many substitutions differed among parallel viral lineages (Perales et al., 2013). It is expected that viruses find higher genetic and phenotypic barriers to respond to multicomponent antiviral responses than to a single inhibitor with a defined target (Perales et al., 2014). A high genetic barrier is due to the requirement of multiple mutations, while a high phenotypic barrier reflects the fact that fitness decrease due to amino acid substitutions in several proteins will be accentuated by the multifunctionality of most proteins. The observations with HCV in cell culture agree with multiple possible IFN-α-resistance mutations identified in clinical practice (reviewed in Perales et al., 2014).

It is an open question whether the concept of high fitness cost to overcome an interferon response can be extended to other systems. I. S. Novella, J. Holland and colleagues showed that passage of VSV in IFN-treated cells selected only variants of limited IFN resistance (Novella et al., 1996). However, field isolates of VSV appear to contain clones with different capacity to resist or to induce IFN, to the point that interferon induction was used as a quasispecies marker for the virus (Marcus et al., 1998).

Quantification of fitness cost is described in Chapters 5 and 8, and the use of broad-spectrum antiviral agents that promote a multicomponent antiviral state is discussed in Chapter 9.

4.7 ANTIGENIC VARIATION IN THE ABSENCE OF IMMUNE SELECTION

Several cases of antigenic variation of DNA and RNA viruses in the absence of immune selection have been described (Domingo et al., 1993, 2001 and references therein). They have been attributed to two possible mechanisms: (i) tolerance of antigenic sites to accept amino acid replacements by virtue of being relatively free of structural constraints (Section 4.5). Fluctuations of mutant distributions (through selection of an unrelated trait or through random drift) may raise antigenic variants to dominance. (ii) Not mutually exclusive with the previous mechanism, selective forces other than an immune response may result in amino acid replacements at antigenic sites (as in coevolution of receptor recognition specificity and antigenicity, discussed above). Antigenic variation may follow from the hitchhiking of mutations that encode amino acid substitutions at antigenic sites, following selection sweeps

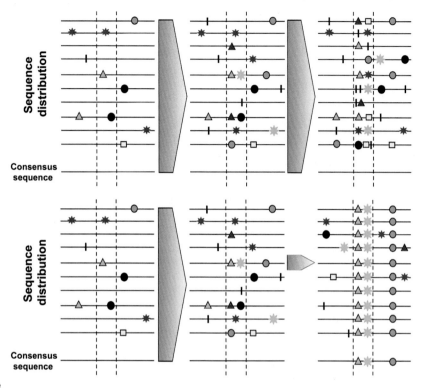

FIGURE 4.8

Antigenic variation in the absence of immune selection. The genomic region encoding an antigenic site is delimited by two discontinuous vertical lines. At the top, extensive replication of the virus leads to accumulation of amino acid substitutions at the antigenic site because of its structural flexibility. At the bottom, a genetic bottleneck or a selection event results in the fixation (hitchhiking) of two amino acid substitutions (green triangle and asterisk) at the antigenic site.

The figure has been modified from Domingo et al. (1993).

(Chapter 3) (Figure 4.8). Therefore, a consequence of quasispecies dynamics is that amino acid substitutions observed at the antigenic sites of viruses are not necessarily the result of immune selection.

4.8 CONSTRAINTS AS A DEMAND ON MUTATION RATE LEVELS

The nature of the selective constraint influences the requirements—in terms of mutation rate and complexity of the mutant spectrum—to be fulfilled by a virus to overcome that specific selective constraint. For a virus to lose sensitivity to an antiviral inhibitor directed to a viral protein, when such a loss depends on a single transition mutation (low genetic barrier), the standard mutation rate, or even a mutation rate lower than the standard, may suffice to generate a resistant mutant. In contrast, when a virus has to adapt to a complex environment, multiple mutations located at different genomic sites might be required for effective replication in the new environment.

The occurrence of multiple mutations in the same viral genome will be favored by low-fidelity polymerases because they generate mutant spectra characterized by larger average number of mutations per genome (higher mutation frequency and nucleotide diversity) than an enzyme with standard copying fidelity. We refer to "broad" or "narrow" mutant spectra to indicate whether they include many types of variants or a restricted repertoire of variants, respectively (Figure 4.9).

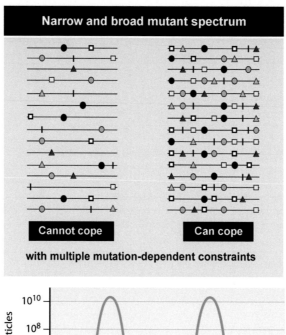

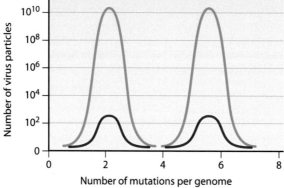

FIGURE 4.9

Two representations of viral populations with different amplitude and size of the mutant spectrum. At the top a narrow mutant spectrum that cannot cope with a complex selective constraint is represented (left), in contrast with a more complex mutant distribution that can cope with a constraint (right). Genomes are depicted as horizontal lines, and mutations as symbols on the lines. At the bottom, mutant distributions of different size (large in the blue and small in the red curves) and complexity (average number of mutations per genome given in the abscissa) are represented.

The bottom part is reproduced from Domingo et al. (2012) with permission from the American Society for Microbiology, Washington DC, USA.

A selective disadvantage due to a narrow mutant spectrum was elegantly documented with a high-fidelity PV mutant and its inability to cause neurological disease in mice (Pfeiffer and Kirkegaard, 2005; Vignuzzi et al., 2006) (see also Chapter 3). Examples of complex environments for a virus—those that often require a broad mutant spectrum for a virus to confront them successfully—are the invasion of tissues or organs to ensure extended replication (that may or may not lead to disease) or survival in face of the immune response of the host, with its multiple (humoral and cellular) branches (Section 4.6).

Arguments occasionally put forward by theoretical biologists that high mutation rates are a consequence of some other parameter (i.e., rate of RNA genome replication) but not a trait directly selected for adaptation, ignore the necessity of viruses to periodically respond to complex (multiple) selective constraints for which high mutation rate should be positively selected, and we know that mutation rate is an evolvable trait. In particular, following a population bottleneck, an event which from all evidence is very frequent in nature (Section 4.1.2), a single infectious virus must repopulate a mutant spectrum to provide a range of phenotypes that can mediate adaptation. The need to rapidly provide a mutant repertoire following population bottlenecks may by itself justify that high mutation rates in RNA viruses have been positively selected (Vignuzzi and Andino, 2010). In addition, mutant viruses displaying higher or lower than standard template-copying fidelity generally display lower fitness than their standard parental genomes. Again, evidence suggests that the observed mutation rate values are a consequence of evolutionary adaptation of viruses.

4.9 MULTIFUNCTIONAL VIRAL PROTEINS IN INTERACTION WITH HOST FACTORS: JOKER SUBSTITUTIONS

Many viral proteins are multifunctional, a feature that may be regarded as an adaptation of compact viral genomes to exploit as much as possible the phenotypic possibilities of each coding nucleotide (a point mentioned in Section 4.6 in connection with functions exerted by proteins that modulate the host immune response). Multifunctionality of proteins is not unique to viruses, since it is increasingly documented for cellular proteins. However, it is in the case of viruses in which the broadest functional repertoires of a single protein have been documented. A few of many examples are briefly mentioned here.

Proteins NS1 and NS2 of orbiviruses play roles in replication, assembly and morphogenesis. NS1 is synthesized abundantly in bluetongue virus-infected cells, the protein assembles as tubular structures, and a single-chain antibody against NS1 expressed intracellularly can lead to reduction of cytopathology and to an increase of virus budding and release from the cells (reviewed in Roy, 2008).

Picornavirus nonstructural protein 2C constitutes another dramatic example of multifunctionality. 2C includes nucleoside-triphosphatase (NTPase) and RNA-binding activities, acts as a chaperone during picornaviral replication, and participates in viral RNA encapsidation, the uncoating of viral particles, and host cell membrane rearrangements required for picornavirus replication (reviewed in Ehrenfeld et al., 2010). Amino acid substitutions in 2C may affect killing of cells in culture (Herrera et al., 2007) or virulence *in vivo* (Sanz-Ramos et al., 2008). A substitution in 2C acted as a compensatory mutation of a defect in cytopathology displayed by deletion mutants of FMDV that lack two copies of 3B, the gene encoding VPg which is the protein involved in initiation of picornaviral RNA replication (Arias et al., 2010). 2C is the target of guanidine hydrochloride, a protein-denaturing agent that at low concentrations is an inhibitor of picornavirus replication. This inhibitor has been instrumental to estimate mutation rates of picornaviruses and to understand the molecular basis of virus extinction by enhanced mutagenesis (Chapter 9).

A specific protein domain can participate in two or more functions. The FMDV polymerase 3D includes in its N-terminal region a nuclear localization signal that acts also to regulate nucleotide incorporation during RNA synthesis (Ferrer-Orta et al., 2015). The nucleoprotein (NP) of arenaviruses is the main structural element of the viral ribonucleoprotein that directs viral RNA synthesis. In addition, NP counteracts the activity of type I IFN, and this function was mapped in the C-terminal region of the protein. This region folds in a way similar to DEDDh family of 3'-5' exoribonuclease, the type of activity that in replicative DNA polymerases and coronaviruses RNA polymerase is responsible for proofreading repair of mismatched nucleotides at the growing 3'-end of the genome (Chapter 2). NP mutants lacking this activity display remarkable decreases in viral fitness (Martinez-Sobrido et al., 2009; reviewed in Grande-Perez et al., 2016). There are many additional examples of multifunctionality of proteins encoded by DNA and RNA viruses, again a manifestation of the need to exploit any coding stretch for completion of the replication cycles.

Multifunctionality of viral proteins has several implications for the modulation of host cell functions during infections, virus evolution, viral genome stability, and the response of viruses to selective constraints. An amino acid substitution needed to overcome a specific selective pressure may also affect unrelated functions performed by the same protein. This may result in a fitness cost, frequently observed with viral mutants selected for their capacity to overcome an inhibitory activity. When the selected mutants are allowed to continue their replication, additional mutations may occur to increase fitness. Multifunctionality may limit the number of possible amino acid substitutions that can be used for such an increase, because the same substitutions that may compensate a limitation of one of the functions may adversely affect other functions exerted by the same protein. Again, *trade-offs* (the virus *modus vivendi*) enter the picture (Section 4.2). Restrictions imposed by multifunctionality may underlie the observation that a single amino acid replacement I248T in protein 2C was selected repeatedly and independently when FMDV had to respond to different environmental demands: in the course of adaptation of a biological clone of the virus to guinea pig (Núñez et al., 2001), or upon replication of the same clone in mice (Sanz-Ramos et al., 2008). Substitutions that serve a virus to increase its fitness under different environmental circumstances are termed "joker" substitutions. Joker substitutions (in the sense of the extra playing card in some games) map at amino acid positions apt to affect positively one or more activities of one protein, leading to a general increase in replication efficiency under different environmental demands. Joker substitutions may compensate for fitness costs endured by the virus, even those due to malfunction of an unrelated viral protein.

The expectation that the genomic nucleotide sequences that belong to overlapping genes (i.e., polynucleotide stretches that contribute to the coding of more than one protein) would be much more conserved than standard open-reading frames for a single protein has not been confirmed, at least to the extent of being able to derive a general conclusion. A reason for the limited differences in conservation of these two categories of coding regions may lie in protein multifunctionality, in combination with the phenotypic involvement of the viral RNA itself, independent of its coding function (Section 4.3 and Chapter 2). Indeed, if the same protein performs different functions and the RNA is also involved in regulatory activities, each nucleotide may be subjected to constraints that may not differ significantly from the constraints operating on the products expressed from overlapping genes (Domingo et al., 2012).

4.10 **ALTERNATING SELECTIVE PRESSURES: THE CASE OF ARBOVIRUSES**

Selective pressures that viruses encounter when infecting their hosts in nature are rarely constant and uniform. The experimental designs in which a virus is subjected to a specific selective pressure in an established cell line in culture are a gross simplification of reality that has nevertheless allowed quantification of the effects of well-defined variables (Chapter 6). However, the first point to note in dealing with the interaction of viral populations with their hosts is the multitude of selective pressures, often conflicting, that viruses must confront. Expressed in a simple way, selective pressures vary in kind and intensity in space and time.

An interesting, biologically relevant case of alternating selective pressures is provided by the animal and plant arboviruses which successfully alternate replication in vertebrate animals or in plants and insect hosts, and have successfully persisted in nature as disease agents (reviewed in different chapters of Morse, 1994; Gibbs et al., 1995; Domingo, 2006). Three strategies have been distinguished regarding the part of the virus life cycle that elapses in the vector: (i) the virus attaches to the vector, usually at its external organs, but does not reach the internal milieu, and does not undergo replication. This style has been termed noncirculative. (ii) The virus enters the vector through specific receptors and multiplies in it. This style is often termed circulative-propagative. (iii) The virus cycles inside the vector, but does not propagate in it. This is the case of the plant nanovirus in its aphid vectors. Interactive styles (i) and (iii) must be distinguished from purely mechanical transmission because inside the vector the virus may meet conditions that alter particle stability relative to the outside environment. During the circulative-propagative style (ii), the virus undergoes several bottleneck events. In mosquito vectors the virus must transit from one compartment into another, and most of the compartments are separated by basal lamina that limits the penetration of viruses (Forrester et al., 2014). Bottlenecks may affect fitness, limit the number of particles that can be transmitted and, in consequence, accentuate the stochasticity in evolutionary outcomes (Hanley and Weaver, 2008; Gutierrez et al., 2012; Forrester et al., 2014).

Zoonotic vector-borne flaviviruses have been extensively studied because of their pathogenic potential for humans. They include yellow fever virus, Dengue virus, Chikungunya virus, and Venezuelan equine encephalitis virus, among others. Several types of insect-mammalian infection cycles have been characterized for these viruses. A sylvatic cycle is defined as the one that involves nonhuman animal host and insects. Phylogenetic evidence suggests that sylvatic Dengue was the precursor of Dengue viruses that infected humans and established enzootic and endemic cycles (Figure 4.10). Here we return to the importance of virus and host population numbers that were discussed in the first section of this Chapter. In the case of Dengue virus it has been estimated that efficient human-to-human transmission required a minimum human population size in the range of 10^4 to 10^6 individuals, a size that was attained only with the advent of urban life (Gubler, 1997). Critical population numbers of hosts and vectors, as well as numbers of infectious particles in viremic hosts, are necessary for the emergence and maintenance of viral diseases (Chapter 7). Dengue virus appears to be highly adaptable to new animal and vector hosts, notably the transition from *Aedes albopictus* to *Aedes aegypti* as mosquito vectors (reviewed in Hanley and Weaver, 2008).

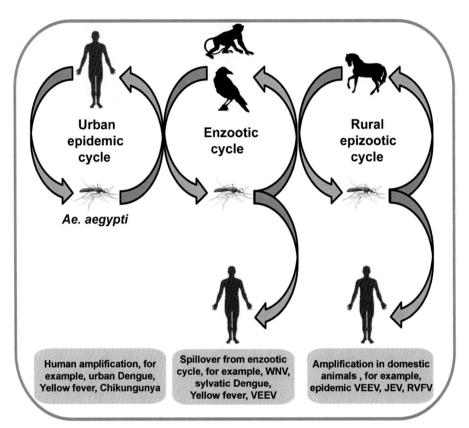

Urban
epidemic
cycle

Enzootic
cycle

Rural
epizootic
cycle

Ae. aegypti

Human amplification, for
example, urban Dengue,
Yellow fever, Chikungunya

Spillover from enzootic
cycle, for example, WNV,
sylvatic Dengue,
Yellow fever, VEEV

Amplification in domestic
animals , for example,
epidemic VEEV, JEV, RVFV

FIGURE 4.10

Simplified representation of three types of infectious cycles (epidemic, enzootic, and rural epizootic) identified for zoonotic arboviruses. Host alternations between vertebrate and mosquito hosts are depicted by curved arrows. Examples of viruses documented to participate in these cycles are shown at the bottom (WNV, West Nile virus; VEEV, Venezuelan equine encephalitis virus; JEV, Japanese encephalitis virus; RVFV, Rift Valley fever virus).

Figure modified from Hanley and Weaver (2008) with permission from the authors.

4.11 OVERVIEW AND CONCLUDING REMARKS

Virus population numbers are orders of magnitude greater than the numbers of individuals in the host species that they infect. Large population numbers, however, are not a constant during the virus life cycles. Viruses undergo periodic bottleneck events, that is, drastic reductions in population size as they spread within individual hosts and during host-to-host transmission. Reductions of population size accentuate the effect of chance in virus evolution. This is because out of a large mutant repertoire that is produced during an infection, only a minority contributes to the next round of progeny production either in the same host (because of a change of compartment) or in a different host individual due to limitation in the number of transmitted viruses. Bottlenecks result also in transient reductions of population heterogeneity (a decrease in the amplitude of mutant spectra) because only a few founder viral particles will originate the progeny repertoire. It has been suggested that one of the reasons why high mutation rates are maintained in viruses is the need to restore population heterogeneity following a population bottleneck. This proposal is in line with the adaptive value of complex mutant spectra documented in Chapters 2 and 3.

Viruses are subjected to several classes of constraints during their multiplication in cellular hosts, an obvious condition for long-term virus-host coexistence given the disparity of their population numbers. Basal constraints are those imposed by the core replicative needs of viruses. Other constraints are due to responses of the individual cells or entire organisms in the event of an infection. Still other constraints are the result of human intervention such as vaccination or the administration of antiviral drugs. Emphasis has been put on codon usage and codon-pair associations as a basal constraint for virus replication, for two reasons. Because it constitutes an example of the nonneutral character of synonymous mutations in viral RNAs, and because codon and codon-pair deoptimization has opened a new prospects for the design of live-attenuated antiviral vaccines.

The modulation and evasion strategies of viruses to overcome immune response have been summarized, and emphasis put on fitness costs and the multifunctionality of viral proteins that act as immune modulators. The studies on evasion strategy, based on selection of antibody- and CTL-resistant mutants have been instrumental for the understanding of quasispecies dynamics. Viruses can use one several receptors that belong to different families of cell surface proteins that mediate virus internalization. Viruses display remarkable flexibility to modify their cell target preference. Alterations of cell tropism and co-evolution of antigenicity and receptor recognition specificity have been outlined as a consequence of quasispecies dynamics and the overlap between antigenic sites and receptor-recognition domains in viruses. The adaptability of viral quasispecies is also reflected in the capacity of viruses to confront changing and alternating selective pressures. The arboviruses that alternate between insect and vertebrate or plant hosts constitute an example of adaptation to multiple hosts as a standard way of life (see Summary Box).

SUMMARY BOX

- Virus population numbers are several orders of magnitude larger than those of their host organisms.
- Viruses are subjected to constraints at multiple levels that limit their replication and contribute to maintaining their biological identity.
- Viruses exploit a variety of molecular mechanisms to overcome external constraints such as therapeutic interventions.
- Codon usage and codon-pair frequencies represent constraints for viral gene expression that can be used to design new generation vaccines.
- Viruses can modify their cellular tropisms through amino acid substitutions in surface protein residues and other mechanisms. Overlap between antigenic sites and receptor recognition sites facilitates coevolution of antigenicity and cellular tropism.
- Viruses use two major strategies to cope with the host immune response: modulation and escape. Escape mutations generally imply a fitness cost for the virus.
- Viruses have evolved means to cope with multiple constraints, including remarkable environmental alternations. This is the case of pathogenic arboviruses whose natural infectious cycles involve replication in vertebrate and insect hosts.

REFERENCES

Acharya, R., Fry, E., Stuart, D., Fox, G., Rowlands, D., et al., 1989. The three-dimensional structure of foot-and-mouth disease virus at 2.9 Å resolution. Nature 337, 709–716.

Alberts, B., Johnson, A., Lewis, J., Raff, M., Roberts, K., et al., 2002. Molecular Biology of the Cell. Garland Science, New York, N.Y.

Alcami, A., 2003. Viral mimicry of cytokines, chemokines and their receptors. Nat. Rev. Immunol. 3, 36–50.

Ali, A., Li, H., Schneider, W.L., Sherman, D.J., Gray, S., et al., 2006. Analysis of genetic bottlenecks during horizontal transmission of Cucumber mosaic virus. J. Virol. 80, 8345–8350.

Aragones, L., Guix, S., Ribes, E., Bosch, A., Pinto, R.M., 2010. Fine-tuning translation kinetics selection as the driving force of codon usage bias in the hepatitis a virus capsid. PLoS Pathog. 6, e1000797.

Arias, A., Perales, C., Escarmis, C., Domingo, E., 2010. Deletion mutants of VPg reveal new cytopathology determinants in a picornavirus. PLoS One 5, e10735.

Arora, R., Priano, C., Jacobson, A.B., Mills, D.R., 1996. cis-acting elements within an RNA coliphage genome: fold as you please, but fold you must!! J. Mol. Biol. 258, 433–446.

Artenstein, M.S., Miller, W.S., 1966. Air sampling for respiratory disease agents in army recruits. Bacteriol. Rev. 30, 571–572.

Bae, Y.S., Yoon, J.W., 1993. Determination of diabetogenicity attributable to a single amino acid, Ala776, on the polyprotein of encephalomyocarditis virus. Diabetes 42, 435–443.

Baldwin, C., Berkhout, B., 2007. HIV-1 drug-resistance and drug-dependence. Retrovirology 4, 78.

Baldwin, C.E., Sanders, R.W., Deng, Y., Jurriaans, S., Lange, J.M., et al., 2004. Emergence of a drug-dependent human immunodeficiency virus type 1 variant during therapy with the T20 fusion inhibitor. J. Virol. 78, 12428–12437.

Ballesteros, M.L., Sanchez, C.M., Enjuanes, L., 1997. Two amino acid changes at the N-terminus of transmissible gastroenteritis coronavirus spike protein result in the loss of enteric tropism. Virology 227, 378–388.

Baranowski, E., Ruíz-Jarabo, C.M., Sevilla, N., Andreu, D., Beck, E., et al., 2000. Cell recognition by foot-and-mouth disease virus that lacks the RGD integrin-binding motif: flexibility in aphthovirus receptor usage. J. Virol. 74, 1641–1647.

Baranowski, E., Ruiz-Jarabo, C.M., Domingo, E., 2001. Evolution of cell recognition by viruses. Science 292, 1102–1105.

Baranowski, E., Ruiz-Jarabo, C.M., Pariente, N., Verdaguer, N., Domingo, E., 2003. Evolution of cell recognition by viruses: a source of biological novelty with medical implications. Adv. Virus Res. 62, 19–111.

Basler, C.F., Garcia-Sastre, A., 2002. Viruses and the type I interferon antiviral system: induction and evasion. Int. Rev. Immunol. 21, 305–337.

Belsham, G.J., Sonenberg, N., 1996. RNA-protein interactions in regulation of picornavirus RNA translation. Microbiol. Rev. 60, 499–511.

Bergelson, J.M., 2010. Receptors. In: Ehrenfeld, E., Domingo, E., Roos, R.P. (Eds.), The Picornaviruses. ASM Press, Washington DC, pp. 73–86.

Betancourt, M., Fereres, A., Fraile, A., Garcia-Arenal, F., 2008. Estimation of the effective number of founders that initiate an infection after aphid transmission of a multipartite plant virus. J. Virol. 82, 12416–12421.

Bhella, D., 2015. The role of cellular adhesion molecules in virus attachment and entry. Philos. Trans. R. Soc. Lond. B Biol. Sci. 370.

Borrego, B., Novella, I.S., Giralt, E., Andreu, D., Domingo, E., 1993. Distinct repertoire of antigenic variants of foot-and-mouth disease virus in the presence or absence of immune selection. J. Virol. 67, 6071–6079.

Borrow, P., Lewicki, H., Wei, X., Horwitz, M.S., Peffer, N., et al., 1997. Antiviral pressure exerted by HIV-1-specific cytotoxic T lymphocytes (CTLs) during primary infection demonstrated by rapid selection of CTL escape virus. Nat. Med. 3, 205–211.

Bosch, A., Mueller, S., Pintó, R.M., 2010. Codon biases and viral fitness. In: Ehrenfeld, E., Domingo, E., Roos, R.P. (Eds.), The Picornaviruses. ASM Press, Washington, DC, pp. 271–283.

Boyd, M.T., Simpson, G.R., Cann, A.J., Johnson, M.A., Weiss, R.A., 1993. A single amino acid substitution in the V1 loop of human immunodeficiency virus type 1 gp120 alters cellular tropism. J. Virol. 67, 3649–3652.

Britten, R.J., 1993. Forbidden synonymous substitutions in coding regions. Mol. Biol. Evol. 10, 205–220.

Budkowska, A., Bedossa, P., Groh, F., Louise, A., Pillot, J., 1995. Fibronectin of human liver sinusoids binds hepatitis B virus: identification by an anti-idiotypic antibody bearing the internal image of the pre-S2 domain. J. Virol. 69, 840–848.

Buenno, L.H., Leme, J., Caricati, C.P., Tonso, A., Rocha, J.C., et al., 2015. Use of uniform designs in combination with neural networks for viral infection process development. Biotechnol. Prog. 31, 532–540.

Bull, R.A., Luciani, F., McElroy, K., Gaudieri, S., Pham, S.T., et al., 2011. Sequential bottlenecks drive viral evolution in early acute hepatitis C virus infection. PLoS Pathog. 7, e1002243.

Burns, C.C., Shaw, J., Campagnoli, R., Jorba, J., Vincent, A., et al., 2006. Modulation of poliovirus replicative fitness in HeLa cells by deoptimization of synonymous codon usage in the capsid region. J. Virol. 80, 3259–3272.

Campbell, C.L., Smith, D.R., Sanchez-Vargas, I., Zhang, B., Shi, P.Y., et al., 2014. A positively selected mutation in the WNV 2K peptide confers resistance to superinfection exclusion in vivo. Virology 464–465, 228–232.

Cantara, W.A., Olson, E.D., Musier-Forsyth, K., 2014. Progress and outlook in structural biology of large viral RNAs. Virus Res. 193, 24–38.

Carson, S.D., Chapman, N.M., Hafenstein, S., Tracy, S., 2011. Variations of coxsackievirus B3 capsid primary structure, ligands, and stability are selected for in a coxsackievirus and adenovirus receptor-limited environment. J. Virol. 85, 3306–3314.

Centlivre, M., Sommer, P., Michel, M., Ho Tsong Fang, R., Gofflo, S., et al., 2005. HIV-1 clade promoters strongly influence spatial and temporal dynamics of viral replication in vivo. J. Clin. Invest. 115, 348–358.

Chamary, J.V., Hurst, L.D., 2005. Biased codon usage near intron-exon junctions: selection on splicing enhancers, splice-site recognition or something else? Trends Genet. 21, 256–259.

Charlesworth, B., 2009. Fundamental concepts in genetics: effective population size and patterns of molecular evolution and variation. Nat. Rev. 10, 195–205.

Cheng, B.Y., Ortiz-Riano, E., Nogales, A., de la Torre, J.C., Martinez-Sobrido, L., 2015. Development of live-attenuated arenavirus vaccines based on codon deoptimization. J. Virol. 89, 3523–3533.

Ciurea, A., Klenerman, P., Hunziker, L., Horvath, E., Senn, B.M., et al., 2000. Viral persistence *in vivo* through selection of neutralizing antibody-escape variants. Proc. Natl. Acad. Sci. U. S. A. 97, 2749–2754.

Clarke, D.K., Duarte, E.A., Elena, S.F., Moya, A., Domingo, E., et al., 1994. The red queen reigns in the kingdom of RNA viruses. Proc. Natl. Acad. Sci. U. S. A. 91, 4821–4824.

Co, M.S., Gaulton, G.N., Fields, B.N., Greene, M.I., 1985. Isolation and biochemical characterization of the mammalian reovirus type 3 cell-surface receptor. Proc. Natl. Acad. Sci. U. S. A. 82, 1494–1498.

Coffin, J.M., 1995. HIV population dynamics in vivo: implications for genetic variation, pathogenesis, and therapy. Science 267, 483–489.

Coleman, J.R., Papamichail, D., Skiena, S., Futcher, B., Wimmer, E., et al., 2008. Virus attenuation by genome-scale changes in codon pair bias. Science 320, 1784–1787.

Colston, E.M., Racaniello, V.R., 1995. Poliovirus variants selected on mutant receptor-expressing cells identify capsid residues that expand receptor recognition. J. Virol. 69, 4823–4829.

Connor, R.J., Kawaoka, Y., Webster, R.G., Paulson, J.C., 1994. Receptor specificity in human, avian, and equine H2 and H3 influenza virus isolates. Virology 205, 17–23.

Connor, R.I., Sheridan, K.E., Ceradini, D., Choe, S., Landau, N.R., 1997. Change in coreceptor use correlates with disease progression in HIV-1-infected individuals. J. Exp. Med. 185, 621–628.

Consigli, R.A., Griffith, G.R., Marriott, S.J., Ludlow, J.W., 1986. Biochemical characterization of poliomavirus-receptor interactions. In: Crowell, R.L., Lonberg-Holm, K. (Eds.), Virus Attachment and Entry into Cells. American Society for Microbiology, Washington, DC, pp. 44–53.

Costafreda, M.I., Perez-Rodriguez, F.J., D'Andrea, L., Guix, S., Ribes, E., et al., 2014. Hepatitis A virus adaptation to cellular shutoff is driven by dynamic adjustments of codon usage and results in the selection of populations with altered capsids. J. Virol. 88, 5029–5041.

Coulon, P., Ternaux, J.P., Flamand, A., Tuffereau, C., 1998. An avirulent mutant of rabies virus is unable to infect motoneurons in vivo and in vitro. J. Virol. 72, 273–278.

Curry, S., Fry, E., Blakemore, W., Abu-Ghazaleh, R., Jackson, T., et al., 1996. Perturbations in the surface structure of A22 Iraq foot-and-mouth disease virus accompanying coupled changes in host cell specificity and antigenicity. Structure 4, 135–145.

de Borba, L., Villordo, S.M., Iglesias, N.G., Filomatori, C.V., Gebhard, L.G., et al., 2015. Overlapping local and long range RNA-RNA interactions modulate dengue virus genome cyclization and replication. J. Virol. 89, 3430–3437.

De Cock, K.M., Jaffe, H.W., Curran, J.W., 2012. The evolving epidemiology of HIV/AIDS. AIDS 26, 1205–1213.

de la Torre, J.C., 2002. Molecular biology of Borna disease virus and persistence. Front. Biosci. 7, d569–d579.

de la Torre, J.C., Wimmer, E., Holland, J.J., 1990. Very high frequency of reversion to guanidine resistance in clonal pools of guanidine-dependent type 1 poliovirus. J. Virol. 64, 664–671.

de Los Santos, T., de Avila Botton, S., Weiblen, R., Grubman, M.J., 2006. The leader proteinase of foot-and-mouth disease virus inhibits the induction of beta interferon mRNA and blocks the host innate immune response. J. Virol. 80, 1906–1914.

Domingo, E. (Ed.), 2006. Quasispecies: Concepts and Implications for Virology. Curr. Top. Microbiol. Immunol., vol. 299. Springer, Berlin.

Domingo, E., 2011. Paradoxical interplay of viral and cellular functions. Viruses 3, 272–277.

Domingo, E., Díez, J., Martínez, M.A., Hernández, J., Holguín, A., et al., 1993. New observations on antigenic diversification of RNA viruses. Antigenic variation is not dependent on immune selection. J. Gen. Virol. 74, 2039–2045.

Domingo, E., Biebricher, C., Eigen, M., Holland, J.J., 2001. Quasispecies and RNA Virus Evolution: Principles and Consequences. Landes Bioscience, Austin.

Domingo, E., Sheldon, J., Perales, C., 2012. Viral quasispecies evolution. Microbiol. Mol. Biol. Rev. 76, 159–216.

dos Reis, M., Savva, R., Wernisch, L., 2004. Solving the riddle of codon usage preferences: a test for translational selection. Nucleic Acids Res. 32, 5036–5044.

Ehrenfeld, E., Domingo, E., Ross, R.P., 2010. The Picornaviruses. ASM Press, Washington, DC.

Eigen, M., 1992. Steps Towards Life. Oxford University Press, Oxford.

Erickson, A.K., Pfeiffer, J.K., 2013. Dynamic viral dissemination in mice infected with yellow fever virus strain 17D. J. Virol. 87, 12392–12397.

Escarmís, C., Dopazo, J., Dávila, M., Palma, E.L., Domingo, E., 1995. Large deletions in the 5'-untranslated region of foot-and-mouth disease virus of serotype C. Virus Res. 35, 155–167.

Este, J.A., Cabrera, C., Blanco, J., Gutierrez, A., Bridger, G., et al., 1999. Shift of clinical human immunodeficiency virus type 1 isolates from X4 to R5 and prevention of emergence of the syncytium-inducing phenotype by blockade of CXCR4. J. Virol. 73, 5577–5585.

Eyre-Walker, A., 1996. Synonymous codon bias is related to gene length in Escherichia coli: selection for translational accuracy? Mol. Biol. Evol. 13, 864–872.

Fabre, F., Moury, B., Johansen, E.I., Simon, V., Jacquemond, M., et al., 2014. Narrow bottlenecks affect Pea seedborne mosaic virus populations during vertical seed transmission but not during leaf colonization. PLoS Pathog. 10, e1003833.

Ferrer-Orta, C., De la Higuera, I., Caridi, F., Sanchez-Aparicio, M.T., Moreno, E., et al., 2015. Multifunctionality of a picornavirus polymerase domain: nuclear localization signal is involved in nucleotide recognition. J. Virol. 89, 6849-6859.

Forrester, N.L., Coffey, L.L., Weaver, S.C., 2014. Arboviral bottlenecks and challenges to maintaining diversity and fitness during mosquito transmission. Viruses 6, 3991–4004.

Foy, B.D., Myles, K.M., Pierro, D.J., Sanchez-Vargas, I., Uhlirova, M., et al., 2004. Development of a new Sindbis virus transducing system and its characterization in three Culicine mosquitoes and two Lepidopteran species. Insect Mol. Biol. 13, 89–100.

Fry, E.E., Stuart, D.I., 2010. Virion structure. In: Ehrenfeld, E., Domingo, E., Roos, R.P. (Eds.), The Picornaviruses. ASM Press, Washington, DC, pp. 59–72.

Fry, E.E., Lea, S.M., Jackson, T., Newman, J.W., Ellard, F.M., et al., 1999. The structure and function of a foot-and-mouth disease virus- oligosaccharide receptor complex. EMBO J. 18, 543–554.

Fung, A., Jin, Z., Dyatkina, N., Wang, G., Beigelman, L., Deval, J., 2014. Efficiency of incorporation and chain termination determines the inhibition potency of 2'-modified nucleotide analogs against hepatitis C virus polymerase. Antimicrobial Agents and Chemotherapy 58, 3636–3645.

Futuyma, D.J., Slatkin, M. (Eds.), 1983. Coevolution. Sinauer Associates Inc., Sunderland, Massachusetts.

Gaulton, G., Co, M.S., Greene, M.I., 1985. Anti-idiotypic antibody identifies the cellular receptor of reovirus type 3. J. Cell. Biochem. 28, 69–78.

Gebauer, F., de la Torre, J.C., Gomes, I., Mateu, M.G., Barahona, H., et al., 1988. Rapid selection of genetic and antigenic variants of foot-and-mouth disease virus during persistence in cattle. J. Virol. 62, 2041–2049.

Gerone, P.J., Couch, R.B., Keefer, G.V., Douglas, R.G., Derrenbacher, E.B., et al., 1966. Assessment of experimental and natural viral aerosols. Bacteriol. Rev. 30, 576–588.

Gibbs, A.J., Calisher, C.H., García-Arenal, F. (Eds.), 1995. Molecular Basis of Virus Evolution. Cambridge University Press, Cambridge.

Gitlin, L., Hagai, T., LaBarbera, A., Solovey, M., Andino, R., 2014. Rapid evolution of virus sequences in intrinsically disordered protein regions. PLoS Pathog. 10, e1004529.

Gloor, G.B., Martin, L.C., Wahl, L.M., Dunn, S.D., 2005. Mutual information in protein multiple sequence alignments reveals two classes of coevolving positions. Biochemistry 44, 7156–7165.

Gomez, P., Ashby, B., Buckling, A., 2015. Population mixing promotes arms race host-parasite coevolution. Proc. Biol. Sci. 282, 20142297.

Grande-Perez, A., Martin, V., Moreno, H., de la Torre, J.C., 2016. Arenaviruses quasispecies and their biological implications. In: Domingo, E., Schuster, P. (Eds.), Quasispecies: From Theory to Experimental Systems. In: Current Topics in Microbiology and Immunology, in press.

Gubler, D.J., 1997. Dengue and dengue hemorrhagic fever: its history and resurgence as a global public health problem. In: Gubler, D.J., Kuno, G. (Eds.), Dengue and Dengue Hemorrhagic Fever. CAB International, New York.

Gutierrez, S., Michalakis, Y., Blanc, S., 2012. Virus population bottlenecks during within-host progression and host-to-host transmission. Curr. Opin. Virol. 2, 546–555.

Haaland, R.E., Hawkins, P.A., Salazar-Gonzalez, J., Johnson, A., Tichacek, A., et al., 2009. Inflammatory genital infections mitigate a severe genetic bottleneck in heterosexual transmission of subtype A and C HIV-1. PLoS Pathog. 5, e1000274.

Hall, C.B., Douglas Jr., R.G., Geiman, J.M., Meagher, M.P., 1979. Viral shedding patterns of children with influenza B infection. J. Infect. Dis. 140, 610–613.

Halstead, S.B., 1980. Immunological parameters of Togavirus disease syndromes. In: Schlesinger, R.W. (Ed.), The Togaviruses. Biology, Structure, Replication. Academic Press, New York, pp. 107–173.

Hanham, C.A., Zhao, F., Tignor, G.H., 1993. Evidence from the anti-idiotypic network that the acetylcholine receptor is a rabies virus receptor. J. Virol. 67, 530–542.

Hanley, K.H., Weaver, S.C., 2008. Arbovirus evolution. In: Domingo, E., Parrish, C.R., Holland, J.J. (Eds.), Origin and Evolution of Viruses second ed. Elsevier, Oxford, pp. 351–392.

Harber, J., Bernhardt, G., Lu, H.H., Sgro, J.Y., Wimmer, E., 1995. Canyon rim residues, including antigenic determinants, modulate serotype-specific binding of polioviruses to mutants of the poliovirus receptor. Virology 214, 559–570.

He, M., An, T.Z., Teng, C.B., 2014. Evolution of mammalian and avian bornaviruses. Mol. Phylogenet. Evol. 79, 385–391.

Herrera, M., Garcia-Arriaza, J., Pariente, N., Escarmis, C., Domingo, E., 2007. Molecular basis for a lack of correlation between viral fitness and cell killing capacity. PLoS Pathog. 3, e53.

Herrmann Jr., E.C., Herrmann, J.A., 1977. A working hypothesis—virus resistance development as an indicator of specific antiviral activity. Ann. N. Y. Acad. Sci. 284, 632–637.

Hertogs, K., Depla, E., Crabbe, T., De Bruin, W., Leenders, W., et al., 1994. Spontaneous development of anti-hepatitis B virus envelope (anti-idiotypic) antibodies in animals immunized with human liver endonexin II or with the F(ab')2 fragment of anti-human liver endonexin II immunoglobulin G: evidence for a receptor-ligand-like relationship between small hepatitis B surface antigen and endonexin II. J. Virol. 68, 1516–1521.

Hewat, E.A., Neumann, E., Conway, J.F., Moser, R., Ronacher, B., et al., 2000. The cellular receptor to human rhinovirus 2 binds around the 5-fold axis and not in the canyon: a structural view. EMBO J. 19, 6317–6325.

Hiramatsu, K., Tadano, M., Men, R., Lai, C.J., 1996. Mutational analysis of a neutralization epitope on the dengue type 2 virus (DEN2) envelope protein: monoclonal antibody resistant DEN2/DEN4 chimeras exhibit reduced mouse neurovirulence. Virology 224, 437–445.

Ho, D.D., 1995. Time to hit HIV, early and hard. N. Engl. J. Med. 333, 450–451.

Ho, D.D., Neumann, A.U., Perelson, A.S., Chen, W., Leonard, J.M., et al., 1995. Rapid turnover of plasma virions and CD4 lymphocytes in HIV-1 infection. Nature 373, 123–126.

Hsu, E.C., Sarangi, F., Iorio, C., Sidhu, M.S., Udem, S.A., et al., 1998. A single amino acid change in the hemagglutinin protein of measles virus determines its ability to bind CD46 and reveals another receptor on marmoset B cells. J. Virol. 72, 2905–2916.

Huang, T., Campadelli-Fiume, G., 1996. Anti-idiotypic antibodies mimicking glycoprotein D of herpes simplex virus identify a cellular protein required for virus spread from cell to cell and virus-induced polykaryocytosis. Proc. Natl. Acad. Sci. U. S. A. 93, 1836–1840.

Huang, S., Reddy, V., Dasgupta, N., Nemerow, G.R., 1999. A single amino acid in the adenovirus type 37 fiber confers binding to human conjunctival cells. J. Virol. 73, 2798–2802.

Hueffer, K., Parrish, C.R., 2003. Parvovirus host range, cell tropism and evolution. Curr. Opin. Microbiol. 6, 392–398.

Hueffer, K., Parker, J.S., Weichert, W.S., Geisel, R.E., Sgro, J.Y., et al., 2003. The natural host range shift and subsequent evolution of canine parvovirus resulted from virus-specific binding to the canine transferrin receptor. J. Virol. 77, 1718–1726.

Hunt, R.C., Simhadri, V.L., Iandoli, M., Sauna, Z.E., Kimchi-Sarfaty, C., 2014. Exposing synonymous mutations. Trends Genet. 30, 308–321.

Iorio, R.M., Glickman, R.L., Riel, A.M., Sheehan, J.P., Bratt, M.A., 1989. Functional and neutralization profile of seven overlapping antigenic sites on the HN glycoprotein of Newcastle disease virus: monoclonal antibodies to some sites prevent viral attachment. Virus Res. 13, 245–261.

Jacob, F., 1977. Evolution and tinkering. Science 196, 1161–1166.

Jennings, A.D., Gibson, C.A., Miller, B.R., Mathews, J.H., Mitchell, C.J., et al., 1994. Analysis of a yellow fever virus isolated from a fatal case of vaccine-associated human encephalitis. J. Infect. Dis. 169, 512–518.

Jnaoui, K., Michiels, T., 1998. Adaptation of Theiler's virus to L929 cells: mutations in the putative receptor binding site on the capsid map to neutralization sites and modulate viral persistence. Virology 244, 397–404.

Jung, Y.T., Kozak, C.A., 2000. A single amino acid change in the murine leukemia virus capsid gene responsible for the Fv1(nr) phenotype. J. Virol. 74, 5385–5387.

Katze, M.G., He, Y., Gale Jr., M., 2002. Viruses and interferon: a fight for supremacy. Nat. Rev. Immunol. 2, 675–687.

Keay, S., Baldwin, B., 1991. Anti-idiotype antibodies that mimic gp86 of human cytomegalovirus inhibit viral fusion but not attachment. J. Virol. 65, 5124–5128.

Keay, S., Merigan, T.C., Rasmussen, L., 1989. Identification of cell surface receptors for the 86-kilodalton glycoprotein of human cytomegalovirus. Proc. Natl. Acad. Sci. U. S. A. 86, 10100–10103.

Keck, Z.Y., Angus, A.G., Wang, W., Lau, P., Wang, Y., et al., 2014. Non-random escape pathways from a broadly neutralizing human monoclonal antibody map to a highly conserved region on the hepatitis C virus E2 glycoprotein encompassing amino acids 412–423. PLoS Pathog. 10, e1004297.

Keele, B.F., Giorgi, E.E., Salazar-Gonzalez, J.F., Decker, J.M., Pham, K.T., et al., 2008. Identification and characterization of transmitted and early founder virus envelopes in primary HIV-1 infection. Proc. Natl. Acad. Sci. U. S. A. 105, 7552–7557.

Koel, B.F., Burke, D.F., Bestebroer, T.M., van der Vliet, S., Zondag, G.C., et al., 2013. Substitutions near the receptor binding site determine major antigenic change during influenza virus evolution. Science 342, 976–979.

Koel, B.F., van der Vliet, S., Burke, D.F., Bestebroer, T.M., Bharoto, E.E., et al., 2014. Antigenic variation of clade 2.1 H5N1 virus is determined by a few amino acid substitutions immediately adjacent to the receptor binding site. mBio 5 (3), e01070–14.

Korber, B.T., Farber, R.M., Wolpert, D.H., Lapedes, A.S., 1993. Covariation of mutations in the V3 loop of human immunodeficiency virus type 1 envelope protein: an information theoretic analysis. Proc. Natl. Acad. Sci. U. S. A. 90, 7176–7180.

Krakauer, D.C., Jansen, V.A., 2002. Red queen dynamics of protein translation. J. Theor. Biol. 218, 97–109.

Kubo, H., Takase-Yoden, S., Taguchi, F., 1993. Neutralization and fusion inhibition activities of monoclonal antibodies specific for the S1 subunit of the spike protein of neurovirulent murine coronavirus JHMV c1-2 variant. J. Gen. Virol. 74 (Pt 7), 1421–1425.

Kubo, H., Yamada, Y.K., Taguchi, F., 1994. Localization of neutralizing epitopes and the receptor-binding site within the amino-terminal 330 amino acids of the murine coronavirus spike protein. J. Virol. 68, 5403–5410.

Kuss, S.K., Etheredge, C.A., Pfeiffer, J.K., 2008. Multiple host barriers restrict poliovirus trafficking in mice. PLoS Pathog. 4, e1000082.

Laeeq, S., Smith, C.A., Wagner, S.D., Thomas, D.B., 1997. Preferential selection of receptor-binding variants of influenza virus hemagglutinin by the neutralizing antibody repertoire of transgenic mice expressing a human immunoglobulin mu minigene. J. Virol. 71, 2600–2605.

Lanza, A.M., Curran, K.A., Rey, L.G., Alper, H.S., 2014. A condition-specific codon optimization approach for improved heterologous gene expression in Saccharomyces cerevisiae. BMC Syst. Biol. 8, 33.

Lavner, Y., Kotlar, D., 2005. Codon bias as a factor in regulating expression via translation rate in the human genome. Gene 345, 127–138.

Le Nouen, C., Brock, L.G., Luongo, C., McCarty, T., Yang, L., et al., 2014. Attenuation of human respiratory syncytial virus by genome-scale codon-pair deoptimization. Proc. Natl. Acad. Sci. U. S. A. 111, 13169–13174.

Lea, S., Hernández, J., Blakemore, W., Brocchi, E., Curry, S., et al., 1994. The structure and antigenicity of a type C foot-and-mouth disease virus. Structure 2, 123–139.

Lea, S., Abu-Ghazaleh, R., Blakemore, W., Curry, S., Fry, E., et al., 1995. Structural comparison of two strains of foot-and-mouth disease virus subtype O1 and a laboratory antigenic variant, G67. Structure 3, 571–580.

Lee, P., Knight, R., Smit, J.M., Wilschut, J., Griffin, D.E., 2002. A single mutation in the E2 glycoprotein important for neurovirulence influences binding of sindbis virus to neuroblastoma cells. J. Virol. 76, 6302–6310.

Lerner, D.L., Elder, J.H., 2000. Expanded host cell tropism and cytopathic properties of feline immunodeficiency virus strain PPR subsequent to passage through interleukin-2-independent T cells. J. Virol. 74, 1854–1863.

Li, H., Roossinck, M.J., 2004. Genetic bottlenecks reduce population variation in an experimental RNA virus population. J. Virol. 78, 10582–10587.

Li, J.S., Tong, S.P., Wands, J.R., 1996. Characterization of a 120-Kilodalton pre-S-binding protein as a candidate duck hepatitis B virus receptor. J. Virol. 70, 6029–6035.

Li, W., Zhang, C., Sui, J., Kuhn, J.H., Moore, M.J., et al., 2005. Receptor and viral determinants of SARS-coronavirus adaptation to human ACE2. EMBO J. 24, 1634–1643.

Liebermann, H., Mentel, R., Bauer, U., Pring-Akerblom, P., Dolling, R., et al., 1998. Receptor binding sites and antigenic epitopes on the fiber knob of human adenovirus serotype 3. J. Virol. 72, 9121–9130.

Lonberg-Holm, K., Gosser, L.B., Shimshick, E.J., 1976. Interaction of liposomes with subviral particles of poliovirus type 2 and rhinovirus type 2. J. Virol. 19, 746–749.

Lopez-Bueno, A., Rubio, M.P., Bryant, N., McKenna, R., Agbandje-McKenna, M., et al., 2006. Host-selected amino acid changes at the sialic acid binding pocket of the parvovirus capsid modulate cell binding affinity and determine virulence. J. Virol. 80, 1563–1573.

Lucas-Hourani, M., Dauzonne, D., Jorda, P., Cousin, G., Lupan, A., et al., 2013. Inhibition of pyrimidine biosynthesis pathway suppresses viral growth through innate immunity. PLoS Pathog. 9, e1003678.

Manning, T., Sleator, R.D., Walsh, P., 2013. Naturally selecting solutions: the use of genetic algorithms in bioinformatics. Bioengineered 4, 266–278.

Maraia, R.J., Iben, J.R., 2014. Different types of secondary information in the genetic code. RNA 20, 977–984.

Marcus, P.I., Rodriguez, L.L., Sekellick, M.J., 1998. Interferon induction as a quasispecies marker of vesicular stomatitis virus populations. J. Virol. 72, 542–549.

Markowitz, M., Louie, M., Hurley, A., Sun, E., Di Mascio, M., et al., 2003. A novel antiviral intervention results in more accurate assessment of human immunodeficiency virus type 1 replication dynamics and T-cell decay in vivo. J. Virol. 77, 5037–5038.

Martínez, M.A., Dopazo, J., Hernandez, J., Mateu, M.G., Sobrino, F., et al., 1992. Evolution of the capsid protein genes of foot-and-mouth disease virus: antigenic variation without accumulation of amino acid substitutions over six decades. J. Virol. 66, 3557–3565.

Martinez, M.A., Verdaguer, N., Mateu, M.G., Domingo, E., 1997. Evolution subverting essentiality: dispensability of the cell attachment Arg-Gly-Asp motif in multiply passaged foot-and-mouth disease virus. Proc. Natl. Acad. Sci. U. S. A. 94, 6798–6802.

Martinez-Salas, E., Regalado, M.P., Domingo, E., 1996. Identification of an essential region for internal initiation of translation in the aphthovirus internal ribosome entry site and implications for viral evolution. J. Virol. 70, 992–998.

Martinez-Sobrido, L., Emonet, S., Giannakas, P., Cubitt, B., Garcia-Sastre, A., et al., 2009. Identification of amino acid residues critical for the anti-interferon activity of the nucleoprotein of the prototypic arenavirus lymphocytic choriomeningitis virus. J. Virol. 83, 11330–11340.

Martrus, G., Nevot, M., Andres, C., Clotet, B., Martinez, M.A., 2013. Changes in codon-pair bias of human immunodeficiency virus type 1 have profound effects on virus replication in cell culture. Retrovirology 10, 78.

Mateu, M.G., 1995. Antibody recognition of picornaviruses and escape from neutralization: a structural view. Virus Res. 38, 1–24.

Matsuzaki, M., Sugawara, K., Adachi, K., Hongo, S., Nishimura, H., et al., 1992. Location of neutralizing epitopes on the hemagglutinin-esterase protein of influenza C virus. Virology 189, 79–87.

Maury, W., Thompson, R.J., Jones, Q., Bradley, S., Denke, T., et al., 2005. Evolution of the equine infectious anemia virus long terminal repeat during the alteration of cell tropism. J. Virol. 79, 5653–5664.

Maynard Smith, J., Burian, R., Dauffman, S., Alberch, P., Campbell, J., 1985. Developmental constraints and evolution. Quart. Rev. Biol. 60, 265–287.

McMichael, A.J., Phillips, R.E., 1997. Escape of human immunodeficiency virus from immune control. Annu. Rev. Immunol. 15, 271–296.

Mendelsohn, C.L., Wimmer, E., Racaniello, V.R., 1989. Cellular receptor for poliovirus: molecular cloning, nucleotide sequence, and expression of a new member of the immunoglobulin superfamily. Cell 56, 855–865.

Minocha, H.C., Xue, W., Reddy, J.R., 1997. A 50 kDa membrane protein from bovine kidney cells is a putative receptor for bovine viral diarrhea virus (BVDV). Adv. Exp. Med. Biol. 412, 145–148.

Morse, S.S., 1994. The Evolutionary Biology of Viruses. Raven Press, New York.

Mosier, D.E., Picchio, G.R., Gulizia, R.J., Sabbe, R., Poignard, P., et al., 1999. Highly potent RANTES analogues either prevent CCR5-using human immunodeficiency virus type 1 infection in vivo or rapidly select for CXCR4-using variants. J. Virol. 73, 3544–3550.

Moury, B., Fabre, F., Senoussi, R., 2007. Estimation of the number of virus particles transmitted by an insect vector. Proc. Natl. Acad. Sci. U. S. A. 104, 17891–17896.

Murcia, P.R., Baillie, G.J., Daly, J., Elton, D., Jervis, C., et al., 2010. Intra- and interhost evolutionary dynamics of equine influenza virus. J. Virol. 84, 6943–6954.

Murphy, B.R., Webster, R.G., 1985. Influenza viruses. In: Fields, B.N. (Ed.), Virology. Raven Press, New York, pp. 1179–1239.

Murray, M.G., Bradley, J., Yang, X.F., Wimmer, E., Moss, E.G., et al., 1988. Poliovirus host range is determined by a short amino acid sequence in neutralization antigenic site I. Science 241, 213–215.

Neumann, A.U., Lam, N.P., Dahari, H., Gretch, D.R., Wiley, T.E., et al., 1998. Hepatitis C viral dynamics in vivo and the antiviral efficacy of interferon-alpha therapy. Science 282, 103–107.

Neurath, A.R., Kent, S.B., Strick, N., Parker, K., 1986. Identification and chemical synthesis of a host cell receptor binding site on hepatitis B virus. Cell 46, 429–436.

Nice, T.J., Strong, D.W., McCune, B.T., Pohl, C.S., Virgin, H.W., 2013. A single-amino-acid change in murine norovirus NS1/2 is sufficient for colonic tropism and persistence. J. Virol. 87, 327–334.

Nogales, A., Baker, S.F., Ortiz-Riano, E., Dewhurst, S., Topham, D.J., et al., 2014. Influenza A virus attenuation by codon deoptimization of the NS gene for vaccine development. J. Virol. 88, 10525–10540.

Novella, I.S., Cilnis, M., Elena, S.F., Kohn, J., Moya, A., et al., 1996. Large-population passages of vesicular stomatitis virus in interferon-treated cells select variants of only limited resistance. J. Virol. 70, 6414–6417.

Nowak, M.A., Bonhoeffer, S., Hill, A.M., Boehme, R., Thomas, H.C., et al., 1996. Viral dynamics in hepatitis B virus infection. Proc. Natl. Acad. Sci. U. S. A. 93, 4398–4402.

Núñez, J.I., Baranowski, E., Molina, N., Ruiz-Jarabo, C.M., Sánchez, C., et al., 2001. A single amino acid substitution in nonstructural protein 3A can mediate adaptation of foot-and-mouth disease virus to the guinea pig. J. Virol. 75, 3977–3983.

Olsthoorn, R.C., Licis, N., van Duin, J., 1994. Leeway and constraints in the forced evolution of a regulatory RNA helix. EMBO J. 13, 2660–2668.

Ortiz-Riano, E., Ngo, N., Devito, S., Eggink, D., Munger, J., et al., 2014. Inhibition of arenavirus by A3, a pyrimidine biosynthesis inhibitor. J. Virol. 88, 878–889.

Palstra, F.P., Ruzzante, D.E., 2008. Genetic estimates of contemporary effective population size: what can they tell us about the importance of genetic stochasticity for wild population persistence? Mol. Ecol. 17, 3428–3447.

Parrish, C.R., Kawaoka, Y., 2005. The origins of new pandemic viruses: the acquisition of new host ranges by canine parvovirus and influenza A viruses. Annu. Rev. Microbiol. 59, 553–586.

Paul, A.V., 2002. Possible unifying mechanism of picornavirus genome replication. In: Semler, B.L., Wimmer, E. (Eds.), Molecular Biology of Picornaviruses. ASM Press, Washington, DC, pp. 227–246.

Pauli, G., Ludwig, H., 1985. Increase of virus yields and releases of Borna disease virus from persistently infected cells. Virus Res. 2, 29–33.

Pawlak, E.N., Dikeakos, J.D., 2015. HIV-1 Nef: a master manipulator of the membrane trafficking machinery mediating immune evasion. Biochim. Biophys. Acta 1850, 733–741.

Perales, C., Beach, N.M., Gallego, I., Soria, M.E., Quer, J., et al., 2013. Response of hepatitis C virus to long-term passage in the presence of alpha interferon: multiple mutations and a common phenotype. J. Virol. 87, 7593–7607.

Perales, C., Beach, N.M., Sheldon, J., Domingo, E., 2014. Molecular basis of interferon resistance in hepatitis C virus. Curr. Opin. Virol. 8, 38–44.

Perelson, A.S., Neumann, A.U., Markowitz, M., Leonard, J.M., Ho, D.D., 1996. HIV-1 dynamics in vivo: virion clearance rate, infected cell life-span, and viral generation time. Science 271, 1582–1586.

Petit, M.A., Capel, F., Dubanchet, S., Mabit, H., 1992. PreS1-specific binding proteins as potential receptors for hepatitis B virus in human hepatocytes. Virology 187, 211–222.

Pfeiffer, J.K., Kirkegaard, K., 2005. Increased fidelity reduces poliovirus fitness under selective pressure in mice. PLoS Pathog. 1, 102–110.

Philadelpho, N.A., Rubbenstroth, D., Guimaraes, M.B., Piantino Ferreira, A.J., 2014. Survey of bornaviruses in pet psittacines in Brazil reveals a novel parrot bornavirus. Vet. Microbiol. 174, 584–590.

Pleij, C.W.A., 1994. RNA pseudoknots. Curr. Opin. Struct. Biol. 4, 337–344.

Quer, J., Esteban, J.I., Cos, J., Sauleda, S., Ocana, L., et al., 2005. Effect of bottlenecking on evolution of the nonstructural protein 3 gene of hepatitis C virus during sexually transmitted acute resolving infection. J. Virol. 79, 15131–15141.

Quer, J., Martell, M., Rodriguez, A., Bosch, A., Jardi, R., et al., 2008. The impact of rapid evolution of hepatitis viruses. In: Domingo, E., Parrish, C., Holland, J.J. (Eds.), Origin and Evolution of Viruses. Elsevier, Oxford, pp. 303–350.

Ramratnam, B., Bonhoeffer, S., Binley, J., Hurley, A., Zhang, L., et al., 1999. Rapid production and clearance of HIV-1 and hepatitis C virus assessed by large volume plasma apheresis. Lancet 354, 1782–1785.

Reeve, R., Smith, E., Wallace, B., 1990. Components of fitness become effectively neutral in equilibrium populations. Proc. Natl. Acad. Sci. U. S. A. 87, 2018–2020.

Remmert, H., 1980. Ecology. Springer-Verlag, Berlin Heidelberg.

Richards, O.C., Ehrenfeld, E., 1990. Pliovirus RNA replication. Curr. Top. Microbiol. Immunol. 161, 90–119.

Richman, D.D., Wrin, T., Little, S.J., Petropoulos, C.J., 2003. Rapid evolution of the neutralizing antibody response to HIV type 1 infection. Proc. Natl. Acad. Sci. U. S. A. 100, 4144–4149.

Robertson, H.D., 1975. Functions of replicating RNA in cells infected by RNA bacteriophages. In: RNA Phages. Cold Spring Harbor Laboratory, New York, pp. 113–115.

Robertson, J.S., Bootman, J.S., Newman, R., Oxford, J.S., Daniels, R.S., et al., 1987. Structural changes in the haemagglutinin which accompany egg adaptation of an influenza A(H1N1) virus. Virology 160, 31–37.

Rodriguez-Frías, F., Tabernero, D., Quer, J., Esteban, J.I., Ortega, I., et al., 2012. Ultra-deep pyrosequencing detects conserved genomic sites and quantifies linkage of drug-resistant amino acid changes in the hepatitis B virus genome. PLoS One 7, e37874.

Rogers, G.N., Paulson, J.C., Daniels, R.S., Skehel, J.J., Wilson, I.A., et al., 1983. Single amino acid substitutions in influenza haemagglutinin change receptor binding specificity. Nature 304, 76–78.

Roossinck, M.J., 2008. Mutant clouds and bottleneck events in plant virus evolution. In: Domingo, E., Parrish, C., Holland, J.J. (Eds.), Origin and Evolution of Viruses. Elsevier, Oxford, pp. 251–258.

Rouzine, I.M., Coffin, J.M., Weinberger, L.S., 2014. Fifteen years later: hard and soft selection sweeps confirm a large population number for HIV in vivo. PLoS Genet. 10, e1004179.

Roy, P., 2008. Molecular dissection of bluetongue virus. In: Mettenleiter, T.C., Sobrino, F. (Eds.), Animal Viruses. Molecular Biology. Cister Academic Press, Norfolk, U.K., pp. 305–353.

Ruiz-Jarabo, C.M., Pariente, N., Baranowski, E., Dávila, M., Gómez-Mariano, G., et al., 2004. Expansion of host-cell tropism of foot-and-mouth disease virus despite replication in a constant environment. J. Gen. Virol. 85, 2289–2297.

Rustagi, A., Gale Jr., M., 2014. Innate antiviral immune signaling, viral evasion and modulation by HIV-1. J. Mol. Biol. 426, 1161–1177.

Sa-Carvalho, D., Rieder, E., Baxt, B., Rodarte, R., Tanuri, A., et al., 1997. Tissue culture adaptation of foot-and-mouth disease virus selects viruses that bind to heparin and are attenuated in cattle. J. Virol. 71, 5115–5123.

Sacristan, S., Diaz, M., Fraile, A., Garcia-Arenal, F., 2011. Contact transmission of Tobacco mosaic virus: a quantitative analysis of parameters relevant for virus evolution. J. Virol. 85, 4974–4981.

Saha, K., Zhang, J., Gupta, A., Dave, R., Yimen, M., et al., 2001. Isolation of primary HIV-1 that target CD8+ T lymphocytes using CD8 as a receptor. Nat. Med. 7, 65–72.

Sala, M., Centlivre, M., Wain-Hobson, S., 2006. Clade-specific differences in active viral replication and compartmentalization. Curr. Opin. HIV AIDS 1, 108–114.

Salazar-Gonzalez, J.F., Salazar, M.G., Keele, B.F., et al., 2009. Genetic identity, biological phenotype, and evolutionary pathways of transmitted/founder viruses in acute and early HIV-1 infection. J. Exp. Med. 206, 1273–1289.

Salemi, M., Vandamme, A.M., 2004. The Phylogenetic Handbook. A Practical Approach to DNA and Protein Phylogeny. Cambridge University Press, Cambridge.

Sánchez, G., Bosch, A., Pinto, R.M., 2003. Genome variability and capsid structural constraints of hepatitis a virus. J. Virol. 77, 452–459.

Sanz-Ramos, M., Diaz-San Segundo, F., Escarmis, C., Domingo, E., Sevilla, N., 2008. Hidden virulence determinants in a viral quasispecies in vivo. J. Virol. 82, 10465–10476.

Sato, S., Zhang, L., Kim, J., Jakob, J., Grant, R.A., et al., 1996. A neutralization site of DA strain of Theiler's murine encephalomyelitis virus important for disease phenotype. Virology 226, 327–337.

Scholle, F., Girard, Y.A., Zhao, Q., Higgs, S., Mason, P.W., 2004. Trans-packaged West Nile virus-like particles: infectious properties in vitro and in infected mosquito vectors. J. Virol. 78, 11605–11614.

Schuster, P., 2011. Mathematical modeling of evolution. Solved and open problems. Theory Biosci. 130, 71–89.

Seet, B.T., Johnston, J.B., Brunetti, C.R., Barrett, J.W., Everett, H., et al., 2003. Poxviruses and immune evasion. Annu. Rev. Immunol. 21, 377–423.

Sellers, R.F., 1971. Quantitative aspects of the spread of foot-and-mouth disease. Vet. Bull. 41, 431–439.

Sevilla, N., Kunz, S., Holz, A., Lewicki, H., Homann, D., et al., 2000. Immunosuppression and resultant viral persistence by specific viral targeting of dendritic cells. J. Exp. Med. 192, 1249–1260.

Sevilla, N., Domingo, E., de la Torre, J.C., 2002. Contribution of LCMV towards deciphering biology of quasispecies in vivo. Curr. Top. Microbiol. Immunol. 263, 197–220.

Shackelton, L.A., Parrish, C.R., Truyen, U., Holmes, E.C., 2005. High rate of viral evolution associated with the emergence of carnivore parvovirus. Proc. Natl. Acad. Sci. U. S. A. 102, 379–384.

Shackelton, L.A., Parrish, C.R., Holmes, E.C., 2006. Evolutionary basis of codon usage and nucleotide composition bias in vertebrate DNA viruses. J. Mol. Evol. 62, 551–563.

Sharma, N.K., Sherker, A.H., 2009. Epidemiology, risk factors, and natural history of chronic hepatitis C. In: Shetty, K., Wu, G.Y. (Eds.), Clinical Gastroenterology. Chronic Viral Hepatitis. Diagnosis and Therapeutics. Humana Press, Springer Science + Business media, LLC, New York, pp. 33–70.

Shepard, C.W., Finelli, L., Alter, M.J., 2005. Global epidemiology of hepatitis C virus infection. Lancet Infect. Dis. 5, 558–567.

Simmonds, P., Tuplin, A., Evans, D.J., 2004. Detection of genome-scale ordered RNA structure (GORS) in genomes of positive-stranded RNA viruses: implications for virus evolution and host persistence. RNA 10, 1337–1351.

Skehel, J.J., Wiley, D.C., 2000. Receptor binding and membrane fusion in virus entry: the influenza hemagglutinin. Annu. Rev. Biochem. 69, 531–569.

Smelt, S.C., Borrow, P., Kunz, S., Cao, W., Tishon, A., et al., 2001. Differences in affinity of binding of lymphocytic choriomeningitis virus strains to the cellular receptor alpha-dystroglycan correlate with viral tropism and disease kinetics. J. Virol. 75, 448–457.

Smith, T.J., Cheng, R.H., Olson, N.H., Peterson, P., Chase, E., et al., 1995. Putative receptor binding sites on alphaviruses as visualized by cryoelectron microscopy. Proc. Natl. Acad. Sci. U. S. A. 92, 10648–10652.

Smith, T.J., Chase, E.S., Schmidt, T.J., Olson, N.H., Baker, T.S., 1996. Neutralizing antibody to human rhinovirus 14 penetrates the receptor- binding canyon. Nature 383, 350–354.

Smith, D.R., Adams, A.P., Kenney, J.L., Wang, E., Weaver, S.C., 2008. Venezuelan equine encephalitis virus in the mosquito vector Aedes taeniorhynchus: infection initiated by a small number of susceptible epithelial cells and a population bottleneck. Virology 372, 176–186.

Spear, P.G., Eisenberg, R.J., Cohen, G.H., 2000. Three classes of cell surface receptors for alphaherpesvirus entry. Virology 275, 1–8.

Stewart, P.L., Nemerow, G.R., 1997. Recent structural solutions for antibody neutralization of viruses. Trends Microbiol. 5, 229–233.

Stoll, R., Hobom, G., Muller, H., 1994. Host restriction in the productive cycle of avian polyomavirus budgerigar fledgling disease virus type 3 depends on a single amino acid change in the common region of structural proteins VP2/VP3. J. Gen. Virol. 75 (Pt 9), 2261–2269.

Strauss, J.H., Wang, K.S., Schmaljohn, A.L., Kuhn, R.J., Strauss, E.G., 1994. Host-cell receptors for Sindbis virus. Arch. Virol. 9, 473–484.

Strauss, M., Filman, D.J., Belnap, D.M., Cheng, N., Noel, R.T., et al., 2015. Nectin-like interactions between poliovirus and its receptor trigger conformational changes associated with cell entry. J. Virol. 89, 4143–4157.

Sunyach, C., Rollier, C., Robaczewska, M., Borel, C., Barraud, L., et al., 1999. Residues critical for duck hepatitis B virus neutralization are involved in host cell interaction. J. Virol. 73, 2569–2575.

Switzer, W.M., Salemi, M., Shanmugam, V., Gao, F., Cong, M.E., et al., 2005. Ancient co-speciation of simian foamy viruses and primates. Nature 434, 376–380.

Taboga, O., Tami, C., Carrillo, E., Núñez, J.I., Rodríguez, A., et al., 1997. A large-scale evaluation of peptide vaccines against foot-and-mouth disease: lack of solid protection in cattle and isolation of escape mutants. J. Virol. 71, 2606–2614.

Takeuchi, Y., Akutsu, M., Murayama, K., Shimizu, N., Hoshino, H., 1991. Host range mutant of human immunodeficiency virus type 1: modification of cell tropism by a single point mutation at the neutralization epitope in the env gene. J. Virol. 65, 1710–1718.

Tami, C., Taboga, O., Berinstein, A., Nuñez, J.I., Palma, E.L., et al., 2003. Evidence of the coevolution of antigenicity and host cell tropism of foot-and-mouth disease virus in vivo. J. Virol. 77, 1219–1226.

Thaker, S.R., Stine, D.L., Zamb, T.J., Srikumaran, S., 1994. Identification of a putative cellular receptor for bovine herpesvirus 1. J. Gen. Virol. 75 (Pt 9), 2303–2309.

Thulke, S., Radonic, A., Nitsche, A., Siegert, W., 2006. Quantitative expression analysis of HHV-6 cell receptor CD46 on cells of human cord blood, peripheral blood and G-CSF mobilised leukapheresis cells. Virol. J. 3, 77.

Tong, S., Li, J., Wands, J.R., 1995. Interaction between duck hepatitis B virus and a 170-kilodalton cellular protein is mediated through a neutralizing epitope of the pre-S region and occurs during viral infection. J. Virol. 69, 7106–7112.

Tseng, Y.S., Gurda, B.L., Chipman, P., McKenna, R., Afione, S., et al., 2015. Adeno-associated virus serotype 1 (AAV1)- and AAV5-antibody complex structures reveal evolutionary commonalities in parvovirus antigenic reactivity. J. Virol. 89, 1794–1808.

Tucker, P.C., Griffin, D.E., 1991. Mechanism of altered Sindbis virus neurovirulence associated with a single-amino-acid change in the E2 Glycoprotein. J. Virol. 65, 1551–1557.

Ubol, S., Griffin, D.E., 1991. Identification of a putative alphavirus receptor on mouse neural cells. J. Virol. 65, 6913–6921.

Umetsu, Y., Sugawara, K., Nishimura, H., Hongo, S., Matsuzaki, M., et al., 1992. Selection of antigenically distinct variants of influenza C viruses by the host cell. Virology 189, 740–744.

Vahlenkamp, T.W., Verschoor, E.J., Schuurman, N.N., van Vliet, A.L., Horzinek, M.C., et al., 1997. A single amino acid substitution in the transmembrane envelope glycoprotein of feline immunodeficiency virus alters cellular tropism. J. Virol. 71, 7132–7135.

Van Valen, L., 1973. A new evolutionary law. Evol. Theory 1, 1–30.

Varthakavi, V., Minocha, H.C., 1996. Identification of a 56 kDa putative bovine herpesvirus 1 cellular receptor by anti-idiotype antibodies. J. Gen. Virol. 77 (Pt 8), 1875–1882.

Ventoso, I., 2012. Adaptive changes in alphavirus mRNA translation allowed colonization of vertebrate hosts. J. Virol. 86, 9484–9494.

Verdaguer, N., Mateu, M.G., Andreu, D., Giralt, E., Domingo, E., et al., 1995. Structure of the major antigenic loop of foot-and-mouth disease virus complexed with a neutralizing antibody: direct involvement of the Arg-Gly-Asp motif in the interaction. EMBO J. 14, 1690–1696.

Verdaguer, N., Ferrero, D., Murthy, M.R., 2014. Viruses and viral proteins. IUCrJ 1, 492–504.

Verschoor, E.J., Boven, L.A., Blaak, H., van Vliet, A.L., Horzinek, M.C., et al., 1995. A single mutation within the V3 envelope neutralization domain of feline immunodeficiency virus determines its tropism for CRFK cells. J. Virol. 69, 4752–4757.

Vignuzzi, M., Andino, R., 2010. Biological implications of picornavirus fidelity mutants. In: Ehrenfeld, E., Domingo, E., Roos, R.P. (Eds.), The Picornaviruses. ASM Press, Washington, DC, pp. 213–228.

Vignuzzi, M., Stone, J.K., Arnold, J.J., Cameron, C.E., Andino, R., 2006. Quasispecies diversity determines pathogenesis through cooperative interactions in a viral population. Nature 439, 344–348.

Villarreal, L.P., 2005. Viruses and the Evolution of Life. ASM Press, Washington, DC.

Villordo, S.M., Filomatori, C.V., Sanchez-Vargas, I., Blair, C.D., Gamarnik, A.V., 2015. Dengue virus RNA structure specialization facilitates host adaptation. PLoS Pathog. 11, e1004604.

Virgin, H.W., 2014. The virome in mammalian physiology and disease. Cell 157, 142–150.

Wang, K.S., Schmaljohn, A.L., Kuhn, R.J., Strauss, J.H., 1991. Antiidiotypic antibodies as probes for the Sindbis virus receptor. Virology 181, 694–702.

Weber, F., Kochs, G., Haller, O., 2004. Inverse interference: how viruses fight the interferon system. Viral Immunol. 17, 498–515.

Webster, B., Ott, M., Greene, W.C., 2013. Evasion of superinfection exclusion and elimination of primary viral RNA by an adapted strain of hepatitis C virus. J. Virol. 87, 13354–13369.

Wei, X., Ghosh, S.K., Taylor, M.E., Johnson, V.A., Emini, E.A., et al., 1995. Viral dynamics in human immunodeficiency virus type 1 infection. Nature 373, 117–122.

Weiner, A., Erickson, A.L., Kansopon, J., Crawford, K., Muchmore, E., et al., 1995. Persistent hepatitis C virus infection in a chimpanzee is associated with emergence of a cytotoxic T lymphocyte escape variant. Proc. Natl. Acad. Sci. U. S. A. 92, 2755–2759.

Whalley, S.A., Murray, J.M., Brown, D., Webster, G.J., Emery, V.C., et al., 2001. Kinetics of acute hepatitis B virus infection in humans. J. Exp. Med. 193, 847–854.

Whitbeck, J.C., Muggeridge, M.I., Rux, A.H., Hou, W., Krummenacher, C., et al., 1999. The major neutralizing antigenic site on herpes simplex virus glycoprotein D overlaps a receptor-binding domain. J. Virol. 73, 9879–9890.

Whitley, D., Watson, J.P., 2005. Complexity theory and the no free lunch theorem. In: Burke, E.K., Kendall, G. (Eds.), Search Methodologies. Introductory Tutorials in Optimization and Decision Support Techniques. Springer Science + Business Media, LLC, New York, pp. 317–339.

Williams, W.V., Guy, H.R., Rubin, D.H., Robey, F., Myers, J.N., et al., 1988. Sequences of the cell-attachment sites of reovirus type 3 and its anti-idiotypic/antireceptor antibody: modeling of their three-dimensional structures. Proc. Natl. Acad. Sci. U. S. A. 85, 6488–6492.

Williams, W.V., Weiner, D.B., Greene, M.I., 1989. Development and use of antireceptor antibodies to study interaction of mammalian reovirus type 3 with its cell surface receptor. Methods Enzymol. 178, 321–341.

Williams, W.V., Kieber-Emmons, T., Weiner, D.B., Rubin, D.H., Greene, M.I., 1991. Contact residues and predicted structure of the reovirus type 3-receptor interaction. J. Biol. Chem. 266, 9241–9250.

Wommack, K.E., Bhavsar, J., Polson, S.W., Chen, J., Dumas, M., et al., 2012. VIROME: a standard operating procedure for analysis of viral metagenome sequences. Stand. Genomic Sci. 6, 427–439.

Woolhouse, M.E.J., Webster, J.P., Domingo, E., Charlesworth, B., Levin, B.R., 2002. Biological and biomedical implications of the coevolution of pathogens and their hosts. Nat. Genet. 32, 569–577.

Wu, Z., Asokan, A., Grieger, J.C., Govindasamy, L., Agbandje-McKenna, M., et al., 2006. Single amino acid changes can influence titer, heparin binding, and tissue tropism in different adeno-associated virus serotypes. J. Virol. 80, 11393–11397.

Xu, G., Wilson, W., Mecham, J., Murphy, K., Zhou, E.M., et al., 1997. VP7: an attachment protein of bluetongue virus for cellular receptors in Culicoides variipennis. J. Gen. Virol. 78 (Pt 7), 1617–1623.

Xue, W., Minocha, H.C., 1993. Identification of the cell surface receptor for bovine viral diarrhoea virus by using anti-idiotypic antibodies. J. Gen. Virol. 74 (Pt 1), 73–79.

Yamada, A., Brown, L.E., Webster, R.G., 1984. Characterization of H2 influenza virus hemagglutinin with monoclonal antibodies: influence of receptor specificity. Virology 138, 276–286.

Yang, Z., Nielsen, R., 2008. Mutation-selection models of codon substitution and their use to estimate selective strengths on codon usage. Mol. Biol. Evol. 25, 568–579.

Yang, H., Carney, P.J., Chang, J.C., Guo, Z., Villanueva, J.M., et al., 2015. Structure and receptor binding preferences of recombinant human A(H3N2) virus hemagglutinins. Virology 477C, 18–31.

Yeates, T.O., Jacobson, D.H., Martin, A., Wychowski, C., Girard, M., et al., 1991. Three-dimensional structure of a mouse-adapted type 2/type 1 poliovirus chimera. EMBO J. 10, 2331–2341.

Ying, T., Du, L., Ju, T.W., Prabakaran, P., Lau, C.C., et al., 2014. Exceptionally potent neutralization of Middle East respiratory syndrome coronavirus by human monoclonal antibodies. J. Virol. 88, 7796–7805.

VIRAL FITNESS AS A MEASURE OF ADAPTATION

5

CHAPTER CONTENTS

ABBREVIATIONS

A	adenine
AIDS	acquired immune deficiency syndrome
BHK	baby hamster kidney
CHO	Chinese hamster ovary
CVB	coxsackievirus group B
FMDV	foot-and-mouth disease virus
G	guanine
HBV	hepatitis B virus
HCV	hepatitis C virus
HIV-1	human immunodeficiency virus type 1
IV	influenza virus
LCMV	lymphocytic choriomeningitis virus

Virus as Populations. http://dx.doi.org/10.1016/B978-0-12-800837-9.00005-8

MOI	multiplicity of infection
NGS	next generation sequencing
PBMC	peripheral blood mononuclear cell
PV	poliovirus
RTP	ribavarin-triphosphate
RT-PCR	reverse transcription followed by polymerase chain reaction
TCID50	tissue culture infectious dose needed to kill 50% of cells
tRNA	transfer RNA
U	uracil

5.1 ORIGIN OF THE FITNESS CONCEPT AND ITS RELEVANCE TO VIRUSES

Viral fitness, also termed selective or adaptive value, is an application to viruses of the concept of Darwinian fitness. From the early foundations of evolutionary theory, genetic variation, fitness, and evolutionary change were considered connected concepts. The relative reproductive efficiency that we recapitulate with the term "fitness," correlates with the rate of evolutionary change promoted by natural selection (Fisher, 1930). In practical terms, fitness refers to the relative contributions to the next generation made by the different individuals that compose a population. This definition is adequate for viruses because it can be adopted to mean the relative replication capacity of the members of a mutant spectrum, despite being modified by internal intrapopulation interactions (Chapter 3). Fitness quantifies the ability of a virus population (irrespective of its complexity) to produce infectious progeny, usually compared with a reference viral clone, in a defined environment (Domingo and Holland, 1997; Quinones-Mateu and Arts, 2006; Martinez-Picado and Martinez, 2008; Wargo and Kurath, 2012).

There is no general agreement on the precise meaning of fitness for biological entities in general, and how it should be measured. Some authors prefer to restrict the definition to the ability of an organism to transfer its genes to the next generation. Others favor to measure it as the number of progeny in the second or future generations. In the case of viruses, by emphasizing viral production in a specific environment, it is assumed that fitness measured as the progeny in an infection will be maintained in the next generations (next rounds of infection) provided the environment remains invariant. This assumption should, however, be taken with caution because of the continuous variation of mutant spectrum composition due to stochastic occurrence of mutations, and consequent uncertainties inherent to quasispecies evolution, even in a constant environment. Different points of view have been expressed regarding whether fitness should be considered an absolute or relative parameter, and if fitness is exactly equivalent to a selection coefficient (Maree et al., 2000; Wargo and Kurath, 2011). We will not dwell on these issues, we will use the term fitness rather than selection coefficient, and for convenience we consider that fitness is a relative value. This solution facilitates the comparison and interpretation of fitness values and differences, as determined by growth-competition experiments between two viruses, the standard approach to quantify fitness.

5.1.1 MEASUREMENT OF VIRAL FITNESS

Competition aimed at determining a relative fitness value is achieved by coinfecting a host (either a primary cell culture, an established cell line, an explanted tissue, an animal, or a plant), with a mixture containing known proportions of the virus to be tested and a reference variant or population of the same virus. For higher reliability, different initial proportions of the two competitors should be assayed. The progeny virus obtained in the first infection is then used to infect fresh cells (or the chosen host) under

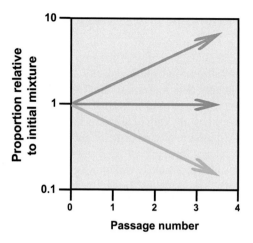

FIGURE 5.1

Representation of fitness vectors. An equal mixture of the mutant and population to be tested and the reference virus is subjected to four serial passages in a host. The amount of the two competing viruses is determined at each passage. The logarithm of the proportion of the two viruses relative to the initial mixture is plotted as a function of passage number. In the upper, orange vector, the mutant or population displays a fitness advantage over the reference virus. In the horizontal, blue vector, both viruses have equal fitness. In the downward, green vector, the mutant virus displays a selective disadvantage relative to the reference. See text for studies where numerical quantifications of fitness can be found.

the same conditions, and the proportion of the two viruses, relative to their proportion in the initial mixture, is determined. The logarithm of the ratio is plotted as a function of passage number. This yields a fitness vector which is adjusted to an exponential equation. The antilogarithm of the vector slope quantitates the relative fitness between the reference and competitor virus (Holland et al., 1991; Duarte et al., 1992; Clarke et al., 1994; Escarmís et al., 1999; Domingo et al., 2001; Domingo, 2006; Quinones-Mateu and Arts, 2006; Sheldon et al., 2014) (Figure 5.1). The two competing viruses can be distinguished by a phenotypic or a genotypic marker. Phenotypic markers can be any measurable trait that differs between the two viruses. They include resistance to a neutralizing monoclonal antibody or to an inhibitor so that the two viruses can be quantified by plating the virus at each passage in the presence and absence of the selective agent. Markers independent of a selective agent that can be used for cytopathic viruses in cell culture are plaque morphology or size, provided the difference of these traits between the competing viruses is clear and stable; relying on plaque size has potential pitfalls derived from its dependence on environmental factors, some of which may elude control. The typical genotypic marker is a nucleotide sequence difference between the genome of the reference virus and of the virus to be tested. The proportion of the two viruses can be estimated by consensus nucleotide sequence of the population at each competition passage, by integrating the amounts of the nucleotides that distinguish the two viruses using the sequencing peak output, or through new of generation sequence (NGS) methods. Alternatively, specific primers can be designed to discriminate the two viral genomes by selective amplification on the basis of the nucleotide difference, using real time RT-PCR for quantification (Yuste et al., 1999; Herrera et al., 2007; Wargo and Kurath, 2011; Sheldon et al., 2014). Fitness may be determined for an entire virus, or for a viral gene or regulatory region inserted into an infectious clone, with a preselected genetic background (Centlivre et al., 2005; Domingo, 2006).

5.1.2 **POWER AND LIMITATIONS OF FITNESS MEASUREMENTS**

Comparison of fitness values is an important tool in viral genetics because fitness captures the overall capacity of a virus to replicate, independently of the step in the virus life cycle and the molecular mechanisms that sustain an overall multiplication capacity. Replicative capacity of a virus *in vivo* is relevant medically because it largely determines the viral load in infected individuals, and viral load is a major influence in the progression of viral infections toward disease manifestations (Chapter 8). The fitness level may be influenced by any step in the virus life cycle, namely, entry into the host cells, uncoating of the viral genome, expression of viral proteins and genome replication (or transcription and replication), particle assembly, release form the cell, and stability of the free viral particle. Elucidation of which of the steps is responsible for a fitness variation is obviously of great interest regarding the understanding of virus-cell interactions. Identification of the critical step that determines a fitness level is a prerequisite to design biochemical experiments to define the molecular interactions that explain fitness differences. For some purposes, however, the fitness value by itself may be sufficiently informative to assist in understanding virus behavior.

A number of limitations must be taken into consideration in performing and interpreting fitness measurements by growth-competition experiments in cell culture or *in vivo*. A major limitation stems from the fact that a fitness value depends on the environment in which fitness is measured, and an entire host represents a heterogeneous, highly compartmentalized environment. Compartmentalization played an important role in the origin of complex life forms on earth (Chapter 1), and it has been maintained as an essential attribute of all differentiated organisms. For this reason, an array of environmental conditions is unavoidably presented to all viruses during an infection process. A virus may display high relative fitness in one cell type and low fitness in another cell type. A cell-dependent fitness difference was used to demonstrate that low fitness is an important determinant of viral extinction through the transition into error catastrophe: the same clonal population of foot-and-mouth disease virus (FMDV) could be easily extinguished in a cell line in which its fitness was low, but not in a cell line in which its fitness was high (Pariente et al., 2001) (Chapter 9).

The environment-dependent nature of fitness means that a ranking of fitness values established in cell culture may not reflect fitness differences in the natural hosts of the virus. This limitation of cell culture or *ex vivo* measurements to reflect fitness differences *in vivo* has been amply recognized, particularly in several studies that have addressed human immunodeficiency virus type 1 (HIV-1) fitness *in vivo* [reviewed in (Quinones-Mateu and Arts, 2006; Martinez-Picado and Martinez, 2008; Clementi and Lazzarin, 2010)]. For similar reasons, relative fitness of HIV-1 isolates may yield a different ranking when fitness is measured in peripheral blood mononuclear cells (PBMCs) or in established cell lines. Effort has been put in understanding HIV-1 fitness cost as a barrier to escape from antiretroviral agents, with the development of computational models proposed to assist in treatment planning (Gopalakrishnan et al., 2014) (fitness cost of drug resistance is discussed in Chapter 8).

The fundamental limitations of fitness values *in vivo* due to environmental heterogeneity are not the only concern regarding fitness determinations. Even in cell culture experiments several key variables may also affect fitness values. If cells are infected at low multiplicity of infection (MOI), initially each of the two viruses involved in the competition for fitness measurement will replicate in separate cells. However, as the infection proceeds, if the yield of the two viruses is sufficiently high and the infection is not interrupted, a second round of high multiplicity infection will occur in the same virus passage. The consequence of this infection design is that there will be two waves of replication: a first wave without intracellular competition between the reference and competitor viruses, and a second wave of intracellular competition between the two viruses in need of exploiting cell resources (ribosomes and tRNA for viral protein expression, structures where replication complexes and replication factories are located, cellular proteins needed for viral RNA replication, nucleotide substrates and other intracellular metabolites, etc.). Some of these resources may not be limiting or display a preference for any of the two competing viruses, but other resources may favor replication of some viral mutants (or genome subpopulations) over others. When this happens, an advantage of one viral population over another may be manifested with more intensity when both replicate in the same cell than when they replicate in separate cells.

Viruses can adapt to the MOI (Chapter 3) thus rendering essential the control of MOI values in comparative growth-competition assays. The situation can become even more complex in the case in which the yield of one virus is higher than the other as a result of the first infection wave, and if the extent of intracellular competition in the second infection wave affects the relative replication capacity of the two viruses. Furthermore, intracellular competition may involve complementation or interference between the mutant clouds originated by the two viruses (Novella, 2003; García-Arriaza et al., 2004; Wilke et al., 2004; Domingo, 2006; Perales et al., 2010) (see Chapter 3).

There are alternative assay designs to minimize replicative phases at different MOI. Viral vectors and subgenomic replicon constructs can be designed to limit replication to a single cycle, avoiding subsequent, second wave cell infections [as examples, see (Mansky, 1998; Scholle et al., 2004)]. For natural viral isolates or laboratory viruses or clones that should be studied unchanged, one possibility is to carry out parallel, independent infections with each virus separately. The virus titer at a defined time (or at multiple times) postinfection can be used as a surrogate fitness value, avoiding intracellular competition between the viruses under comparison. This procedure was used to compare the fitness of FMDV clones subjected to plaque transfers since comparative experiments showed that the individual viral yield led to fitness values comparable to those obtained by growth-competition experiments [compare procedures in (Escarmís et al., 1999, 2002)]. An alternative possibility is to restrict each of the serial infections with the competing viruses to a single round of high or low MOI infection, in the latter case by limiting the duration of infection to prevent the high MOI phase. This and other alternatives pose technical difficulties that have been previously discussed (Wilke et al., 2004; Domingo, 2006; Wargo and Kurath, 2012).

The above considerations render essential for the growth-competition assays to define as precisely as possible the experimental conditions (MOI, number of cells and their passage history, cell density in the case of cell suspensions or confluence status in the case of cell monolayers, the time postinfection at which the virus is collected at each passage, etc.), and to use the same conditions for comparative fitness measurements. Furthermore, since unavoidably mutation and recombination events occur stochastically in the course of the competitions, it is necessary to carry out triplicate fitness determinations in parallel, and advisable to limit the number of passages to four, under a controlled set of experimental conditions. In the case of fitness determinations *in vivo*, well-characterized host organisms should be used; when possible, inbred, sex-, and age-matched pathogen-screened animals kept under controlled conditions, etc. These precautions should minimize confounding effects resulting from random genetic changes and environmental differences in the course of the competitions (Box 5.1).

5.1.3 DISSECTION OF FITNESS DETERMINANTS

The usual aim of fitness determinations is to establish relative replication capacities of variant viruses, but fitness assays may also be designed to test differences in efficacy of specific steps in the virus life

---

BOX 5.1 MAIN POINTS ABOUT VIRAL FITNESS

- It measures the replication capacity of a virus relative to a reference isolate, mutant or population of the same virus.
- It is important because it captures an overall performance of the virus, and it can be used to plan virological and biochemical tests to identify the basis of fitness increase or decrease.
- A fitness value is relative to an environment. It is not a universal value for a viral mutant or population.
- Fitness differences *in vivo* are useful to compare the replicative or pathogenic potential of a virus. Fitness *in vivo* is a combined outcome of multiple influences due to the heterogeneous environment provided by differentiated organisms.

---

cycle (virus entry by determining the kinetics of internalization of two mutant viruses, synthesis of viral proteins, release of virus from cells, etc.). An example of fitness measurements on the basis of single-step competitions (for entry, replication, and release of virus form cells) was the elucidation that an increased thermal stability of the viral particle harboring RNA genomes with internal deletions contributed to the selective advantage of a segmented genome version of FMDV over its corresponding nonsegmented form (Ojosnegros et al., 2011). Segmentation took place when a biological clone of FMDV was subjected to more than 200 serial infections in baby hamster kidney (BHK-21) cells at high MOI (see section 2.11 in Chapter 2). The increased stability of the bipartite virus version was probably associated with a lower packaging density of the genomes with deletions inside the geometrically constrained capsid. The fitness decrease inflicted by thermal instability of the particles with full length genomes coincided with the fitness decrease determined by competition between the two viral forms (Ojosnegros et al., 2011). Experiments can be designed to identify which of the several steps of the viral infectious cycle is responsible of one virus displaying higher fitness than a related mutant.

For viral quasispecies, the term environment regarding its influence on fitness has to be taken in its broadest sense, with multiple participating factors. Physical parameters such as temperature, as well as biological factors contributed by the cells are part of the environment. If two competing viruses display different thermal stability, because they differ in amino acids at the capsid or surface proteins, temperature will affect fitness, and the effect will be critically dependent on the time that the assembled viral particles virus spend in the extracellular culture medium relative to the time in the intracellular environment, prior to exit from the cell. Time in the extracellular environment must be included as an additional step of the replication cycle that can affect a fitness value.

The influences on fitness extend to the virus itself since the mutant spectrum of a virus (or two competing viruses) must be considered part of the environment; its composition (subjected to indeterminacies derived from the random occurrence of mutation and recombination) may modulate the behavior of the entire quasispecies (Chapter 3). A viral population can modify its fitness without alteration of the consensus sequence (González-López et al.,2005; Grande-Pérez et al., 2005) (see also Chapter 9).

5.2 THE CHALLENGE OF FITNESS *IN VIVO*

The limitations of fitness determinations are accentuated in the assays performed *in vivo* because of the multiple steps and changing environments involved in the progress of a viral infection in an entire organism, be it an animal or a plant. A relevant factor is the effect of chance in the form of genetic drift that in the course of a fitness assay may modify the composition of the viral population relative to the composition it would have if selection were the only evolutionary force (section 3.2 in Chapter 3). Experiments with FMDV by F. Sobrino and colleagues constituted the first direct experimental measurements of fitness *in vivo*, and documented fitness variations among components of the mutant spectrum of a virus replicating in an animal. FMDV variants isolated during an infection of a pig displayed different capacities to compete with a reference clone in coinfections carried out in pigs (Carrillo et al., 1998). FMDV variants found at low frequency in the viral quasispecies *in vivo* were competed with the parental, standard isolate in pigs. Different proportions of the two competitors were found in different animal lesions (aphthae), thus documenting for the first time the fitness variations among components of FMDV quasispecies *in vivo*. The results provided evidence that each individual aphtha harbors virus with a different mutant spectrum composition, thus accentuating the stochastic component of FMDV transmission. Replacements of subsets of variants by others have been reported that suggest preferential colonization of specific organs [(Sanz-Ramos et al., 2008) among other examples]. Coxsackievirus with deletions in the 5′-terminal region of the genome can persist in the pancreas of mice where they displace the corresponding parental genomes (Tracy et al., 2015).

Fitness assays in cell culture may encourage extending the determinations *in vivo*. This is the case with hepatitis C virus (HCV). The availability of a cell culture system for HCV (Lindenbach et al., 2005; Wakita et al., 2005; Zhong et al., 2005) opened the way to experimental studies on HCV evolution in cell culture. For the first time it has been possible to obtain HCV from the same evolutionary lineages displaying fitness differences exceeding twofold (Perales et al., 2013; Sheldon et al., 2014). High HCV fitness was associated with significant resistance to multiple anti-HCV inhibitors, including inhibitors that target a cellular function (Sheldon et al., 2014). This observation and the likely possibility that fitness may also affect treatment response *in vivo*, renders extremely relevant to find means to measure or at least estimate fitness of HCV replicating in infected patients. Unless methods to grow natural HCV isolates *ex vivo* are found, it would be desirable to find *in vivo* parameters that can be used as surrogate fitness values, and attempts to reach this goal are now in progress, including computations based on NGS data (Seifert and Beerenwinkel, 2016). Model studies have suggested great relevance for HCV fitness regarding antiviral treatments, but the challenge to find fitness parameters that characterize HCV in patients remains.

Despite obvious difficulties in their completion and interpretation, fitness determinations have been (and continue being) of great value in the studies of viral population dynamics in cell culture, *ex vivo* and *in vivo*. Concepts relevant to medical research and to clinical practice such as the connection between fitness and viral load, fitness costs of drug resistance mutations, identification of compensatory mutations for fitness gain, and several other questions, are now amenable to quantification thanks to fitness assays (Quinones-Mateu and Arts, 2006; Martinez-Picado and Martinez, 2008; Iyidogan and Anderson, 2014) (Box 5.1) (Chapter 8).

5.3 FITNESS LANDSCAPES

Fitness landscapes (for historical origins and theoretical developments see (Schuster, 2016)) are a pictorial display in the form of mountains and valleys that represent high fitness values at the top of mountains and low fitness values down the valleys (Figure 5.2; a mountain landscape has been chosen as a symbol for viral fitness in the cover picture of this book). This description was used by S. Wright (Wright, 1931), and it has been extensively applied to represent transitions of viral populations from a high

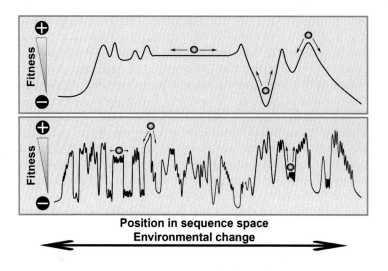

FIGURE 5.2

Putative schematic fitness landscapes for viruses. In both panels, high and low fitness values are represented as mountain peaks and valleys, respectively. Spheres and thin arrows indicate possible fitness variations (increases by upward movements, and decreases by downward movements) of a virus. Note the presence of a plateau at the center of the upper landscape where movements do not entail a fitness cost. The two parameters that here we consider that can mediate a fitness change are the position in sequence space of the virus (due to genetic modifications) or an environmental change (assuming a fixed position in sequence space). The two variables are indicated with a double head, thick arrow at the bottom of the scheme. Examples of the effect of these two parameters on fitness levels, and evidence that the second, highly rugged landscape is a more faithful representation of fitness variation in viruses are presented in the text. Note that despite ruggedness, the same fitness value is shared by many points in the abscissa.

fitness value (high replicative and survival capacity; location at top of a mountain) to a low fitness value (a point down in a valley). In the representations in two dimensions, fitness values are plotted in ordinate, but it is not always clear what is plotted in the abscissa. Here, we will consider that the variable in the abscissa is either the genomic nucleotide sequence or the environment (Figure 5.2). Genomic sequence may signify the consensus sequence of the population or the position of a mutant cloud in sequence space. According to this representation, a virus that acquires a mutation that leads to an amino acid that directly or indirectly affects the active site of the polymerase will most probably drive the virus from a fitness peak down toward a valley. In a specific case involving the FMDV polymerase, several substitutions in the enzyme, that were selected by passaging the virus in the presence of the mutagenic purine nucleoside analog ribavirin, produced a fitness decrease. In a related experiment, the mutant cloud that was produced by treating a clonal population of FMDV with the mutagenic pyrimidine base analog 5-fluorouracil displayed a sixfold reduction in fitness relative to a nonmutagenized control. In addition, the cloud exerted an interfering activity on the replication of standard FMDV RNA (González-López et al., 2004; González-López et al., 2005). The mutagenic treatment did not result in any variation of the consensus sequence of the population. Therefore, in this case the element responsible of the downhill movement in the fitness landscape was an altered cloud that kept its consensus sequence invariant.

The alternative plot is to represent an environmental change applied to a sequence or cloud at a defined position in sequence space. Again, continuing with FMDV as model, a clonal population displayed high fitness in BHK-21 cells, but low fitness in Chinese hamster ovary (CHO) cells (we refer to this difference in Section 5.1.2 in connection with fitness-dependent extinction). In this case, the downward movement in the landscape refers to the same sequence (a clonal population with limited cloud amplitude) transported to a different cell type (from BHK-21 to Chinese hamster ovary cells). The downhill movement may also be triggered by an intracellular environmental change within the same cellular type. If a given genomic sequence aptly replicating inside a cell is suddenly surrounded by an inhibitor of viral replication, its replicative fitness will decrease because of the presence of the inhibitor, and the virus will move from a fitness peak down to a valley. Conversely, if the virus acquires an inhibitor resistance mutation, its fitness in the presence of the inhibitor will increase, and the virus will move from a valley to a peak (Coffin, 1995; Quinones-Mateu and Arts, 2006) (Figure 5.2). Fitness is relative to an environment, and the presence or absence of a drug represents an obvious and important change in the environment. Thus, in analyzing evolutionary transitions, the experimentalist must be aware that both position in sequence space and environment can affect the fitness value of a virus. Both factors (and perhaps even additional influences) may actually act conjointly rather than independently. It is noteworthy that environment-driven fitness changes can occur without modifications of the consensus sequence. This may come about by at least two different mechanisms: (i) an effect of the mutant spectrum composition per se, not translated in dominance of alternative genomes (Lorenzo-Redondo et al., 2011) (Chapter 3), or (ii) transitions of a mutant spectrum without the selection of specific genomes. This is the case of the "melting" of information during the transition into error catastrophe promoted by mutagenic agents (Chapter 9). How high fitness peaks or how deep fitness valleys are during evolutionary transitions, or how the tie between cloud composition and environmental change is manifested, will have to wait for multiple fitness measurements and comparisons of many virus-host systems, a field which is only in its infancy.

5.3.1 JUSTIFICATION OF RUGGEDNESS IN FITNESS LANDSCAPES FOR VIRUSES

Very simple and unrealistic fitness landscapes are frequently used to depict fitness. A single peak landscape is often assumed for mathematical tractability in theoretical studies, implying a constant environment, and fitness variation considered a function of genomic nucleotide sequence. Of the two types of fitness landscapes for viruses depicted in Figure 5.2, the second rugged landscape appears to be the one that best describes evolving viruses. The main argument is that fitness variations are the norm as a result of displacements in sequence space or environmental modifications, despite acknowledging that interactions among mutations (rescuing effect of compensatory mutations also termed positive epistasis) may buffer or subvert the negative effects of individual mutations. In a fitness plateau such as the one depicted toward the middle of the first landscape in Figure 5.2, there is little or no fitness variation despite a movement in sequence space or an environmental modification. We favor the view that flat landscapes (or robust viral populations and clones; see Section 5.7) are not the norm for viruses, and that rugged landscapes are the ones that best represent reality while evolution is at play. A different matter is that well-settled products of some optimization process, such as those viral genomes that are seen as well adapted, might not have taken advantage of epistatic interactions to reach some point of fitness stability that facilitates their continued presence. We prefer focusing on the process before reaching stability points; what we start witnessing when applying next generation sequencing (NGS) surveys is that some evolutionary intermediates are established, but many others are not. Those that are not established now are, however, the raw material for potential alternative or future establishments. The beauty—but at the same time annoying complexity—is that what is fleeting a given time can be semipermanent another time, and vice versa. These considerations justify that we are not only concerned with genomes that work in comfortable regions of fitness landscapes, but also in those that had a fleeting existence.

The nascent evolutionary events are those in which fitness landscapes manifest their ruggedness. The evidence for preferring rugged fitness landscapes for viruses is severalfold, and it actually recapitulates several features of virus evolution portrayed in others chapters of this book. Briefly, the main experimental arguments are as follows: (i) many point mutations are deleterious when performance is measured under a defined environment. It means that just moving one position in sequence space often changes drastically the replicative and survival capacity of a virus. Moving several positions should be even more deleterious. Again, positive epistasis may compensate deleterious effects of individual mutations, but reaching them may imply considerable searching in sequence space; the focus is in the process. (ii) Interfering genomes that are responsible for the loss of viral infectivity can cease to be interfering just by one additional mutation. (iii) Variants from the same virus isolated from different organs of an infected host often display different fitness. (iv) Finally, as emphasized in previous sections of this chapter, fitness is extremely environment dependent. The reader will find many literature references in several chapters of the book that support the above statements. From current experimental evidence, rugged landscapes are the ones most suitable to depict virus evolution, although simple landscapes are needed for progress in theoretical models.

5.4 POPULATION FACTORS ON FITNESS VARIATIONS: COLLECTIVE FITNESS AND PERTURBATIONS BY ENVIRONMENTAL HETEROGENEITY

Fitness of a viral population is influenced by the average replicative capacity of the individuals that compose the population, modified by modulatory effects of mutant spectra (Chapter 3). Some features of population dynamics quite reproducibly may have an effect on fitness: serial infections involving passages with large numbers of cells and viruses tend to result in fitness gain while repeated bottleneck events lead to average fitness decreases (Figure 5.3) (Chao, 1990; Duarte et al., 1992; Novella et al., 1995; Escarmis et al., 1996; Escarmis et al., 1999; Yuste et al., 2000, 2005). The most straightforward interpretation of this result is that in the course of large population passages the Darwinian principles of genetic variation, competition, and selection (Chapter 3), will generously operate to progressively enrich the population in those subpopulations capable of increased infectious progeny production. In contrast, repeated bottleneck events acting on highly mutable entities results unavoidably in the accumulation of mutations reflected in a modification of the consensus sequence. Since, on average, mutations tend to be deleterious rather than beneficial, the result of mutation accumulation will be fitness decrease. The several experimental evolution studies that have led to this conclusion are described in Chapter 6. Here we deal with the main point regarding fitness: depending on the passage regime, a virus can move from high to low fitness values in a rugged fitness landscape (compare Figures 5.2 and 5.3). Evidence of these passage regimen-dependent fitness transitions *in vivo* is limited due to technical difficulties such as the control of viral population size *in vivo* and the likely and unavoidable existence of stages with large and low population sizes during the infection. The occurrence of bottlenecks *in vivo* is explained in Chapter 4.

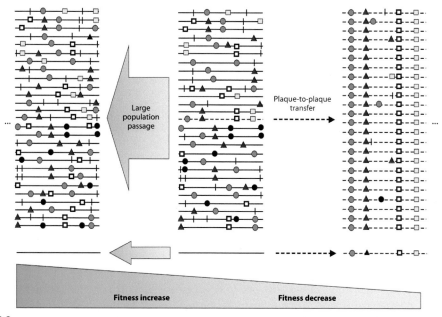

FIGURE 5.3

The effect of passage conditions on fitness variation. A virus subjected to large, unrestricted population passages (large arrow) tends to increase its fitness due to quasispecies optimization in the considered environment. In contrast, a viral population subjected to repeated bottleneck events (plaque-to-plaque transfers in the laboratory, discontinuous arrow) tends to decrease its fitness through accumulation of mutations, reflected in a modification of the consensus sequence (bottom horizontal lines). Experimental evidence in support of this scheme is given in the text and in Chapter 6.

    Fitness dynamics provides an interpretation of Haldane's dilemma (Haldane, 1937), when the concept is applied to viruses. Haldane calculated that if an individual has a lower than optimal genotype at *n loci*, and such deviation of optimality is reflected in a $(1-S_i)$ decrease in fitness, its resulting fitness will be $(1-S_1)(1-S_2) \ldots (1-Sn)$ times that of the optimal. Since many *loci* are polymorphic and selection acts on many *loci*, average fitness values for individuals must be very low, which seems to be contrary to observations with natural populations. A numerical example (Williams, 1992) illustrates the problem. If 1000 *loci* have an average fitness loss of 1%, the average individual will display a fitness of 10^{-5} relative to the fitness of the best possible genotype. This will result in a thoroughly inadequate gene pool if a good proportion of fitness components interact multiplicatively. For asexual populations subjected to quasispecies dynamics, mutant distributions are ranked according to fitness (Eigen and Schuster, 1979) (Chapter 3). Most individuals in the distributions have suboptimal fitness (below that of the master genome) due to similar arguments put forward by Haldane. Yet, many individuals which die (go extinct) leave their place to other, temporarily more fit individuals, that, despite being suboptimal, have an acceptable mutational load for survival, and give rise to new distributions in a continuous reiteration of the process. That is, new individuals will again generate progeny that will be suboptimal at many *loci*, reinitiating the competition-displacement cycle. Haldane's dilemma in general genetics is largely obsolete, but its consideration for virus evolution leads to the realization that the continuous replenishment of nascent viral subpopulations and the relative nature of fitness values, aided by accelerated evolutionary events, explains a continuous biological competence of highly variable viruses.

The major conclusions on the factors that promote increases or decreases of viral fitness (recapitulated in Figure 5.3) have been established with viral populations and clones in the laboratory or with natural isolates *ex vivo*. Many interesting phenomena and principles of general genetics have been experimentally tested with viruses, based on fitness variations following highly biased population regimes or competitions among viruses. Muller's ratchet, Red Queen hypothesis, or the competitive exclusion principle are examples. Since the experiments that probed such fundamental concepts in genetics fall into the area of experimental evolution, they are described in Chapter 6.

5.5 QUASISPECIES MEMORY AND FITNESS RECOVERY

Quasispecies memory means the presence in the mutant spectrum of a specific subpopulation of genomes by virtue of their having been dominant at an early evolutionary phase of the same virus lineage (Ruiz-Jarabo et al., 2000). The basis of this class of molecular memory is fitness gain during the process of viral replication in the presence of a selective agent. This is the reason to describe it here once the concepts of selection and fitness have been covered. Memory was first described using two markers of FMDV: resistance to neutralization by a monoclonal antibody—a phenotypic marker related to a set of specific mutations—and an elongated internal oligoadenylate tract in the FMDV genome, generated as a result of repeated plaque-to-plaque transfers (Chapter 6). The experiments intended to show the presence of quasispecies memory were inspired in immunological memory, and more generally in the behavior of the immune system regarded as a complex adaptive system (Frank, 1996). The course of the observations on quasispecies memory was as follows: monoclonal antibody-escape mutants were selected by multiplying FMDV in the presence of the antibody, and they became dominant in the population, as expected. When the mutants were passaged in the absence of antibody, true revertants arose, and became dominant in the population, which is expected if the escape mutants had endured some fitness cost. The key finding was that the frequency of escape mutants that remained in the population after revertants had taken over increased from a basal level of 4.10^{-3} to 5.10^{-4} to a memory level of 1.10^{-2} to 5.10^{-1}, and the difference was statistically significant (Ruiz-Jarabo et al., 2000) (Figure 5.4).

A similar observation was made with the genetic modification that consisted of an elongation of four adenylate residues located between the two functional AUG protein synthesis initiation codons in the FMDV genome. The number of adenylates increased by 2 and up to 23 residues in individual clones that had been subjected up to 30 serial plaque transfers, with considerable intraclone heterogeneity (Escarmis et al., 1996). This oligoadenylate elongation has never been seen in natural isolates of FMDV, and the likely reason why it repeatedly occurred in independent FMDV clones is that the viral polymerase has an inherent tendency to slippage or misalignment mutagenesis when copying homopolymeric template residues (Chapter 2). Usually, negative selection eliminates such altered genomes because they display alterations in protein expression as the molecular basis of fitness decrease (Escarmis et al., 1996). For reasons discussed in Chapter 6, negative selection is attenuated during plaque transfers of virus. Therefore, when clones with the elongated oligoadenylate tract were subjected to large population passages, they were overgrown by genomes with the four

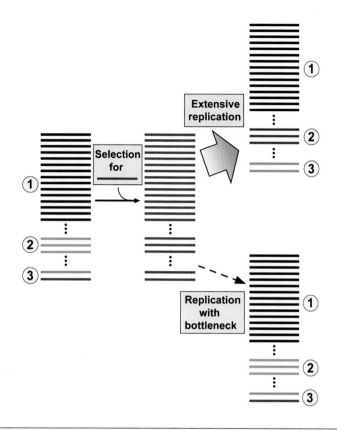

FIGURE 5.4

A schematic representation of quasispecies memory and the effect of a population bottleneck on memory. For clarity, the different scattered symbols on lines used to represent mutations in other figures of this book, have been omitted. Instead, frequency levels of genome subpopulations are depicted by horizontal lines with different colors. Level indicated by circled number 1 includes the most frequent genomes in the quasispecies (black horizontal lines in the distribution drawn on the left). A second and third frequency level is indicated with numbers 2 and 3, and green, blue and red genomes. When a low frequency, red genome is selected, it dominates the population (middle distribution). When replicated in the absence of the selective constraint (extensive replication, big arrow), revertants become dominant and occupy frequency level 1 (black lines in the upper right quasispecies). Because in the process of selective amplification the red genomes have attained higher fitness than the parental red genomes in the initial distribution, they occupy level 2 rather than level 1 (compare first distribution on the left with the upper distribution on the right). Level 2 is termed memory level. When replication is allowed, and then a bottleneck intervenes, the result is the dominance of revertants, together with exclusion of low frequency genomes, leading to memory loss. A distribution similar to the original one will be produced (discontinuous arrow and bottom right distribution). The red genome is now present again at its basal frequency which depends on the mutational pressure operating in the system. See text for experimental justification of this scheme.

standard adenylates, as the population recovered fitness (compare with the concept summarized in Figure 5.3). Here again, the frequency of genomes with the elongated oligonucleotide remained at memory levels when a clone with an elongated oligoadenylate had been subjected to 50 large population passages. Clones with a higher than standard number of adenylate residues were found in 11% to 17% of the molecular and biological clones analyzed. In contrast, none out of 40 molecular clones from the control population devoid of memory deviated from the standard sequence (Ruiz-Jarabo et al., 2000; Arias et al., 2001), and again the difference was statistically significant. (Figure 2.4(b) in Section 2.3 of Chapter 2 depicts FMDV genomes with the internal oligoadenylate elongation as a consequence of misalignment mutagenesis).

The interpretation of the presence of a molecular memory in FMDV is that in the process of being selected, those genomes harboring the memory markers increased their replicative fitness so that when they went back to the status of minority genome subpopulations, they reached a higher frequency dictated by their fitness level (remember that mutant spectrum components are ranked according to the fitness as explained in Chapter 3) (Figure 5.4). Once established, the memory genomes can gain replicative fitness in parallel with the dominant components of the quasispecies (Arias et al., 2004). Consistent with the proposed mechanism of memory implementation, it was shown that memory levels were dependent on the relative fitness of the parental populations that underwent selection of memory markers. A 7.6-fold higher fitness resulted in 30- to 100-fold higher memory level (Ruiz-Jarabo et al., 2002). The frequency of memory genomes decreased with the number of serial passages in the course of FMDV replication in a process akin to "diffusion" in sequence space (see Section 3.7 in Chapter 3). The decay was strikingly reproducible in parallel viral lineages (Ruiz-Jarabo et al., 2003). This reproducible behavior is one of several findings that denote the potential of viral quasispecies to display a deterministic behavior as discussed in Section 3.6.2 of Chapter 3.

The demonstration of quasispecies memory was extended to HIV-1 *in vivo* by C. Briones and colleagues who distinguished two classes of memory. One is the replicative memory that we have just examined in the case of FMDV. The second class of memory is termed reservoir, cellular, or anatomical memory, which is due to proviral DNA integration in the target cells. The viral DNA replicates as part of the cellular chromosome, which provides a reservoir of ancestral sequences. The reservoir memory was identified by a phylogenetic analysis of consensus sequences obtained at different times after the initiation of retroviral therapy in HIV-1-infected patients, while replicative memory was evidenced by an extensive clonal analysis of HIV-1 samples from patients at different times of treatment, using multidrug resistance markers [detailed in (Briones et al., 2003) and reviewed in (Briones and Domingo, 2008)]. The two classes of memory were supported both by experimental findings and a theoretical model that considered both replicative memory and integration of ancient sequences as proviral DNA in the cellular genome (Briones et al., 2003; Briones and Bastolla, 2005). These early studies on memory have permitted an interpretation of the reemergence of ancestral sequences during virus evolution, and several observations are consistent with virus behavior being influenced by memory genomes (Borrow et al., 1997; Wyatt et al., 1998; Karlsson et al., 1999; Briones et al., 2000; Lau et al., 2000; García-Lerma et al., 2001; Imamichi et al., 2001; Kijak et al., 2002; Charpentier et al., 2004). Box 5.2. summarizes the main features of quasispecies memory and its implications. After being established with viruses, the memory concept was extended to *Drosophila melanogaster*, thus reinforcing its applicability to various types of biological systems (Teotonio et al., 2009).

> **BOX 5.2 MAIN FEATURES OF QUASISPECIES MEMORY**
> - The presence of memory genomes in a viral lineage is an expected consequence of quasispecies dynamics, and of the fitness gain of a viral population while subjected to a selective amplification.
> - Memory levels are dependent on the relative fitness of the viral genomes destined to become memory genomes.
> - Upon replication, memory genomes gain fitness in parallel with the entire population that harbors them.
> - Memory genomes can modify the response of viral quasispecies to selective constraints directed to traits embodied in the memory genomes.
> - Population bottlenecks erase memory because bottlenecks exclude low frequency genomes and memory genomes are present at low frequency.

5.5.1 IMPLICATIONS OF QUASISPECIES MEMORY: HARBINGER MUTATIONS

Fitness-dependent memory is a collective property of the entire quasispecies and not of its individual components, as shown by memory loss when the population was subjected to a bottleneck event (Ruiz-Jarabo et al., 2000; Briones and Domingo, 2008) (Figure 5.4 and Box 5.2). Memory has a number of theoretical and practical implications that extend to medical practice. On the theoretical side, a way of expressing the collective behavior of a viral quasispecies is that its response to a selective constraint may depend on whether the population has gone through a bottleneck or not. Memory can facilitate the response of a collectivity of genomes to selective constraints that occur intermittently in the life cycle of an evolutionary lineage. Examination of the types of selective constraints that often operate in nature suggests that many of them are not exceptional or unique, but rather, they are repeated occasionally or even regularly (physiological responses dependent on circadian rhythms, cycles of temperature variations, exposures to the same species of predators, etc.). In the case of viruses, if a response to recurrent constraints involves selection of genome subpopulations, once the selective force wanes, the storage of the responsive subpopulation at memory levels guarantees preparedness to confront the next identical or related constraint. The preparedness effect is, however, abolished whenever a bottleneck event excludes the memory genomes from the evolving population (Figure 5.4 and Box 5.2). Since bottlenecks are increasingly recognized as abundant in the life cycles of viruses, it is uncertain how frequently memory of the type described here can indeed have an influence in virus behavior. It may have its major effect during prolonged chronic infections involving large viral loads, known also to be an important determinant of treatment response and disease progression (Chapter 8).

A practical application of quasispecies memory may be found in treatment planning and treatment modifications for chronically infected patients, for example, HIV-1-, HBV-, or HCV-infected patients. When an antiviral treatment is interrupted because the virus in the patient is dominated by mutants resistant to the inhibitor (or inhibitors) in use, inhibitor-sensitive subpopulations will replenish the mutant spectrum if the resistance mutations inflicted a fitness cost. Since the resistant mutants increased their fitness in the course of selection, they will not return to basal levels but they will occupy a memory level. When time elapses, if no alternative treatments are available for the patient, and the previous inhibitor is administered again, the virus may find a way to resist the new round of treatment because of the presence of the relevant resistant genomes at memory levels. Many, but not all, studies have indicated that the presence in mutant spectra of low frequency resistance mutations may jeopardize treatment efficacy. In cases in which minority mutations may confer partial resistance to inhibitors, it is important to avoid a treatment that includes again the same drugs used in the previous treatment period (Briones and Domingo, 2008). This point is discussed again in Chapter 8 in relation to the relevance of quasispecies dynamics in disease control.

Some mutations that are maintained at memory levels can be a predictor of later dominance in the same evolutionary lineage. Such mutations are termed "harbinger" mutations. They were reported during a clonal analysis of FMDV during a process of fitness gain (Arias et al., 2001). The presence of harbinger mutations is likely to become more evident as successive samples of an evolving viral population are examined through next generation sequencing (NGS) methodology, since virus evolution consists of successive replacements of some viral genome subpopulations by others Thus, quasispecies dynamics can have as one of its consequences the presence of a molecular memory of past selection events, and memory levels are decisively influenced by fitness values.

5.6 THE RELATIONSHIP BETWEEN FITNESS AND VIRULENCE

Virulence of a virus (or of any parasite) is the ability to induce pathology in a host. Viruses generally encode one or several virulence determinants either in regulatory regions or in viral proteins. Pathological manifestations can be due to the direct action of viral gene products on cells of specific tissues or organs, or indirectly through the host response to the invading virus that can trigger immunopathological processes. Increasing evidence suggests that during their epidemic spread most viruses may undergo genetic modifications that can either enhance or attenuate their virulence. For example, the Ebola virus that caused the African outbreak of 2014-2015, termed strain Guinea-2014 (with cases of contagion outside Africa) was derived from strain Zaire-1976 (Gire et al., 2014). While mortality of the Zaire-1976 strain was around 90%, the estimated mortality of the Guinea-2014 strain was about 50% (although caution should be exercised concerning precision of these estimates in areas with poor health infrastructure and presence of confounding factors). The decrease of mortality probably conferred the virus an epidemiological fitness advantage (see Section 5.9) by favoring virus spread among the susceptible human population. While the previous Ebola outbreaks in the twentieth century caused infection in hundreds of individuals, the Guinea-2014 outbreaks infected tens of thousands individuals, representing a higher risk of disease expansion and reemergence in distant areas. For virtually any virus there have been studies that document that genetic change, sometimes a minimal change, can modify virus virulence.

Virulence, as fitness, is relative to a biological environment. The response to an infection is influenced by the host immune response, and multitudes of largely unknown influences. About 99% of poliovirus (PV) infections are asymptomatic, and this was also true at the peak of the poliomyelitis epidemics in the middle of the twentieth century (Koike and Nomoto, 2010). Yet, in a small proportion of infected individuals invasion of the central nervous system and neurological disease is produced, likely influenced by viral and host determinants. Since the time that infectious clones became available to express altered viruses, it was shown that one or a few mutations engineered in the clones could alter virulence and that excessive attenuation rendered putative vaccines ineffective. This is one of the reasons why very few engineered vaccines have replaced the classical inactivated and attenuated antiviral vaccines. Fitness must be adjusted to attain the required degree of attenuation of a live virus.

Regarding the influence of the immune system, a debilitated immune response may favor viral infections for obvious reasons: the arms race between the virus and the host is tilted in favor of the virus. Furthermore, immunocompromised individuals (because of inherited or acquired immuno-deficiency) may shed large amounts of PV, and they are termed "supershedders" or "superspread-ers"; they may contribute to the spread of PV, and may constitute a factor of disease maintenance or reemergence (Martin et al., 2000; Martin, 2006). Similar observations had previously been made with an immunodeficient child persistently infected by influenza virus (Rocha et al., 1991) (Section 7.7.1 in Chapter 7). In these cases there is an environment-dependent increase of viral fitness.

Attenuation of a virus or any pathogen to produce a live-attenuated vaccine, using the classical multipassage in an alternate host, can be viewed as a process of accumulation of mutations that alters relative fitness values. Indeed, such extensive passage renders the virus fit to replicate in the alterna-tive environment. When this adaptation results in the adequate degree of fitness loss in the authentic host, the virus may no longer manifest disease potential. The derivation of an attenuated virus by serial passage of the pathogenic counterpart in an alternative host (cell culture or animal) is an example of a virus increasing its fitness in a nonnatural environment concomitantly with decrease in fitness and virulence for the authentic host. The serial passages of rabies virus in rabbits, a process termed "lapinization," carried out by L. Pasteur are an example of preparation of a vaccine for human use. Tissue culture-adapted PV can be highly attenuated for humans and monkeys, rendering a live vaccine virus which is highly fit in cell culture (at least more fit than the initial virulent isolate), but unable to replicate in the host cell subsets associated with causing pathology. This does not mean that the tissue culture-adapted PV is not capable of replicating in *any* host cells. In fact, it must replicate at least to the extent of evoking a protective immune response if the virus has a value as a live vaccine. Thus, fitness is not only host dependent, but organ-, tissue-, or cell population dependent. The decrease in disease potential will often be associated with a decrease of viral fitness in those cells associated with disease manifestations, or those cells needed to reach the ones involved in disease. Studies are needed to quantify organ-dependent variation in viral fitness.

Some models of evolution of virulence propose that viruses that would naturally cause harm and death in their main natural hosts would be selected against because depletion of the host population would prevent long-term survival of the virus. Several lines of evidence support the occurrence of virulent viral strains mainly as a result of selection of variant viruses from essentially avirulent popula-tions. Cardiovirulent coxsackievirus B (CVB) may arise commonly in infected neonates with an im-mature immune system. In this specific environment some CVB mutant may gain access to the heart where they may adapt to replicate efficiently, causing myocarditis (Tracy et al., 2006). This view has a conceptual parallelism with the mechanisms that lead to the emergence of a new human viral disease caused by a virus from some natural animal reservoir in which the virus is harmless or nearly so. Out of many variants generated in replicating viral quasispecies and that can potentially become established in a new host species, only the chance contact with the new host in which a new quasispecies can be selected, can lead to a viral disease emergence (compare the proposed mechanism for CVB cardioviru-lence with suggested mechanisms of viral disease emergence in Chapter 7). The scenario of an efficient replication of a viral subpopulation in a specific organ, tissue, or cell subset as being associated with disease is supported in the case of CVB by the finding of 10^8 TCID50 (tissue culture infectious dose needed to kill 50% of cells) per gram of pediatric heart tissue obtained at death (Chapman et al., 2008).

The observations just summarized raise the issue of the relationship between viral virulence and fitness, as defined in Section 5.1, that is, the relative capacity of a virus to produce progeny in a given environment. Models of virulence have been surrounded by controversy perhaps because of the difficulties of capturing the complex set of evolutionary and biochemical ingredients that result in what we term virulence. In particular, the connection between viral fitness and virulence has been a much-debated matter, involving experimentalists and theoreticians alike. Some models assume a direct relationship between fitness and capacity to produce disease (Lenski and May, 1994; Poulin and Combes, 1999; Brown et al., 2006). Experimental studies with viruses and other types of pathogens have provided evidence either in favor or against such direct relationship between fitness and virulence. Among other examples, the diabetogenic phenotype of CVB correlates with replication efficiency (Kanno et al., 2006). In a comparative study of two divergent strains of the whispovirus white spot syndrome virus, virulence—measured by cumulative mortality rates of its shrimp host—correlated with replicative fitness *in vivo* (Marks et al., 2005). A direct correlation between the level of RNA replication and cytopathology was also observed in engineered alphavirus replicons (Frolov et al., 1999). In other cases, a direct connection between fitness and virulence has been assumed, but not proven experimentally. A greater replicative fitness of historical versus current HIV-1 isolates was taken as evidence of attenuation of HIV-1 as the AIDS pandemic advanced (Arien et al., 2005).

Other studies favor the view that fitness and virulence do not necessarily correlate. Despite similar replication of R5-tropic and X4-tropic clones of HIV-1 in mitogen-activated T cells, X4 tropic clones were transferred more efficiently than R5-tropic clones to CD4(+) T cells. The increased transfer could contribute to the competitive advantage of X4 viruses in AIDS patients (Arien et al., 2006). In a related finding in a different biological setting, nucleopolyhedrovirus that was transmitted early to its host, the moth *Lymantriadispar*, was more virulent than virus transmitted late, despite the latter being more productive because it could use more host tissue for replication (Cooper et al., 2002). Lack of correlation between fitness and virulence was also observed with some plant viruses (Escriu et al., 2003) and in some fungal infections (Nielsen et al., 2005).

An observation with FMDV has unveiled one of the possible molecular mechanisms of a lack of correlation between fitness and virulence. A biological clone of FMDV acquired high fitness and virulence, measured in this case as the cell-killing capacity of populations that had been subjected to large population passages in BHK-21 cells. This was the expected result of competitive optimization of the replicating quasispecies. Subsequent plaque-to-plaque transfers of the clone (serial bottleneck events; Figure 5.5 and Chapter 6) resulted in profound fitness loss, but only a minimal decrease of virulence (Herrera et al., 2007). Comparison of the behavior of constructed chimeric viruses mapped the fitness-decreasing mutations throughout the genome, while virulence determinants were concentrated on some genomic regions. Thus, virulence was more robust than fitness to the effects of mutations that accumulate upon serial plaque transfers (Escarmis et al., 1996; Escarmis et al., 2002). It must be emphasized that in this study FMDV fitness and virulence was measured in the same biological environment provided by BHK-21 cells. Thus, depending on the passage regime viral fitness and virulence can follow different evolutionary trajectories (Herrera et al., 2007) (Figure 5.5).

A distribution of fitness-enhancing mutation throughout the viral genome has been also observed in HCV passaged in human hepatoma cells in culture (Perales et al., 2013). Incidentally, the observation that fitness determinants are scattered throughout a viral genome is expected from the

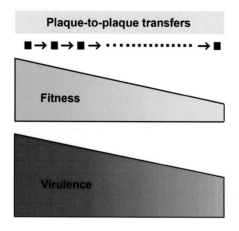

FIGURE 5.5

Evolution of fitness and virulence of a FMDV clone subjected to multiple plaque-to-plaque transfers. Fitness was measured by growth-competition assays as described in Section 5.1.1, and virulence as the capacity of the virus to kill the host BHK-21 cells. Fitness determinants were scattered throughout the viral genome, and decreased considerably as a result of the transfers. In contrast, virulence determinants were concentrated on more specific genomic regions, and decreased to a lower extent than fitness as a result of plaque transfers. The drawing is a summary of a study by (Herrera et al., 2007). See text for experimental justification and additional references on the relationship between fitness and virulence.

fact underlined in Chapters 2 and 4 that most mutations in viral genomes are not neutral. A minimal change can deoptimize a virus, as reflected in a fitness decrease in the considered environment. The results of the link between fitness and virulence carried out with FMDV imply also that in theoretical models of virulence it is not justified to equate virulence with replicative capacity, again an example of an experimental finding that should be entered into theoretical studies to render them more realistic. It is likely that the connection between fitness and virulence will continue being a subject of debate.

5.7 FITNESS LANDSCAPES FOR SURVIVAL: THE ADVANTAGE OF THE FLATTEST

The first fitness landscape depicted in Figure 5.2 includes a flat plateau with a virus whose movement does not entail any fitness variation. If we take position in sequence space as the variable in the horizontal axis, what this part of the figure means is that a virus can accept mutations without its fitness being affected. The term robustness has been used to denote the invariance of a phenotypic trait despite the occurrence of mutations or, in more general terms, the preservation of the phenotype in the face of perturbations (de Visser et al., 2003; Elena, 2012). *In silico* results with digital organisms documented a selective advantage of those organism that live in a flat fitness surface, an event that in terms of fitness landscapes is known as "the advantage of the flattest" (Wilke et al., 2001).

Despite evidence summarized in Section 5.3.1 that a rugged version of fitness landscapes is more realistic for viruses (bottom landscape in Figure 5.2) one can envision areas of sequence space in which a virus is more resistant than average to the effect of mutations, or that a virus type has evolved to be more resistant to mutations. Several possible mechanisms of virus robustness have been proposed (Elena et al., 2006). Possibilities include evolution toward a genomic base composition or a codon usage that does not change the frequency of misincorporations, but limits the proportion of amino acid substitutions prone to functional alterations, among many other possibilities one could imagine. The main problem is that there is a deficit of experimental descriptions of robustness with a precise supportive molecular mechanism. Coxsackievirus B3 (CVB3) exhibited increased vulnerability than PV to lethal mutagenesis by ribavirin, that could not be attributed to increased incorporation of ribavirin-triphosphate (RTP) (Graci et al., 2012). CVB3 manifested increased sensitivity to the random introduction of a pyrimidine analog to the viral RNA, with the major effect being at the level of the encoded proteins rather than at the level of RNA structure, with lower tolerance of nonsynonymous mutations. How differences in robustness might relate to the life style of viruses and their ability to respond to constraints is largely unknown.

High fidelity mutants due to a faithful polymerase cannot be considered a robustness mechanism because what such viruses do is to incorporate a lower number of mutations, not that phenotypic traits are unaltered despite acquisition of mutations. There are several experimental observations that are consistent with some virus mutants being more robust than others regarding mutation tolerance. Early evidence that a standard VSV virus may accept more mutations than one of its neutral derivatives will be presented in Chapter 6 in a discussion of the multiple contributions made by J. Holland and his colleagues on the translation to virology of concepts of population genetics, using VSV as the model system. Viral robustness has also been suggested as a possible mechanism to decrease the efficacy of lethal mutagenesis treatments (Sanjuan et al., 2007), with a seeming agreement with the results of Graci and colleagues on reduced mutational robustness of CVB3 (Graci et al., 2012). However, a distinction is in place here. One issue is that a virus naturally occupies a position of higher robustness than another virus (in the previous example, PV versus CVB3), and a different issue is that within the time frame of a transition toward extinction by a virus-specific mutagen a virus can stably move into a position of sequence space at which robustness is significantly enhanced. This second possibility is, in my view, extremely unlikely, and it was not sustained by experiments on lethal mutagenesis of lymphocytic choriomeningitis virus (LCMV) (Martin et al., 2008). Furthermore, a theoretical study by C. Wilke and colleagues (O'Dea et al., 2009) did not support the concept that enhanced mutagenesis can displace viruses toward regions of increased robustness [reviewed by (Elena, 2012)]. Both experiments and a theoretical model suggest that the prediction of a robustness-mediated failure of lethal mutagenesis was probably stretching too much the "robustness of robustness" so to speak, and this is discussed in Chapter 9 in connection with lethal mutagenesis as an antiviral approach.

Although difficult to prove, there might be instances in nature that a transient high fitness peak can be reached by a virus that has as a cost a high sensitivity to additional mutations, exactly the opposite to advantage of the flattest that we could call "disadvantage of the sharp." It is tempting to speculate that an example of such an occurrence was a virulent influenza virus (IV) whose hemagglutinin gene acquired a fragment of ribosomal RNA by nonhomologous recombination (Khatchikian et al., 1989). No such mosaic influenza viruses have been perpetuated in nature suggesting that such constructs, when they occur, may drive the virus toward extinction. They may become less tolerant of additional genetic modifications.

5.8 FITNESS AND FUNCTION

Molecular virology has developed very important tools to dissect viral functions by engineering infectious molecular clones with preselected modifications. The derivations of reverse genetics first developed by C. Weissmann with the bacteriophage Qβ system (see Chapter 3) have permeated virology to the point that, with the exception of a few viruses, reverse genetics has been applied to most of them, be they RNA or DNA viruses. Here fitness comes as a relevant parameter concerning the difference between a virus having the attribute of "viable" as such versus "viable" for evolutionary success. At least part of the constructs that are currently investigated with viruses (numerous and readily accessible to any literature search) yield infectious viruses that will have only a minimal chance to survive in an evolutionarily competitive scenario. What an infectious construct tells us is that basic functional performance of a virus is operating, and this is extremely informative of the basic molecular requirements for a virus to express minimally its replicative program. We should be aware, however, that an infectious construct may be too debilitated to have prospects of its being established as a competitive biological element in nature or even in cell culture or in *ex vivo* experiments. In support of this assertion is the fact that replication of many such constructs leads to variant form that display fitness increases and that have accumulated multiple mutations in their genomes. Such fitness optimization through accumulation of mutations is observed with entire viruses and subgenomic replicons. The advantage conferred by a mutant spectrum over its individual components may be one of the reasons of poor performance of specific constructs. Therefore, what is functional biochemically or in a single round of infection may be nonfunctional from the point of view of evolutionary prospects.

5.9 EPIDEMIOLOGICAL FITNESS

Epidemiological fitness is a semiquantitative concept based on observations such as the prevalence of some viral genomic sequences over others in some geographical areas, based on diagnostic surveys, and the spread of a particular variant virus into neighbor territories. A high epidemiological fitness may be due to a combination of parameters such as the average rate of virus replication in different infected host individuals, the immune status of the host population, virus transmissibility, virus particle stability in the environment, etc. that altogether confer a high reproductive ratio (R_0, number of infected contacts per infected host; Chapter 7). For example, a FMDV of serotype O termed O PanAsia was first isolated in India in 1990, and in the following years the virus displaced other FMDV strains as it spread to other continents (Knowles and Samuel, 2003). At that time, FMDV O PanAsia displayed high epidemiologic fitness, for reasons that are not understood. Several sublineages of a successor virus termed O PanAsia-II expanded in subsequent decades. Also, in 2009 a new H1N1 IV that had originated in swine spread into the human population. It partially displaced other H1N1 and H3N1 viruses that were circulating at the time, although in subsequent years the new virus coexisted with other, previously dominant influenza virus genotypes. In this case, the high epidemiologic fitness displayed by the new H1N1 virus when it irrupted into the human population was likely associated with having acquired a new genomic composition and antigenic determinants that conferred the virus high replicative capacity in a human population that had a partially protective immune response addressed to other antigenic forms of IV (see Chapter 7 on antigenic shift and drift as evolutionary events). In summary, the fitness of a virus (or any pathogen) *in vivo* is determined basically by the complete genome of the pathogen and its expression products in interaction with the physical and biological environments provided by the host cell or organism.

5.10 OVERVIEW AND CONCLUDING REMARKS

Fitness is an important parameter to describe the survival potential of a virus (be it a clone, population, mutant, or natural isolate) when confronted with related competitors in a given environment. Fitness, as a measure of relative replication capacity, can influence the progress of viral infections *in vivo* and, in consequence, the disease manifestations in the infected organisms. It is not surprising that, despite several technical difficulties, efforts are being put in its determination. In this area, application of NGS should provide new prospects for fitness quantification and, equally important, the fine molecular details of dominance of some variant subsets in competition with many others.

The argument has been made that fitness landscapes, a classic representation of fitness values and variations in the form of mountains and valleys, should be very rugged for viruses as they replicate in their natural environments. One of the reasons is the frequent deleterious effects of individual mutations, evidenced by many of the examples included in different chapters of this book. There is a general debate in biology on the relationship between fitness and virulence, and it is unlikely that this debate is solved solely by studies with viruses. There are several elements that complicate attempts at clarification, the least trivial being that the very exact meaning of both fitness and virulence is itself object of debate.

One of the consequences of the fitness gain that accompanies a selection event is the presence of memory genomes in viral quasispecies. Many of the implications of memory as part of quasispecies dynamics, and also for the planning at antiviral treatments, have not been fully appreciated.

Even more complex than the standard fitness concept for a virus are attempts to define in quantitative terms fitness in the field, the so-called epidemiological fitness. Again, multiple factors participate in conferring a virus an epidemiological advantage, and several such factors are probably unrelated to the fitness parameters we determine in the laboratory. This is a key point in dealing with viruses as populations that will again be treated in Chapter 7 in relation to the emergence of viral disease agents (see Summary Box).

SUMMARY BOX

- Fitness is a measure of the relative efficacy of two viruses to produce progeny in a given environment.
- Fitness landscapes are a pictorial representation of variation of fitness values as a function of the genomic sequence or the environment. They serve to visualize fitness gains, fitness losses, the advantage of the flattest, and the disadvantage of the sharpest.
- Molecular evidence suggests that fitness landscapes for viruses are extremely rugged.
- Fitness of a viral population may vary without any modification of the consensus sequence.
- A consequence of fitness increase during selection of viral subpopulations is the establishment of memory genomes in the population. Memory levels are occupied by those genomes that were dominant at an earlier evolutionary phase of the same lineage. The presence of a molecular memory in viral quasispecies has several theoretical and medical implications.
- Epidemiological fitness is a semiquantitative parameter that reflects the capacity of a virus to become dominant in the field.

REFERENCES

Arias, A., Lázaro, E., Escarmís, C., Domingo, E., 2001. Molecular intermediates of fitness gain of an RNA virus: characterization of a mutant spectrum by biological and molecular cloning. J. Gen. Virol. 82, 1049–1060.

Arias, A., Ruiz-Jarabo, C.M., Escarmis, C., Domingo, E., 2004. Fitness increase of memory genomes in a viral quasispecies. J. Mol. Biol. 339, 405–412.

Arien, K.K., Troyer, R.M., Gali, Y., Colebunders, R.L., Arts, E.J., et al., 2005. Replicative fitness of historical and recent HIV-1 isolates suggests HIV-1 attenuation over time. AIDS 19, 1555–1564.

Arien, K.K., Gali, Y., El-Abdellati, A., Heyndrickx, L., Janssens, W., et al., 2006. Replicative fitness of CCR5-using and CXCR4-using human immunodeficiency virus type 1 biological clones. Virology 347, 65–74.

Borrow, P., Lewicki, H., Wei, X., Horwitz, M.S., Peffer, N., et al., 1997. Antiviral pressure exerted by HIV-1-specific cytotoxic T lymphocytes (CTLs) during primary infection demonstrated by rapid selection of CTL escape virus. Nat. Med. 3, 205–211.

Briones, C., Bastolla, U., 2005. Protein evolution in viral quasispecies under selective pressure: a thermodynamic and phylogenetic analysis. Gene 347, 237–246.

Briones, C., Domingo, E., 2008. Minority report: hidden memory genomes in HIV-1 quasispecies and possible clinical implications. AIDS Rev. 10, 93–109.

Briones, C., Mas, A., Gomez-Mariano, G., Altisent, C., Menendez-Arias, L., et al., 2000. Dynamics of dominance of a dipeptide insertion in reverse transcriptase of HIV-1 from patients subjected to prolonged therapy. Virus Res. 66, 13–26.

Briones, C., Domingo, E., Molina-París, C., 2003. Memory in retroviral quasispecies: experimental evidence and theoretical model for human immunodeficiency virus. J. Mol. Biol. 331, 213–229.

Brown, N.F., Wickham, M.E., Coombes, B.K., Finlay, B.B., 2006. Crossing the line: selection and evolution of virulence traits. PLoS Pathog. 2, e42.

Carrillo, C., Borca, M., Moore, D.M., Morgan, D.O., Sobrino, F., 1998. *In vivo* analysis of the stability and fitness of variants recovered from foot-and-mouth disease virus quasispecies. J. Gen. Virol. 79, 1699–1706.

Centlivre, M., Sommer, P., Michel, M., Ho Tsong Fang, R., Gofflo, S., et al., 2005. HIV-1 clade promoters strongly influence spatial and temporal dynamics of viral replication in vivo. J. Clin. Invest. 115, 348–358.

Chao, L., 1990. Fitness of RNA virus decreased by Muller's ratchet. Nature 348, 454–455.

Chapman, N.M., Kim, K.S., Drescher, K.M., Oka, K., Tracy, S., 2008. 5' terminal deletions in the genome of a coxsackievirus B2 strain occurred naturally in human heart. Virology 375, 480–491.

Charpentier, C., Dwyer, D.E., Mammano, F., Lecossier, D., Clavel, F., et al., 2004. Role of minority populations of human immunodeficiency virus type 1 in the evolution of viral resistance to protease inhibitors. J. Virol. 78, 4234–4247.

Clarke, D.K., Duarte, E.A., Elena, S.F., Moya, A., Domingo, E., et al., 1994. The red queen reigns in the kingdom of RNA viruses. Proc. Natl. Acad. Sci. U. S. A. 91, 4821–4824.

Clementi, M., Lazzarin, A., 2010. Human immunodeficiency virus type 1 fitness and tropism: concept, quantification, and clinical relevance. Clin. Microbiol. Infect. 16, 1532–1538.

Coffin, J.M., 1995. HIV population dynamics in vivo: implications for genetic variation, pathogenesis, and therapy. Science 267, 483–489.

Cooper, V.S., Reiskind, M.H., Miller, J.A., Shelton, K.A., Walther, B.A., et al., 2002. Timing of transmission and the evolution of virulence of an insect virus. Proc. Biol. Sci. 269, 1161–1165.

de Visser, J.A., Hermisson, J., Wagner, G.P., Ancel Meyers, L., Bagheri-Chaichian, H., et al., 2003. Perspective: evolution and detection of genetic robustness. Evolution 57, 1959–1972.

Domingo, E., 2006. Quasispecies: concepts and implications for virology. Curr. Top. Microbiol. Immunol. 299, 83–140.

Domingo, E., Holland, J.J., 1997. RNA virus mutations and fitness for survival. Annu. Rev. Microbiol. 51, 151–178.

Domingo, E., Biebricher, C., Eigen, M., Holland, J.J., 2001. Quasispecies and RNA Virus Evolution: Principles and Consequences. Landes Bioscience, Austin.

Duarte, E., Clarke, D., Moya, A., Domingo, E., Holland, J., 1992. Rapid fitness losses in mammalian RNA virus clones due to Muller's ratchet. Proc. Natl. Acad. Sci. U. S. A. 89, 6015–6019.

Eigen, M., Schuster, P., 1979. The Hypercycle. A principle of natural self-organization, Springer, Berlin.

Elena, S.F., 2012. RNA virus genetic robustness: possible causes and some consequences. Curr. Opin. Virol. 2, 525–530.

Elena, S.F., Carrasco, P., Daros, J.A., Sanjuan, R., 2006. Mechanisms of genetic robustness in RNA viruses. EMBO Rep. 7, 168–173.

Escarmis, C., Davila, M., Charpentier, N., Bracho, A., Moya, A., et al., 1996. Genetic lesions associated with Muller's ratchet in an RNA virus. J. Mol. Biol. 264, 255–267.

Escarmís, C., Dávila, M., Domingo, E., 1999. Multiple molecular pathways for fitness recovery of an RNA virus debilitated by operation of Muller's ratchet. J. Mol. Biol. 285, 495–505.

Escarmís, C., Gómez-Mariano, G., Dávila, M., Lázaro, E., Domingo, E., 2002. Resistance to extinction of low fitness virus subjected to plaque-to- plaque transfers: diversification by mutation clustering. J. Mol. Biol. 315, 647–661.

Escriu, F., Fraile, A., Garcia-Arenal, F., 2003. The evolution of virulence in a plant virus. Evolution 57, 755–765.

Fisher, R.A., 1930. The Genetical Theory of Natural Selection. Oxford University Press, Oxford.

Frank, S.A., 1996. The design of natural and artificial adaptive systems. In: Rose, M.R., Lauder, G.V. (Eds.), Adaptation. Academic Press, San Diego, pp. 451–505.

Frolov, I., Agapov, E., Hoffman Jr., T.A., Pragai, B.M., Lippa, M., et al., 1999. Selection of RNA replicons capable of persistent noncytopathic replication in mammalian cells. J. Virol. 73, 3854–3865.

García-Arriaza, J., Manrubia, S.C., Toja, M., Domingo, E., Escarmís, C., 2004. Evolutionary transition toward defective RNAs that are infectious by complementation. J. Virol. 78, 11678–11685.

García-Lerma, J.G., Nidtha, S., Blumoff, K., Weinstock, H., Heneine, W., 2001. Increased ability for selection of zidovudine resistance in a distinct class of wild-type HIV-1 from drug-naive persons. Proc. Natl. Acad. Sci. U. S. A. 98, 13907–13912.

Gire, S.K., Goba, A., Andersen, K.G., Sealfon, R.S., Park, D.J., et al., 2014. Genomic surveillance elucidates Ebola virus origin and transmission during the 2014 outbreak. Science 345, 1369–1372.

González-López, C., Arias, A., Pariente, N., Gómez-Mariano, G., Domingo, E., 2004. Preextinction viral RNA can interfere with infectivity. J. Virol. 78, 3319–3324.

González-López, C., Gómez-Mariano, G., Escarmís, C., Domingo, E., 2005. Invariant aphthovirus consensus nucleotide sequence in the transition to error catastrophe. Infect. Genet. Evol. 5, 366–374.

Gopalakrishnan, S., Montazeri, H., Menz, S., Beerenwinkel, N., Huisinga, W., 2014. Estimating HIV-1 fitness characteristics from cross-sectional genotype data. PLoS Comput. Biol. 10, e1003886.

Graci, J.D., Gnadig, N.F., Galarraga, J.E., Castro, C., Vignuzzi, M., et al., 2012. Mutational robustness of an RNA virus influences sensitivity to lethal mutagenesis. J. Virol. 86, 2869–2873.

Grande-Pérez, A., Gomez-Mariano, G., Lowenstein, P.R., Domingo, E., 2005. Mutagenesis-induced large fitness variations with an invariant arenavirus consensus genomic nucleotide sequence. J. Virol. 79, 10451–10459.

Haldane, J.B.S., 1937. The effect of variation on fitness. Am. Nat. 71, 337–349.

Herrera, M., Garcia-Arriaza, J., Pariente, N., Escarmis, C., Domingo, E., 2007. Molecular basis for a lack of correlation between viral fitness and cell killing capacity. PLoS Pathog. 3, e53.

Holland, J.J., de la Torre, J.C., Clarke, D.K., Duarte, E., 1991. Quantitation of relative fitness and great adaptability of clonal populations of RNA viruses. J. Virol. 65, 2960–2967.

Imamichi, H., Crandall, K.A., Natarajan, V., Jiang, M.K., Dewar, R.L., et al., 2001. Human immunodeficiency virus type 1 quasi species that rebound after discontinuation of highly active antiretroviral therapy are similar to the viral quasi species present before initiation of therapy. J. Infect. Dis. 183, 36–50.

Iyidogan, P., Anderson, K.S., 2014. Current perspectives on HIV-1 antiretroviral drug resistance. Viruses 6, 4095–4139.

Kanno, T., Kim, K., Kono, K., Drescher, K.M., Chapman, N.M., et al., 2006. Group B coxsackievirus diabetogenic phenotype correlates with replication efficiency. J. Virol. 80, 5637–5643.

Karlsson, A.C., Gaines, H., Sallberg, M., Lindback, S., Sonnerborg, A., 1999. Reappearance of founder virus sequence in human immunodeficiency virus type 1-infected patients. J. Virol. 73, 6191–6196.

Khatchikian, D., Orlich, M., Rott, R., 1989. Increased viral pathogenicity after insertion of a 28S ribosomal RNA sequence into the haemagglutinin gene of an influenza virus. Nature 340, 156–157.

Kijak, G.H., Simon, V., Balfe, P., Vanderhoeven, J., Pampuro, S.E., et al., 2002. Origin of human immunodeficiency virus type 1 quasispecies emerging after antiretroviral treatment interruption in patients with therapeutic failure. J. Virol. 76, 7000–7009.

Knowles, N.J., Samuel, A.R., 2003. Molecular epidemiology of foot-and-mouth disease virus. Virus Res. 91, 65–80.

Koike, S., Nomoto, A., 2010. Poliomyelitis. In: Ehrenfeld, E., Domingo, E., Roos, R.P. (Eds.), The Picornaviruses. ASM Press, Washington DC, pp. 339–352.

Lau, D.T., Khokhar, M.F., Doo, E., Ghany, M.G., Herion, D., et al., 2000. Long-term therapy of chronic hepatitis B with lamivudine. Hepatology 32, 828–834.

Lenski, R.E., May, R.M., 1994. The evolution of virulence in parasites and pathogens: reconciliation between two competing hypotheses. J. Theor. Biol. 169, 253–265.

Lindenbach, B.D., Evans, M.J., Syder, A.J., Wolk, B., Tellinghuisen, T.L., et al., 2005. Complete replication of hepatitis C virus in cell culture. Science 309, 623–626.

Lorenzo-Redondo, R., Borderia, A.V., Lopez-Galindez, C., 2011. Dynamics of in vitro fitness recovery of HIV-1. J. Virol. 85, 1861–1870.

Mansky, L.M., 1998. Retrovirus mutation rates and their role in genetic variation. J. Gen. Virol. 79 (Pt 6), 1337–1345.

Maree, A.F., Keulen, W., Boucher, C.A., De Boer, R.J., 2000. Estimating relative fitness in viral competition experiments. J. Virol. 74, 11067–11072.

Marks, H., van Duijse, J.J., Zuidema, D., van Hulten, M.C., Vlak, J.M., 2005. Fitness and virulence of an ancestral white spot syndrome virus isolate from shrimp. Virus Res. 110, 9–20.

Martin, J., 2006. Vaccine-derived poliovirus from long term excretors and the end game of polio eradication. Biologicals 34, 117–122.

Martin, J., Dunn, G., Hull, R., Patel, V., Minor, P.D., 2000. Evolution of the Sabin strain of type 3 poliovirus in an immunodeficient patient during the entire 637-day period of virus excretion. J. Virol. 74, 3001–3010.

Martin, V., Grande-Perez, A., Domingo, E., 2008. No evidence of selection for mutational robustness during lethal mutagenesis of lymphocytic choriomeningitis virus. Virology 378, 185–192.

Martinez-Picado, J., Martinez, M.A., 2008. HIV-1 reverse transcriptase inhibitor resistance mutations and fitness: a view from the clinic and ex vivo. Virus Res. 134, 104–123.

Nielsen, C., Keena, M., Hajek, A.E., 2005. Virulence and fitness of the fungal pathogen *Entomophaga maimaiga* in its host *Lymantria dispar*, for pathogen and host strains originating from Asia, Europe, and North America. J. Invertebr. Pathol. 89, 232–242.

Novella, I.S., 2003. Contributions of vesicular stomatitis virus to the understanding of RNA virus evolution. Curr. Opin. Microbiol. 6, 399–405.

Novella, I.S., Duarte, E.A., Elena, S.F., Moya, A., Domingo, E., et al., 1995. Exponential increases of RNA virus fitness during large population transmissions. Proc. Natl. Acad. Sci. U. S. A. 92, 5841–5844.

O'Dea, E.B., Keller, T.E., Wilke, C.O., 2009. Does mutational robustness inhibit extinction by lethal mutagenesis in viral populations? PLoS Comput. Biol. 6, e1000811.

Ojosnegros, S., Garcia-Arriaza, J., Escarmis, C., Manrubia, S.C., Perales, C., et al., 2011. Viral genome segmentation can result from a trade-off between genetic content and particle stability. PLoS Genet. 7, e1001344.

Pariente, N., Sierra, S., Lowenstein, P.R., Domingo, E., 2001. Efficient virus extinction by combinations of a mutagen and antiviral inhibitors. J. Virol. 75, 9723–9730.

Perales, C., Lorenzo-Redondo, R., López-Galíndez, C., Martínez, M.A., Domingo, E., 2010. Mutant spectra in virus behavior. Futur. Virol. 5, 679–698.

Perales, C., Beach, N.M., Gallego, I., Soria, M.E., Quer, J., et al., 2013. Response of hepatitis C virus to long-term passage in the presence of alpha interferon: multiple mutations and a common phenotype. J. Virol. 87, 7593–7607.

Poulin, R., Combes, C., 1999. The concept of virulence: interpretations and implications. Parasitol. Today 15, 474–475.

Quinones-Mateu, M.E., Arts, E.J., 2006. Virus fitness: concept, quantification, and application to HIV population dynamics. Curr. Top. Microbiol. Immunol. 299, 83–140.

Rocha, E., Cox, N.J., Black, R.A., Harmon, M.W., Harrison, C.J., et al., 1991. Antigenic and genetic variation in influenza A (H1N1) virus isolates recovered from a persistently infected immunodeficient child. J. Virol. 65, 2340–2350.

Ruiz-Jarabo, C.M., Arias, A., Baranowski, E., Escarmís, C., Domingo, E., 2000. Memory in viral quasispecies. J. Virol. 74, 3543–3547.

Ruiz-Jarabo, C.M., Arias, A., Molina-París, C., Briones, C., Baranowski, E., et al., 2002. Duration and fitness dependence of quasispecies memory. J. Mol. Biol. 315, 285–296.

Ruiz-Jarabo, C.M., Miller, E., Gomez-Mariano, G., Domingo, E., 2003. Synchronous loss of quasispecies memory in parallel viral lineages: a deterministic feature of viral quasispecies. J. Mol. Biol. 333, 553–563.

Sanjuan, R., Cuevas, J.M., Furio, V., Holmes, E.C., Moya, A., 2007. Selection for robustness in mutagenized RNA viruses. PLoS Genet. 3, e93.

Sanz-Ramos, M., Diaz-San Segundo, F., Escarmis, C., Domingo, E., Sevilla, N., 2008. Hidden virulence determinants in a viral quasispecies in vivo. J. Virol. 82, 10465–10476.

Scholle, F., Girard, Y.A., Zhao, Q., Higgs, S., Mason, P.W., 2004. Trans-packaged West Nile virus-like particles: infectious properties in vitro and in infected mosquito vectors. J. Virol. 78, 11605–11614.

Schuster, P., 2016. Quasispecies on fitness landscapes. Current Topics in Microbiology and Immunology, in press.

Seifert, D., Beerenwinkel, N., 2016. Estimating fitness of viral quasispecies from next generation sequencing data. Curr. Top. Microbiol. Immunol.

Sheldon, J., Beach, N., Moreno, E., Gallego, I., Piñeiro, D., et al., 2014. Increased replicative fitness can lead to decreased drug sensitivity of hepatitis C virus. J. Virol. 88, 12098–12111.

Teotonio, H., Chelo, I.M., Bradic, M., Rose, M.R., Long, A.D., 2009. Experimental evolution reveals natural selection on standing genetic variation. Nat. Genet. 41, 251–257.

Tracy, S., Chapman, N.M., Drescher, K.M., Kono, K., Tapprich, W., 2006. Evolution of virulence in picornaviruses. Curr. Top. Microbiol. Immunol. 299, 193–209.

Tracy, S., Smithee, S., Alhazmi, A., Chapman, N., 2015. Coxsackievirus can persist in murine pancreas by deletion of 5' terminal genomic sequences. J. Med. Virol. 87, 240–247.

Wakita, T., Pietschmann, T., Kato, T., Date, T., Miyamoto, M., et al., 2005. Production of infectious hepatitis C virus in tissue culture from a cloned viral genome. Nat. Med. 11, 791–796.

Wargo, A.R., Kurath, G., 2011. In vivo fitness associated with high virulence in a vertebrate virus is a complex trait regulated by host entry, replication, and shedding. J. Virol. 85, 3959–3967.

Wargo, A.R., Kurath, G., 2012. Viral fitness: definitions, measurement, and current insights. Curr. Opin. Virol. 2, 538–545.

Wilke, C.O., Wang, J.L., Ofria, C., Lenski, R.E., Adami, C., 2001. Evolution of digital organisms at high mutation rates leads to survival of the flattest. Nature 412, 331–333.

Wilke, C.O., Reissig, D.D., Novella, I.S., 2004. Replication at periodically changing multiplicity of infection promotes stable coexistence of competing viral populations. Evolution 58, 900–905.

Williams, G.C., 1992. Natural Selection. Domains, Levels and Challenges. Oxford University Press, Oxford, New York.

Wright, S., 1931. Evolution in Mendelian populations. Genetics 16, 97–159.

Wyatt, C.A., Andrus, L., Brotman, B., Huang, F., Lee, D.H., et al., 1998. Immunity in chimpanzees chronically infected with hepatitis C virus: role of minor quasispecies in reinfection. J. Virol. 72, 1725–1730.

Yuste, E., Sanchez-Palomino, S., Casado, C., Domingo, E., Lopez-Galindez, C., 1999. Drastic fitness loss in human immunodeficiency virus type 1 upon serial bottleneck events. J. Virol. 73, 2745–2751.

Yuste, E., López-Galíndez, C., Domingo, E., 2000. Unusual distribution of mutations associated with serial bottleneck passages of human immunodeficiency virus type 1. J. Virol. 74, 9546–9552.

Yuste, E., Bordería, A.V., Domingo, E., López-Galíndez, C., 2005. Few mutations in the 5' leader region mediate fitness recovery of debilitated human immunodeficiency type 1 viruses. J. Virol. 79, 5421–5427.

Zhong, J., Gastaminza, P., Cheng, G., Kapadia, S., Kato, T., et al., 2005. Robust hepatitis C virus infection in vitro. Proc. Natl. Acad. Sci. U. S. A. 102, 9294–9299.

CHAPTER

VIRUS POPULATION DYNAMICS EXAMINED WITH EXPERIMENTAL MODEL SYSTEMS

6

CHAPTER CONTENTS

ABBREVIATIONS

BHK	baby hamster kidney cells
DI	defective interfering
FMDV	foot-and-mouth disease virus
HIV-1	human immunodeficiency virus type 1
IRES	internal ribosome entry site
IV	influenza virus

miRNA micro-RNA
MAb   monoclonal antibody
MOI   multiplicity of infection
NCCR   noncoding control region
NGS   next generation sequencing
PBMCs   peripheral blood mononuclear cells
PV   poliovirus
VSV   vesicular stomatitis virus

6.1 VALUE OF EXPERIMENTAL EVOLUTION

There are good reasons to suspect that we are not aware of many of the influences that guide the evolution of viruses in nature, and that only a few of the influences can be properly quantified. A rapid review of the major concepts expressed in previous chapters (the chance origin of mutations, multiple selective forces and random events that alter population compositions in an almost incessant manner, potential fidelity modifications of viral polymerases, rapid diversification of viruses even within an individual host, fitness variations due to changes in genomic sequences or in the environment, etc.) forces us to realize that the interpretation of how and why viral populations evolve in nature must be based on largely indirect evidence and a considerable number of assumptions. An understanding of the basic principles that preside over the generation of viral diversity and how some viral forms replace others should be based not only on indirect evidence, but also on the design of experiments in which a limited number of variables can be examined. Ideally, we need to investigate the operation of one variable at a time in an evolutionary outcome, and then try to extend the simplified approach to integrate and interpret the effect of multiple variables acting conjointly.

Some of the concepts explained in previous chapters can indeed be translated into one defined variable. If we ignore intrinsic population heterogeneity, we can compare the effect of viral population size by examining the outcome of an infection of a cell culture or an animal with 0, 1, 10, 10^2, or 10^3 infectious units of a virus. We can fix the amount of virus to 10^2 and examine the influence of prior treatment with 0, 1, 10, 10^2, or 10^3 doses of any compound that might modify the virus, the cell, or the virus-cell interaction with the aim of investigating the virus life cycle or finding a new antiviral target. Parallel replicas may help controlling the possible effects of differences in virus population composition. We may also consider physico-chemical variables such as temperature or ionic composition of the cell culture medium. A diagram of some experimental possibilities with viruses that can grow in a cell culture is depicted in Figure 6.1. For some virus-cell systems, persistent infections are readily established because the virus displays limited cytopathology. In other cases, persistence is only established from cells that survive killing by the virus (Section 6.4). Remarkable insights have been obtained during the last four decades by subjecting viral populations to defined constraints in the laboratory, inspired by the types of constraints likely to operate *in vivo*. This is one of the major objectives of experimental evolution in virology. The studies included in this chapter serve the dual purpose of identifying the experimental origin of some of conclusions drawn in other chapters, and of providing additional complementary information on basic concepts of virus evolution.

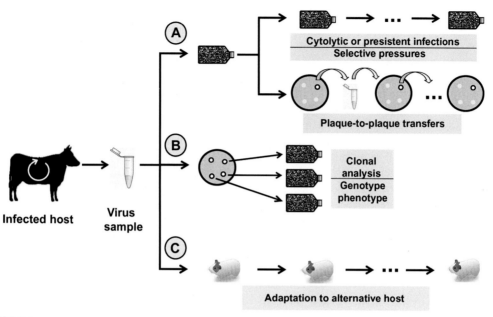

FIGURE 6.1

Scheme of possible laboratory experiments with a virus that can grow in cell culture. A virus sample from an infected host can be used to infect a cell culture (upper branch A) and the progeny can be passaged serially in cytolytic or persistent infections under different selective pressures. The virus can be diluted and plated to characterize biological clones from the population (middle branch B). The virus may also be adapted to an alternative host by serially passaging the progeny virus produced in the new host (branch C, at the bottom). These and additional possible designs form the basis of experimental evolution, and several examples are discussed in the text, with literature references.

6.2 EXPERIMENTAL SYSTEMS IN CELL CULTURE AND *IN VIVO*

For viruses that grow in cell culture, the experimental designs may involve serial passages of cytolytic viruses or of persistently infected cells under different passage conditions (i.e., virus or cell population numbers) or environmental alterations (absence or presence of drugs or antibodies). Passage of a cytolytic virus involves the infection of fresh (uninfected) cells at each passage, implying that no cell evolution can take place. In contrast, passage of persistently infected cells involves successive rounds of cell multiplication with the virus replicating in the cells. The number of duplications that the cells undergo depends on the number of cells seeded on the plate and their number when they reach confluence or are taken for the next passage. In this type of system both the cells and the resident virus may evolve (Section 6.4). When bottlenecks are introduced (meaning a reduction of the number of cells per passage), they will affect both viruses and cells, with consequences for the evolutionary outcome.

During cytolytic infections in which only the virus can evolve, population size in experimental designs has two components: the multiplicity of infection (MOI, or number of infectious particles added per cell) or the total amount of virus used in each infection. In the case of plaque-to-plaque transfers, first performed by L. Chao using bacteriophage φ6 (Chao, 1990), the population size is reduced to one in each transfer, representing an extreme passage regime that has been highly informative of the profound molecular alterations associated with fitness decrease underwent by the components of mutant spectra (Section 6.5). The viral population size has yet another relevant influence that can be explained with a numerical example. In standard virus passages (not plaque-to-plaque transfers), infection of 10^3 cells with 10^4 infectious particles yields the same MOI than infection of 10^6 cells with 10^7 particles (in both cases the MOI is 10 infectious particles per cell). Yet, the two infections may lead to different outcomes because the number of genomes that initiate infection is 1000-fold larger in the second case. Specifically, any relevant mutant present at a frequency of 10^{-4} in the viral populations will have a high probability of exclusion from participation in the next infection round if the latter involves 10^3 infectious particles. In contrast, the mutant will be included in infections started with 10^7 particles. [The same concept applies to the loss of memory genomes in viral quasispecies due to intervening bottlenecks that exclude subsets of minority and memory genomes for the next infection rounds (Section 5.5 in Chapter 5).]

The presence and maintenance of a specific mutant in a viral population depends on the virus population size and, therefore, it is an "extrinsic" property of a viral population, as opposed to the "intrinsic" properties that are independent of the population size (Domingo and Perales, 2012) ("intrinsic" vs. "extrinsic" properties of mutant spectra were discussed in Chapter 3 in connection with biological complexity as applied to viruses). A natural isolate of a virus may also be adapted to some alternative hosts *in vivo* depending on the barriers to be confronted by the virus in the potential new host (Figure 6.1 and Chapter 4). The basic designs depicted in Figure 6.1 can be extended and modified including complex scenarios such as environmental heterogeneity (alternations or migrations among different cell lines, or between cell lines and host organism, etc.), including the alternancy between mammalian and insect cells in the study of arbovirus evolution (Coffey and Vignuzzi, 2011; Coffey et al., 2008; Novella et al., 1999a, 2007, among other investigations). In a study with vesicular stomatitis virus (VSV) populations passaged in different cell lines, virus adaptation was cell specific in the absence of cell flow, and fitness in all environments decreased with migration rate (Cuevas et al., 2003). The scope of possibilities of designed experiments to learn about virus evolution is truly remarkable. However, we have to be aware that viruses change continuously in the course of experiments.

6.2.1 "TO CULTURE IS TO DISTURB"

The transfer of a viral isolate into an alternative host (either a cell culture, an explant, or an intact organism) implies a perturbation regarding the representation of the parental quasispecies in the new host for two reasons: (i) the involvement of a bottleneck event whose intensity depends on the amount of virus in the biological sample relative to the total amount in the infected parental host, and (ii) the change of biological environment from the donor into the recipient host (Figure 6.2). Concerning (i), a virus sample contains only a subset of the genomes present in the infected host, and the relative fitness of subsets may not represent the fitness of the entire population, if it were feasible to introduce the entire population into the recipient host. Both chance and selection events will modify the genome composition that will enter subsequent rounds of multiplication (Figure 6.2). This important point was first noted by A. Meyerhans, S. Wain-Hobson, and their colleagues in a comparison of the *tat* gene from sequential human immunodeficiency virus type 1 (HIV-1) isolates, and from peripheral blood mononuclear cells (PBMCs) infected with the natural isolates. The study unveiled the difficulties of defining HIV-1 infections in molecular terms, and the authors coined the following sentence that became popular in virus evolution: "to culture is to disturb" (Meyerhans et al., 1989). It gives a very pertinent image of what is hidden behind quasispecies dynamics.

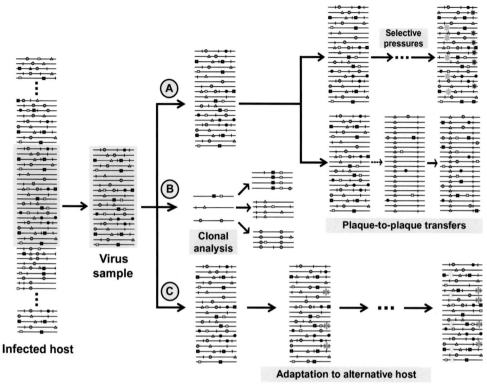

FIGURE 6.2

The implication of mutant spectra and population size in experimental designs. This scheme is parallel to that shown in Figure 6.1, except that infected objects have been replaced by mutant spectra in those same objects. From left to right: the infected host has a huge mutant spectrum whose depiction would occupy millions of columns as the one drawn on the left. The virus sample taken from the infected host includes only a subset of genomes (even if they amount to hundred-thousands), here shadowed in gray. Upper branch A: When this genomic subset is passaged in cells under a selective pressure, the mutant spectrum will be modified, and genomes with specific mutations (green and red asterisks) will increase in dominance. If the same subset is subjected to plaque-to-plaque transfers, mutations accumulate beginning with a founder genome that will hitchhike two initial mutations (red triangle and horizontal line). In the middle branch B the three biological clones will have different initial mutations, and heterogeneity will increase upon expansion of the population. In the bottom branch C, adaptation to a new host will imply increasing dominance of new mutations (yellow and blue asterisks). Many examples of genetic variation due to the types of effects shown here in a diagrammatic form are discussed in the text and in other chapters of the book.

Important genetic and phenotypic differences between natural isolates and their cell culture-adapted counterparts have been observed in several viruses, including DNA viruses. Polyomaviruses are widespread in humans, but they are rarely pathogenic except for immunocompromised individuals. One of the regions of the polyomavirus genome, the noncoding control region (or NCCR), is quite conserved among natural, usually nonpathogenic isolates. When these viruses are adapted to cell culture, the

NCCR undergoes rearrangements involving deletions and duplications. Rearrangements are observed in variants that cause pathology, and they can be readily grown in cell culture. An interesting possibility is that NCCR variants display high replication rates associated with pathogenic potential (Gosert et al., 2010). The connection between replication rate, viral load, and disease progression is discussed in Chapter 8. Genome rearrangements were also identified in regions with repeated sequences in cell culture-adapted cytomegaloviruses (Murphy and Shenk, 2008). These observations with DNA viruses, as well as their relevance for viral persistence and the possible implication of micro-RNA (miRNA) expression have been reviewed (see Imperiale and Jiang, 2015, and references therein). It appears as virtually impossible to maintain a virus population invariant when it enters a different environment.

6.2.2 EXPERIMENTAL EVOLUTION *IN VIVO*

The field of experimental evolution includes designed experiments with viruses in their natural or alternative potential hosts. The objective is to probe concepts of viral evolution and pathogenesis, which are viewed as increasingly interconnected. There is an extensive literature with plant and animal viruses that has contributed to the major concepts discussed in this book. Because the studies are very numerous, here we will discuss some selected examples, again with the objective of underlining some general conclusions.

Studies that rose considerable controversy relate to the elucidation of amino acid substitutions in avian influenza virus (IV) that mediated pathogenicity and human-to-human transmission potential of the virus (reviews in Bouvier and Lowen, 2010; Schrauwen et al., 2014). The ferret is a useful animal model for IV research since it produces respiratory symptoms similar to those in humans. The use of ferret as animal model showed that amino acid substitutions in the receptor-binding domain of the hemagglutinin and the polymerase PB2 were important for transmission of an avian IV among ferrets. In one of the experiments to unveil critical determinants for human-to-human transmission, an avian H9N2 IV was adapted to replication in mammals by serially passaging the avian IV in ferrets. Considerable IV heterogeneity that was diagnostic of active population dynamics in the animals was recorded. The ferret-adapted virus was transmitted efficiently among ferrets, indicating, that, not surprisingly, an avian IV can adapt to be transmissible among humans (Sorrell et al., 2009). Likewise, a number of different mutations, including those that affect receptor preference are needed for effective airborne transmission of an avian H5N1 IV (Herfst et al., 2012). These and other experiments to unveil transmissibility of avian IV in humans open the possibility of next generation sequencing (NGS) screening of avian IVs in search for mutations that may approximate avian viruses to represent a zoonotic threat for humans. The controversy surrounding the need to carry out these types of experiments ("gain of function" experiments) has, in my view, two major components: the uncertainties derived from specifying and publicizing mutations that may render avian IV a biological weapon, and the danger of laboratory escape of a modified IV that can be highly pathogenic and transmissible among humans, causing a devastating disease, in the shadow of the 1918 influenza pandemic. With regard to the first concern, unfortunately, there are many biological weapons available without the need of new ones, and it is very likely that pathogen-enhancing mutations in general (and for IV in particular) do not have a universal value and may change even for closely related isolates of a virus (even, we may speculate, in the course of preparing new stocks for ill-intentioned purposes). Concerning the second concern, indeed, high containment facilities and strict protocols must be used for experiments on directed changes of virus virulence and transmissibility.

A virus that naturally displays a broad host range, and that has contributed examples of biological modifications *in vivo* is foot-and-mouth disease virus (FMDV). In Chapter 4, a molecular analysis of adaptation of a swine FMDV to the guinea pig was used as an example of the influence of substitutions in nonstructural viral proteins in host range (Section 4.4.1). An earlier analysis of FMDV evolution *in vivo* involved a collaboration between Madrid, Rio de Janeiro, and Buenos Aires teams. It consisted in the genetic and antigenic analysis of sequential FMDV samples extracted from cattle during a persistent infection established experimentally with a clonal virus population derived from a cattle isolate (Gebauer et al., 1988). The reason to bring this study here is not only because it is an interesting and informative example of experimental evolution *in vivo*, but also because its results were published at the time when the extensive genetic diversity of HIV-1 was being discovered. Many virologists regarded variation of HIV-1 as unusual, even unique, perhaps only paralleled by that of IV, with the concept of antigenic shift and drift of IV well established at the time (Chapter 7). H. Temin took the analysis of variation of FMDV reported by F. Gebauer and colleagues to emphasize to his retrovirologist colleagues that HIV-1 was not "unique but merely different" (Temin, 1989). In the study with FMDV, virus was recovered from the esophageal-pharyngeal area, the site of FMDV persistence in ruminants, and the virus was examined for up to 539 days postinfection. Despite the infection originating from a biological clone, the sequential samples displayed genetic heterogeneity and dominance of viral subpopulations. Moreover, the persistent virus evolved at rates as high as 0.9×10^{-2} to 7.4×10^{-2} substitutions per nucleotide and year (s/nt/y), which is as high or even higher than the rate calculated for HIV-1 [10^{-2} to 3.7×10^{-4} s/nt/y according to several studies (Hahn et al., 1986; Korber et al., 2000; Shankarappa et al., 1999)]. The mutations were certainly not neutral since several of them affected the reactivity of the virus with antibodies. This controlled experiment showed that FMDV underwent extensive genetic and antigenic variation during persistence in cattle. Thus, H. Temin could emphasize that HIV-1 is not unique concerning variation potential. If a virus is confronted with a focused selective pressure *in vivo*—as is the case of FMDV in the pharyngeal region where an active local mucosal immune response is triggered—it can reach remarkable rates of evolution that are 10^{6}- to 10^{7}-fold higher than average values for cellular genes (Holland et al., 1982) (Chapter 7).

In agreement with the model study by F. Gebauer and colleagues with FMDV, persistent viral infections in hosts that display an active immune response (albeit insufficient to clear the virus), may constitute a source of antigenic variants that may occasionally be transmitted to new hosts, or may remain essentially confined to the persistently infected individual. In the latter case, successive waves of variants may be selected to prevent virus clearance by the immune system thus contributing to persistence (Clements et al., 1988; Narayan et al., 1981; Pawlotsky, 2006; Richman et al., 2003; Sponseller et al., 2007, among other examples). Given the potential coevolution of antigenic sites and receptor-recognition domains (Section 4.5 in Chapter 4), persistent infections in animals are a potential threat for the zoonotic emergence of human pathogens, accentuated by the possibility of recombination between the persistent and a related virus from an external source.

6.3 VIRAL DYNAMICS IN CONTROLLED ENVIRONMENTS: ALTERATIONS OF VIRAL SUBPOPULATIONS

Experiments that revealed a sustained heterogeneity in replicating viral populations have spanned four decades and have been based on widely different technologies: the very early RNA T1 oligonucleotide fingerprint used to analyze biological clones of Qβ RNA (Domingo et al., 1978) and the CirSeq design for next generation sequencing (NGS) applied to poliovirus (PV) populations (Acevedo et al., 2014), with many studies in between (Andino and Domingo, 2015).

The initial findings with bacteriophage Qβ illustrate how difficult it is for a virus to reach a true population equilibrium. The phage was multiplied for many years in its host *Escherichia coli* since its isolation from Kyoto feces in 1961 by I. Watanabe and colleagues (Miyake et al., 1967). It was passaged in the laboratories of I. Watanabe, S. Spiegelman (in Urbana), and C. Weissmann (in New York, 1965-1967 and in Zürich, 1967-1974). There is no record that the virus had been biologically cloned during these multiple passages in *E. coli*. In 1974, a clone termed A.S. (from A. Shapira who was working with M. Billeter in Zürich) was isolated and its T1 oligonucleotide fingerprint was compared with that of a reconstructed stock of the uncloned Qβ population. The comparison revealed that the uncloned stock was heterogeneous since one oligonucleotide that was present in full molar amount in clone A.S. was present in submolar amounts in the uncloned stock, and had been replaced by a mutant oligonucleotide (explained in Domingo et al., 1978). The resolved oligonucleotides that represented about 10% of the genome had been sequenced and mapped by M. Billeter, although the results were published years later (Billeter, 1978). These very early comparisons reinforced a suspicion that was frequently commented in the discussions held in Zürich: that bacteriophage Qβ, and other RNA bacteriophages, were probably highly heterogeneous [see Weissmann et al. (1973) for statements about potential heterogeneity at the time that nucleotide sequencing was slow and cumbersome]. Earlier indirect evidence of genetic instability of RNA viruses (abundance of temperature sensitive mutants in virus stocks, frequent reversion of phenotypic markers, etc.) was reviewed (Domingo and Holland, 1988). The observations with bacteriophage Qβ prompted the discovery of viral quasispecies (as described in Chapter 3), but the reason to bring them here is to emphasize that even after extensive passage without cloning, no population equilibrium with one dominant genome type was reached in the Qβ population.

From the current knowledge of quasispecies dynamics we can interpret that one of the reasons why no equilibrium with a defined consensus sequence was produced lies in the multiple possibilities of exploration of sequence space, open to any virus during replication as a result of the stochastic generation of mutations. Nonequilibrium might have been favored by the mutual effects (positive or negative, sometimes termed epistatic effects) among mutations in the same genome, and interactions among different genomes, or sampling effects during passages among other factors. It is worth (albeit not easy) picturing that a tendency toward equilibrium (meaning a trend toward a steady distribution of mutant forms) has to be based on the number and types of mutants available to the viral population at a given time. When new mutants are generated stochastically, the pathway toward equilibrium may change. A very large population will generate a larger number of alternatives toward equilibrium to choose from, as compared to a small population. This is why deterministic features of quasispecies are more likely to be observed with large viral populations (Section 3.6.2 in Chapter 3). No equilibrium can be assumed (and much less so the absence of mutations!) even for a virus with a long history of multiplication in the same environment (Box 6.1).

The initial FMDV passage experiments in cell culture that provided evidence of quasispecies dynamics for an animal virus were carried out with a biological clone of a swine isolate (Sobrino et al., 1983). The viral genome displayed increases and decreases in the molar proportion of T1 oligonucleotides, with heterogeneity levels estimated in an average of 2-8 mutations per genome in the mutant spectrum, as compared with the corresponding consensus sequence. Passage resulted in adaptation of the virus to the culture cells, as documented by an increase of infectious progeny production. Many experimental studies by J.J. Holland and colleagues contributed greatly to what we know about viral quasispecies, starting with pioneer experiments on the generation of defective interfering (DI) particles

---

BOX 6.1 SOME TEACHINGS OF EXPERIMENTAL VIRUS EVOLUTION

- Mutant viruses are continuously arising in viral populations. Even prolonged passage in the same environment does not mean that a population equilibrium has been attained. An invariant consensus sequence does not imply absence of mutations. It means that the genomes are mutating continuously to yield the same consensus.
- Virus evolution consists in the replacement of some viral subpopulation by others, due to random events or in response to selective constraints.
- The reorganization of viral subpopulations is profoundly altered by bottleneck events.
- The model studies predict that the effect of a bottleneck depends on its size and the fitness of the parental population.
- Work conducted under the guidance of J.J. Holland has permitted testing experimentally several theoretical proposals, hypotheses, and principles of general genetics. They include:
  - Muller's ratchet
  - Competitive Exclusion principle
  - Red Queen hypothesis

---

and their competition with standard virus (Holland et al., 1979, 1982). The studies carried out in J.J. Holland's laboratory in San Diego are summarized in several chapters of this book (Novella, 2003).

Application of standard molecular cloning and sequencing, and NGS to the analysis of viral populations during experimental infections has supported the view that viral populations are composed of many genome subpopulations, and that their evolution is best described as the replacement of some viral subpopulations by others (Acevedo et al., 2014; Baccam et al., 2003; Sobrino et al., 1983). Because conditions are far from population equilibrium, such replacements occur even in the absence of an externally applied selective pressure. A selective pressure acts as a guiding force to decant viral subpopulations in favor of those that best respond to the constraint, often after many transient, abortive attempts, as also observed in natural infections (Cale et al., 2011; Fischer et al., 2010; Kortenhoeven et al., 2015; Tsibris et al., 2009) (Figure 6.3). In other terms, different areas of sequence space are dynamically occupied prior to the occurrence of selective constraints as well as in response to selective demands (Chapters 3 and 7).

6.4 PERSISTENT INFECTIONS IN CELL CULTURE: VIRUS-CELL COEVOLUTION

Persistent infections *in vivo* may result from failure of immune surveillance systems to clear a virus, from infection of cells that exert functions related to the immune response, or from other mechanisms that limit viral population numbers and cell killing. Persistent infections with or without pathology may involve integration of genetic material of the virus into host cells or maintenance of the virus replicating by its standard mechanisms, but with modulation of viral population numbers. Persistent infections are abundant in most biological phyla that have been examined [review of mechanisms and biological consequences of viral persistence in (Ahmed and Chen, 1999; Nash et al., 2015; Nathanson and Gonzalez-Scarano, 2007; Oldstone, 2006; Roossinck, 2014)]. A feature of persistent infections is that the population numbers of infectious particles remain limited and constrained to a specific

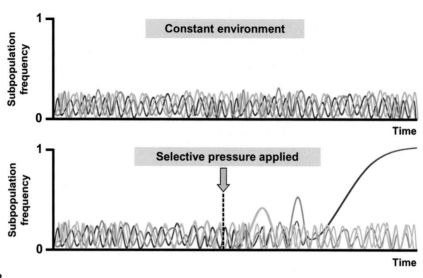

FIGURE 6.3

Diagram of dynamic genome subpopulations in the course of virus replication. Four different colors have been chosen to depict fluctuations of four genomic classes. In a real population thousands of genomes may be involved in each infected cell. When a selective pressure is applied, multiple transient selection events occur (green and blue outstanding waves) only to be finally displaced by a winning subset of genomes here represented in red. The time frame in the abscissa is left without units on purpose, because the diagram can represent events in a few infected cells during minutes or in a persistently infected organism during years. See text for examples and references.

environmental domain where viruses are enclosed (liver, kidney, brain, etc.). Under these circumstances the *r* strategy typical of overt virus infections (success based on rapid reproduction to cope with different environments) is transformed into one closer to the *K* strategy (limited progeny adapted to a specific environment) which is typical of large animals (compare with Section 4.1 in Chapter 4).

Cell culture systems have provided excellent tools to address the mechanisms of persistence of DNA and RNA viruses, including herpesviruses, polyomaviruses, parvoviruses, picornaviruses, reoviruses, flaviviruses, and retroviruses (Herrera et al., 2008; Imperiale and Jiang, 2015; Olivares et al., 2013, among other examples). Here we will analyze some of these systems that have provided new insights into the interaction of cells and viruses in a controlled environment, and in particular the evidence of virus-host coevolution during persistence. Takemoto and Habel reported the first evidence of virus-cell coevolution with the picornavirus Coxsackievirus A9 (Takemoto and Habel, 1959). R. Ahmed, B.N. Fields, and colleagues documented a role of the cells and coevolution of reovirus and L cells during a persistent infection in cell culture (Ahmed et al., 1981). These early studies were followed by experiments with the lymphotropic minute virus of mice (Ron and Tal, 1985, 1986), reovirus (Chiarini et al., 1983), poliomavirus (Delli Bovi et al., 1984), and FMDV (de la Torre et al., 1988) (several systems are summarized in Table 6.1).

Persistent infectious of FMDV were established in BHK-21 and IBRS-2 cells by growing the cells that survived a cytolytic infection (the frequency of survivors was 10^{-3} to 10^{-4}). During persistence the resident virus changed genetically and phenotypically: the heterogeneity of the resident viral genomes

Table 6.1 Examples of Virus-Cell Coevolution in Cell Culture

Virus Cell	Main Findings	References
Coxsackievirus A9 HeLa	Carrier cells showed increased resistance to virus. Virus underwent antigenic variation, plaque size reduction, increased virulence for HeLa cells and decreased virulence for suckling mice	Takemoto and Habel (1959)
Reovirus L	Mutant cells selected during persistence displayed partial resistance to reinfection with the parental virus, and readily reestablished persistence. Mutant virus selected during persistence overgrew the parental virus	Ahmed et al. (1981)
Bovine rotavirus AU-BEK	Persistence was dependent on the presence of fetal calf serum. Cells evolved to be highly resistant to the virus	Chiarini et al. (1983)
Polyomavirus (Py) Friend erythroleukemic (FL)	Viral genomes integrate at high temperature (large T-inactive). Viral genomes rescued at permissive temperature. Coevolution with selection of virus-free Py-resistant cells or cells shedding Py variants	Delli Bovi et al. (1984)
Lymphotropic minute virus of mice L	Host range virus mutants and virus-resistant cells were selected. Reconstruction of the persistently infected culture required mutant virus and mutant cells. Resistance to infection was due to an intracellular block that affected synthesis of viral single-stranded DNA	Ron and Tal (1985)
Herpes simplex virus type 1 Human lymphoblastoid CEM	Cured cells were partially resistant to parental virus. Virus isolated during persistence displayed increased virulence for the parental CEM cells	Cummings and Rinaldo (1989)
Reovirus L	Mutation in cells and virus affect an early step of the virus replication cycle. Amino acid substitutions that alter virus sigma 1 protein oligomerization mediate infection of virus-resistant L cells. The capsid is a determinant of reovirus adaptation	Dermody et al. (1993) and Wilson et al. (1996)
Poliovirus HEp-2c	Persistently infected cells cured of the resident poliovirus (PV) displayed selective permissivity to the parental PV (Mahoney strain) but partial resistance to several PV mutants. Restriction was mapped at an early phase (adsorption or uncoating) of the infection	Calvez et al. (1995)
Mouse hepatitis DBT	Murine astrocytoma (DBT) cells resistant to mouse hepatitis virus (MHV) were selected; resistance included diminished expression of the MHV receptor. MHVs with increased avidity for the receptor—and that recognized additional receptors—were selected	Chen and Baric (1996)
Rotavirus MA104	Virus rescued from persistently infected cells produced higher viral yield than the parental virus in persistently infected cells that were cured of the virus. Mutations in virus and cells affected virus entry	Mrukowicz et al. (1998)
Hepatitis C virus Human hepatoma cells	Virus variants with accelerated replication kinetics and higher peak titers were selected during persistence. The altered phenotype was associated with a substitution in the envelope E2 protein. Cells with decreased permissivity to the virus due to a block of virus entry, RNA replication, or both were selected	Zhong et al. (2006)

was estimated in 5×10^{-4} substitutions per nucleotide, and the virus displayed temperature sensitivity and small plaque morphology (de la Torre et al., 1985, 1988). During FMDV persistence in BHK-21 cells the virus became increasingly virulent for the parental BHK-21 cells, and, in turn, the cells became progressively more resistant to the parental FMDV (de la Torre et al., 1985, 1988). The resistance of the cells to the parental virus was not due to an impairment of virus attachment, penetration, or uncoating, but to some intracellular block as evidenced by virological and cell fusion experiments (de la Torre et al., 1988, 1989a).

The changes of FMDV and its host cells adhere to the concept of virus-cell coevolution, meaning a mutual influence in the evolution of two interacting biological entities (trait coevolution is discussed in Section 4.5 of Chapter 4). The coevolutionary process involved both viral and cellular heterogeneity. J.C. de la Torre and colleagues analyzed a total of 248 stable BHK-21 cell lineages established from individual cells isolated by limiting dilution from the persistently infected cultures (de la Torre et al., 1989a,b). They distinguished six cell categories based on morphology, duplication rate, and resistance to the FMDV that initiated persistence. The study indicated that the coexistence of the cell with the virus led to selection of cells with genetic modifications that resulted in altered phenotypes. In particular, increased cell transformation (evidenced by rapid cell duplication, loss of contact inhibition, and growth in semisolid agar medium) correlated with resistance to FMDV, although a possible mechanism that could link the two phenotypic traits was not investigated. Contrary to expectations, it was not the virus but the initial capacity of BHK-21 cells to vary that permitted the establishment of the persistent infection with FMDV (Martin Hernandez et al., 1994). Mutual influences that lead to both cell and virus modifications provide long-term stability to the system without either the cells or the virus eliminating its partner (except for documented "crises of cytopathology" during which high virus titers are reached, suggesting a transient imbalance in favor of the virus) (Herrera et al., 2008). The generality of these proposals are supported by observations with several virus-host systems (Table 6.1).

6.4.1 BACK AGAIN 4000 MILLION YEARS: CONTINGENCY IN EVOLUTION

In an excursion toward the question of contingency in evolution, M. Herrera and colleagues repeated again the entire process of establishment and maintenance of FMDV persistence 20 years later, starting from frozen virus and cell stocks (Herrera et al., 2008). The main conclusion with two new persistent BHK-21-FMDV lineages established in parallel was that the virus-cell coevolution displayed features very similar to those observed in the first persistent lineage, although there were differences in some of the mutations in FMDV. However, two amino acid substitutions which affected residues around a pore located at the capsid 5-fold axis were selected in the three persistent lineages. One of the substitutions (D9 → A in capsid protein VP3) was later selected in other FMDV serotypes that having originated in acute infections in animals, were passaged in evolved BHK-21 cells that displayed partial resistance to the virus (de la Torre et al., 1988; Díez et al., 1990; Escarmis et al., 1998). This substitution belongs to the category of "joker" substitutions in the sense of providing a selective advantage to the virus in a variety of sequence contexts and environments (Section 4.9 in Chapter 4).

The two newly established BHK-21-FMDV persistent lineages displayed strikingly parallel features in the variation of progeny production and imbalances in the ratio of positive versus negative strand viral RNA, that were found at the same passage number (Herrera et al., 2008). This parallel behavior reflects a deterministic feature of virus evolution, that has been encountered with other virus-host systems (see Section 6.7.1 and the general concept of deterministic vs. stochastic quasispecies in Section 3.6.2 of

Chapter 3). Interestingly, a deterministic behavior should be generally favored by large viral population sizes, but in this case a drastic decrease in population size preceded its manifestation. This unexpected association was interpreted as deterministic behavior being due to strict selective demands that guide the virus toward one (or very few) biological solutions to respond to the constraint (Herrera et al., 2008). The reproducibility of the main features of FMDV persistence in independent establishment events was suggested to mean that when a biological system is highly constrained, there may be limited room for contingency. Thus, we can now consider the question: If we could rewind the tape of evolution and play it again, would it turn out to be similar or different from what we know? According to S.J. Gould if we could rewind the tape to the remote past, evolution would turn out entirely different (Gould, 1989). The answer proposed by M. Herrera and colleagues was that the tape would turn out equally, differently or "in between" depending on the constraints operating at the time we start to retape. If the tape was situated in the RNA world around 3.6×10^9 to 4.2×10^9 years ago or perhaps prior to the Cambrian explosion, about 5×10^8 years ago (Conway Morris, 1998; Gould, 1989) there would be ample room for divergence due to equally valid alternative solutions. In contrast, once the constraints inherent to functional cells and viruses are in place, the system should be forced to a seemingly deterministic behavior (see Chapter 1 for the time frame in which the increasingly chemical and precellular complexity evolved on Earth).

It is tempting to make a connection between virus-cell coevolution in cell culture and the models of virus origins that contemplated vesicle-wrapped primitive cellular and viral entities (Section 1.5.5 in Chapter 1). What the experiments of viral persistence in cell culture show is that as part of coevolutionary mechanisms, a cell may diminish the expression of surface proteins that act as receptors for the coevolving virus or even alter fundamental properties such as the rate of cellular multiplication. Cells persistently infected with PV express mutated forms of the PV receptor CD155 (Gosselin et al., 2003; Pavio et al., 2000), and persistently infected BHK-21 cells increased their degree of cellular transformation (de la Torre et al., 1988, 1989b). [As review of picornavirus persistence, see Colbère-Garapin and Lipton (2010).] It has been proposed that in the primitive biosphere early vesicles evolved to limit their invasion by virus-like elements, and they did so by building protective walls that culminated in primitive receptor-dependent virus entry into increasingly autonomous precellular entities. The insights provided by current cell-virus coevolution models indicate that such kind of mutual influences occur with naked or enveloped viruses without the need for the virus to integrate its genome (or part of it) into the genetic material of the coevolving cell (Colbère-Garapin and Lipton, 2010; de la Torre et al., 1988). During persistent infectious in cell culture, virus titers change in ways partially dictated by their own quasispecies dynamics and coevolving carrier cells.

An area of experimental evolution that puts its emphasis on the virus rather than the host cells consists in subjecting the virus to repeated bottlenecks that are experimentally realized through plaque-to-plaque transfers (Figures 6.1 and 6.2). In this design, the cell obviously plays the essential role of hosting viral replication, but it cannot evolve since fresh (uninfected) cells are used for each virus transfer. We have learned a lot from this class of experiments.

6.5 TEACHINGS FROM PLAQUE-TO-PLAQUE TRANSFERS

Bottlenecks increase the stochasticity of evolutionary events and may modify the course of selection (Chapter 3). In the present section we review evidence that population bottlenecks may have additional and even more profound influences in the composition of viral quasispecies, by permitting hidden minority genomes to surface in populations. This important line of investigation of viral genetics was initiated by L. Chao working with the tripartite double-stranded RNA bacteriophage φ6 (Chao, 1990). He demonstrated average fitness decrease in φ6 clones subjected to serial plaque transfers in its host bacterium *Pseudomonas phaseolicola*. The results constituted the first experimental support for the operation of Muller's ratchet, a concept from theoretical biology explained in Section 6.5.1. However, the molecular basis of fitness decrease was not investigated in this first study.

The experimental design of L. Chao was extended by J.J. Holland and his colleagues to the animal virus VSV growing in mammalian cells (Duarte et al., 1992). The results agreed with those of L. Chao and documented variable, but in some cases severe fitness drops by subjecting VSV to only 20 serial plaque-to-plaque transfers. A relevant observation was the large difference in the extent of fitness decrease among biological VSV clones (from a common parental population) subjected to the same passage regime, which is itself a reflection of the remarkable heterogeneity within a VSV population. A second highly significant observation was that VSV clones isolated from a population with a long history of passage in BHK-21 cells displayed a more pronounced fitness decrease when subjected to plaque-to-plaque transfers in the alternative hosts HeLa or MDCK cells than in BHK-21 cells to which the virus was better adapted due to its prior passage history. This difference probably reflects the fact that when the virus replicated in a less adequate environment provided by the unfamiliar host cell, a larger proportion of genetic variants occupied the most frequent class of genomes. The genomes forced to be more represented are those more likely to be picked at random from individual plaques to enter the following transfer. J.J. Holland and colleagues extended these studies with VSV to show fitness decreases at different plating temperatures, and contrasted the fitness loss associated with plaque transfers with fitness gain upon large population passages in the same host cells. Clones that had attained low fitness due to plaque transfers rapidly regained fitness when they were passaged as large populations (Clarke et al., 1993). A study of alternation between passage regimes evidenced that the fitness increase that occurs during two successive large population passages was not sufficient to overcome the decrease produced by a single bottleneck passage (Duarte et al., 1993).

The model studies of J.J. Holland, I.S. Novella, and colleagues with VSV represented great progress in the understanding of virus evolution, and some of the information provided by the results is not (even today!) sufficiently considered in interpreting the consequences of bottlenecks and population expansions *in vivo* (Chapter 5). A pertinent study was the quantification of the effect of the bottleneck size (how many infectious particles participate in the bottleneck passage) on viral fitness. This effect depends on the fitness of the population subjected to the bottleneck (Novella et al., 1995) (Figure 6.4). When the starting population has low fitness (point L in the middle and bottom schemes of Figure 6.4), few particles per transfer are sufficient for the passaged population to gain fitness. In contrast, when the initial population has high fitness (point H in the middle and bottom schemes of Figure 6.4), a large number of particles is needed just to maintain fitness. An even larger number is necessary to increase fitness above the initial level of point H.

If these observations on fitness dependence of fitness evolution were operative in the patchy environments provided by nature (still to be proven) they could contribute to maintaining equilibrium between population numbers of viruses and their hosts. Large fitness increases of viruses would be prevented since whenever a bottleneck of any size is reached, it will act to limit fitness to an extent which is commensurate with the fitness value already attained by the relevant population (see also Section 6.6). This modulating effect is credible in view of the evidence of the frequent occurrence of bottlenecks in nature (compare Figure 6.4 with the evidence of bottlenecks during the arbovirus life cycle described in Section 4.10 of Chapter 4). It is tempting to consider that the equilibrium between

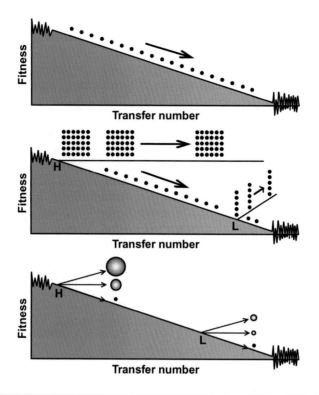

FIGURE 6.4

The effect of bottleneck size on fitness is dependent on the fitness of the population subjected to the bottleneck. At the top, plaque-to-plaque transfers (bottleneck size 1, black dots) lead to fitness decrease dependent on the number of transfers (downward arrow). The middle diagram shows that when the starting population has high fitness (point H in the graph) a bottleneck size of 30 (pool of virus from thirty plaques) is necessary to just maintain the fitness value (horizontal arrow). In contrast, when the starting point has low fitness (point L in the graph) a bottleneck size of 5 is sufficient to increase fitness (upward arrow). In the bottom diagram bottleneck sizes are depicted as spheres of different size to summarize schematically the bottleneck size requirements to maintain or increase fitness depending on the initial fitness (H, high; L, low). The concept expressed in this diagram is based on studies with VSV by I.S. Novella, J.J. Holland, and colleagues, as described in the text. The fluctuations in fitness values at very high and very low-fitness values have been observed experimentally, and their interpretation and implications are also discussed in the text.

viral and host populations numbers that had to exist from ancestral times (Chapter 1) is not only due to interaction and escape strategies between viruses and host immune systems (Chapter 4), but also due to in-built self-regulatory mechanisms inherent to viral genome dynamics that limit excessive replication capacity in the face of a far more inflexible cellular world. Bottleneck size may be an evolvable trait that has contributed to some balance between virus and host population numbers.

Repeated bottlenecks not only decrease fitness as summarized in previous paragraphs, but, in addition, they may have long-term detrimental effects on virus adaptability. Even when a bottlenecked VSV population had the same fitness than a virus maintained under large population passages, it displayed lower adaptability (Novella, 2004). This result means that it is not fitness *per se* that determines the adaptive potential of a virus, but it is the part of sequence space that it occupies that does. Different areas of sequence space may provide a mutant cloud with comparable fitness, but with different capacity to respond to perturbations (see Section 6.7.1 in this chapter and Section 3.7 in Chapter 3). A similar, also rather counterintuitive conclusion is reached when comparing mutation frequencies (or other complexity measurements) of viral populations: the same high mutation frequency may either permit viral survival or drive the virus toward extinction. It is not the mutation frequency value *per se* that counts, but the context in which the mutations occur (the experimental values that justify such an assertion are discussed in Chapter 9).

Once more we come to the conclusion of the importance of viral population numbers in fitness variations and in guiding evolutionary episodes. It is obvious that some nuances must be introduced into the rather simplified view that bottlenecks lead to fitness loss (Chapter 5). There is no absolute population size value that guarantees avoidance of fitness decrease. The bottleneck size that leads to decrease, maintenance, or increase of fitness is dependent on the initial fitness and on the position of the virus in sequence space. It is quite clear that in this type of studies, experimental evolution has gone far ahead than field observations, partly due to the difficulties in following fitness evolution and in controlling bottleneck sizes *in vivo*. With these clarifications, we are now in a position to begin to approach the general question (or dilemma) between asexual and sexual modes of reproduction, as a precedent to approach clonality versus nonclonality in biological evolution to be addressed in the closing Chapter 10.

6.5.1 MULLER'S RATCHET AND THE ADVANTAGE OF SEX

Fitness loss due to bottleneck events (with all its participating parameters discussed in the previous section) provides experimental support to a theoretical proposal made by H.J. Muller, known as Muller's ratchet (Maynard-Smith, 1976; Muller, 1964; Nowak and Schuster, 1989). The proposal is that small populations of asexual organisms that display high mutation rates will tend to incorporate mutations (the majority deleterious) in an irreversible, ratchet-type mechanism, unless recombination can yield the initial type of genome devoid of mutations (often termed the zero mutation class of genomes). Fitness decrease due to accumulation of mutations proposed by H.J. Muller in general terms is expected to be accentuated in the case of viral populations with continuous generation of new mutant genomes. In mutant spectra of viruses the least mutated class of genomes will be the one displaying highest replicative fitness and will correspond to the master sequence that dominates the population (Chapter 3). With the plaque-to-plaque transfer design there is a probability that in each plating (a step or click in the ratchet) the least mutated class of genomes is lost. Therefore, the viral population is forced to regenerate a distribution from which again the least mutated class will be lost in the next ratchet click. The system is doomed to rapid deterioration and extinction, unless mechanisms to restore fitter genomes operate.

The operation of Muller's ratchet appears as rather general in viruses since, in addition to bacteriophage φ6 and the animal virus VSV (studies summarized in Section 6.5), other viral systems have been used to document fitness loss associated with serial bottlenecks (de la Iglesia and Elena, 2007; Escarmís et al., 1996; Jaramillo et al., 2013; Yuste et al., 1999). The studies by C. Escarmís and colleagues with FMDV contributed decisively to define the molecular basis of fitness loss in an RNA virus, and to unveil unusual genetic lesions (with phenotypic consequences) that hide in mutant spectra. For these reasons the studies by C. Escarmís and colleagues are detailed in Section 6.5.2. Muller's ratchet operates not only in viruses but also in bacteria, protozoa, plants, fish, and mitochondrial DNA (Allen et al., 2009; Andersson and Hughes, 1996; Bell, 1988; Coates, 1992; Engelstadter, 2008; Leslie and Vrïjenhoek, 1980; Loewe, 2006; Moran, 1996).

The concept that sex conferred an advantage to organisms by providing new gene combinations for adaptation dates back to A. Weismann (Bell, 1982; Weismann, 1889-1892). Avoidance of the detrimental effects of Muller's ratchet is believed to have been one of the driving forces to introduce genetic recombination and sex in the reproduction and evolution of living systems (Agrawal, 2006; Barton and Charlesworth, 1998; Maynard Smith and Szathmáry, 1999; Maynard-Smith, 1976). In the words of G. Bell: "Sex acts as an editor which detects serious copying errors and enables the genetic message to be transmitted without contamination" (Bell, 1988). It is worth comparing this sentence by G. Bell with the molecular evidence discussed in Chapters 2 and 3 that copying errors during virus replication are not corrected because of the absence of proofreading exonucleases (or related) activities in RNA viruses. Lack of correcting activities is remarkable from the point of view of an evolution rich in lateral gene transfers because proofreading-repair activities are available not only from DNA polymerases but also from the coronavirus polymerases (and likely other large RNA genomes to be discovered). This raises the issue of clonal evolution in viruses, an important feature of long-term virus evolution (Chapter 10). Now we return to plaque-to-plaque transfers because there is more to learn from them.

6.5.2 MOLECULAR BASIS OF FITNESS DECREASE: DEEP FLUCTUATIONS, MASSIVE EXTINCTIONS, AND RARE SURVIVORS

The genomic RNA of 19 biological clones of FMDV that had undergone fitness decrease as a consequence of 30 serial plaque-to-plaque transfers was sequenced and the sequences compared with those of the parental clones, and to the parental populations subjected to large population passages. Many mutations that had never been found in natural isolates or laboratory clones and populations of FMDV were found. A total of 69 mutations affecting the internal ribosome entry site (IRES), L-protease, P1- and polymerase-coding regions were identified (Escarmís et al., 1996). Notably, 9 out of 19 clones showed an extension of an internal tract of four adenylate residues that precedes the second functional AUG. Depending on the clone, the extension was of one, or several residues and up to 23 additional adenylates in one of the clones. This elongation affected translation, was one of the determinants of fitness decrease, and reverted when the clones were subjected to large population passages. It was proposed that the elongation, that constitutes a hot spot for variation, was prompted by slippage mutagenesis of the polymerase. This genetic lesion in the FMDV genome was given as an example of mutational instruction in Section 2.3 of Chapter 2, and it served as one of the genetic markers for quasispecies memory when it reverted upon large population passages of the altered clones (Section 5.5 in Chapter 5). The molecular instability associated with this oligoadenylate elongation is illustrated by the fact that genomes with different numbers of adenylate residues coexisted in the same viral plaque.

Another rare lesion was a point deletion located between the two functional AUG protein synthesis initiation codons, that led to a predicted stop codon for the proteins synthesized from the first AUG. A significant finding is that of the total number of mutations that had accumulated at the end of the 30 transfers, one-third were already present in the first transfer, although a rather steady accumulation of mutations was observed when the clones were subjected to many additional transfers. A possible limitation in the accumulation of mutations is examined again in connection with limited tolerance to mutations and the concept of contingent neutrality (Section 6.7.1).

FMDV clones were subjected to additional (up to 409) plaque-to-plaque transfers that unveiled additional rare genetic lesions, unusual phenotypes, and a remarkable resistance to extinction (Escarmís et al., 2002, 2008). The most salient phenotypic change is that after more than 100 transfers the virus became noncytolytic, was unable to form visible plaques, and it could readily establish a persistent, non-cytocidal infection that normally would be established only from the few cells that survived a cytolytic infection (see Section 6.4). Thus, strikingly, a fundamental property of the virus-host interaction such as the capacity to kill cells [a very marker of virus virulence! (see Section 5.6 in Chapter 5)], was totally altered as a result of repeated bottleneck passages. In fact, if infectivity is judged by the capacity to produce plaques, these multiply transferred clones underwent decreases of specific infectivity (the ratio between the amount of infectivity and that of viral RNA) of at least 140-fold relative to their corresponding parental biological clone. Such a reduction is enormous, and it should serve to compare the decrease of specific infectivity that accompanies virus extinction by lethal mutagenesis. (Decreases of specific infectivity is mentioned in Section 4.3 of Chapter 4 as one of the consequences of suboptimal codon usage, and are revisited in Chapter 9 as a diagnostic parameter of lethal mutagenesis.) A proof that pursuing molecular analysis of clones subjected to many bottleneck transfers may provide new information was documented by the discovery of an amino acid substitution in capsid protein VP1 that produced virus thermosensitivity, and exerted an effect at a distance in the processing of the FMDV polyprotein (Escarmís et al., 2009), unveiling a new feature of picornavirus protein processing (Martínez-Salas and Ryan, 2010). Unusual FMDVs isolated after plaque transfers are summarized in Box 6.2.

Several mechanisms have been proposed to explain why so many unusual, often unique, detrimental mutations can be rescued in clones subjected to many plaque transfers. One model emphasizes that negative selection is attenuated during plaque transfers in the sense that no truly competitive optimization of the mutant spectrum is allowed, except for intracellular competition in each individual cell. The competitive mixing of genomes is limited in the cell monolayer whose individual cells become infected by the wave of progeny virus that had its initial focus in the cell hit by the virus in the corresponding transfer. Even accepting that intraplaque MOI might be high, no competition among many of the newly arising genomes can take place in any of the cells. Therefore, the displacement of unfit genomes by fitter ones is limited. A necessity to explain the frequency and types of mutations observed is that the virus must be subjected to high mutational pressure. Although mutation rates for RNA viruses are discussed in Chapter 2, it must be stated here that many of the studies that allowed a quantification of high mutation rates came from experimental evolution designs. The fact that one of the clones at transfer 409 differed from its parental clone in 122 mutations, implies a mutation frequency of 1.5×10^{-2} s/nt (Escarmís et al., 2008). Remarkably when FMDV is subjected to mutagenesis, a tenfold lower frequency can drive the virus to extinction (Chapter 9). The observed mutability during plaque transfers renders perfectly understandable multiple reversion events scored in picornavirus genomes (de la Torre et al., 1992; Domingo et al., 2010).

Another, not mutually exclusive, mechanism is that many extinction events take place in the course of the transfers and that the low-fitness survivors acquire compensatory mutations that still allow them

---

BOX 6.2 UNUSUAL GENOTYPES AND PHENOTYPES IN FOOT-AND-MOUTH DISEASE VIRUS SUBJECTED TO PLAQUE-TO-PLAQUE TRANSFERS

- Mutations never found in other populations of the same virus subjected to other passage regimes, including a rare one nucleotide deletion.
- Noncytocidal mutants that can establish a persistent infection in cell culture without intervening cell killing.
- Amino acid substitutions in the capsid that can affect thermal stability and polyprotein processing.

to form a plaque, unless a noncytolytic phenotype that still allows intracellular RNA replication to take place is produced.

C. Carrillo, D.L. Rock, and colleagues applied an *in vivo* protocol of 20 serial swine-to-swine contact transmission that resembles the plaque-to-plaque transfer design (Carrillo et al., 2007). Interestingly, profound phenotypic changes occurred in the virus after several transfers, including reduction of virulence and establishment of a carrier state in pigs, previously thought to be typical of ruminants. Several mutations accumulated in the viral genome, suggestive of the operation of Muller's ratchet *in vivo*. This study proves the feasibility of serial bottleneck passages in animals, that may be highly informative of potential viral alterations associated with transmission events.

6.6 LIMITS TO FITNESS GAIN AND LOSS

A key issue in the studies of fitness evolution is whether fitness of a virus population can grow indefinitely or it has a limit, reaching a plateau value. This question is of theoretical and experimental interest. Limitations in the capacity to occupy sequence space suggest that there must be a limit to fitness gain and that, in the case of viruses this limit may be imposed by the viral population size. Several experimental studies of experimental evolution and theoretical predictions support that either increases or decreases of fitness reach a plateau, proposed to be the result of mutational effects and the ratio of beneficial to deleterious mutations (average mutations will increase fitness of a very low-fitness population but decrease fitness of a high-fitness population) (Silander et al., 2007).

The situation may not be so simple. I.S. Novella and colleagues demonstrated that exponential increase of VSV fitness has a limit but that when the limit is reached, stochastic fitness fluctuations occur (Novella et al., 1999b). Variations are represented by the zig zag lines at the upper left side of the fitness diagrams shown in Figure 6.4. Since the viral population size needed to increase viral fitness is larger the higher the fitness value, the results of I.S. Novella and colleagues suggest that VSV reached a fitness value whose further increase could not be guaranteed by the replicating population size attained. An area of sequence space is reached where the newly arising mutations are not steadily incorporated, but rather they produce unpredictable fitness jumps.

A similar fluctuation of fitness values was observed in the case of FMDV subjected to many plaque-to-plaque transfers (Lázaro et al., 2003). Variations are represented by the zig zag lines at the lower right side of the fitness diagrams shown in Figure 6.4. An example with four FMDV subclones is presented in Figure 6.5. The fluctuating pattern at low-fitness values followed a Weibull statistical distribution which was taken to mean that complex virus-cell interactions contribute to the level of progeny in each individual plaque. The results reinforce the concept of extreme resistance of viruses to extinction at the population level despite many extinctions at the individual level. The reason is that when fitness is very low the probability of stochastic occurrence of beneficial mutations can rescue subpopulations from their fate toward extinction (Escarmís et al., 2002).

Fitness instability might underlie the transition toward genome segmentation underwent by FMDV upon extended high MOI replication in cells, an experiment intended to see if FMDV reached a fitness plateau (García-Arriaza et al., 2004). In the course of hundreds of passages, the virus diversified in two distinct subpopulations that exhibited a competition-colonization dynamics previously shown by D. Tilman to operate in classical ecological systems (Tilman, 1994). The viruses diversified into colonizers which were efficient in killing cells, and competitors that modulated cell killing. Thus, internal quasispecies interactions modulated virus virulence (Ojosnegros et al., 2010). This balance was maintained

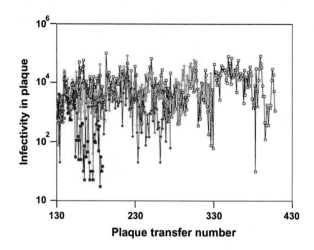

FIGURE 6.5

The fluctuation pattern of fitness values (infectivity in plaque) of FMDV subclones subjected to a maximum of 400 serial plaque-to-plaque transfers. Results for four different subclones (color-coded) for passage 130–400 are shown. Only one of the subclones (identified in black color) survived to produce plaques to the end of the transfer series. The other clones (coded red, blue, and green) became noncytopathic at different passages.

Modified from Escarmís et al. (2008), with permission of the authors.

until around passage 260 at which the viral genome became segmented: the monopartite genome was converted into two segments that complemented each other to replicate and kill cells (García-Arriaza et al., 2004, 2006; Moreno et al., 2014; Ojosnegros et al., 2011). This transition was discussed in Section 2.11 of Chapter 2 in connection with a mutation-driven genome transition involving RNA recombination. The transition toward segmentation might have been favored by fitness instability when fitness increase reached its population size-dependent limits.

6.7 COMPETITIVE EXCLUSION PRINCIPLE AND RED QUEEN HYPOTHESIS

J.J. Holland and colleagues examined the fate of two competing VSV clones that displayed approximately equal fitness. The two clones coexisted for several generations until one of the populations rapidly outgrew the other (Clarke et al., 1994). This observation is schematically depicted in Figure 6.5, and agrees with the Competitive Exclusion principle of population genetics (Gause, 1971). This principle states that one competing species will always outcompete the other provided no niche differentiation exists between them. In a related formulation, the resource-based competition theory asserts that two consumers that share a single, limiting resource cannot stably coexist in a spatially and temporally homogeneous habitat. It is assumed that there are no other intervening ecological factors. In testing the Competitive Exclusion principle experimentally, the precise definition of "limiting resource" is essential. In a series of classical experiments, F.J. Ayala showed that two species of *Drosophila* could coexist for many generations in competition for limited resources (Ayala, 1971). The genetic and phenotypic inflexibility of a differentiated organism such as *Drosophila* may explain its very different behavior as compared with two VSV quasispecies. The genetic and phenotypic constraints of *Drosophila* within the time frame of the experiment contrasts with the much more voluble and dynamic nature of two competing VSV mutant clouds. Indeed, the eventual exclusion of one VSV quasispecies by the other is expected as a consequence of the random, infrequent occurrence of advantageous mutations in a genome of one of the populations and not in the other. During the competition process, both the winners and the losers gained fitness at similar rates, in agreement with the Red Queen hypothesis: "No species can ever win and new adversaries grinningly replace the losers" (Van Valen, 1973). Among competing viral quasispecies that have approached some population equilibrium in which the mutant spectrum can modulate the behavior of the ensemble, infrequently arising, superior mutants are likely to perturb the equilibrium in such a way as to exclude or maintain at low levels all other mutants present in the competing quasispecies. Until such exclusion occurs, most members of the quasispecies gain replicative fitness, including minority members present at "memory" levels (Section 5.5 in Chapter 5). As expressed by the Red Queen in Lewis Carrol's *Through the Looking Glass*: "It takes all the running you can do to stay in the same place." It appears as if the mutational background associated with some competitive optimization of quasispecies affected both competing populations in a similar way, while the stochastic occurrence of a saliently beneficial mutation disrupted coexistence. The results imply also that the frequency of mutations that are advantageous enough to upset the coexistence of the two populations is low. This is expected from the fact that detrimental and lethal mutations are far more frequent than advantageous mutations (Figure 6.6).

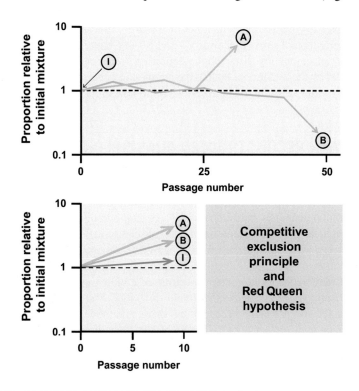

FIGURE 6.6

Schematic representation of the Competitive Exclusion principle and the Red Queen hypothesis applied to RNA viruses. At the top, two populations of fitness equal to a control I (initial population), were serially passaged in cell culture. After many passages in which both populations coexisted at about the same frequency, suddenly one of the populations (A, yellow arrow, or B, green arrow) displaced the other. This behavior is in agreement with the Competitive Exclusion principle of population genetics. Below, both populations A (winner) and B (looser) gained fitness relative to the control I. This observation agrees with the Red Queen hypothesis. See text for implications and references.

6.7.1 CONTINGENT NEUTRALITY IN VIRUS

The behavior of competing VSV populations was not always unpredictable. In parallel competitions of a wild-type VSV and a surrogate marked subclone of equal fitness, a predictable nonlinear behavior of the two populations was characterized (Quer et al., 1996). After nearly constant times, the two viruses competing in a constant cell culture environment followed different trajectories. There was a reproducible tendency of the wild type to gain fitness at a higher rate than the surrogate mutant. Thus, despite the stochastic occurrence of mutations, a nearly deterministic evolutionary behavior was observed (Tsimring et al., 1996). A number of environmental perturbations (presence of DIs, increased temperature during viral replication, or limited, enhanced mutagenesis) led to an accelerated dominance of the wild type over the mutant (Quer et al., 2001). Comparison of consensus nucleotide sequences of the entire VSV genomes showed that the mutant genome had a number of mutations with respect to the wild type, and suggested that neutrality of the mutant relative to the wild type was maintained provided the environment was not perturbed. The behavior of the mutant relative to the wild type was described as being of "contingent neutrality." The presence of mutations rendered the virus less robust to accept additional mutations. In terms of fitness landscapes, the results can be interpreted as if that the wild type lies on a relatively flat (or less rugged) fitness surface that preserves its replicative efficacy despite occurrence of mutations. In contrast, the mutant lies on a sharper fitness peak prone to fitness decreases upon mutation or environmental stress (Sections 5.3 and 5.7 in Chapter 5).

Thus, the behavior of competing viral subpopulations, a frequent occurrence in quasispecies dynamics may be affected by the mutational load in the competing genomes to the point of accelerating a selective advantage by virtue of limited tolerance to acquire new mutations.

6.8 STUDIES WITH RECONSTRUCTED QUASISPECIES

Experimental evolution offers the possibility—that has so far been exploited only minimally—of reconstructing complex quasispecies swarms with specific mutants or viral subpopulations, to examine their behavior during replication under different environmental conditions. These types of experiments in which population evolution can now be followed by NGS methodologies, would actually represent an important contribution to the study of complexity, a field of science in need of experimental approaches.

A FMDV quasispecies was reconstructed with 19 antigenic variants of the virus, each identified by an amino acid substitution at the major antigenic site that conferred resistance to a monoclonal antibody (MAb). The mutants were added to the mutant spectrum of a biological clone of FMDV, at a concentration typical of the antibody-resistant mutants found in FMDV. The reconstructed quasispecies was allowed to replicate in the absence and presence of the MAb, and the resulting populations analyzed. In the populations passaged in the presence of the antibody, but not in the control population, 10 out of the 19 mutant introduced became dominant, indicating selection of a mutant cloud that shared the required phenotype (Perales et al., 2005). In a subsequent study, the FMDV quasispecies was reconstructed with matched pairs of distinguishable MAb-escape mutants of the same antigenic site. Each mutant of a pair differed from the other in 11- or 33-fold in fitness. The analysis of the populations subjected to antibody selection revealed dominance of the corresponding high-fitness mutants. Thus, relative viral fitness can influence significantly the mutant repertoire selected by neutralizing antibody (Martin et al., 2006). Such fitness effects are likely to underlie the response to any selective constraint, and should be considered in the interpretation of the effect of selective forces acting on complex viral populations.

6.9 QUASISPECIES DYNAMICS IN CELL CULTURE AND *IN VIVO*

Some authors have argued that quasispecies dynamics is valid for viruses replicating in cell culture, but that its basic principles are not adequate to understand virus behavior *in vivo*. A reexamination of the recent evidence derived from application of NGS to *in vivo* systems renders the suggestion of a fundamental difference between cell culture and *in vivo* quasispecies dynamics untenable. One can argue that *in vivo* systems offer highly complex, compartmentalized environment that delays a possible approach to population equilibrium, as compared with a uniform and constant environment. What we have learnt, however, from the several model experimental studies summarized in this chapter, both using cell culture systems and animals, is that there are no fundamental differences in the quasispecies behavior of viruses in cell culture and *in vivo*. Perhaps a significant example is provided by the accumulation of mutations in the FMDV genomes and the profound phenotypic changes observed both in cell culture plaque-to-plaque transfers and in serial bottleneck transmission of the virus in swine (Section 6.5.2). Evidence in support of the value of quasispecies to interpret viral population dynamics *in vivo* is increasing as the new tools of NGS are applied to evolving viral population.

6.10 OVERVIEW AND CONCLUDING REMARKS

The possibilities of experimental evolution to gain new insights into the mechanisms of virus evolution are enormous, and they remain largely unexploited. In this chapter we have summarized cell culture and *in vivo* designs that have revealed fundamental features of virus evolution, including some that have provided experimental confirmation of some concepts of population biology.

Coevolution of cells and viruses appears to be quite common during persistent infections in cell culture. Several reproducible traits of virus and cell variation are shared by widely different viral pathogens and cell types. It is tempting to speculate that coevolutionary interactions reflect an inheritance of old-time relationships between primitive forms of viruses and cells.

Experimental evolution opens the possibility to examine the effects of extreme population regimes in fitness evolution, notably the result of massive infections versus the most profound bottleneck restriction: serial infections limited to one infectious particle per passage. Teachings of plaque-to-plaque transfers include the evidence of highly unusual mutations, and phenotypes that contradict textbook types of viral properties. In particular, the recognized cytopathic nature of FMDV is turned into noncytopathic in clones rescued from the low-frequency levels of viral quasispecies. In a similar note, repeated bottleneck passages of FMDV in swine produced viruses that established a carrier state in swine, a type of virus-host interaction previously thought to occur only in ruminants.

Muller's ratchet, Competitive Exclusion principle, Red Queen hypothesis, and contingent neutrality are concepts that have been established in experimental designs based on competitive, large population passages of marked mutant mixtures. These types of experiments, and those using reconstructed quasispecies, face a promising research time in which the application of NGS should clarify the mechanisms by which some variants overgrow others, opening new avenues for the understanding of viruses at the population level. The potential implications for the diagnosis of viral disease and considerations for treatment options are very evident, and they will be discussed in coming chapters (see Summary Box).

SUMMARY BOX

- Experimental evolution permits the establishment of many fundamental concepts of virus evolution that are facilitating the interpretation of observations in the far more complex natural scenarios.
- Designs intended to reproduce extreme passage regimes have unveiled many features of mutant spectra that are hidden to standard analyses based exclusively on consensus sequences. In particular, multiple plaque-to-plaque transfers have revealed that extremely unusual mutations can be found in low-fitness viruses that result in extreme phenotypes. Such unusual viral subpopulations replicate thanks to the presence of compensatory mutations that rescue a few genomes out of a great majority that are extinguished.
- The experimental studies have revealed the fitness dependence of fitness variation, a field of research still to be applied *in vivo*. Fitness increases and decreases have a limit imposed either by an insufficient population size or by extremely low replicative fitness. In such extreme scenarios, unexpected evolutionary transitions may be favored.
- Several concepts of population genetics have found experimental support in work with viruses, notably Muller's ratchet, Competitive Exclusion principle, and the Red Queen hypothesis.
- The dynamics of competition between viral populations has defined the concept of contingent neutrality in virus evolution.
- The capacity to reconstruct quasispecies with selected types of mutants opens new research avenues to understand viral population dynamics under controlled conditions.
- It will be extremely interesting to reexamine the population dynamics in the different designs described in this chapter with the new tool of next generation sequencing.

REFERENCES

Acevedo, A., Brodsky, L., Andino, R., 2014. Mutational and fitness landscapes of an RNA virus revealed through population sequencing. Nature 505, 686–690.

Agrawal, A.F., 2006. Evolution of sex: why do organisms shuffle their genotypes? Curr. Biol. 16, R696–R704.

Ahmed, R., Chen, S.Y. (Eds.), 1999. Persistent Viral Infections. John Wiley and Sons, New York.

Ahmed, R., Canning, W.M., Kauffman, R.S., Sharpe, A.H., Hallum, J.V., et al., 1981. Role of the host cell in persistent viral infection: coevolution of L cells and reovirus during persistent infection. Cell 25, 325–332.

Allen, J.M., Light, J.E., Perotti, M.A., Braig, H.R., Reed, D.L., 2009. Mutational meltdown in primary endosymbionts: selection limits Muller's ratchet. PLoS One 4, e4969.

Andersson, D.I., Hughes, D., 1996. Muller's ratchet decreases fitness of a DNA-based microbe. Proc. Natl. Acad. Sci. U. S. A. 93, 906–907.

Andino, R., Domingo, E., 2015. Viral quasispecies. Virology 479–480, 46–51.

Ayala, F.J., 1971. Competition between species: frequency dependence. Science 171, 820–824.

Baccam, P., Thompson, R.J., Li, Y., Sparks, W.O., Belshan, M., et al., 2003. Subpopulations of equine infectious anemia virus Rev coexist *in vivo* and differ in phenotype. J. Virol. 77, 12122–12131.

Barton, N.H., Charlesworth, B., 1998. Why sex and recombination? Science 281, 1986–1990.

Bell, G., 1982. The Master Piece of Nature: The Evolution and Genetics of Sexuality. Croom-Helm, Ltd., London.

Bell, G., 1988. Sex and Death in Protozoa: The History of an Obsession. Cambridge University Press, Cambridge.

Billeter, M.A., 1978. Sequence and location of large RNase T1 oligonucleotides in bacteriophage Qbeta RNA. J. Biol. Chem. 253, 8381–8389.

Bouvier, N.M., Lowen, A.C., 2010. Animal models for influenza virus pathogenesis and transmission. Virus 2, 1530–1563.

Cale, E.M., Hraber, P., Giorgi, E.E., Fischer, W., Bhattacharya, T., et al., 2011. Epitope-specific CD8+ T lymphocytes cross-recognize mutant simian immunodeficiency virus (SIV) sequences but fail to contain very early evolution and eventual fixation of epitope escape mutations during SIV infection. J. Virol. 85, 3746–3757.

Calvez, V., Pelletier, I., Couderc, T., Pavio-Guedo, N., Blondel, B., et al., 1995. Cell clones cured of persistent poliovirus infection display selective permissivity to the wild-type poliovirus strain Mahoney and partial resistance to the attenuated Sabin 1 strain and Mahoney mutants. Virology 212, 309–322.

Carrillo, C., Lu, Z., Borca, M.V., Vagnozzi, A., Kutish, G.F., et al., 2007. Genetic and phenotypic variation of foot-and-mouth disease virus during serial passages in a natural host. J. Virol. 81, 11341–11351.

Chao, L., 1990. Fitness of RNA virus decreased by Muller's ratchet. Nature 348, 454–455.

Chen, W., Baric, R.S., 1996. Molecular anatomy of mouse hepatitis virus persistence: coevolution of increased host cell resistance and virus virulence. J. Virol. 70, 3947–3960.

Chiarini, A., Arista, S., Giammanco, A., Sinatra, A., 1983. Rotavirus persistence in cell cultures: selection of resistant cells in the presence of fetal calf serum. J. Gen. Virol. 64, 1101–1110.

Clarke, D.K., Duarte, E.A., Moya, A., Elena, S.F., Domingo, E., et al., 1993. Genetic bottlenecks and population passages cause profound fitness differences in RNA viruses. J. Virol. 67, 222–228.

Clarke, D.K., Duarte, E.A., Elena, S.F., Moya, A., Domingo, E., et al., 1994. The red queen reigns in the kingdom of RNA viruses. Proc. Natl. Acad. Sci. U. S. A. 91, 4821–4824.

Clements, J.E., Gdovin, S.L., Montelaro, R.C., Narayan, O., 1988. Antigenic variation in lentiviral diseases. Annu. Rev. Immunol. 6, 139–159.

Coates, D.J., 1992. Genetic consequences of a bottleneck and partial genetic structure in the triggerplant *Stylidium coroniforme* (Stylidiaceae). Heredity 69, 512–520.

Coffey, L.L., Vignuzzi, M., 2011. Host alternation of chikungunya virus increases fitness while restricting population diversity and adaptability to novel selective pressures. J. Virol. 85, 1025–1035.

Coffey, L.L., Vasilakis, N., Brault, A.C., Powers, A.M., Tripet, F., et al., 2008. Arbovirus evolution *in vivo* is constrained by host alternation. Proc. Natl. Acad. Sci. U. S. A. 105, 6970–6975.

Colbère-Garapin, F., Lipton, H.L., 2010. Persistent infections. In: Ehrenfeld, E., Domingo, E., Roos, R.P. (Eds.), The Picornaviruses. ASM Press, Washington, DC, pp. 321–338.

Conway Morris, S., 1998. The Crucible of Creation. The Burgess Shale and the Rise of Animals. Oxford University Press, Oxford.

Cuevas, J.M., Moya, A., Elena, S.F., 2003. Evolution of RNA virus in spatially structured heterogeneous environments. J. Evol. Biol. 16, 456–466.

Cummings, P.J., Rinaldo Jr., C.R., 1989. Coevolution of virulent virus and resistant cells as a mechanism of persistence of herpes simplex virus type 1 in a human T lymphoblastoid cell line. J. Gen. Virol. 70 (Pt 1), 97–106.

de la Iglesia, F., Elena, S.F., 2007. Fitness declines in tobacco etch virus upon serial bottleneck transfers. J. Virol. 81, 4941–4947.

de la Torre, J.C., Davila, M., Sobrino, F., Ortin, J., Domingo, E., 1985. Establishment of cell lines persistently infected with foot-and-mouth disease virus. Virology 145, 24–35.

de la Torre, J.C., Martínez-Salas, E., Diez, J., Villaverde, A., Gebauer, F., et al., 1988. Coevolution of cells and viruses in a persistent infection of foot-and-mouth disease virus in cell culture. J. Virol. 62, 2050–2058.

de la Torre, J.C., de la Luna, S., Diez, J., Domingo, E., 1989a. Resistance to foot-and-mouth disease virus mediated by trans-acting cellular products. J. Virol. 63, 2385–2387.

de la Torre, J.C., Martínez-Salas, E., Díez, J., Domingo, E., 1989b. Extensive cell heterogeneity during persistent infection with foot-and-mouth disease virus. J. Virol. 63, 59–63.

de la Torre, J.C., Giachetti, C., Semler, B.L., Holland, J.J., 1992. High frequency of single-base transitions and extreme frequency of precise multiple-base reversion mutations in poliovirus. Proc. Natl. Acad. Sci. U. S. A. 89, 2531–2535.

Delli Bovi, P., De Simone, V., Giordano, R., Amati, P., 1984. Polyomavirus growth and persistence in Friend erythroleukemic cells. J. Virol. 49, 566–571.

Dermody, T.S., Nibert, M.L., Wetzel, J.D., Tong, X., Fields, B.N., 1993. Cells and viruses with mutations affecting viral entry are selected during persistent infections of L cells with mammalian reoviruses. J. Virol. 67, 2055–2063.

Díez, J., Dávila, M., Escarmís, C., Mateu, M.G., Dominguez, J., et al., 1990. Unique amino acid substitutions in the capsid proteins of foot-and-mouth disease virus from a persistent infection in cell culture. J. Virol. 64, 5519–5528.

Domingo, E., Holland, J.J., 1988. High error rates population equilibrium, and evolution of RNA replication systems. In: Domingo, E., Holland, J.J., Ahlquist, P. (Eds.), RNA Genetics. CRC Press, Boca Raton, FL, pp. 3–36.

Domingo, E., Perales, C., 2012. From quasispecies theory to viral quasispecies: how complexity has permeated virology. Math. Model. Nat. Phenom. 7, 32–49.

Domingo, E., Sabo, D., Taniguchi, T., Weissmann, C., 1978. Nucleotide sequence heterogeneity of an RNA phage population. Cell 13, 735–744.

Domingo, E., Perales, C., Agudo, R., Arias, A., Ferrer-Orta, C., et al., 2010. Mutation, quasispecies and lethal mutagenesis. In: Ehrenfeld, E., Domingo, E., Roos, R.P. (Eds.), The Picornaviruses. ASM Press, Washington, DC, pp. 197–211.

Duarte, E., Clarke, D., Moya, A., Domingo, E., Holland, J., 1992. Rapid fitness losses in mammalian RNA virus clones due to Muller's ratchet. Proc. Natl. Acad. Sci. U. S. A. 89, 6015–6019.

Duarte, E.A., Clarke, D.K., Moya, A., Elena, S.F., Domingo, E., et al., 1993. Many-trillion fold amplification of single RNA virus particles fails to overcome the Muller's ratchet effect. J. Virol. 67, 3620–3623.

Engelstadter, J., 2008. Muller's ratchet and the degeneration of Y chromosomes: a simulation study. Genetics 180, 957–967.

Escarmís, C., Dávila, M., Charpentier, N., Bracho, A., Moya, A., et al., 1996. Genetic lesions associated with Muller's ratchet in an RNA virus. J. Mol. Biol. 264, 255–267.

Escarmis, C., Carrillo, E.C., Ferrer, M., Arriaza, J.F., Lopez, N., et al., 1998. Rapid selection in modified BHK-21 cells of a foot-and-mouth disease virus variant showing alterations in cell tropism. J. Virol. 72, 10171–10179.

Escarmís, C., Gómez-Mariano, G., Dávila, M., Lázaro, E., Domingo, E., 2002. Resistance to extinction of low fitness virus subjected to plaque-to-plaque transfers: diversification by mutation clustering. J. Mol. Biol. 315, 647–661.

Escarmís, C., Lazaro, E., Arias, A., Domingo, E., 2008. Repeated bottleneck transfers can lead to non-cytocidal forms of a cytopathic virus: implications for viral extinction. J. Mol. Biol. 376, 367–379.

Escarmís, C., Perales, C., Domingo, E., 2009. Biological effect of Muller's Ratchet: distant capsid site can affect picornavirus protein processing. J. Virol. 83, 6748–6756.

Fischer, W., Ganusov, V.V., Giorgi, E.E., Hraber, P.T., Keele, B.F., et al., 2010. Transmission of single HIV-1 genomes and dynamics of early immune escape revealed by ultra-deep sequencing. PLoS One 5, e12303.

García-Arriaza, J., Manrubia, S.C., Toja, M., Domingo, E., Escarmís, C., 2004. Evolutionary transition toward defective RNAs that are infectious by complementation. J. Virol. 78, 11678–11685.

García-Arriaza, J., Ojosnegros, S., Dávila, M., Domingo, E., Escarmis, C., 2006. Dynamics of mutation and recombination in a replicating population of complementing, defective viral genomes. J. Mol. Biol. 360, 558–572.

Gause, G.F., 1971. The Struggle for Existence. Dover, New York.

Gebauer, F., de la Torre, J.C., Gomes, I., Mateu, M.G., Barahona, H., et al., 1988. Rapid selection of genetic and antigenic variants of foot-and-mouth disease virus during persistence in cattle. J. Virol. 62, 2041–2049.

Gosert, R., Kardas, P., Major, E.O., Hirsch, H.H., 2010. Rearranged JC virus noncoding control regions found in progressive multifocal leukoencephalopathy patient samples increase virus early gene expression and replication rate. J. Virol. 84, 10448–10456.

Gosselin, A.S., Simonin, Y., Guivel-Benhassine, F., Rincheval, V., Vayssiere, J.L., et al., 2003. Poliovirus-induced apoptosis is reduced in cells expressing a mutant CD155 selected during persistent poliovirus infection in neuroblastoma cells. J. Virol. 77, 790–798.

Gould, S.J., 1989. Wonderful Life: The Burgess Shale and the Nature of History. W.W. Norton, New York.

Hahn, B.H., Shaw, G.M., Taylor, M.E., Redfield, R.R., Markham, P.D., et al., 1986. Genetic variation in HTLV-III/LAV over time in patients with AIDS or at risk for AIDS. Science 232, 1548–1553.

Herfst, S., Schrauwen, E.J., Linster, M., Chutinimitkul, S., de Wit, E., et al., 2012. Airborne transmission of influenza A/H5N1 virus between ferrets. Science 336, 1534–1541.

Herrera, M., Grande-Perez, A., Perales, C., Domingo, E., 2008. Persistence of foot-and-mouth disease virus in cell culture revisited: implications for contingency in evolution. J. Gen. Virol. 89, 232–244.

Holland, J.J., Grabau, E.A., Jones, C.L., Semler, B.L., 1979. Evolution of multiple genome mutations during long-term persistent infection by vesicular stomatitis virus. Cell 16, 495–504.

Holland, J.J., Spindler, K., Horodyski, F., Grabau, E., Nichol, S., et al., 1982. Rapid evolution of RNA genomes. Science 215, 1577–1585.

Imperiale, M.J., Jiang, M., 2015. What DNA viral genomic rearrangements tell us about persistence. J. Virol. 89, 1948–1950.

Jaramillo, N., Domingo, E., Munoz-Egea, M.C., Tabares, E., Gadea, I., 2013. Evidence of Muller's ratchet in herpes simplex virus type 1. J. Gen. Virol. 94, 366–375.

Korber, B., Muldoon, M., Theiler, J., Gao, F., Gupta, R., et al., 2000. Timing the ancestor of the HIV-1 pandemic strains. Science 288, 1789–1796.

Kortenhoeven, C., Joubert, F., Bastos, A., Abolnik, C., 2015. Virus genome dynamics under different propagation pressures: reconstruction of whole genome haplotypes of West Nile viruses from NGS data. BMC Genomics 16, 118.

Lázaro, E., Escarmis, C., Perez-Mercader, J., Manrubia, S.C., Domingo, E., 2003. Resistance of virus to extinction on bottleneck passages: study of a decaying and fluctuating pattern of fitness loss. Proc. Natl. Acad. Sci. U. S. A. 100, 10830–10835.

Leslie, J.F., Vrijenhoek, R.C., 1980. Consideration of Muller's ratchet mechanism through studies of genetic linkage and genomic compatibilities in clonally reproducing *Poeciliopsis*. Evolution 34, 1105–1115.

Loewe, L., 2006. Quantifying the genomic decay paradox due to Muller's ratchet in human mitochondrial DNA. Genet. Res. 87, 133–159.

Martin Hernandez, A.M., Carrillo, E.C., Sevilla, N., Domingo, E., 1994. Rapid cell variation can determine the establishment of a persistent viral infection. Proc. Natl. Acad. Sci. U. S. A. 91, 3705–3709.

Martin, V., Perales, C., Davila, M., Domingo, E., 2006. Viral fitness can influence the repertoire of virus variants selected by antibodies. J. Mol. Biol. 362, 44–54.

Martínez-Salas, E., Ryan, M.D., 2010. Translation and protein processing. In: Ehrenfeld, E., Domingo, E., Roos, R.P. (Eds.), The Picornaviruses. ASM Press, Washington, DC, pp. 141–164.

Maynard Smith, J., Szathmáry, E., 1999. The Origins of Life: From the Birth of Life to the Origins of Language. Oxford University Press, Oxford.

Maynard-Smith, J., 1976. The Evolution of Sex. Cambridge University Press, Cambridge.

Meyerhans, A., Cheynier, R., Albert, J., Seth, M., Kwok, S., et al., 1989. Temporal fluctuations in HIV quasispecies *in vivo* are not reflected by sequential HIV isolations. Cell 58, 901–910.

Miyake, T., Shiba, T., Watanabe, I., 1967. Grouping of RNA phages by a Milipore filtration method. Jpn. J. Microbiol. 11, 203–211.

Moran, N.A., 1996. Accelerated evolution and Muller's rachet in endosymbiotic bacteria. Proc. Natl. Acad. Sci. U. S. A. 93, 2873–2878.

Moreno, E., Ojosnegros, S., Garcia-Arriaza, J., Escarmis, C., Domingo, E., et al., 2014. Exploration of sequence space as the basis of viral RNA genome segmentation. Proc. Natl. Acad. Sci. U. S. A. 111, 6678–6683.

Mrukowicz, J.Z., Wetzel, J.D., Goral, M.I., Fogo, A.B., Wright, P.F., et al., 1998. Viruses and cells with mutations affecting viral entry are selected during persistent rotavirus infections of MA104 cells. J. Virol. 72, 3088–3097.

Muller, H.J., 1964. The relation of recombination to mutational advance. Mutat. Res. 1, 2–9.

Murphy, E., Shenk, T., 2008. Human cytomegalovirus genome. Curr. Top. Microbiol. Immunol. 325, 1–19.

Narayan, O., Clements, J.E., Griffin, D.E., Wolinsky, J.S., 1981. Neutralizing antibody spectrum determines the antigenic profiles of emerging mutants of visna virus. Infect. Immun. 32, 1045–1050.

Nash, A.A., Dalziel, R.G., Fitzgerald, J.R., 2015. Mims' Pathogenesis of Infectious Disease. Elsevier Science & Technology, Edinburgh.

Nathanson, N., Gonzalez-Scarano, F., 2007. Viral persistence. In: Nathanson, N. (Ed.), Viral Pathogenesis and Immunity. Academic Press, San Diego, pp. 130–145.

Novella, I.S., 2003. Contributions of vesicular stomatitis virus to the understanding of RNA virus evolution. Curr. Opin. Microbiol. 6, 399–405.

Novella, I.S., 2004. Negative effect of genetic bottlenecks on the adaptability of vesicular stomatitis virus. J. Mol. Biol. 336, 61–67.

Novella, I.S., Elena, S.F., Moya, A., Domingo, E., Holland, J.J., 1995. Size of genetic bottlenecks leading to virus fitness loss is determined by mean initial population fitness. J. Virol. 69, 2869–2872.

Novella, I.S., Hershey, C.L., Escarmis, C., Domingo, E., Holland, J.J., 1999a. Lack of evolutionary stasis during alternating replication of an arbovirus in insect and mammalian cells. J. Mol. Biol. 287, 459–465.

Novella, I.S., Quer, J., Domingo, E., Holland, J.J., 1999b. Exponential fitness gains of RNA virus populations are limited by bottleneck effects. J. Virol. 73, 1668–1671.

Novella, I.S., Ebendick-Corpus, B.E., Zarate, S., Miller, E.L., 2007. Emergence of mammalian cell-adapted vesicular stomatitis virus from persistent infections of insect vector cells. J. Virol. 81, 6664–6668.

Nowak, M.A., Schuster, P., 1989. Error thresholds of replication in finite populations mutation frequencies and the onset of Muller's ratchet. J. Theor. Biol. 137, 375–395.

Ojosnegros, S., Beerenwinkel, N., Antal, T., Nowak, M.A., Escarmis, C., et al., 2010. Competition-colonization dynamics in an RNA virus. Proc. Natl. Acad. Sci. U. S. A. 107, 2108–2112.

Ojosnegros, S., Garcia-Arriaza, J., Escarmis, C., Manrubia, S.C., Perales, C., et al., 2011. Viral genome segmentation can result from a trade-off between genetic content and particle stability. PLoS Genet. 7, e1001344.

Oldstone, M.B., 2006. Viral persistence: parameters, mechanisms and future predictions. Virology 344, 111–118.

Olivares, I., Sanchez-Jimenez, C., Vieira, C.R., Toledano, V., Gutierrez-Rivas, M., et al., 2013. Evidence of ongoing replication in a human immunodeficiency virus type 1 persistently infected cell line. J. Gen. Virol. 94, 944–954.

Pavio, N., Couderc, T., Girard, S., Sgro, J.Y., Blondel, B., et al., 2000. Expression of mutated poliovirus receptors in human neuroblastoma cells persistently infected with poliovirus. Virology 274, 331–342.

Pawlotsky, J.M., 2006. Hepatitis C virus population dynamics during infection. Curr. Top. Microbiol. Immunol. 299, 261–284.

Perales, C., Martin, V., Ruiz-Jarabo, C.M., Domingo, E., 2005. Monitoring sequence space as a test for the target of selection in viruses. J. Mol. Biol. 345, 451–459.

Quer, J., Huerta, R., Novella, I.S., Tsimring, L., Domingo, E., et al., 1996. Reproducible nonlinear population dynamics and critical points during replicative competitions of RNA virus quasispecies. J. Mol. Biol. 264, 465–471.

Quer, J., Hershey, C.L., Domingo, E., Holland, J.J., Novella, I.S., 2001. Contingent neutrality in competing viral populations. J. Virol. 75, 7315–7320.

Richman, D.D., Wrin, T., Little, S.J., Petropoulos, C.J., 2003. Rapid evolution of the neutralizing antibody response to HIV type 1 infection. Proc. Natl. Acad. Sci. U. S. A. 100, 4144–4149.

Ron, D., Tal, J., 1985. Coevolution of cells and virus as a mechanism for the persistence of lymphotropic minute virus of mice in L-cells. J. Virol. 55, 424–430.

Ron, D., Tal, J., 1986. Spontaneous curing of a minute virus of mice carrier state by selection of cells with an intracellular block of viral replication. J. Virol. 58, 26–30.

Roossinck, M.J., 2014. Metagenomics of plant and fungal viruses reveals an abundance of persistent lifestyles. Front. Microbiol. 5, 767.

Schrauwen, E.J., de Graaf, M., Herfst, S., Rimmelzwaan, G.F., Osterhaus, A.D., et al., 2014. Determinants of virulence of influenza A virus. Eur. J. Clin. Microbiol. Infect. Dis. 33, 479–490.

Shankarappa, R., Margolick, J.B., Gange, S.J., Rodrigo, A.G., Upchurch, D., et al., 1999. Consistent viral evolutionary changes associated with the progression of human immunodeficiency virus type 1 infection. J. Virol. 73, 10489–10502.

Silander, O.K., Tenaillon, O., Chao, L., 2007. Understanding the evolutionary fate of finite populations: the dynamics of mutational effects. PLoS Biol. 5, e94.

Sobrino, F., Dávila, M., Ortín, J., Domingo, E., 1983. Multiple genetic variants arise in the course of replication of foot-and-mouth disease virus in cell culture. Virology 128, 310–318.

Sorrell, E.M., Wan, H., Araya, Y., Song, H., Perez, D.R., 2009. Minimal molecular constraints for respiratory droplet transmission of an avian-human H9N2 influenza A virus. Proc. Natl. Acad. Sci. U. S. A. 106, 7565–7570.

Sponseller, B.A., Sparks, W.O., Wannemuehler, Y., Li, Y., Antons, A.K., et al., 2007. Immune selection of equine infectious anemia virus env variants during the long-term inapparent stage of disease. Virology 363, 156–165.

Takemoto, K.K., Habel, K., 1959. Virus-cell relationship in a carrier culture of HeLa cells and Coxsackie A9 virus. Virology 7, 28–44.

Temin, H.M., 1989. Is HIV unique or merely different? J. Acquir. Immune Defic. Syndr. 2, 1–9.

Tilman, D., 1994. Competition and biodiversity in spatially structured habitats. Ecology 75, 2–16.

Tsibris, A.M., Korber, B., Arnaout, R., Russ, C., Lo, C.C., et al., 2009. Quantitative deep sequencing reveals dynamic HIV-1 escape and large population shifts during CCR5 antagonist therapy *in vivo*. PLoS One 4, e5683.

Tsimring, L.S., Levine, H., Kessler, D.A., 1996. RNA virus evolution via a fitness-space model. Phys. Rev. Lett. 76, 4440–4443.

Van Valen, L., 1973. A new evolutionary law. Evol. Theory 1, 1–30.

Weismann, A., 1891-1892. Essays upon Heredity and Kindred Subjects. vols. I and II. Clarendon Press, Oxford.

Weissmann, C., Billeter, M.A., Goodman, H.M., Hindley, J., Weber, H., 1973. Structure and function of phage RNA. Annu. Rev. Biochem. 42, 303–328.

Wilson, G.J., Wetzel, J.D., Puryear, W., Bassel-Duby, R., Dermody, T.S., 1996. Persistent reovirus infections of L cells select mutations in viral attachment protein sigma 1 that alter oligomer stability. J. Virol. 70, 6598–6606.

Yuste, E., Sánchez-Palomino, S., Casado, C., Domingo, E., López-Galíndez, C., 1999. Drastic fitness loss in human immunodeficiency virus type 1 upon serial bottleneck events. J. Virol. 73, 2745–2751.

Zhong, J., Gastaminza, P., Chung, J., Stamataki, Z., Isogawa, M., et al., 2006. Persistent hepatitis C virus infection in vitro: coevolution of virus and host. J. Virol. 80, 11082–11093.

LONG-TERM VIRUS EVOLUTION IN NATURE

CHAPTER CONTENTS

ABBREVIATIONS

CAT	colonization-adaptation trade-off
cccDNA	covalently closed circular DNA
CTL	cytotoxic T lymphocyte
CRF	circulating recombinant forms
ELISA	enzyme-linked immunosorbent assay
EMCV	encephalomyocarditis virus
ET	evolutionary trace
FAO	Food and Agriculture Organization of the United Nations
FMD	foot-and-mouth disease
FMDV	foot-and-mouth disease virus
HAV	hepatitis A virus
HBV	hepatitis B virus
HCV	hepatitis C virus
HIV-1	human immunodeficiency virus type 1

Virus as Populations. http://dx.doi.org/10.1016/B978-0-12-800837-9.00007-1

HLA	human leukocyte antigen
HTLV-1	human T-cell lymphotropic virus type 1
HTLV-2	human T-cell lymphotropic virus type 2
HRV	human rhinovirus
ICTV	International Committee on Taxonomy of Viruses
IV	influenza virus
MARM	monoclonal antibody-resistant mutant
ML	maximum likelihood
MV	measles virus
NGS	next-generation sequencing
PAM	percent accepted mutation
PIR	protein information resource
PV	poliovirus
RV	rabies virus
SARS	severe acute respiratory syndrome
SIVcpz	chimpanzee simian immunodeficiency virus
s/nt/y	substitutions per nucleotide and year
URF	unique recombinant form
URL	uniform resource locator
WNV	west Nile virus

7.1 INTRODUCTION TO THE SPREAD OF VIRUSES. OUTBREAKS, EPIDEMICS, AND PANDEMICS

Intrahost virus replication and evolution are the first steps in the process of virus diversification that continues with successive virus transmission events that are a condition for long-term survival in nature. Viruses are perpetuated as a consequence of many rounds of persistent or acute infections, with possible extracellular stages in which genomes remain basically invariant. Despite lacking direct evidence, we presume that multitudes of successive transmissions have allowed viruses to survive at least for thousands of years, probably undergoing continuous genetic change. Picornavirologists are familiar with an Egyptian stela dated 1550-1333 B.C. (18th Egyptian dynasty) that portrays the image of a man with an atrophic leg probably a consequence of infection with poliovirus (PV) or a related virus (Eggers, 2002). In this chapter we deviate from the focus on how viral population numbers affect short-term survival and evolution, and we turn to features of viruses as they infect successive hosts to persist in nature.

Viruses can be transmitted vertically or horizontally. Vertical transmission occurs from parental organisms to their offspring, and it includes infection through the germ line in animals and plants, from the mother to the embryo during fetal development, and also postnatal transmission to the newborn via blood, milk, or contact (Mims, 1981; Nash et al., 2015). In the horizontal transmission, a virus spreads from infected individuals to susceptible recipients. This is the type of transmission we are most familiar with. It frequently gives rise to disease outbreaks (infection episodes localized in space and time that affect a few individuals), epidemics (that affect an ample geographical area and are often extended in time), and pandemics (that affect most areas of our planet), typically the periodic influenza pandemics.

All transmission modes have probably contributed to the maintenance of viruses in our biosphere. Persistent infections are likely to have played a major role when the number of individual humans or animals living in close contact was limited throughout the preagricultural era, earlier than 10,000 years ago. A favorable climate change during the Holocene (the geological epoch that began at the end of the

Pleistocene, around 12,000 years before present; compare with Chapter 1) was probably an important driver toward large-scale domestication of plants and animals, 10,000-7000 years B.C. From the behavior of current viruses, there might have always been a dynamics of virus change for adaptability within individual hosts, and transmissions among animals or plants. The probability of transmission increased as host population numbers rose with agricultural practices and urban life in the last several thousand years. Intensive agriculture must have contributed to accelerated sequence space exploration by viruses with consequences for the emergence of viral disease (Section 7.7). There has been probably a continuous dynamics of viral emergences, reemergences, and extinctions with patterns that may be parallel to those observed with present-day viruses. Virology has existed as an organized scientific discipline with the possibility to isolate, store, and study viruses only for about one century. The challenge to reconstruct the events that might have led to viruses similar to the ones we isolate today was addressed in Chapter 1, with a critical first question being if viruses originated 4000 million years ago, or "only" 2000 million years ago (diagrams in Figures 1.3 and 1.4, and Section 1.5 in Chapter 1). In this chapter, we are more modest in our aspirations and we will analyze with the tools of genomics what happens when viruses evolve for months or years in what we call interhost virus evolution.

Unfortunately viruses have not left any fossil record (at least that we can uncover with the available tools), since according to current paleontology nitrogen- and phosphorus-rich molecules are unlikely to be protected in fossils older than 1 million years. At most, decades-old biological samples containing viruses (such as those from lung specimens or a frozen body infected with the incorrectly called "Spanish" influenza of 1918) have been preserved, and sequences have been retrieved. Fortunately, there is a different type of "molecular fossil" record of viral genomes in the DNA of differentiated organisms, in the form of integrated virus-like genetic elements. This research area is termed Paleovirology (Aswad and Katzourakis, 2012). The presence of recognizable viral genomic sequences in cellular DNA suggests a history of long-term interaction between viruses and cells, in support of some models of virus origins that propose a long coevolution between precellular and cellular entities with virus-like elements (Chapter 1). Despite the current capacity to amplify tiny amounts of viral nucleic acids for nucleotide sequence determinations, proposals on how long-term viral evolution might proceed have to be based mainly on the comparison of viral genomes and the structure of viral proteins from modern representatives of different virus groups. First we should understand the basic concepts related to virus transmission, keeping in mind viral population numbers and the complexity of viral populations.

7.2 REPRODUCTIVE RATIO AS A PREDICTOR OF EPIDEMIC POTENTIAL. INDETERMINACIES IN TRANSMISSION EVENTS

The basic reproductive ratio (R_0) is the average number of infected contacts per infected individual. At a population level, a value of R_0 larger than 1 means that a virus will continue its propagation among susceptible hosts, if no environmental changes or external influences intervene. A R_0 value lower that 1 means that the virus is doomed to extinction at the epidemiological level under those specific circumstances. The basic models of infection dynamics were developed by R.M. Anderson, R.M. May, and M.A. Nowak, with inclusion of the following key parameters: rate k at which uninfected hosts enter the population of susceptibles (x), their normal death rate (u) (so that the equilibrium abundance of uninfected hosts is k/u), number of infected hosts (y), mortality due to infection (v) (so that $1/u+v$ is the average lifetime of an infected host), a rate constant (β) that characterizes parasite infectivity (so that βx is the rate of new infections and βxy is the rate at which infected hosts transmit the virus to uninfected

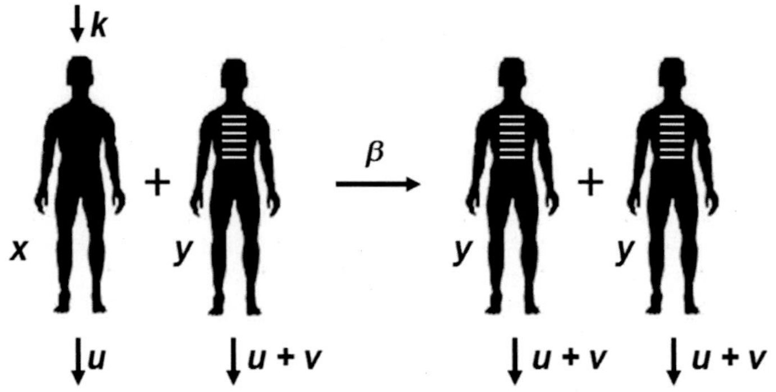

$$\frac{dx}{dt} = k - ux - \beta xy$$

$$\frac{dy}{dt} = y(\beta x - u - v)$$

$$R_0 = \frac{\beta}{u + v} \cdot \frac{k}{u}$$

R_0 = Basic reproductive ratio

FIGURE 7.1

A schematic representation of the main parameters of viral dynamics that enter the equations that predict the rate of variation of uninfected and infected (internal horizontal lines in the human figure) individuals (shaded box on the left) and the R_0 value (shaded box on the right). The meaning of parameters and literature references are given in the text.

hosts). These parameters are schematically indicated in Figure 7.1 and they provide a theoretical value for R_0 (Anderson and May, 1991; Nowak and May, 2000; Nowak, 2006).

R_0 values are not a universal constant for viruses because, as discussed in Chapters 3 and 4, virus variation may affect viral fitness and viral load in infected individuals, and the latter, in turn, may influence the amount of virus that surfaces in a host to permit transmission. Despite uncertainties, consistent R_0 values have been estimated for different viral pathogens based on field observations. Values of R_0 for human immunodeficiency virus type 1 (HIV-1) and severe acute respiratory syndrome (SARS) corona-virus range from 2 to 5, for PV from 5 to 7, for Ebola virus from 1.5 to 2.5. For measles virus (MV), which is one of the most contagious viruses described to date, the R_0 reaches 12-18 (Heffernan et al., 2005; Althaus, 2014). Most isolates of the SARS coronavirus that circulated months after the emergence of this human pathogen had modest R_0 values, and this is consistent with SARS not having reached the pandemic proportions feared immediately following its emergence. In contrast, MV is highly transmissible, thus explaining frequent outbreaks as soon as a sizable population stops vaccinating its infants. Since some of the parameters that enter the basic equations of viral dynamics depend on the nucleotide sequence of the viral genome, mutations may alter R_0 values, allowing some virus variants to overtake those that were previously circulating in the population (Figure 7.2). Viral replication, fitness, load, transmissibility, and virulence are all interconnected factors that contribute to virus persistence in its broader sense of virus being perpetuated in nature. These parameters can affect both disease progression in an infected individual and transmissibility at the epidemiological level.

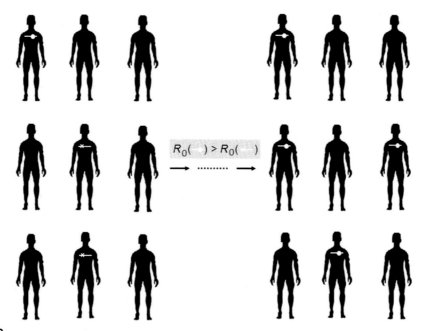

FIGURE 7.2

Displacement of a virus variant by another, by virtue of the latter displaying a higher R_0 value. The competing viruses are depicted as horizontal lines with a distinctive symbol. Differences in R_0 recapitulate part of the determinants of epidemiological fitness (Section 5.9 in Chapter 5). Concepts of competition among clones or populations within infected host organisms or cell cultures, treated in previous chapters, can be extended at the epidemiological level, with the appropriate choice of the key parameters. References are given in the text.

The difference between the numbers of infectious particles that participate in transmission versus the total number of virus in an infected, donor organism provides a first picture of the indeterminacies involved in viral transmissions. The larger the population size and genetic heterogeneity of the virus in an infected individual, the higher will be the likelihood that independent transmission events have different outcomes. Individual susceptible hosts will receive subsets of related but nonidentical genomes. In a bright article that emphasized the molecular evidence and medical implications of quasispecies in viruses, J.J. Holland and colleagues wrote the following statement: "Therefore, the acute effects and subtle chronic effects of infections will differ not only because we all vary genetically, physiologically, and immunologically, but also because we all experience a different array of quasispecies challenges. These facts are easily overlooked by clinicians and scientists because disease syndromes are often grossly similar for each type of virus, and because it would appear to make no difference in a practical sense. However, for the person who develops Guillain-Barré syndrome following a common cold, or for the individual who remains healthy despite many years of HIV-1 infection, for example, it may make all the difference in the world" (Holland et al., 1992). Indeterminacies in the process of virus spread can be viewed as an extension of the diversification due to bottleneck events in the case of virus transmission, as visualized in Figures 6.1 and 6.2 in Chapter 6, when dealing with the limitations of the virus samples retrieved from an infected host as the starting material for experimental evolution approaches.

7.3 RATES OF VIRUS EVOLUTION IN NATURE

Despite a necessarily approximate and imprecise knowledge of how many and which types of genomes participate in successive horizontal and vertical transmissions, we can obtain an overall estimate of the rate at which viruses evolve in nature. This is commonly done by comparing consensus genomic nucleotide sequences of viruses isolated at different times during an outbreak, epidemic, or pandemic.

The rate of evolution (also termed rate of fixation or rate of accumulation of mutations) is generally expressed as substitutions per nucleotide and year (s/nt/y). The term fixation is not the most adequate when dealing with virus evolution given that the term refers to a consensus that, in addition to being an average of the real sequences, has a fleeting dominance. Yet, the term is frequently used in the literature of general genetics and virus evolution. The rate of evolution is calculated from genetic distances between consensus viral genomic sequences of successive viral samples from a single persistently or acutely infected host, or from different host individuals infected at different times. Rates of evolution are only indirectly related to mutation rates and mutation frequencies that do not include a time factor in them. There have been several comparisons of rates of evolution for viruses that document the differences between RNA and DNA viruses (Jenkins et al., 2002; Hanada et al., 2004; Domingo, 2007). A few comparative values are given in Table 7.1.

Herpes simplex virus constitutes an example of a complex DNA virus for which, despite uncertainties (Firth et al., 2010) a calculated rate of evolution was 10^{-8} s/nt/y (Sakaoka et al., 1994), which is actually closer to the rate estimated for cellular genes than for most viruses. Yet, its mutation frequencies, measured by independent procedures are in the range of 7×10^{-3} to 1×10^{-5} (see also Section 7.4.2). The latter values may result from the selective agent targeting a replicating herpes simplex virus that has produced multiple variants, while the overall slow rate of evolution may be influenced by periods of latency. Slow evolution is expected for retroviruses such as human T-cell lymphotropic virus types 1 and 2 (HTLV-1 and HTLV-2) whose life cycles are dominated by the integrated provirus stage, with the viruses following the clonal expansion of their host cells (Melamed et al., 2014). Some single-stranded DNA viruses display rates of evolution typical of the rapidly evolving RNA viruses (Table 7.1).

Different genes of the same virus set may show different rates of evolution (i.e., the polymerase and other nonstructural proteins may evolve more slowly than structural proteins). Thus, a rate of evolution is far from being a universal feature of a virus. A comparison of rates of synonymous substitutions (under the assumption that synonymous substitutions do not affect protein function; see Chapter 2 for limitations of considering synonymous mutations as neutral) for several RNA

Table 7.1 Some Representative Rates of Virus Evolution in Nature

Virus or Organism	Range of Values
RNA viruses (riboviruses)	10^{-2} to 10^{-5}
Retroviruses	10^{-1} to 10^{-5}
Single-stranded DNA viruses	10^{-3} to 10^{-4}
Double-stranded DNA viruses	10^{-7} to 10^{-8}
Cellular genes of host organisms	10^{-8} to 10^{-9}

Values are expressed as substitutions per nucleotide and year. The range of values is based on several studies. Values depend on the virus under study, the genomic region analyzed, and several factors discussed in the text.

viruses, yielded a range of evolutionary rates of 6×10^{-2} to 1×10^{-7} synonymous substitutions per synonymous site per year (Hanada et al., 2004). The values were recalculated from primary phylogenetic data using maximum likelihood (ML) (Section 7.6), under the assumption of the molecular clock, and inference of the ancestral nucleotide sequences at the tree nodes. The five orders of magnitude variation were attributed mainly to the degree of virus replication rather than to differences in error rate. We will deal with the molecular clock hypothesis (constant rate of accumulation of mutations) in Section 7.3.3, but the major features of virus evolution studied in previous chapters (mainly those typical of mutant swarm-forming RNA and DNA viruses) should make us skeptical of similar evolution rates in different biological contexts. Rate variations were documented with HIV-1 subpopulations in different compartments of the human brain (Salemi et al., 2005). The data did not fit a "global" molecular clock for the virus in the brain, and "local" clocks showed that meninges and temporal lobe HIV-1 subpopulations evolved 30 and 100 times faster, respectively, than other HIV-1 populations in the brain. It is believed that these differences were due to random drift rather than selection. An additional complication is that even restricting virus isolations to the same biological material in a standard epidemiological setting, several measurements indicated discontinuities in evolutionary rates. The discontinuities had at least two origins: the nonlinear effect of time, and some unique features of evolution occurring inside an infected host. These points are examined next.

7.3.1 INFLUENCE OF THE TIME OF SAMPLING

Noncumulative sequence changes in the hemagglutinin of influenza virus (IV) type C were found in an early study by Buonagurio et al. (1985). The authors proposed a cocirculation of variants that belonged to different evolutionary lineages. If multiple evolutionary pathways coexist in a given geographical area, and they establish a network of lineages that evolve with time, variations of calculated rates of evolution are expected, and they may distort the rate of evolution of individual lineages.

A second early observation was made during an episode of foot-and-mouth disease (FMD) in Spain. Estimates of the rate of evolution of the virus ranged from $<4 \times 10^{-4}$ to 4×10^{-2} s/nt/y, depending on the genomic region analyzed, and the time period between isolations (Sobrino et al., 1986). Cocirculation of multiple heterogeneous foot-and-mouth disease virus (FMDV) samples ("evolving quasispecies") was proposed. The result to be emphasized here is that the calculated rates of evolution were extremely high (higher than 10^{-2} s/nt/y) if the two FMDVs compared were isolated at close time points, while lower values were calculated when the viruses were sampled from different animals at distant time points.

The dependence of the calculated rate of evolution during the epidemic spread of the virus on the time interval between virus isolations for sequence determination is expected for viruses that need not be transmitted by direct contact between an infected and a susceptible host. Some viruses remain infectious in the environment for prolonged time periods, until they reach a susceptible host in which to initiate replication rounds. This is the case of viruses transmitted by the fecal-oral route such as enteroviruses. FMDV can adhere and remain infectious on many objects (fomites), including dust particles, food products with neutral pH, or insects that can transport the virus mechanically. Infectious FMDV can traverse long distances (many kilometers) on dust particles, people, trains, and the like. Even if some infectivity is lost, a few infectious particles are sufficient to infect an animal (Sellers, 1971, 1981). There are some classic examples of long-distance transport of FMDV, a virus subjected to close scrutiny due to its economic impact. One is the spread of SAT1 and A22 FMDV during the 1960s in Turkey along the railway line from the cattle raising region of Lake Van to slaughterhouses in Istambul [this and other examples are described in (Brooksby, 1981)]. Computer models have been developed to explain and predict possible airborne FMDV transmission in different geographical areas (Sorensen et al., 2000). [As an anecdote, in my experience as a member of the Research Group of the Standing Technical Committee for the Control of FMD of Food and Agriculture Organization of the United Nations (FAO) in the 1980s, FMD outbreaks in any country always came from somewhere else.].

For viruses that can remain infectious outside their hosts, and that do not need donor-recipient host contacts to perpetuate transmission chains, the time between isolations will influence the calculated rate of evolution based on genomic nucleotide sequences. The reason is that during the extracellular stages, the virus will not undergo genetic change, at least to the extent of variation during intracellular replication (possible mutations due to chemical damage in viral genomes is indicated in Section 2.2. of Chapter 2). The effect of nonreplicative time intervals in the rate of evolution is illustrated in Figure 7.3.

Some complications should be considered in the interpretation of the analyses depicted in Figure 7.3: (i) the consensus sequences determined to characterize the virus shed by each animal is a simplification of the real genome composition of the virus. (ii) Individual animals vary in physiological and immunological status, and, obviously, they are not in line waiting to be infected; they move, gather around water and food sources, some are isolated, others in close contact with their peer, and

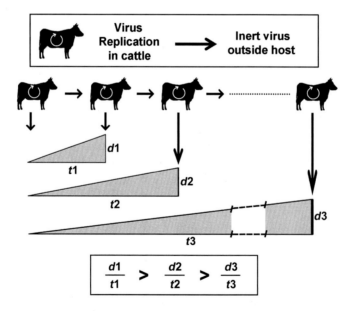

FIGURE 7.3

Illustration of the inverse correlation between the time between viral isolations for consensus sequence determination, and the calculated rate of evolution. Animals sustain the replication (inside curved arrow) of virus that will be transmitted to a susceptible animal. The time that the virus spends outside an animal (absence of replication) is depicted by a horizontal arrow. The time between virus isolation is given by $t1$, $t2$, and $t3$. The number of nucleotide differences in the virus isolates relative to the sequence of the initial (reference) animal is given by $d1$, $d2$, and $d3$ (vertical arrows). Because of the increasing periods of stasis (addition of horizontal arrows), calculated rates of evolution given by the d/t ratio will be higher the shorter the time interval between isolations. See text for references.

so on. (iii) In this case, virus transport is assumed to be mechanical (on dust particles carried by wind, aerosols, insects, etc.) without additional viral replication during transport. Yet, subpopulations of the most environment-resistant particles, or particles that adhere best to the transporter object, may bias the composition of the virus that will reach an animal to pursue replication. Such events, occurring for ten to hundred rounds of host infections, render the appalling virus diversity described in Chapter 1 a bit less appalling. Since several additional environmental circumstances are changeable and unpredictable, it is unlikely that rates of viral evolution in nature can remain invariant on the basis of some internal principle of constant mutation occurrence (as if accumulation of mutations was as monotonous as radioactive decay!).

7.3.2 INTERHOST VERSUS INTRAHOST RATE OF EVOLUTION

Additional observations against constant mutational input with time have been made with HIV-1 and human and avian hepatitis B virus (HBV). The main finding is that interhost rates of evolution are lower than intrahost rates, even under a comparable set of epidemiological parameters. Several proposals have been made to account for this difference. A.J. Leslie and colleagues described cytotoxic T lymphocyte (CTL)-escape mutants of HIV-1 from infected patients. Some of the mutants reverted to the wild-type sequence after transmission to individuals negative for the human leukocyte antigen (HLA) alleles associated with long-term HIV-1 control (Leslie et al., 2004). Strong intrahost selective pressures and reversion of part of the selected mutations upon transmission to a susceptible individual is one of the possible mechanisms behind diminished evolutionary rates when viruses from multiple hosts are compared (Figure 7.4, Box 7.1).

J.T. Herbeck, J.I. Mullins, and colleagues systematically observed lower nucleotide sequence divergence between HIV-1 isolates from different individuals sampled in primary infection than between isolates from individuals with advanced illness. HIV-1 regained some ancestral features when infecting a new host, again explaining a higher intrahost than interhost evolutionary rate (Herbeck et al., 2006). In a study of HIV-1 transmission between several pairs of individuals over an 8-year period, A.D. Redd and colleagues reported that the viral populations found in the newly infected recipients were more closely related to ancestral sequences from the donor than to the sequences found in the donor near the time of transmission (Redd et al., 2012). Preferential transmission of ancestral sequences may also contribute to lower interhost than intrahost rates of evolution (Box 7.1).

K.A. Lythgoe and C. Fraser provided evidence that cycling of HIV-1 through long-lived memory CD4+T cells is probably the main contributing factor to slower HIV-1 evolution at the epidemic level (Lythgoe and Fraser, 2012). Ancestral sequences of HIV-1 in infected individuals may arise by the activation of proviral sequences kept in the form of quasispecies memory. In this case, it is the type of molecular memory that we defined as reservoir, anatomical, or cellular memory in Section 5.5 of Chapter 5. A related type of reservoir memory is found in HBV, in the form of covalently closed circular DNA (cccDNA) that persists in the nuclei of infected hepatocytes, and acts as a template for the synthesis of pregenomic RNA and viral mRNAs (Kay and Zoulim, 2007). In this case, a record of ancient sequences is registered in the cccDNA. It should be noted that memory levels are dependent on fitness values, as evidenced experimentally with FMDV and expected from the theoretical basis of memory implementation (Chapter 5). In consequence, the most abundant memory genomes established early in an infection might be those displaying the highest fitness early in infection, and they might be better adapted to initiate infections than to sustain them (Figure 7.4).

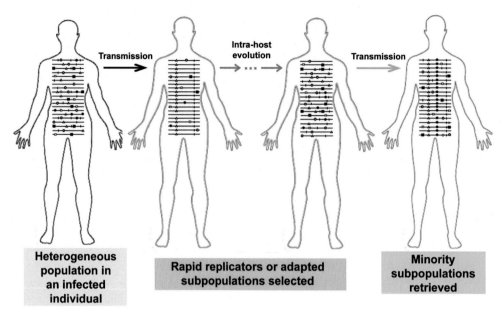

FIGURE 7.4

A possible mechanism for a faster intrahost than interhost rate of evolution. Transmission events are represented by long arrows, and intrahost evolution by short arrows. The virus in the person on the left (black outline) has evolved to generate a complex mutant spectrum. However, only a subset of the genomes are efficiently transmitted to the recipient person (brown outline). The virus in the recipient person evolves toward a complex mutant spectrum. Again, in this new mutant spectrum only a subset of genomes that resemble the ones in the first transmission are efficiently transmitted to the third person (green outline). The net result is that because at each transmission the genomes related to those that first entered the previous host have an advantage, rates of evolution will appear as slower than those within each host. Boxes at the bottom summarize the major event at each step. See text for additional related mechanisms and references.

BOX 7.1 MODELS FOR NONLINEAR RATES OF EVOLUTION

- For viruses that remain infectious in the extracellular environment, stasis due to absence of replication will result in rates of evolution inversely correlated with the time between the isolation of the compared viruses.
- Adapt and revert. Mutants that permit adaptation to a new host individual revert upon transmission.
- Preferential transmission of ancestral sequences. Despite diversification in any host, ancestral sequences have a selective advantage in transmission. They may be retrieved from cellular memory (integrated provirus in HIV-1 or cccDNA in HBV).
- Colonization-adaptation trade-off. Sequential changes in the intensity of the host immune response favor dominance of some genome subpopulations over others. Upon transmission, ancestral minority subpopulations may become dominant.

Additional mechanisms for the time dependence of evolutionary rates have been suggested for HBV. In a 17 years follow-up of several patients, HBV diversity increased during periods of active host immune response, and viral copy numbers decreased. When the immune response was weak, viral genome diversity decreased and viral copy numbers increased; these periods are expected to be those of high transmissibility (Wang et al., 2010).

Endogenous hepadnaviruses are present in the genomes of several organisms. There is evidence that some of the integration events in avian hosts are at least 19 million years old. These integrated hepadnaviruses maintain about 75% nucleotide sequence identity with present-day hepadnaviruses, and the comparisons suggest that the long-term substitution rates are 10^3-fold lower than those for circulating avian HBVs (Gilbert and Feschotte, 2010). Permanence of viral genomic sequences in cellular DNA is a mechanism of evolutionary stasis, as it was emphasized in Chapter 3 with the comparison of the evolutionary rate of the retroviral v-*mos* gene and its cellular counterpart c-*mos* (Gojobori and Yokoyama, 1985), among other evidence. Considerable evolutionary stasis is also observed by comparing isolates of HTLV-1 and HTLV-2 whose replication displays preference for maintaining its integration in cellular DNA (Melamed et al., 2014). For viruses that have a dual potential of error-prone replication and of cellular DNA-like stasis, the permanence in cellular DNA may also contribute to reduced long-term evolutionary rates.

HBV quasispecies dynamics was examined in virus that infected members of the same family that presumably acquired the virus through mother-to-infant transmission (Lin et al., 2015). Again, intrahost evolutionary rate was higher than interhost rate, and the latter decreased with the number of transmissions. The differences were mainly due to nonsynonymous substitutions at limited sites. These observations were interpreted as a rapid switch of HBV between colonization (invasion of new host) and adaptation (quasispecies optimization in the new host). The authors referred to the colonization-adaptation trade-off (CAT) model, or alternations of virus facing an environment marked by a limited host immune response followed by a period of active immune response. In the former environment, viruses displaying rapid replication are selected, while in the latter environment, HBV escape mutants with lower productivity are selected. In each transmission, when the virus reaches a new host, the previously adapted subpopulations are overgrown by the rapidly replicating ones. cccDNA can serve as a reservoir of ancient sequences.

In agreement with these proposals, rates of evolution measured in a single infected individual persistently infected with a continuously replicating virus tend to be higher than those observed with the same viruses isolated from different individuals (Morse, 1994; Domingo et al., 2001; Domingo, 2006). Slowly evolving viral genes may nevertheless undergo episodes of rapid evolution and, vice versa, a rapidly evolving gene may be transiently static. This should be considered in statistical approaches to evolution (Gaucher et al., 2002).

7.3.3 RATE DISCREPANCIES AND THE CLOCK HYPOTHESIS

Several not mutually exclusive mechanisms can account for the difference between intra- and interhost rates of evolution as well as the inverse correlation between time of viral isolation and rate of evolution in a scenario of viral disease outbreaks or epidemics (Box 7.1). Some of the mechanisms involve viral population numbers and competition among subpopulations of mutant spectra as critical ingredients. The molecular clock hypothesis dwindles as a conceptual framework because from all evidence, virus evolution is far from being dictated by a steady accumulation of mutations in viral genomes. The major event is "replacement of subpopulations" rather than "accumulation of mutations in genomes." This conceptual change is as important for the long-term evolution of viruses as it was the consideration of the wild type as a cloud of mutants in the definition of a viral population (Chapter 3).

There are additional arguments against the operation of a molecular clock in virus evolution. According to the clock hypothesis, the rate of accumulation of mutations coincides with the rate at which mutations arise in the infected individuals. This holds for neutral mutations and, as documented in several chapters of this book, very few mutations occurring in highly compact viral genomes are truly neutral (with no functional consequences in any environment; see Section 2.3 in Chapter 2). Even for neutral evolution, spatial asymmetries in populations are sufficient to perturb the molecular clock rate, as documented with a theoretical model of broad applicability (Allen et al., 2015). Yet, quasispecies and the operation of a clock are not totally irreconcilable. An epidemiological study with FMDV suggested that viral quasispecies could produce a transient molecular clock due to the periodic sampling of components of the mutant spectrum in transmission (Villaverde et al., 1991). In this case, both time differences between transmissions and spatial heterogeneities would blur the transiently observed regularity.

7.4 LONG-TERM ANTIGENIC DIVERSIFICATION OF VIRUSES

Viruses can change their antigenic properties gradually, in a process termed antigenic drift, or suddenly, in a process termed antigenic shift. The distinction between antigenic drift and shift was established with IV (Gething et al., 1980; Webster, 1999; Parrish and Kawaoka, 2005). Shift in IV is due to genome segment reassortment that incorporates new hemagglutinin or neuraminidase genes. In monopartite viruses, the difference between gradual and drastic antigenic change has also been established (Martínez et al., 1991b).

The antigenic diversification of one FMDV serotype was examined over a six-decade period by comparing amino acid sequences of the major antigenic sites of the virus isolated in three continents (Martínez et al., 1992). The evolution of the capsid genes was associated with linear accumulation of synonymous mutations, but not of amino acid substitutions. Remarkably, the antigenic variation over six decades was due to fluctuations among limited combinations of amino acid residues without net accumulation of amino acid substitutions over time (Figure 7.5). This result suggests that constraints at the protein level may maintain a long-term virus identity at the antigenic level. In a related observation

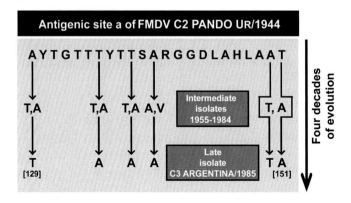

FIGURE 7.5

Evolution of a major antigenic site of FMDV over four decades. The sequence at the top is that of amino acid residues 129-151 of capsid protein VP1 of FMDV C2 Pando, isolated in Uruguay in 1944. Key amino acid positions did not diverge in a linear fashion and isolates over four decades displayed only two types of amino acid residues at each position. See text for possible mechanisms and references.

on the inter-host evolution of HIV-1 mentioned in Section 7.3.2. HIV-1 recovered ancestral features when infecting a new host (Herbeck et al., 2006). Thus, multiple constraints in viruses may limit the rate and mode of long-term diversification, resulting in different numbers of circulating serotypes among related viruses.

7.4.1 WIDELY DIFFERENT NUMBER OF SEROTYPES AMONG GENETICALLY VARIABLE VIRUSES

A puzzling question in evolutionary virology is that despite sharing high mutation rates, some viruses display extensive antigenic diversity in nature reflected in multiple serotypes, while other viruses maintain a relatively invariant antigenic structure, with only one serotype recorded. For the latter group of viruses, the same vaccine can maintain its efficacy over many decades; examples are rabies virus (RV) and MV, two RNA viruses that show remarkable genetic diversity in nature, and estimates of mutation rates and frequencies comparable to other RNA viruses. Antigenic constancy versus variation is also a determinant of long-lasting immunity after infection or vaccination. MV infection produces lifelong immunity (probably as a result of several factors) while patients that have cleared hepatitis C virus (HCV) can be re-infected by the same virus. Cases of patients infected with HCVs of different genotypes are increasingly identified, as more refined diagnostic tests are utilized. No correlation between virus structure (or morphotype) and antigenic diversity has been found. Among structurally closely related viruses, differences in antigenic diversity are apparent. A dramatic case is that of the picornaviruses since encephalomyocarditis virus (EMCV) or hepatitis A virus (HAV) have a single serotype, while human rhinoviruses (HRVs) have been divided into more than 100 serotypes. Other picornaviruses have intermediate numbers of serotypes: three in the case of PV and seven in the case of FMDV. Although it may seem that a diverse antigenic structure may predict a broad host range, this is actually not the case. HAV is highly specialized for the human host while EMCV infects more than 30 species, including mammals, birds, and invertebrates (Knowles et al., 2010).

Several, not mutually exclusive models, have been proposed to account for differences in the antigenic stability (number of serotypes) among viruses:

- Differences in mutation rates, either the average value for the entire genome, or the local mutation rate at the genomic sites that encode antigenic determinants.
- The presence of some dominant and invariant antigenic sites that evoke long-lasting antibodies in the infected hosts, and that obscure other antigenic sites that produce different antibodies that have a limited impact on the antigenic profile of the virus.
- Differences among the assays used for serotype classification. If a universal and standard procedure to classify virus isolates in different serotypes were applied, differences among viruses would be largely lost.
- Difference in the history of virus circulation. Ancient viruses that undergo many rounds of genome replication in each infected host have had an opportunity to diversify antigenically in a manner not possible with viruses that have a more limited history of circulation among susceptible hosts. Antigenic diversification of some viruses currently viewed as antigenically invariant will take place during the next hundreds of years if their circulation continues.
- Some viruses have antigenic sites that cannot vary because they are under severe constraints to accept amino acid substitutions. Antigenic variants may exist as low-fitness subpopulations, but their frequency is too low to modify the results of the diagnostic tests used for serological classifications.

Consideration of these possibilities requires examining some experimental data on virus antigenicity. First, as a conceptual precision, we assume that the number of serotypes is essentially determined by amino acid sequences located in the virus particle, and that either directly or indirectly can affect the interaction of virus with antibodies. Neutralizing and nonneutralizing antibodies may contribute to serological distinctions, depending on the assays performed for serotyping. Serum neutralization tests will identify differences in sensitivity to neutralization while enzyme-linked immunosorbent assay (ELISA) tests will capture reactivity by all raised antibodies. Antibodies can be obtained from infected natural hosts, or from some laboratory animals which are not a natural host for the virus. An ensemble of amino acid residues forms an antigenic determinant which is usually composed of multiple epitopes [defined here as a unit of interaction with a monoclonal antibody (MAb)]. Epitopes can be either continuous (also termed linear) or discontinuous (also termed structured). Continuous epitopes are those whose primary amino acid sequence has the information to react with the cognate antibody. Discontinuous epitopes are those whose reactive residues come from distant positions of the same protein or from residues of different proteins. Many overlapping epitopes can be found within the same antigenic site. Epitopes can include modified amino acid residues such as glycosylated amino acids. Reactivity of discontinuous epitopes with the cognate antibody is generally lost as a consequence of denaturation of the proteins that form the epitope.

With these introductory clarifications, we can now examine the different possibilities listed above.

There is no correlation between limited antigenic diversity and low average mutation rate. Mutation rates and frequencies for RNA viruses fall in the range of 10^{-5} to 10^{-3} substitutions per nucleotide (Chapter 2). However, mutation rates along a viral genome are not uniform, as evidenced by the occurrence of hot spots for variation. Influences such as nucleotide sequence context or RNA structure may conceivably alter mutation rates. It was proposed that a predicted double-stranded RNA at the region encoding the major antigenic site of FMDV might increase locally the polymerase error rate and give rise to multiple amino acid substitutions (Weddell et al., 1985). While at some specific sites polymerases may be more error prone than average, subsequent evidence for FMDV indicated that antigenic variation is due to amino acid substitutions at different antigenic sites and that even variation at the major site can be mediated by distant amino acids on the viral capsid (Rowlands et al., 1983; Geysen et al., 1984; Mateu et al., 1990; Feigelstock et al., 1996). Later molecular studies have not provided evidence that viruses may have a large number of serotypes because their polymerases are more error prone when copying regions encoding amino acids that belong to antigenic sites. Therefore, the possibility that differences in mutation rates determine the number of serotypes is highly unlikely.

Most viruses include multiple antigenic sites, and antibodies are raised against several surface proteins to produce an array of neutralizing and nonneutralizing antibody molecules. Taking again picornaviruses as an example, the number of antigenic domains (each composed of multiple epitopes) varies between one and four (Mateu, 1995). There is no evidence that a restriction in the number of sites or epitopes or that the expression of a salient class of antibody molecule may explain a 100-fold difference in the number of serotypes among picornavirus genera. Thus, the second proposal is unlikely to be correct.

The difference among classification assays argument does not have an easy response. Indeed, there is no universal procedure used to classify viruses serologically, and, therefore, strictly speaking, there in the possibility that different numbers may be obtained using alternative classification procedures. FMDV is a pertinent example. Its seven serotypes are defined on the basis of a very stringent test that cannot be performed with a human virus for obvious reasons: absence of cross protection resulting from vaccination or infection with a given FMDV. Infection or vaccination with FMDV of one serotype does not confer protection against FMDV of a different serotype. In contrast, the subtype classification of FMDV was based on serological assays such as cross-neutralization or complement fixation tests, usually using sera raised in guinea pigs. These assays allowed classification of FMDV in more than 65 serological subtypes. Subtyping was stopped when it was realized that using increasingly discriminatory assays such as reactivity with MAbs, virtually any new isolate could define a new subtype [(Mateu et al., 1988), see Domingo et al., 1990; Sobrino and Domingo, 2004 for review of serotype and subtype classification of FMDV]. Despite these considerations, it is unlikely that serological assays using *in vitro* tests would be responsible for a 100-fold difference between two human pathogens such as HRV and HAV. Thus, it does not seem justified to attribute antigenic constancy to an artifact derived from diagnostic procedures.

More extensive virus circulation will favor genetic and antigenic diversification, as repeatedly justified in several chapters of this book. Obviously, following hundreds of additional years of circulation of a virus, a single serotype may diversify in multiple serotypes. What we are describing today may be a snapshot of an evolving process. Genotype differentiation is actually being witnessed during the expansion of HCV pandemics, partly due to a true genetic diversification of the virus as it circulated over the last decades, and partly due to increasing capacity of virus surveillance, and of molecular and phylogenetic tools for genome analysis. The reader can find an illustration of this point by comparing the expanded phylogenetic HCV tree from six to seven genotypes and the subtype ramifications, published by P. Simmonds and colleagues in 1993 and 2014 [compare (Simmonds et al., 1993) and (Smith et al., 2014)]. Although it cannot be excluded that time might tend to equalize the number of serotypes among viruses, current evidence does not justify blaming unknowns of long-term evolution to settle this issue.

We come to constraints at antigenic sites that limit the number of accepted amino acid substitutions as a model for antigenic invariance. It is the preferred model of molecular virologists. The initial concept was proposed by M.G. Rossmann, in his canyon hypothesis (Rossmann, 1989), based on studies with HRV14. A canyon in the virus preserves the receptor-binding site inside the canyon, while permitting amino acid substitutions that affect antigenicity, without consequences for receptor recognition. A physical and functional separation between receptor and antibody binding allows extensive antigenic variation. Clearly, in many viruses there is an overlap between antigenic and receptor recognition sites (Section 4.5 in Chapter 4), that could limit antigenic variation. A difference in constraints is also supported by a structural comparison carried out by J.M. Casasnovas and his colleagues of the interaction of PV and HRV16 with their respective cellular receptors, that revealed a receptor-binding site more accessible in PV than in HVR16 (Xing et al., 2000). This would render HRV the picornavirus most prone to antigenic variation, as indeed found in nature. Thus, constraints imposed by the requirement to interact with the cellular receptor may explain limited capacity for antigenic diversification, and perhaps with the contribution of other possible influences, the puzzle of widely different antigenic types despite similar high genome mutability. Additional structural and functional studies with viruses of different families are necessary to substantiate this proposal.

7.4.2 SIMILAR FREQUENCIES OF MONOCLONAL ANTIBODY-ESCAPE MUTANTS IN VIRUSES DIFFERING IN ANTIGENIC DIVERSITY

If some viruses have a limitation in accepting amino acid substitutions at their antigenic sites, they are expected to yield low frequencies of MAb-escape mutants [monoclonal antibody-resistant mutant (MARM) frequencies] in laboratory experiments. Comparison of MARM frequencies of different viruses shows that this is not the case (Table 7.2). In particular, the cardiovirus Mengo virus (one serotype) displays similar MAR frequencies than HRV (hundred serotypes). In fact, none of the RNA and DNA viruses listed in Table 7.2 deviate from a broad range of MARM frequencies of 10^{-3} to 10^{-5}, except for substitutions at some discontinuous epitopes of FMDV (Lea et al., 1994). In several of the studies listed, the stability of the selected escape mutants was tested after a few passages in cell culture, but in other studies, lack of reversion of the antigenic change was not ascertained. Two FMDV escape mutants showed a selective disadvantage over the parental wild-type virus (fitness decrease); upon continued replication, the mutants acquired fitness-enhancing mutations without reversion of the antigenic change (Martínez et al., 1991a).

Unless the escape mutations are selectively neutral, the expectation is that MARM frequencies may be an underestimate of the real rate at which the amino acid substitutions occur. Thus, it is possible that following selection by an antibody, some mutants may decrease in frequency due to a fitness cost, or that their level is maintained due to additional compensatory mutations acquired by the replicating genomes (Figure 7.6). Viruses that are highly constrained for antigenic variation may be diagnosed through fitness decrease of MARM mutants despite them occurring at similar rates as those that affect unconstrained sites. This is a concept similar to the distinction between fitness and function that we made in Section 5.8 of Chapter 5. That is, the occurrence of an antigenic change does not guarantee that the change will be perpetuated in nature and contribute to natural antigenic diversification. Again, fitness should be considered as a relevant parameter, and fitness effects on antigenic stability have been largely unexplored.

Table 7.2 Frequency of Monoclonal Antibody-resistant Mutants (MARMs) for Some Viruses

Virus	Monoclonal Antibody-resistant Mutant (MARM) Frequencies	References
Poliovirus	10^{-4} to 10^{-5}	Emini et al. (1982) and Minor et al. (1983, 1986)
Mengovirus	3×10^{-3} to 5×10^{-5}	Boege et al. (1991)
Foot-and-mouth disease virus	10^{-4} to 10^{-5} (continuous epitopes)	Martínez et al. (1991a)
	10^{-4} to 10^{-7} (discontinuous epitopes)	Lea et al. (1994)
Rhinovirus	10^{-4} to 10^{-5}	Sherry et al. (1986)
Hepatitis A virus	3×10^{-3}	Stapleton and Lemon (1987)
Vesicular stomatitis virus	0.5×10^{-4} to 1×10^{-4}	Holland et al. (1990)
Rabies virus	10^{-4}	Wiktor and Koprowski (1980)
Measles virus	9×10^{-5}	Schrag et al. (1999)
Sindbis virus	10^{-3} to 10^{-5}	Stec et al. (1986)
Canine parvovirus	$10^{-3.4}$ to $10^{-5.4}$	Smith and Inglis (1987)
Herpes simplex virus	1×10^{-5}	Smith and Inglis (1987)

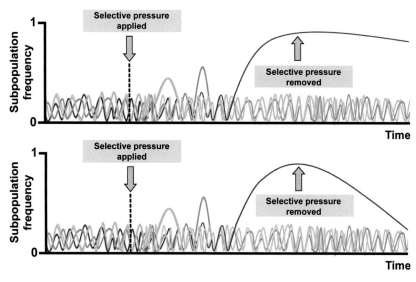

FIGURE 7.6

Stability of selected mutant subpopulations when selective pressure is removed. This scheme is the same portrayed in Figure 6.3 of Chapter 3, but shifted toward the right of the time scale. Five different colors have been chosen to depict fluctuations of four genomic classes. In a real population, thousands of genomes may be involved in each infected cell. In the present diagram, the red line represents genomes selected for their resistance to neutralizing monoclonal or polyclonal antibodies. Once the antibody pressure is removed, the mutant genomes may remain dominant either because the relevant substitutions do not affect viral fitness or because compensatory mutations have been acquired (top diagram). In contrast, upon removal of the antibody pressure, the proportion of selected genomes may fade with time due to a fitness cost and absence of compensatory mutations (bottom diagram). In the latter case, the antibody-resistant mutant will not contribute to the long-term antigenic diversification of the virus.

7.5 COMPARING VIRAL GENOMES. SEQUENCE ALIGNMENTS AND DATABASES

The likely multiple origin of viruses, followed by extended events of interaction with evolving host organisms of all phyla have produced the myriads of viral particles that at the present time outnumber cells by a factor of 10 (Chapter 1). The way to put order into such diversity is to classify viruses as done periodically by the International Committee on Taxonomy of Viruses (ICTV) (http://www.ictvonline.org/).

The computational procedures developed to study phylogenetic relationships in evolutionary biology are routinely applied to virology to establish relationships among closely or distantly related viruses (Page and Holmes, 1998; Hall, 2001; Felsenstein, 2004; Salemi and Vandamme, 2004; Yang, 2006). No phylogenetic tree that connects the viruses that have been characterized to date can be derived in a reliable way, not even a tree for DNA or for RNA viruses. What we can do is to produce trees for related viruses that probably share a common ancestor. Many data banks are available for viruses to retrieve sequences for comparison with new isolates. Despite the fact that data banks are periodically updated, some are listed in Table 7.3, and can serve as the starting point to reach the desired uniform

Table 7.3 Information on Nucleotide and Amino Acid Sequence Alignment Programs, Data Banks and Phylogenetic Procedures

Identification	URL	Contents
EMBL Nucleotide Sequence Database	http://www.ebi.ac.uk/ena	All reported sequences. General database
GenBank, the NIH Genetic Sequence Database	http://www.ncbi.nlm.nih.gov/genbank	All reported sequences. General database
DNA Data Bank of Japan	http://www.ddbj.nig.ac.jp	All reports sequences. General database
UNIPROT	http://www.ebi.ac.uk/uniprot	Protein sequences
Protein Data Bank (PDB)	http://www.rcsb.org	Protein structure data
Virus Particle Explorer	http://viperdb.scripps.edu	Virus structures and structure-derived properties: capsid interactions, residue contributions to protein-protein interactions. Links to sequences and taxonomic information
Viral Genomes Project	http://www.ncbi.nlm.nih.gov/genome/viruses	Complete or nearly complete viral genome sequences. Additional information. Includes Pairwise Sequence Comparisons (PASC) within viral families
The Influenza Sequence Database	http://www.fludb.org	Sequences, tools for the analysis of hemagglutinin and neuraminidase sequences
Picornavirus Sequence Database	http://www.viprbrc.org/brc/home.spg?decorator=picorna	Sequence and specific references for different picornavirus isolates
Plant viruses	http://www.dpvweb.net	Description of Plant Viruses (DPV). Expanding data bank of viruses, viroids and satellites of plants, fungi and protozoa
Potyvirus Database	http://www.danforthcenter.org/iltab/potyviridae/	Taxonomy, references and sequence databases of members of the *Potyviridae* family
Calicivirus Sequence Database	http://www.viprbrc.org/brc/home.spg?decorator=calici	Sequences, information and specific references for different calicivirus isolates
HPV Sequence Database	http://pave.niaid.nih.gov/	Human papilomavirus, sequences, analysis and alignment tools
HIV Sequences Database	http://www.hiv.lanl.gov http://www.hiv.lanl.gov/content/sequence/RESDB/	Sequences, drug resistance. Molecular immunology and vaccine trials. Analysis tools
HIV Drug Resistance Database	http://hivdb.stanford.edu	Sequence, correlations genotype-phenotype and genotype-antiretroviral treatment. Sequence analysis tools
Hepatitis C virus Database	http://hcv.lanl.gov http://www.hcvdb.org	Sequences and genome analysis tools
Hepatitis virus Database	http://s2as02.genes.nic.ac.jp/	Hepatitis B, C, and E virus sequences
Poxvirus Bioinformatics Research Center	http://www.poxvirus.org	Poxvirus genomes
Viral Bioinformatics Resource Center	http://athena.bioc.uvic.ca/	Large DNA viruses (Poxviruses, African Swine Fever Viruses, Iridoviruses, and Baculoviruses)

Continued

Table 7.3   Information on Nucleotide and Amino Acid Sequence Alignment Programs, Data Banks and Phylogenetic Procedures—cont'd

Identification	URL	Contents
Human Endogenous Retroviruses Database	http://herv.img.cas.cz	Human endogenous retroviruses, and genome analysis tools
VIDA, Virus Database at University College London	http://www.biochem.ucl.ac.uk/bsm/virus_database/VIDA.html	Homologous protein families from herpes, pox, papilloma, corona and arteriviruses
Subviral RNA Database	http://subviral.med.uottawa.ca/cgi-bin/home.cgi	Sequences and prediction of RNA secondary structures
Vir Oligo Compilation Lab	http://viroligo.okstate.edu/	Database of oligonucleotides used in virus detection and identification. Technical information and several links to original sequence information
ViPR Virus Pathogen Resource	http://viprbrc.org	Sequences of multiple virus families
Chromas	http://technelysium.com.au/?page_id=13	Software for DNA sequencing
FORMAT CONVERSION	http://hcv.lanl.gov/content/sequence/FORMAT_CONVERSION/form.htm	Program that converts the sequence(s) to a different user-specified format
BEAST, Tracer	http://beast.bio.ed.ac.uk/Tracer	Phylogenetic inferences. Bayesian methods
MODELTEST	http://darwin.uvigo.es/software/modeltest.html	Determination of nucleotide substitution model. Phylogenetic derivations
PHYLIP package	http://evolution.genetics.washington.edu/phylip.html	Programs for inferring phylogenies
PAUP	http://paup.csit.fsu.edu/	Software for inference of evolutionary trees
BiBiServ	http://bibiserv.techfak.uni-bielefeld.de/splits	Bielefeld Bioinformatics Service
Bowtie2	http://bowtie-bio.sourceforge.net/bowtie2/index.shtml	It is a tool for aligning sequencing reads to long reference sequences
SAM tools (Sequence Alignment/Map)	http://samtools.sourceforge.net/	A generic format for storing large nucleotide sequence alignments
Free Bayes	https://github.com/ekg/freebayes	FreeBayes is a Bayesian genetic variant detector designed to find small polymorphisms
EMBOSS	http://www.ebi.ac.uk/Tools/emboss/	Sequence analysis
T-Coffee	http://www.tcoffee.org/	Tools for computing, evaluating and manipulating multiple alignments
IGV, Integrative Genomics Viewer	http://www.broadinstitute.org/igv/	The Integrative Genomics Viewer (IGV) is a high-performance visualization tool for interactive exploration of large, integrated genomic datasets. It supports a wide variety of data types, including array-based and next-generation sequence data, and genomic annotations

resource locator (URL) to implement a procedure for genome characterization. Prior to any comparative study of nucleotide or amino acid sequences (not only to establish phylogenetic relationships, but also to calculate genetic distances, to identify regulatory regions, functional domains, and structural motifs, to design oligonucleotide primers for amplifications, or other applications) it is essential to align sequences accurately, and some programs for sequence alignments are also given in Table 7.3.

Databases differ in format and contents, which may include prediction of traits derived from sequence information (RNA secondary structures, antiviral drug sensitivity levels, assignments to homologous protein families, etc.). Some of them offer a link with the *web* page of the ICTV thus providing background information to assign newly determined sequences to current taxonomic groups. A structure-based amino acid sequence alignment of protein homologues can be carried out based on three-dimensional structures of proteins. Such types of amino acid sequence alignments may help in the identification of relevant structural and functional motifs. Sequence variability among a set of aligned sequences can be quantitated by the number of variable sites, mean pairwise diversity, mutation frequency, and other estimators (i.e., the Watterson's estimator) (Page and Holmes, 1998; Mount, 2004; Salemi and Vandamme, 2004) (see also Chapter 3 for parameters used to quantify mutant spectrum complexity). Relevant information on protein evolution can be derived from alignment of the protein sequence of related viruses (or of isolates from one virus, or for components of the same mutant spectrum) and analyzing the statistical acceptability of the divergent amino acids at each position. Statistical acceptability derives in part from the chemical nature and shape of the amino acid side chains and from the limitations that the genetic code imposes on amino acid replacements (Porto et al., 2005). The basic assumption is that the more conserved the amino acid sequences, and the more similar are the variant amino acids, the more likely is that the proteins derived from a common ancestor. M. Dayhoff pioneered the early comparison of protein sequences establishing a protein information resource (PIR) in the middle of the twentieth century. Tables named PAM (percent accepted mutation) were constructed and several evolved versions such as BLOSUM matrices, based on the BLOCKS database, are used to compare protein sequences. The BLOSUM62 amino acid substitution matrix groups amino acids according to their chemical structure and provides a probability of occurrence of each amino acid replacement: zero, amino acid replacement expected by chance; positive number, replacement found more often than by chance; and negative number, replacement found less often than by chance.

7.6 PHYLOGENETIC RELATIONSHIPS AMONG VIRUSES. EVOLUTIONARY MODELS

The URLs listed in Table 7.3 give access to computational analyses that allow sequence alignments and derivation of phylogenetic trees which are extremely informative of middle- and long-term evolutionary change of viruses (Page and Holmes, 1998; Notredame et al., 2000; Mount, 2004; Salemi and Vandamme, 2004; Holmes, 2008, 2009). Application of phylogenetic methods to virus evolution requires careful consideration of the evolutionary models to be used, including probabilities of the different types of nucleotide and amino acid replacements, and the rates at which they may occur. Statistical methods (i.e., likelihood ratio tests) are available to select an adequate models for a given data set (Salemi and Vandamme, 2004). At the nucleotide sequence level, it is often assumed that when transitions are more frequent than transversions in a set of related sequences, no saturation of mutation took place. In contrast, when transversions are more frequent than transitions, saturation is presumed

(Xia and Xie, 2001). Parameter α applied to amino acid sequence alignments (e.g. using program AAml from PAML package, version 3.14) takes into account multiple amino acid replacements per site, as well as unequal substitution rates among sites (Yang et al., 2000). Parameter α can be calculated using the amino acid replacement matrix WAG available in the program MODELTEST (Posada and Crandall, 1998). Despite their obvious utility, it is unlikely that these statistical procedures which were developed on the assumption of successions of defined sequences (rather than mutant clouds) can capture the complexities underlying long-term evolution of viruses in nature.

Phylogenetic reconstructions based on nucleotide (and deduced amino acid) sequence alignments are generally possible with selected genes of relatively close viruses (i.e., that belong to the same family). The main methods used to derive evolutionary trees are: maximum parsimony, distance, ML, Bayesian methods of phylogenetic inference, and splits-tree analysis [reviewed in (Eigen, 1992; Page and Holmes, 1998; Mount, 2004; Salemi and Vandamme, 2004; Sullivan, 2005; Holmes, 2008)] (Table 7.3).

Maximum parsimony predicts the minimal mutation steps needed to produce the observed sequences from ancestor sequences. It is most suitable for closely related sequences. Often, all possible trees are examined before a consensus tree is produced, and, therefore, the method is time consuming. Most programs based on maximum parsimony assume the operation of a molecular clock, with the limitations that were discussed in Section 7.3.

Distance methods are based on the calculation of genetic distances between any two sequences of a multiple sequence alignment. Large genetic distances require a correction for multiple mutational steps (i.e., Kimura 2-parameter distance). Most distance methods can handle large numbers of sequences, and results are relatively reliable even when a molecular clock does not operate. Commonly applied distance methods include neighbor joining (NJ) (that does not assume a molecular clock and yields an unrooted tree), several variant versions of NJ, and the unweighted pair group method with arithmetic mean (UPGMA, a clustering method that assumes a molecular clock and produces a rooted tree). The software package TREECON was developed to derive NJ trees.

ML methods use probability calculations to derive a branching pattern from the mutations at different positions of the nucleic acids under study. They can estimate both distances and the most accurate mutational pathway between sequences. Generally, supercomputers are needed when many sequences are compared since all possible trees are examined. ML methods are included in several programs listed in Table 7.3. Bayesian methods (based on conditional probabilities derived by Baye's rule) (Huelsenbeck et al., 2001; Ronquist and Huelsenbeck, 2003; Huelsenbeck and Dyer, 2004) have the advantage of increased speed of data processing, but they still require time to avoid incorrect inferences.

Splits-tree procedures are based on split-decomposition theory or statistical geometry, and they provide a geometrical representation of the distance relationships in sequence space (Eigen, 1992; Dopazo et al., 1993; Salemi and Vandamme, 2004) (Chapter 3). The procedure has been used to analyze rapidly evolving viral sequences (http://bibiserv.techfak.uni-bielefeld.de/splits/); methods that allow the inclusion of insertions and deletions have been adapted to the splits-tree program (Cheynier et al., 2001). Phylogenetic trees can be presented as rooted trees (with a reference out-group) and unrooted trees.

When possible it is advisable to apply different phylogenetic procedures to compare tree topologies. Resampling methods (i.e., bootstrapping, jackknifing, etc.) are used to assess the statistical reliability of the trees (Page and Holmes, 1998; Salemi et al., 1998; Mount, 2004; Salemi and Vandamme, 2004). A tree defines clades or lineages of a virus attending to groupings by relatedness. Different tree topologies can be obtained when analyzing different genes of the same virus set. Discordant phylogenetic positions of two different genes of the same virus is suggestive of recombination, that should be evaluated statistically

(Worobey, 2001; Salemi and Vandamme, 2004; Martin et al., 2005). Recombination is very frequently in viruses, and in some of them is intimately linked to the replication mechanism (Chapters 2 and 10).

The more conserved genes (i.e., the polymerase and other nonstructural protein-coding genes) may permit the establishment of phylogenetic relationships among some distant virus groups. Examples are the clustering of a number of animal and plant RNA viruses as supergroups (Morse, 1994). Families of DNA-dependent DNA polymerases group some bacterial and bacteriophage DNA polymerases with some eukaryotic polymerases (Morse, 1994; Villarreal, 2005), in support of active exchange of modules during coevolution of viruses and their hosts (Botstein, 1980, 1981; Zimmern, 1988). In contrast to conserved genes, variable genes (typically capsid proteins and surface glycoproteins) serve to establish short-term evolutionary relationships within the same virus group, including the survey of virus variation during outbreaks, epidemics, and pandemics (Gorman et al., 1992; Martínez et al., 1992; Morse, 1994; Gavrilin et al., 2000).

Distantly related viruses, with no discernible nucleotide or amino acid sequence identity, can sometimes be grouped on the basis of the three-dimensional structures of viral proteins. The evolutionary trace (ET) clustering method combines phylogenetic partition of sequences with structural information (Chakravarty et al., 2005), and it may help identifying functionally relevant domains shared by divergent isolates in particular highly variable capsid and surface viral proteins. ET can be applied to proteins and nucleic acids, and its clustering features may reveal conserved structures that are overlooked when all sequences are compared together. As explained in Chapter 1, the great diversity of amino acid sequences recorded among viral structural proteins (several URL links in Table 7.3) are actually reduced to a limited number of morphotypes at the structural level. In another approach, the probabilities of equivalence between pairs of residues in viral proteins are converted into evolutionary distances (Bamford et al., 2005; Ravantti et al., 2013). The structure-based classification has grouped the coat protein of icosahedral viruses in separate classes, each of which, interestingly, embraces different domains of life (Archaea, Bacteria, and Eukarya). A lineage of structurally related viruses includes tailed bacteriophages and the herpesviruses, suggesting that parts of the genomes of complex viruses may have a very ancient origin. They might have belonged to viruses that infected primitive cells, before the latter diverged into the domains of life that we identify in our biosphere (Bamford et al., 2005; Villarreal, 2005) (compare with models of virus origins in Chapter 1).

Viral clades may cluster with clades of their host species, suggesting either virus-host coadaptation or an extended parasite-host relationship, with limited possibilities of jumping the host barrier (Section 7.7). Hantaviruses and their rodent hosts (Plyusnin and Morzunov, 2001), lyssaviruses and bat species, spumaviruses and their primate hosts, and herpesviruses and their vertebrate hosts, are some among other examples of long-term host-virus coevolution (Mc Geoch and Davison, 1999; Woolhouse et al., 2002; Switzer et al., 2005). (See, however, a discussion on time scale discrepancies of coevolutionary rates (Sharp and Simmonds, 2011), and compare with section 7.3.3)

7.7 EXTINCTION, SURVIVAL, AND EMERGENCE OF VIRAL PATHOGENS. BACK TO THE MUTANT CLOUDS

The viral groups defined by phylogenetic methods may or may not occupy a defined geographical location. It will depend on whether viral vectors or infected individuals carry the virus over long distances or not. A defined phylogenetic group may include viruses that produce similar or different pathology. This is because the capacity of a virus to cause disease may depend on modest genetic change (i.e., one or a few amino acid substitutions) that does not alter its position in a phylogenetic tree. It is important to emphasize that, independently of the time frame considered, the tips of phylogenetic trees are a cloud of mutants, that genomes within the cloud are the origin of future diversification pathways, and that individual cloud components may differ in pathogenic potential. Figure 7.7 summarizes the diversification of HIV-1 since it entered the human population. Once HIV-1 originated from multiple introductions of a chimpanzee simian immunodeficiency virus (SIVcpz), the four major HIV-1 groups M, O, N, and P were generated, and group M evolved into the multiple subtypes and recombinant forms that circulate at present. Many factors determine the pathogenic potential of any of the HIV-1 subtypes and the newly arising recombinant forms.

The relevance of the mutant cloud in determining viral fitness and survival was documented by comparing five isolates of west Nile virus (WNV) that had identical consensus sequences and differed in the mutant spectrum, as analyzed by next-generation sequencing (NGS) (Kortenhoeven et al., 2015) (Figure 7.8). The study concerned a WNV lineage 2 that circulated in Europe during the beginning of the twenty-first century. Environmental changes modified the haplotype composition while maintaining an invariant consensus sequences, an example of "perturbation" manifested only at the level of the mutant spectrum (see Section 6 in Chapter 6).

HIV-1 is a notorious case of successful emergence of a new viral pathogen from a zoonotic reservoir of a related virus. However, despite limited records, there is also evidence that some viruses that once produced human disease might be now extinct. One example is Economo's disease (also termed lethargic encephalitis or epidemic encephalitis), a degenerative disease of the brain that produced loss of neurons. The disease had an acute phase of variable duration and intensity, followed by a chronic phase, sometimes with a late onset of symptoms. The disease showed a seasonal character with maximum incidence in late winter. The first cases were recorded in Eastern Europe in 1915 and the disease was first described by Baron C. Von Economo in Vienna in 1917. In 1920-1923 the disease attained pandemic proportions, although the number of cases and mortality were limited. It was estimated that between 1917 and 1929 about one hundred thousand cases occurred in Germany and Great Britain and then, mysteriously, the number of cases decreased and the disease disappeared (Ford, 1937). Economo's

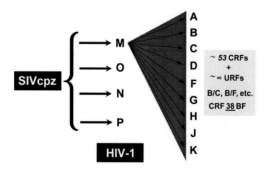

FIGURE 7.7

Diversification of HIV-1 from the time of introduction into the human population of retroviral simian ancestors SIVcpz from chimpanzees. Group M diversified into at least nine subtypes plus about 53 circulating recombinant forms (CRF) (denoted by the identification letter of two parental subtypes), and multitudes of unique recombinant forms that have not reached epidemiological relevance (box on the right). Genetic and antigenic diversifications are discussed in the text.

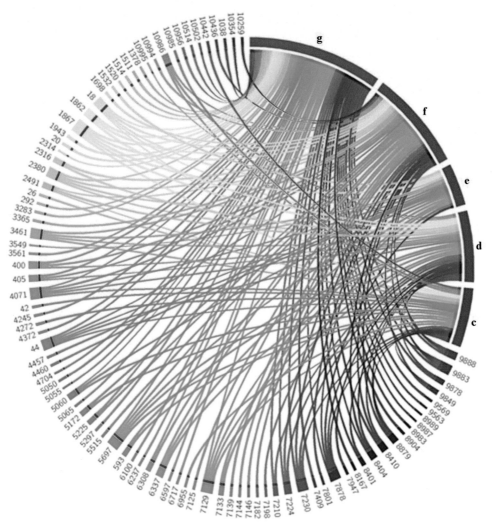

FIGURE 7.8

Visualization of the complexity of mutant spectra (haplotype composition) of five isolates of WNV denoted by c, d, e, f, and g (magenda bars) that have identical consensus sequence. The color lines connect the genomes where the same single nucleotide change occurs. Similarly color-coded ribbons indicate that the same mutation occurs in two genomes at the same position. Nucleotide positions are numbered next to the outer rim of the circle.

Figure reproduced from Kortenhoeven et al. (2015). BMC Genomics is an open access journal, and the article can be reproduced under the terms of the Creative Commons Attribution. The figure has been reproduced with permission of the authors.

disease, of a likely viral origin, is now extinct. At the time it was suspected that a virus similar to IV or some picornavirus might have been the etiological agent of this disease, but no proof could be provided.

FMDV, the agent of the economically most important disease of cattle and other farm animals circulated until recently as seven different serotypes termed A, O, C, Asia 1, SAT1, SAT2, and SAT3, and each serotype as multiple subtypes and antigenic variants (review in Sobrino and Domingo, 2004). Interestingly, in the 1980s the incidence of serotype C FMDV decreased to the point that at the beginning of the twenty-first century this FMDV serotypes was considered nearly extinct and its eradication feasible. It cannot be totally excluded, however, that type C FMDV is replicating in some persistently infected ruminant in some remote part of our planet and that the virus reemerges again. If not, its ecological niche has been occupied by FMDVs of other serotypes. This is one important issue behind virus eradication (smallpox in the late 1970s or rinderpest in 2011): the possibility that the niche left by an eradicated pathogen is occupied by a related pathogen. A.E. Gorbalenya, E. Wimmer, and colleagues examined the possible evolutionary origin of present-day PV, and that other picornaviruses might occupy the PV niche in the event of its eradication (Jiang et al., 2007). Their phylogenetic analysis suggests that PV could originate from a C-cluster coxsackie A virus through amino acid substitutions in the capsid that led to a change of receptor specificity (other cases are discussed in Chapter 4). They generated chimeras of PV and its putative ancestors, and some of them were viable and pathogenic for transgenic mice expressing the PV receptor. The authors suggest that in a world without anti-PV neutralizing antibodies, coxsackieviruses may mutate to generate a new PV-like agent.

Thus, despite virology being a very recent scientific activity, there is ample evidence of emergence of new viral pathogens, as well as cases of extinctions due to human interventions, and possible extinctions by natural influences. Viruses may evolve with regard to the symptoms they inflict upon their hosts. An increase of severity of Dengue virus infection has been observed in some world areas, consisting of neurological manifestations in patients with dengue fever or dengue hemorrhagic fever (Cam et al., 2001), among other examples of human and veterinary viral diseases. The dynamics of extinction of mutant viruses and their replacement by other forms is a continuous process, as the cycles of birth-death for any organism, but in a highly accelerated fashion.

We now turn to the pressing problem of the emergence and reemergence of viral disease.

7.7.1 FACTORS IN VIRAL EMERGENCE

New human viral pathogens emerge or reemerge at a rate of about one per year, representing an important concern for public health. Emergence is defined as the appearance of a new pathogen for a host, while reemergence often refers to the reappearance of a viral pathogen, following a period of absence. Being a popular topic, the reader will find numerous books and reviews on the subject. It is worth emphasizing that in the twentieth century many authors took the lead in emphasizing the problem of viral emergences, and the need to investigate the underlying mechanisms, notably S.S. Morse and J. Lederberg [see several chapters of Morse (1993, 1994)]. Given the adaptive capacity of viruses, in particular the RNA viruses, the reader will certainly suspect that genetic variation of viruses must be one of the factors involved in viral emergences. Indeed, most of the high-impact new viral diseases recorded recently or historically are due to RNA viruses. A statement by J. Lederberg reflects our vulnerability in the face of the nearly unlimited potential of viruses to vary: "Abundant sources of genetic variation exist for viruses to learn new tricks, not necessarily confined to what happens routinely or even frequently" (Lederberg, 1993). The situation is even more complex because genetic variation of viruses is only one of many ingredients that promote the introduction of new viral pathogens in the

---

BOX 7.2 FACTORS IN THE EMERGENCE OF MICROBIAL DISEASE

- Microbial change and adaptation.
- Human susceptibility to infection: impaired host immunity and malnutrition.
- Climate and weather.
- Changing ecosystems: vector ecology; reservoir abundance; and distribution.
- Human demographics and behavior: population growth; aging; and urbanization.
- Economic development and land use.
- International travel and commerce.
- Technology and industry.
- Breakdown of public health measures.
- Poverty and social inequality.
- War and famine.
- Lack of political will.
- Intent to harm: bioterrorism and agroterrorism.
  Points summarized from Smolinski et al. (2003).

---

human population. A report issued by US Institute of Medicine in 2003 analyzed and documented 13 factors that individually or in combination participate in the emergence of microbial disease. They include a number of sociological, environmental, and ecological influences that act to promote the emergence and reemergence of viruses, bacteria, fungi, and protozoa (Smolinski et al., 2003) (Box 7.2).

Here we will deal briefly with those factors of viral emergence related to the virus and host population numbers, in line with the focus of this book. Other aspects have been covered elsewhere (Antia et al., 2003; Haagmans et al., 2009; Wang and Crameri, 2014; Lipkin and Anthony, 2015; among others). The emergence of a viral disease can be regarded as a consequence of virus adaptation to a new environment, therefore, involving the concepts and mechanisms dissected in previous chapters. In particular, a relevant parameter is the variation of viral fitness in different environments (Domingo, 2010; Wargo and Kurath, 2011).

Fitness can directly or indirectly impact any of the three steps involved in viral disease emergence or reemergence, which can be summarized as follows:

- Introduction of virus into a new host species.
- Establishment of the virus in the new host.
- Dissemination of the virus among individuals of the new host species to produce outbreaks, epidemics, or pandemics.

For the introduction and establishment steps, replicative fitness is critical while for the dissemination step, epidemiological fitness plays the major role (Chapter 5).

Two population numbers are key for the establishment step: the number of viral particles shed by the infected donor host, and the number of potential new hosts that come into contact with the infected donor. We are now aware that even if two viral populations shed by an infected host have an identical number of infectious particles, not all mutant spectra might have the genomes subpopulations to permit the establishment in the human host (Figure 7.9). There is a natural lottery regarding which quasispecies subpopulations will hit which host. In the words of J.J. Holland and his colleagues: "Although new RNA virus diseases of humans will continue to emerge at indeterminate intervals, the viruses

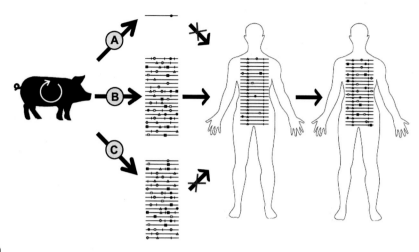

FIGURE 7.9

Relevance of virus population size and mutant spectrum composition in the zoonotic transmission of a virus. Only a subset of the genomes that surface an infected pig may be able to establish an infection in humans. The scheme indicates that a single genome that reached a human was not adequate to establish infection (pathway A). When multiple genomes reached the human (pathways B and C), only those that included a subset that displayed a minimum fitness in humans were able to initiate an infection and expand in the new host (pathway B). For pathways A and B, events are as if the contact between donor and recipient host had not taken place (arrows with cross). See text for implications.

themselves will not really be new, but rather mutated and rearranged to allow infection of new hosts, or to cause new disease patterns. It is important to remember that every quasispecies genome swarm in a infected individual is unique and 'new' in the sense that no identical population of RNA genomes has ever existed before and none such will ever exist again" (Holland et al., 1992).

The higher the number of viral particles shed by an infected host, the higher the probability of transmission to susceptible hosts (Section 7.2), and also of producing an emergence in a new host species. Viral population numbers and the number of transmissible particles can be largely amplified in immunocompromised individuals. Such individuals are termed super-spreaders, and can contribute large amounts of variant viruses to the transmission lottery (Rocha et al., 1991; Paunio et al., 1998; Gavrilin et al., 2000; Khetsuriani et al., 2003; Small et al., 2006; Odoom et al., 2008). Concerning the recipient hosts, the higher the number of potentially susceptible hosts that come into contact with an infected donor, the higher the probability of establishment of an emergent infection. It is likely that the advent of agricultural practices some 10,000 years ago, combined with increased contacts between humans and animals, inaugurated a time of new viral emergences. In the new scenario, viruses could shift from a persistent (low interhost transmission) mode into an acute (high interhost transmission) infection mode.

Not only population numbers are important, the connections between the spatial habitats of potential donor and recipient hosts are also highly relevant (Figure 7.10). As correctly emphasized by S.S. Morse, changes in viral traffic may allow viruses to come near potential new hosts that had never been encountered before. Several sociological and ecological factors that can impact directly or indirectly the accessibility to an infected donor play a role. A typical example that connects several of the points listed in Box 7.2 is provided by the increase of arbovirus vectors during a specially humid season

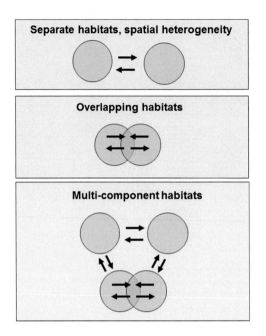

FIGURE 7.10

Types of habitats that may limit or facilitate interaction between hosts that can potentially establish an emergent infection in a local habitat. In separate habitats contacts are restricted while in overlapping habitats contacts are facilitated. In most cases, habitats cannot be reduced to the standard extremes, and are multicomponent habitats with various degrees of complexity. See text for implications for viral emergences.

due to the climate change because insect larvae can proliferate on water reservoirs. Increased travel may put humans infected with arboviruses in contact with the flourishing insect vector population. Climate change may modify the migration routes of some birds, again putting these potential vertebrate hosts in contact with infected animals and insect vectors.

Other points listed in Box 7.2 are worth commenting: close human-to-human contacts are favored by urbanization. In 1975, there were five megacities in the world (meaning cities with more than 10 million human population) while at the time of writing this book the number of megacities exceeds 20. Humans in close contact are, in addition, highly mobile. At present it is possible to go around the world in about 36 h (if you choose the adequate airports…) which represents a 1000-fold increase in spatial mobility of humans relative to the mobility in the year 1800. The 2014-2015 Ebola epidemics in Africa was made worse by the breakdown of public health measures, poverty, and lack of political will of local and international agencies to put efforts in stopping transmission. Underdeveloped countries are a reservoir of viral infections that represent a global threat due to several of the points listed in Box 7.2 [(Smolinski et al., 2003); several chapters of Singh (2014)].

Concerning the establishment and dissemination steps, the molecular mechanisms of quasispecies optimization in the new host environment apply. The underlying events are those presided by the extended Darwinian concepts of variation, competition, and selection, with the perturbations derived from stochastic effects (treated in different chapters of this book).

7.7.2 COMPLEXITY REVISITED

The meanings of complexity in virology were discussed in Section 3.9 of Chapter 3, one of them being the inability to explain a whole as the sum of its parts (Solé and Goodwin, 2000). R.V. Solé and B. Goodwin define the sciences of complexity as "the study of those systems in which there is no simple and predictable relationship between levels, between the properties of parts and of wholes." Several levels of complexity can be identified in the events that give to the emergence of a viral disease (Domingo, 2010; Sáiz et al., 2014). One level of complexity concerns the behavior of viral populations: behavior is often determined by interactions among components of mutant spectra in a way that cannot be predicted by the individual components of the population, even if we knew them!

The second level of complexity that can have an impact in the emergence of viral disease stems from the environmental, sociological, and ecological variables that must converge for a virus from some animal reservoir to come into contact and successfully infect a new host, for example, a human. Despite close surveillance, emergences of viral disease are unpredictable. Experts expect new influenza pandemics to arise somewhere in Asia from the avian reservoirs of IV, yet in 2009 the new influenza pandemic originated in Central America. Paradoxically, despite a general agreement that surveillance of human and zoonotic virus reservoirs should be intensified using new molecular tools (i.e., NGS to go beyond the consensus sequences), the reality is that what we have learnt are the reasons why emergences are unpredictable. The "abundant sources of genetic variation" that was emphasized by J. Lederberg (see Section 7.7.1) should be extended to refer to "abundant sources of complexity in viral emergences." For the time being we have to be ready to react once the emergence has already occurred.

7.8 OVERVIEW AND CONCLUDING REMARKS

Viruses have survived because they have undergone multiple rounds of vertical and horizontal transmission in their host organisms, and because occasionally they have found new suitable hosts where to replicate. Among the many parameters involved, in this chapter we have emphasized the relevance of virus and host population numbers for sustained transmissions and the long-term maintenance of viral entities. A point that is often either ignored or not sufficiently emphasized is that the quasispecies nature of viral populations introduces an element of uncertainty regarding which types of mutants are transmitted to new hosts. Despite being a complication that cannot be easily handled, it is a fact that should stimulate new approaches to the surveillance of virus transmission and the identification of the founder viruses in new infections.

A steady accumulation of mutations during evolution in the field was a proposal that agreed with the neutral theory of molecular evolution developed last century. One suspects that this agreement resulted in a premature preference for a regular clock of steady incorporation of mutations to work in the case of viruses. The evidence, however, is that there are multiple molecular mechanisms that render the operation of a molecular clock for viruses very unlikely, and perhaps fortuitous in some cases. Several possible mechanisms of variable evolutionary rates have been discussed, and further clarification is expected from entire genome sequencing applied to viruses during outbreaks and epidemics. Viruses are probably not the best biological systems to obtain experimental evidence in support of the clock hypothesis.

A puzzling and pendent issue in viral evolution is the interpretation of the widely different number of viral serotypes, despite viruses sharing comparably large mutation rates and frequencies. Different possibilities have been examined, and a slight preference for variable constraints acting on the amino acid residues that determine the antigenic properties of viruses has been expressed. Again, additional work is necessary to solve this interesting problem.

Procedures for sequence alignments and the establishment of phylogenetic relationships among related viruses have been briefly summarized, with some indications to find useful URL sites. The comparison of genomic sequences (and encoded amino acids) of new viral isolates with those of the viruses characterized to date is important given the increasing number of new viruses discovered in natural habitats.

The important problem of viral emergences and reemergences has been treated with emphasis on the concept of complexity. There are multiple interacting influences that converge to produce the emergence or reemergence of a viral pathogen, one of them being the heterogeneity of viral populations at the genetic and phenotypic level. Despite considerable methodological progress we are still in the realm of uncertainty regarding prediction of when and where a new viral pathogen will emerge (see Summary Box).

SUMMARY BOX

- Long-term evolution of viruses is the result of a history of virus transmission among hosts. Basic principles of transmission dynamics must take into consideration sampling effects and the inherent heterogeneity of viral populations.
- Rates of evolution of viruses in nature are extremely high as compared to the estimated rate for their host organisms. Contrary to some tenets of neutral evolution, rates of viral evolution are not constant with time. In particular, several mechanisms explain why intrahost virus evolution is faster than interhost evolution.
- Several procedures for sequence alignments and derivation of phylogenetic trees allow a partial description of virus diversification in nature.
- Antigenic diversification of viruses is subjected to constraints that differ among viruses. Some viruses have a single serotype while others have 100 serotypes. Several possible mechanisms may contribute to this difference.
- The emergence and reemergence of new viral pathogens is a multifactorial event with a clear influence of host and virus population numbers. Several levels of complexity participate in the emergence of a new pathogen, rendering the event highly unpredictable.

REFERENCES

Allen, B., Sample, C., Dementieva, Y., Medeiros, R.C., Paoletti, C., et al., 2015. The molecular clock of neutral evolution can be accelerated or slowed by asymmetric spatial structure. PLoS Comput. Biol. 11, e1004108.

Althaus, C.L., 2014. Estimating the reproduction number of Ebola virus (EBOV) during the 2014 outbreak in West Africa. PLoS Currents Outbreaks. doi:10.1371/currents.outbreaks.91afb5e0f279e7129e7056095255b288.

Anderson, R.M., May, R.M., 1991. Infectious Diseases of Humans. Oxford University Press, Oxford.

Antia, R., Regoes, R.R., Koella, J.C., Bergstrom, C.T., 2003. The role of evolution in the emergence of infectious diseases. Nature 426, 658–661.

Aswad, A., Katzourakis, A., 2012. Paleovirology and virally derived immunity. Trends Ecol. Evol. 27, 627–636.

Bamford, D.H., Grimes, J.M., Stuart, D.I., 2005. What does structure tell us about virus evolution? Curr. Opin. Struct. Biol. 15, 655–663.

Boege, U., Kobasa, D., Onodera, S., Parks, G.D., Palmenberg, A.C., et al., 1991. Characterization of Mengo virus neutralization epitopes. Virology 181, 1–13.

Botstein, D., 1980. A theory of modular evolution for bacteriophages. Ann. N. Y. Acad. Sci. 354, 484–491.

Botstein, D., 1981. A modular theory of virus evolution. In: Fields, B.N., Jaenisch, R., Fox, C.F. (Eds.), Animal Virus Genetics. Academic Press, New York, pp. 363–384.

Brooksby, J.B., 1981. Surveillance and control of virus diseases: Europe, Middle East and Indian sub-continent. In: Gibss, E.P.J. (Ed.), Virus Diseases of Food Animals. In: International Perspectives, vol. I. Academic Press Inc., London, pp. 69–78.

Buonagurio, D.A., Nakada, S., Desselberger, U., Krystal, M., Palese, P., 1985. Noncumulative sequence changes in the hemagglutinin genes of influenza C virus isolates. Virology 146, 221–232.

Cam, B.V., Fonsmark, L., Hue, N.B., Phuong, N.T., Poulsen, A., et al., 2001. Prospective case-control study of encephalopathy in children with dengue hemorrhagic fever. Am. J. Trop. Med. Hyg. 65, 848–851.

Chakravarty, S., Hutson, A.M., Estes, M.K., Prasad, B.V., 2005. Evolutionary trace residues in noroviruses: importance in receptor binding, antigenicity, virion assembly, and strain diversity. J. Virol. 79, 554–568.

Cheynier, R., Kils-Hutten, L., Meyerhans, A., Wain-Hobson, S., 2001. Insertion/deletion frequencies match those of point mutations in the hypervariable regions of the simian immunodeficiency virus surface envelope gene. J. Gen. Virol. 82, 1613–1619.

Domingo, E., 2006. Quasispecies: concepts and implications for virology. Curr. Top. Microbiol. Immunol. 299.

Domingo, E., 2007. Virus evolution. In: Knipe, D.M., Howley, P.M. (Eds.), Fields Virology. fifth ed. Lippincott Williams & Wilkins, Philadelphia, pp. 389–421.

Domingo, E., 2010. Mechanisms of viral emergence. Vet. Res. 41, 38.

Domingo, E., Mateu, M.G., Martínez, M.A., Dopazo, J., Moya, A., et al., 1990. Genetic variability and antigenic diversity of foot-and-mouth disease virus. In: Kurkstak, E., Marusyk, R.G., Murphy, S.A., Van-Regenmortel, M.H.V. (Eds.), Applied Virology Research. Plenum Publishing Co., New York, pp. 233–266.

Domingo, E., Biebricher, C., Eigen, M., Holland, J.J., 2001. Quasispecies and RNA Virus Evolution: Principles and Consequences. Landes Bioscience, Austin.

Dopazo, J., Dress, A., von Haeseler, A., 1993. Split decomposition: a technique to analyze viral evolution. Proc. Natl. Acad. Sci. U. S. A. 90, 10320–10324.

Eggers, H.J., 2002. History of poliomyelitis and poliomyelitis research. In: Semler, B.L., Wimmer, E. (Eds.), Molecular Biology of Picornaviruses. ASM Press, Washington, DC, pp. 3–14.

Eigen, M., 1992. Steps towards life. Oxford University Press, Oxford.

Emini, E.A., Jameson, B.A., Lewis, A.J., Larsen, G.R., Wimmer, E., 1982. Poliovirus neutralization epitopes: analysis and localization with neutralizing monoclonal antibodies. J. Virol. 43, 997–1005.

Feigelstock, D.A., Mateu, M.G., Valero, M.L., Andreu, D., Domingo, E., et al., 1996. Emerging foot-and-mouth disease virus variants with antigenically critical amino acid substitutions predicted by model studies using reference viruses. Vaccine 14, 97–102.

Felsenstein, J., 2004. Inferring Phylogenies. Sinauer Associates, Sunderland, MA.

Firth, C., Kitchen, A., Shapiro, B., Suchard, M.A., Holmes, E.C., et al., 2010. Using time-structured data to estimate evolutionary rates of double-stranded DNA viruses. Mol. Biol. Evol. 27, 2038–2051.

Ford, F.R., 1937. Diseases of the Nervous System in Infancy, Childhood and Adolescence. Charles C. Thomas, Springfield, IL.

Gaucher, E.A., Gu, X., Miyamoto, M.M., Benner, S.A., 2002. Predicting functional divergence in protein evolution by site-specific rate shifts. Trends Biochem. Sci. 27, 315–321.

Gavrilin, G.V., Cherkasova, E.A., Lipskaya, G.Y., Kew, O.M., Agol, V.I., 2000. Evolution of circulating wild poliovirus and of vaccine-derived poliovirus in an immunodeficient patient: a unifying model. J. Virol. 74, 7381–7390.

Gething, M.J., Bye, J., Skehel, J., Waterfield, M., 1980. Cloning and DNA sequence of double-stranded copies of haemagglutinin genes from H2 and H3 strains elucidates antigenic shift and drift in human influenza virus. Nature 287, 301–306.

Geysen, H.M., Meloen, R.H., Barteling, S.J., 1984. Use of peptide synthesis to probe viral antigens for epitopes to a resolution of a single amino acid. Proc. Natl. Acad. Sci. U. S. A. 81, 3998–4002.

Gilbert, C., Feschotte, C., 2010. Genomic fossils calibrate the long-term evolution of hepadnaviruses. PLoS Biol. 8(9). pii:e1000459.

Gojobori, T., Yokoyama, S., 1985. Rates of evolution of the retroviral oncogene of Moloney murine sarcoma virus and of its cellular homologues. Proc. Natl. Acad. Sci. U. S. A. 82, 4198–4201.

Gorman, O.T., Bean, W.J., Webster, R.G., 1992. Evolutionary processes in influenza viruses: divergence, rapid evolution, and stasis. Curr. Top. Microbiol. Immunol. 176, 75–97.

Haagmans, B.L., Andeweg, A.C., Osterhaus, A.D., 2009. The application of genomics to emerging zoonotic viral diseases. PLoS Pathog. 5, e1000557.

Hall, B.G., 2001. Phylogenetic Trees Made Easy: A How-To Manual for Molecular Biologists. Sinauer Association Inc., Sunderland, MA.

Hanada, K., Suzuki, Y., Gojobori, T., 2004. A large variation in the rates of synonymous substitution for RNA viruses and its relationship to a diversity of viral infection and transmission modes. Mol. Biol. Evol. 21, 1074–1080.

Heffernan, J.M., Smith, R.J., Wahl, L.M., 2005. Perspectives on the basic reproductive ratio. J. R. S. Interface R. Soc. 2, 281–293.

Herbeck, J.T., Nickle, D.C., Learn, G.H., Gottlieb, G.S., Curlin, M.E., et al., 2006. Human immunodeficiency virus type 1 env evolves toward ancestral states upon transmission to a new host. J. Virol. 80, 1637–1644.

Holland, J.J., Domingo, E., de la Torre, J.C., Steinhauer, D.A., 1990. Mutation frequencies at defined single codon sites in vesicular stomatitis virus and poliovirus can be increased only slightly by chemical mutagenesis. J. Virol. 64, 3960–3962.

Holland, J.J., de La Torre, J.C., Steinhauer, D.A., 1992. RNA virus populations as quasispecies. Curr. Top. Microbiol. Immunol. 176, 1–20.

Holmes, E.C., 2008. Comparative studies of RNA virus evolution. In: Domingo, E., Parrish, C.R., Holland, J.J. (Eds.), Origin and Evolution of Viruses, second ed. Elsevier, Oxford, pp. 119–134.

Holmes, E.C., 2009. The Evolution and Emergence of RNA Viruses. Oxford Series in Ecology and Evolution. Oxford University Press, New York.

Huelsenbeck, J.P., Dyer, K.A., 2004. Bayesian estimation of positively selected sites. J. Mol. Evol. 58, 661–672.

Huelsenbeck, J.P., Ronquist, F., Nielsen, R., Bollback, J.P., 2001. Bayesian inference of phylogeny and its impact on evolutionary biology. Science 294, 2310–2314.

Jenkins, G.M., Rambaut, A., Pybus, O.G., Holmes, E.C., 2002. Rates of molecular evolution in RNA viruses: a quantitative phylogenetic analysis. J. Mol. Evol. 54, 156–165.

Jiang, P., Faase, J.A., Toyoda, H., Paul, A., Wimmer, E., et al., 2007. Evidence for emergence of diverse polioviruses from C-cluster coxsackie A viruses and implications for global poliovirus eradication. Proc. Natl. Acad. Sci. U. S. A. 104, 9457–9462.

Kay, A., Zoulim, F., 2007. Hepatitis B virus genetic variability and evolution. Virus Res. 127, 164–176.

Khetsuriani, N., Prevots, D.R., Quick, L., Elder, M.E., Pallansch, M., et al., 2003. Persistence of vaccine-derived polioviruses among immunodeficient persons with vaccine-associated paralytic poliomyelitis. J. Infect. Dis. 188, 1845–1852.

Knowles, N.J., Hovi, T., King, A.M.Q., Stanway, G., 2010. Overview of taxonomy. In: Ehrenfeld, E., Domingo, E., Roos, R.P. (Eds.), The Picornaviruses. ASM Press, Washington, DC, pp. 19–32.

Kortenhoeven, C., Joubert, F., Bastos, A., Abolnik, C., 2015. Virus genome dynamics under different propagation pressures: reconstruction of whole genome haplotypes of west Nile viruses from NGS data. BMC Genomics 16, 118.

Lea, S., Hernández, J., Blakemore, W., Brocchi, E., Curry, S., et al., 1994. The structure and antigenicity of a type C foot-and-mouth disease virus. Structure 2, 123–139.

Lederberg, J., 1993. Viruses and humankind: intracellular symbiosis and evolutionary competition. In: Morse, S.S. (Ed.), Emerging Viruses. Oxford University Press, Oxford, pp. 3–9.

Leslie, A.J., Pfafferott, K.J., Chetty, P., Draenert, R., Addo, M.M., et al., 2004. HIV evolution: CTL escape mutation and reversion after transmission. Nat. Med. 10, 282–289.

Lin, Y.Y., Liu, C., Chien, W.H., Wu, L.L., Tao, Y., et al., 2015. New insights into the evolutionary rate of hepatitis B virus at different biological scales. J. Virol. 89, 3512–3522.

Lipkin, W.I., Anthony, S.J., 2015. Virus hunting. Virology 479–480C, 194–199.

Lythgoe, K.A., Fraser, C., 2012. New insights into the evolutionary rate of HIV-1 at the within-host and epidemiological levels. Proc. Biol. Sci. 279, 3367–3375.

Martin, D.P., Williamson, C., Posada, D., 2005. RDP2: recombination detection and analysis from sequence alignments. Bioinformatics 21, 260–262.

Martínez, M.A., Carrillo, C., Gonzalez-Candelas, F., Moya, A., Domingo, E., et al., 1991a. Fitness alteration of foot-and-mouth disease virus mutants: measurement of adaptability of viral quasispecies. J. Virol. 65, 3954–3957.

Martínez, M.A., Hernández, J., Piccone, M.E., Palma, E.L., Domingo, E., et al., 1991b. Two mechanisms of antigenic diversification of foot-and-mouth disease virus. Virology 184, 695–706.

Martínez, M.A., Dopazo, J., Hernandez, J., Mateu, M.G., Sobrino, F., et al., 1992. Evolution of the capsid protein genes of foot-and-mouth disease virus: antigenic variation without accumulation of amino acid substitutions over six decades. J. Virol. 66, 3557–3565.

Mateu, M.G., 1995. Antibody recognition of picornaviruses and escape from neutralization: a structural view. Virus Res. 38, 1–24.

Mateu, M.G., Da Silva, J.L., Rocha, E., De Brum, D.L., Alonso, A., et al., 1988. Extensive antigenic heterogeneity of foot-and-mouth disease virus of serotype C. Virology 167, 113–124.

Mateu, M.G., Martínez, M.A., Capucci, L., Andreu, D., Giralt, E., et al., 1990. A single amino acid substitution affects multiple overlapping epitopes in the major antigenic site of foot-and-mouth disease virus of serotype C. J. Gen. Virol. 71, 629–637.

Mc Geoch, D.J., Davison, A.J., 1999. The molecular evolutionary history of the herpesviruses. In: Domingo, E., Webster, R., Holland, J. (Eds.), Origin and Evolution of Viruses. Academic Press, San Diego, pp. 441–465.

Melamed, A., Witkover, A.D., Laydon, D.J., Brown, R., Ladell, K., et al., 2014. Clonality of HTLV-2 in natural infection. PLoS Pathog. 10, e1004006.

Mims, C.A., 1981. Vertical transmission of viruses. Microbiol. Rev. 45, 267–286.

Minor, P.D., Schild, G.C., Bootman, J., Evans, D.M., Ferguson, M., et al., 1983. Location and primary structure of a major antigenic site for poliovirus neutralization. Nature 301, 674–679.

Minor, P.D., Ferguson, M., Evans, D.M., Almond, J.W., Icenogle, J.P., 1986. Antigenic structure of polioviruses of serotypes 1, 2 and 3. J. Gen. Virol. 67, 1283–1291.

Morse, S.S., 1993. Emerging Viruses. Oxford University Press, Oxford.

Morse, S.S., 1994. The Evolutionary Biology of Viruses. Raven Press, New York.

Mount, D.W., 2004. Bioinformatics. Sequence and Genome Analysis. Cold Spring Harbor Laboratory Press, Cold Spring Harbor, NY.

Nash, A.A., Dalziel, R.G., Fitzgerald, J.R., 2015. Mims' Pathogenesis of Infectious Disease. Elsevier Science & Technology, Edinburgh.

Notredame, C., Higgins, D.G., Heringa, J., 2000. T-Coffee: a novel method for fast and accurate multiple sequence alignment. J. Mol. Biol. 302, 205–217.

Nowak, M.A., 2006. Evolutionary Dynamics. The Belknap Press of Harvard University Press, Cambridge, Massachusetts and London, England.

Nowak, M.A., May, R.M., 2000. Virus Dynamics. Mathematical Principles of Immunology and Virology. Oxford University Press Inc., New York.

Odoom, J.K., Yunus, Z., Dunn, G., Minor, P.D., Martin, J., 2008. Changes in population dynamics during long-term evolution of sabin type 1 poliovirus in an immunodeficient patient. J. Virol. 82, 9179–9190.

Page, R.D.M., Holmes, E.C., 1998. Molecular Evolution. A Phylogenetic Approach. Blackwell Science Ltd., Oxford.

Parrish, C.R., Kawaoka, Y., 2005. The origins of new pandemic viruses: the acquisition of new host ranges by canine parvovirus and influenza A viruses. Annu. Rev. Microbiol. 59, 553–586.

Paunio, M., Peltola, H., Valle, M., Davidkin, I., Virtanen, M., et al., 1998. Explosive school-based measles outbreak: intense exposure may have resulted in high risk, even among revaccinees. Am. J. Epidemiol. 148, 1103–1110.

Plyusnin, A., Morzunov, S.P., 2001. Virus evolution and genetic diversity of hantaviruses and their rodent hosts. Curr. Top. Microbiol. Immunol. 256, 47–75.

Porto, M., Roman, H.E., Vendruscolo, M., Bastolla, U., 2005. Prediction of site-specific amino acid distributions and limits of divergent evolutionary changes in protein sequences. Mol. Biol. Evol. 22, 630–638.

Posada, D., Crandall, K.A., 1998. MODELTEST: testing the model of DNA substitution. Bioinformatics 14, 817–818.

Ravantti, J., Bamford, D., Stuart, D.I., 2013. Automatic comparison and classification of protein structures. J. Struct. Biol. 183, 47–56.

Redd, A.D., Collinson-Streng, A.N., Chatziandreou, N., Mullis, C.E., Laeyendecker, O., et al., 2012. Previously transmitted HIV-1 strains are preferentially selected during subsequent sexual transmissions. J. Infect. Dis. 206, 1433–1442.

Rocha, E., Cox, N.J., Black, R.A., Harmon, M.W., Harrison, C.J., et al., 1991. Antigenic and genetic variation in influenza A (H1N1) virus isolates recovered from a persistently infected immunodeficient child. J. Virol. 65, 2340–2350.

Ronquist, F., Huelsenbeck, J.P., 2003. MrBayes 3: Bayesian phylogenetic inference under mixed models. Bioinformatics 19, 1572–1574.

Rossmann, M.G., 1989. The canyon hypothesis. Hiding the host cell receptor attachment site on a viral surface from immune surveillance. J. Biol. Chem. 264, 14587–14590.

Rowlands, D.J., Clarke, B.E., Carroll, A.R., Brown, F., Nicholson, B.H., et al., 1983. Chemical basis of antigenic variation in foot-and-mouth disease virus. Nature 306, 694–697.

Sáiz, J.C., Sobrino, F., Sevilla, N., Martin, V., Perales, C., et al., 2014. Molecular and evolutionary mechanisms of viral emergence. In: Singh, S.K. (Ed.), Viral Infections and Global Change. John Wiley & Sons Inc., Hoboken, NJ, pp. 297–325.

Sakaoka, H., Kurita, K., Iida, Y., Takada, S., Umene, K., et al., 1994. Quantitative analysis of genomic polymorphism of herpes simplex virus type 1 strains from six countries: studies of molecular evolution and molecular epidemiology of the virus. J. Gen. Virol. 75 (Pt 3), 513–527.

Salemi, M., Vandamme, A.M., 2004. The phylogeny handbook. A Practical Approach to DNA and Protein Phylogeny. Cambridge University Press, Cambridge.

Salemi, M., Vandamme, A.M., Gradozzi, C., Van Laethem, K., Cattaneo, E., et al., 1998. Evolutionary rate and genetic heterogeneity of human T-cell lymphotropic virus type II (HTLV-II) using isolates from European injecting drug users. J. Mol. Evol. 46, 602–611.

Salemi, M., Lamers, S.L., Yu, S., de Oliveira, T., Fitch, W.M., et al., 2005. Phylodynamic analysis of human immunodeficiency virus type 1 in distinct brain compartments provides a model for the neuropathogenesis of AIDS. J. Virol. 79, 11343–11352.

Schrag, S.J., Rota, P.A., Bellini, W.J., 1999. Spontaneous mutation rate of measles virus: direct estimation based on mutations conferring monoclonal antibody resistance. J. Virol. 73, 51–54.

Sellers, R.F., 1971. Quantitative aspects of the spread of foot-and-mouth disease. Vet. Bull. 41, 431–439.

Sellers, R.F., 1981. Factors affecting the geographical distribution and spread of virus diseases of food animals. In: Gibbs, E.P.J. (Ed.), Virus Diseases of Food Animals. In: International Perspectives, vol. I. Academis Press Inc, London, pp. 19–29.

Sharp, P.M., Simmonds, P., 2011. Evaluating the evidence of virus/host co-evolution. Curr Opin Virol. 1, 436–441.

Sherry, B., Mosser, A.G., Colonno, R.J., Rueckert, R.R., 1986. Use of monoclonal antibodies to identify four neutralization immunogens on a common cold picornavirus, human rhinovirus 14. J. Virol. 57, 246–257.

Simmonds, P., Holmes, E.C., Cha, T.A., Chan, S.W., McOmish, F., et al., 1993. Classification of hepatitis C virus into six major genotypes and a series of subtypes by phylogenetic analysis of the NS-5 region. J. Gen. Virol. 74 (Pt 11), 2391–2399.

Singh, S.K., 2014. Viral Infections and Global Change. John Wiley & Sons Inc., Hoboken, NJ.

Small, M., Tse, C.K., Walker, D.M., 2006. Super-spreaders and the rate of transmission of the SARS virus. Physica D 215, 146–158.

Smith, D.B., Inglis, S.C., 1987. The mutation rate and variability of eukaryotic viruses: an analytical review. J. Gen. Virol. 68, 2729–2740.

Smith, D.B., Bukh, J., Kuiken, C., Muerhoff, A.S., Rice, C.M., et al., 2014. Expanded classification of hepatitis C virus into 7 genotypes and 67 subtypes: updated criteria and genotype assignment web resource. Hepatology 59, 318–327.

Smolinski, M.S., Hamburg, M.A., Lederberg, J., 2003. Microbial Threats to Health, Emergence, Detection and Response. The National Academies Press, Washington, DC.

Sobrino, F., Domingo, E., 2004. Foot-and-Mouth Disease: Current Perspectives. Horizon Bioscience, Wymondham, England.

Sobrino, F., Palma, E.L., Beck, E., Dávila, M., de la Torre, J.C., et al., 1986. Fixation of mutations in the viral genome during an outbreak of foot-and-mouth disease: heterogeneity and rate variations. Gene 50, 149–159.

Solé, R., Goodwin, B., 2000. Signs of life. How Complexity Pervades Biology. Basic Books, New York.

Sorensen, J.H., Mackay, D.K., Jensen, C.O., Donaldson, A.I., 2000. An integrated model to predict the atmospheric spread of foot-and-mouth disease virus. Epidemiol. Infect. 124, 577–590.

Stapleton, J.T., Lemon, S.M., 1987. Neutralization escape mutants define a dominant immunogenic neutralization site on hepatitis A virus. J. Virol. 61, 491–498.

Stec, D.S., Waddell, A., Schmaljohn, C.S., Cole, G.A., Schmaljohn, A.L., 1986. Antibody-selected variation and reversion in Sindbis virus neutralization epitopes. J. Virol. 57, 715–720.

Sullivan, J., 2005. Maximum-likelihood methods for phylogeny estimation. Methods Enzymol. 395, 757–779.

Switzer, W.M., Salemi, M., Shanmugam, V., Gao, F., Cong, M.E., et al., 2005. Ancient co-speciation of simian foamy viruses and primates. Nature 434, 376–380.

Villarreal, L.P., 2005. Viruses and the Evolution of Life. ASM Press, Washington, DC.

Villaverde, A., Martínez, M.A., Sobrino, F., Dopazo, J., Moya, A., et al., 1991. Fixation of mutations at the VP1 gene of foot-and-mouth disease virus. Can quasispecies define a transient molecular clock? Gene 103, 147–153.

Wang, L.F., Crameri, G., 2014. Emerging zoonotic viral diseases. Rev. Sci. Tech. (Int. Off. Epiz.) 33, 569–581.

Wang, H.Y., Chien, M.H., Huang, H.P., Chang, H.C., Wu, C.C., et al., 2010. Distinct hepatitis B virus dynamics in the immunotolerant and early immunoclearance phases. J. Virol. 84, 3454–3463.

Wargo, A.R., Kurath, G., 2011. In vivo fitness associated with high virulence in a vertebrate virus is a complex trait regulated by host entry, replication, and shedding. J. Virol. 85, 3959–3967.

Webster, R.G., 1999. Antigenic variation in influenza viruses. In: Domingo, E., Webster, R.G., Holland, J.J. (Eds.), Origin and Evolution of Viruses. Academic Press, San Diego, pp. 377–390.

Weddell, G.N., Yansura, D.G., Dowbenko, D.J., Hoatlin, M.E., Grubman, M.J., et al., 1985. Sequence variation in the gene for the immunogenic capsid protein VP1 of foot-and-mouth disease virus type A. Proc. Natl. Acad. Sci. U. S. A. 82, 2618–2622.

Wiktor, T.J., Koprowski, H., 1980. Antigenic variants of rabies virus. J. Exp. Med. 152, 99–112.

Woolhouse, M.E., Webster, J.P., Domingo, E., Charlesworth, B., Levin, B.R., 2002. Biological and biomedical implications of the co-evolution of pathogens and their hosts. Nat. Genet. 32, 569–577.

Worobey, M., 2001. A novel approach to detecting and measuring recombination: new insights into evolution in viruses, bacteria, and mitochondria. Mol. Biol. Evol. 18, 1425–1434.

Xia, X., Xie, Z., 2001. DAMBE: software package for data analysis in molecular biology and evolution. J. Hered. 92, 371–373.

Xing, L., Tjarnlund, K., Lindqvist, B., Kaplan, G.G., Feigelstock, D., et al., 2000. Distinct cellular receptor interactions in poliovirus and rhinoviruses. EMBO J. 19, 1207–1216.

Yang, Z., 2006. Computational Molecular Evolution. Oxford University Press, Oxford.

Yang, Z., Nielsen, R., Goldman, N., Pedersen, A.M., 2000. Codon-substitution models for heterogeneous selection pressure at amino acid sites. Genetics 155, 431–449.

Zimmern, D., 1988. Evolution of RNA viruses. In: Domingo, E., Holland, J.J., Ahlquist, P. (Eds.), RNA Genetics. CRC Press Inc., FL, pp. 211–240.

QUASISPECIES DYNAMICS IN DISEASE PREVENTION AND CONTROL

CHAPTER CONTENTS

ABBREVIATIONS

Ab	antibody
AIDS	acquired immunodeficiency syndrome
CC$_{50}$	cytotoxic concentration 50
CTL	cytotoxic T cell
FMDV	foot-and-mouth disease virus

Virus as Populations. http://dx.doi.org/10.1016/B978-0-12-800837-9.00008-3

HAV	hepatitis A virus
HBV	hepatitis B virus
HCV	hepatitis C virus
HIV-1	human immunodeficiency virus type 1
IC$_{50}$	inhibitory concentration 50
IFN-α	interferon-alpha
IV	influenza virus
MAb	monoclonal antibody
MOI	multiplicity of infection
NGS	next generation sequencing
NNRTIs	nonnucleotide reverse transcriptase inhibitors
NRTIs	nucleoside/nucleotide reverse transcriptase inhibitors
RT	reverse transcriptase
TI	therapeutic index

8.1 MEDICAL INTERVENTIONS AS SELECTIVE CONSTRAINTS

Medical interventions have dramatically increased over the last century and in the case of infectious diseases the discovery and development of antibiotics and antiviral agents has represented a very powerful external selective constraint imposed upon replicating microbes. Hundreds of antiviral agents have been developed since the second half of the twentieth century, and viruses can generally find evolutionary pathways to continue replication in their presence. The same is true of antibiotic resistance in bacteria. The belief that bacterial diseases were in their way toward extinction was quite widespread in the middle of the twentieth century. Sir M. Burnet wrote in his 1966 textbook: "And since bacterial infections are, with unimportant exceptions, amenable to treatment with one or other of the new drugs, our real problems are likely to be concerned with virus diseases" (Burnet, 1966). In 1932, a prominent Spanish medical doctor, G. Marañón, declared: "In the year 2000 cancer will be a historical disease. Infections will be almost entirely absent as a cause of mortality." (On a personal note, when I joined University of California Irvine in 1969 to work as postdoctoral student with R.C. Warner, I attended some biology courses in which the teachers expressed to students that infectious diseases would disappear in a few decades as a consequence of the use of antibiotics and antiviral agents.) Predictions in science tend to fail.

The optimistic view was not unanimous. A. Fleming, the discoverer of penicillin, recognized the adaptive capacity of bacteria and suggested that bacteria would inevitably find ways of resisting the damage to them caused by antimicrobial drugs [quoted from the document "New antimicrobial drugs" from the European Academies Science Advisory Council, November 2014 (www.easac.eu); see also Chapter 10]. Furthermore, there was early evidence of selection of *Mycobacterium tuberculosis* mutants resistant to streptomycin (Mitchison, 1950). Antibiotic resistance in bacteria has similarities and differences with antiviral resistance in viruses, and they are compared in Chapter 10.

We are now very aware that one of the major problems in antiviral therapy is the nearly systematic selection of drug-resistant virus mutants, which is often associated with treatment failure. Other external influences such as vaccination or immunotherapy, particularly using monoclonal antibodies, can also evoke the selection of viral subpopulations capable of replicating in the presence of those components inherent to an immune response. Thus, selective constraints intended to limit RNA virus replication meet with the broad and dynamic repertoire of variants ingrained in quasispecies dynamics. Two space-time levels of the effects of drugs or vaccines are distinguished in coming sections: (i) short-term consequences for the individual in the form of treatment or vaccination failure and (ii) long-term consequences at the population level in the field, or vaccine-driven evolution of the antigenic properties of viruses.

There are other medical interventions that may alter virus survival. Individuals which are immunocompromised as a consequence of treatment after organ transplantation or those subjected to anti-cancer chemotherapy become particularly vulnerable to viral infections. Enhanced viral replication can favor pathological manifestations in the affected individual as well as the spread of large amounts viruses into the environment, with consequences for the emergence and reemergence of viral disease (Section 7.7 in Chapter 7).

8.2 DIFFERENT MANIFESTATIONS OF VIRUS EVOLUTION IN THE PREVENTION AND TREATMENT OF VIRAL DISEASE

Viral diseases are an important burden for human health and agriculture (Bloom and Lambert, 2003). Virus evolution, through the basic mechanisms exposed in previous chapters, can influence the two major strategies to combat viral infections: prevention by vaccination and treatment by antiviral inhibitors. In considering the design of a new antiviral vaccine, the extent of diversity in the field of the virus to be controlled is critical. The natural evolution of the virus may result in the circulation of one major antigenic type or to the co-circulation of multiple antigenic types. The vaccine composition (independently of the type of vaccine; see Section 8.3.1) must match the antigenic composition of the virus to be controlled. Hepatitis A virus (HAV) circulates as a single serotype while foot-and-mouth disease virus (FMDV) circulates as seven serotypes and diverse subtypes, and the antigenic types are unevenly distributed in different geographical locations. A monovalent vaccine made of the prevailing antigenic type of HAV should be sufficient to confer protection, while a multivalent vaccine composed of several types or subtypes is required to confer protection against FMDV, and the antigenic composition should be selected depending on the circulating viruses. This is why anti-FMD vaccines of different composition are used in different world areas at a given time, and vaccine composition must be periodically updated to maintain its efficacy. Thus, one effect of virus evolution relevant to vaccine design derives from the necessity to prepare a vaccine that mirrors the antigenic composition of the virus to be controlled. In the case of live-attenuated antiviral vaccines, the evolution of the vaccine virus while it replicates in the vaccinee is a risk factor to produce virulent derivatives.

The invasion of a susceptible host by a virus, and the ensuing viral replication, can be regarded as a step-wise process during which the virus must adapt to a series of selective pressures presented by the host, notably the immune response. The outcome can be either viral clearance (elimination of the infection) or virus survival and progression toward an acute or a persistent infection. Administration of antiviral agents is an additional selective constraint that limits viral replication. Evolutionary mechanisms may either succeed in selection of mutants resistant to the antiviral agent that will permit the infection to continue, or fail in sustaining the infection, resulting in the clearing of the virus from the organism.

Treatment planning, one of the aims of the new antiviral pharmacological interventions [motivated largely from information obtained by next generation sequencing (NGS) applied to the virus present in each infected patient] has some parallels with vaccine composition design. For vaccines, the information comes from analyses of antigenic composition of circulating viruses and for antiviral agents, the information comes from the analyses of the quasispecies composition of the virus to be controlled in the infected patient.

8.3 ANTIVIRAL VACCINES AND THE ADAPTIVE POTENTIAL OF VIRUSES

World-wide vaccination campaigns made possible the eradication of human smallpox [with the official declaration by the World Health Organization (WHO) in 1980] and animal rinderpest [with the official declaration by the World Organization for Animal Health, Office International des Epizooties (OIE) in 2011]. The number of new cases has dramatically decreased as a result of vaccination programs against several viral diseases such as measles or hepatitis B (Bloom and Lambert, 2003), and substantial progress has been made toward the eradication of poliomyelitis (Chumakov and Kew, 2010). These facts demonstrate that at least some viral diseases can be controlled on a global basis by vaccination, an unprecedented achievement of human and animal health.

Despite huge economic investment, however, there are important viral diseases such as acquired immunodeficiency syndrome (AIDS), hepatitis C, or viral hemorrhagic fevers for which no effective vaccines are available. For some diseases such as human influenza or animal FMD, vaccines are accessible, but they require periodic updating to approximate the antigenic composition of the vaccine to that of the circulating virus (Section 8.2). In the case of influenza virus (IV), a major change in antigenic composition can occur through antigenic shift, in which the virus acquires new hemagglutinin and neuranimidase genes by genome segment reassortment (Section 7.4 in Chapter 7), with the first evidence obtained by G. Laver as early as 1971 (for the early history of influenza, its causative virus, and vaccine designs, see Beveridge, 1977; Kilbourne, 1987). Antigenic variation of viruses, whatever the mechanism might be, can affect vaccine efficacy and in some cases the extreme rapid intra- and interhost evolution of a virus may render a vaccine unfeasible at least with the current tools of vaccinology. The difficulties for the control of virus disease derived from the adaptive potential of viruses (Domingo, 1989; Domingo and Holland, 1992; Bailey et al., 2004) require the judicious application of existing tools and innovative approaches that are still in their infancy.

8.3.1 SOME REQUIREMENTS FOR THE DESIGN OF VACCINES TO CONTROL HIGHLY VARIABLE VIRUSES

A first basic requisite for the preparation of a vaccine against a viral agent is the understanding of the immune response evoked by the virus when it infects the organism to be protected (activation of B and T lymphocytes for antibody production, cellular responses, and generation of memory cells) and correlates of protection (Bloom and Lambert, 2003; Hagan et al., 2015). For each virus-host system experiments are necessary to try to establish the determinants of protection, which is not a simple issue. The discussions in coming paragraphs are focused on the relevance of virus evolution in vaccine efficacy, irrespective of the type of protection afforded by the vaccine. What we term "protection" may mean total absence of replication of the infecting virus (termed "sterilizing" immunity) or absence of disease manifestations despite infection and some virus replication. As a general initial statement which is widely accepted by vaccinologists, a vaccine is likely to be effective when it evokes an immune response which is similar to the response elicited by the authentic viral pathogen when it produces disease successfully overcome by the infected organism (Evans and Kaslow, 1997; Bloom and Lambert, 2003). We refer to this as the basic principle of vaccinology. When infection by an antigenically constant virus produces lifelong immunity (i.e., measles virus infection) a vaccine is likely to evoke long-lasting protection. In contrast, if a patient cured of a virus can be reinfected by the same (or a closely related) virus (i.e., hepatitis C virus infection) a vaccine—at least one prepared by standard methodology—is unlikely to evoke protection.

Some points to be considered in the design of antiviral vaccines are listed in Box 8.1. They are intended to minimize selection of vaccine-escape mutants and favor the success of vaccination campaigns. Some of the recommendations deserve further comment. First, a basic knowledge of virus evolutionary dynamics and how it affects virus antigenic stability (or lack of) is essential. The fact that a methodology is available (i.e., vectors that can express large amounts of antigens displaying good immunogenicity) does not guarantee vaccine efficacy, and even less if correlates of protection are not understood. The order of efficacy of different vaccine designs proposed in Box 8.1 is justified both by the basic principle of vaccinology and by the mechanisms of selection of antibody (Ab)- and cytotoxic T-cell (CTL)-escape mutants by viruses. Single amino acid substitutions at B- and T-cell epitopes in viral proteins are often sufficient to elude neutralization by the corresponding cognate-specific antibody or to escape recognition by a clonal CTL population. For many viruses, the frequency of monoclonal antibody-escape mutants has been measured in 10^{-4} to 10^{-6}, even in clonal populations obtained under controlled laboratory conditions and that have undergone a limited number of replication rounds (Section 7.4.2 in Chapter 7). The generation of immune-escape variants can result in lack of vaccine efficacy, contribute to viral persistence (Pircher et al., 1990; Weidt et al., 1995; Ciurea et al., 2000, 2001; Richman et al., 2003; Pawlotsky, 2006), and provoke vaccination-induced virus evolution (Section 8.3.2). In human immunodeficiency virus type 1 (HIV-1), antibody-escape variants are incessantly being produced *in vivo* to the point that virus replication continues despite the antibody response (Richman et al., 2003; Bailey et al., 2004).

BOX 8.1 VACCINE DESIGNS AND VACCINATION STRATEGIES FOR ANTIGENICALLY VARIABLE VIRUSES

- Prior to the planning of a vaccine strategy, it is essential to review what is known about genetic and antigenic variation of the virus to be controlled (whether it is a DNA or RNA virus displaying high- or low-fidelity replication, antigenic diversity in the field, location of B- and T-cell epitopes, etc.).
- Carry out research to understand the correlates of protection.
- In keeping with the basic principle of vaccinology, from the point of view of inducing a protective response the preferred order of vaccine types is as follows: live attenuated > whole virus inactivated = empty viral particles > multiple immunogenic viral proteins > a single immunogenic viral protein > mixtures of synthetic peptides, dendrimeric scaffolds, peptide arrays > a single synthetic peptide.
- International vaccination programs should be carried out as quickly as possible.
- Programs to update the antigenic composition of vaccines should be implemented.

Based on Domingo and Holland (1992).

The frequency of selection of mutants that can escape a number (n) of components in which we could hypothetically separate a global immune response is far lower than the frequency of escape to a single (a, b, c, etc.) of the i components of the response. Making a simple mathematical abstraction that is applicable also to antiviral-escape mutants (Section 8.4), the frequency of mutants that escape n components of an immune response is the product of frequencies of escape to each individual component [$10^{-a} \times 10^{-b} \times 10^{-c} \times \ldots 10^{-i} = 10^{-(a+b+c+\ldots i)}$]. This is obviously an oversimplification because it is not realistic to dissect the selective impact of a complex immune response into discrete components. A virus generally includes multiple antigenic sites and each of them is often composed of several overlapping or nonoverlapping epitopes; in addition, a virus has several T-cell epitopes in different structural and nonstructural proteins, and each epitope displays a different degree of relative dominance. The above abstraction reflects, however, the advantage of stimulating the host immune system with a sufficiently broad array of B- and T-cell epitopes to prevent selection of vaccine-escape mutants due to a high genetic and phenotypic barrier (compare with the barrier to drug resistance described in Section 8.4.2). Therefore, selection of vaccine-escape viral mutants is more likely with synthetic peptidic vaccines, than with whole virus-attenuated or inactivated vaccines because the latter present a broad epitopic repertoire to the immune system. New prospects for attenuated vaccines have been opened with the engineering of viruses with suboptimal replication fidelity or deoptimized codon or codon pair usage (Coleman et al., 2008; Vignuzzi et al., 2008; Cheng et al., 2015). Selection of escape-mutants by peptidic vaccines that evoked partial protection of cattle was documented with FMDV (Taboga et al., 1997; Tami et al., 2003). The arguments in favor of multiepitopic presentation are also endorsed by a notorious scarcity of licensed peptidic vaccines for viral diseases despite horrendous economic investments (orders of magnitude greater than investments in quasispecies research!). Use of a complex, multiepitopic vaccine, however, need not prevent long-term selection of antigenic virus variants as a result of vaccine usage, an important still largely underexplored topic discussed in Section 8.3.2.

8.3.2 VACCINATION-INDUCED EVOLUTION

If a virus is allowed to circulate in a population where vaccinated and unvaccinated host individuals coexist, if the vaccine does not induce sterilizing immunity as is often the case, viruses with an altered antigenic profile might be gradually selected. The longer the virus is allowed to replicate in such a scenario, the higher the probability of incorporation of compensatory mutations that yield high-fitness antigenic variants.

These events in the case of vaccines used in veterinary medicine are particularly significant because they may alter the cell tropism and host range of viruses thus increasing the possibilities of zoonotic transmission of viruses into humans (Schat and Baranowski, 2007). Evidence of vaccination-induced DNA and RNA virus evolution is increasing, and it has been documented with bovine respiratory syncytial virus, bovine herpesvirus-1, Marek's disease virus, porcine circovirus 2, and classical swine fever virus, among others (Valarcher et al., 2000; Muylkens et al., 2006; Ji et al., 2014; Kekarainen et al., 2014, reviews in Gandon et al., 2003; Schat and Baranowski, 2007). The timing of dominance of CTL-escape mutants of simian immunodeficiency virus (SIV) was influenced by vaccination, and the process could be analyzed by penetration into the mutant spectra of the relevant viral populations (Loh et al., 2008).

For human viruses, evidence of vaccine-escape mutants has been obtained for hepatitis A and B viruses. Vaccination-associated escape mutants of HAV with substitutions around the immunodominant site of the virus were identified in a cohort of HIV-1, HAV doubly infected individuals (Perez-Sautu et al., 2011). The study suggested that an incomplete vaccination schedule, combined with the HIV-1-produced immunosuppression might have contributed to high-HAV loads thus facilitating the generation and dominance of antigenic variants. In Taiwan, the prevalence of mutants at a major antigenic determinant of the surface antigen of hepatitis B virus (HBV) tripled in one decade, and it has been suggested that this increase of prevalence might be due to the ample vaccination coverage in the region (Hsu et al., 1999).

Vaccines can rarely afford protection to all vaccinated individuals due to many factors that include variations in vaccine receptivity factors due to polymorphisms in genes involved in the adaptive immune response, immunosuppression of the vaccine recipient, insufficient time between vaccination and exposure to the viral pathogen, and antigenic differences between the vaccine strains and circulating viruses. In addition, for massive vaccinations in veterinary medicine, damage to the vaccine and improper administration are additional problems. Vaccination may occasionally promote the selection not only of antigenic variants but also of host cell tropism, host range or virulent variants (Swayne and Kapczynski, 2008; Kirkwood, 2010; Read et al. 2015). To what extent the widespread use of vaccination can contribute to antigenic variation relative to other factors (genetic drift due to genetic bottlenecks, etc.) is not known. However, our current understanding of virus dynamics should encourage investigations on the genetic and antigenic modifications of breakthrough viruses that arise from vaccinated individuals as compared with changes in viruses from unvaccinated host populations.

The above observations with animal and human viruses suggest the following possible scenarios. Reversion of live-attenuated vaccine viruses into virulent forms is a cause of disease derived from the evolutionary potential of viruses. In the case of attenuated Sabin poliovirus vaccine, the rate of vaccine-associated poliomyelitis among those vaccinated for the first time was 1 per 500,000 to 1 per 750,000 vaccinees, and those receiving the second vaccine dose the rate was about 1 in 12 million (reviewed in Rowlands and Minor, 2010). Attenuated anti-FMD vaccines were used in some countries during the second half of the twentieth century, but reversion to virulence forced halting the vaccination programs.

Vaccine-escape mutants may arise due to ineffective vaccines, and concomitant factors such as immunosuppression. The escape mutants may remain confined to the unsuccessfully vaccinated host or may spread to other susceptible individuals, and attain different degrees of epidemiological relevance. Escape mutants may be direct mutants of the infecting virus or may originate by recombination between the infecting virus and other co-infecting related viruses, as observed with poliovirus and bovine herpesvirus-1 [Kew et al., 2002; Thiry et al., 2006; among other studies with these and additional viruses]. Reiteration of vaccine selection and fitness increase processes over many generations of vaccinees (be humans or animals) may result in accelerated virus evolution. Since systematic use of vaccines for humans and animals in intensive production units is relatively recent in terms of evolutionary time (less than 100 years, and in some cases even only a few decades) it is still premature to evaluate whether vaccination is a significant factor in promoting long-term virus evolution.

8.4 RESISTANCE TO ANTIVIRAL INHIBITORS

The first description of virus resistant to an antiviral inhibitor was by J. Barrera-Oro, H.J. Eggers, I. Tamm, and colleagues working with enteroviruses and guanidine hydrochloride and 2-(alpha-hydroxybenzyl)-benzimidazole as inhibitors (Eggers and Tamm, 1961; Melnick et al., 1961). These early results that suggested that antiviral-resistant mutants could be readily selected have been amply confirmed with many viruses and inhibitors in cell culture and *in vivo*. Indeed, the selection of viral mutants resistant to antiviral agents is an extremely frequent occurrence that has been known for decades, although it became widely recognized in the course of development and clinical use of antiretroviral agents to treat HIV-1 infections and AIDS.

The description of drug-escape mutants has been based on three main groups of observations:

- Detection of antiviral-resistant mutants in patients during treatment. When a reverse genetics system is available, the suspected mutation should be introduced in an infectious clone and resistance ascertained and quantified in cell culture or in vitro enzymological assays.
- Selection of resistant mutants in cell culture, by subjecting the viruses to passages in the presence of inhibitors. The viral population size is an important variable in this type of experiment (Section 8.4.1).
- Calculation of the frequency of resistant mutants by plating a virus in the absence and presence of the antiviral agent similarly to the assays to calculate the frequency of monoclonal antibody (MAb)-resistant mutants (described in Chapter 7, Section 7.4.2).

In the three groups of observations, the frequency at which a specific escape mutant is found depends on a number of barriers to resistance (Section 8.4.2).

Traditionally, the fact that a drug can select virus-resistant mutants is regarded as a proof of the selectivity of the drug, as opposed to unspecific or toxic effects on the host cell that indirectly impair virus replication (Herrmann and Herrmann, 1977; Golan and Tashjian, 2011). Selection of viral mutants resistant to antiviral inhibitors is a major problem for the control of viral disease for two main reasons: (i) because it often results in virus breakthrough (increase of viral load) resulting in treatment failure and (ii) because resistant virus variants may become epidemiologically relevant, with the consequent decrease of inhibitor efficacy at the population level (Domingo and Holland, 1992).

Increasing numbers of antiviral agents have been developed based on the three-dimensional structure of viral proteins and their complexes with natural and synthetic ligands, in efforts that have engaged academic institutions and pharmaceutical companies. Antiviral agents may target viral or cellular proteins involved in any step of the virus life cycle. They may interact with virions and inhibit an early step of infection such as the attachment to the host cell, penetration into the cell, or uncoating to liberate the genetic material of the virus inside the cell. Other agents interfere with synthesis of viral nucleic acids or viral protein processing, particle assembly, or virus release from cells. Selection of resistant mutants has been described for virtually any chemical type of antiviral agent directed to any step of the infectious cycle of DNA or RNA viruses, including important pathogens such as herpesviruses, picornaviruses, IV, HBV, and hepatitis C virus (HCV) (several reviews and articles have covered the theoretical basis of drug resistance and descriptions for specific groups of viruses) [see Domingo et al., 2001b and previous versions in Progress in Drug Research; Richman, 1994, 1996; Ribeiro and Bonhoeffer, 2000; Domingo et al., 2001a, 2012; Menendez-Arias, 2010; the 10 articles in the Current Opinion of Virology volume edited by L. Menendez-Arias and D. Richman (Menendez-Arias and Richman, 2014)]. Therefore, the general mechanisms that confer adaptability to viruses are very effective in finding drug-escape pathways, through molecular mechanisms that are summarized in Section 8.5.

8.4.1 REPLICATIVE LOAD AND ANTIVIRAL RESISTANCE

Considering the implications of quasispecies dynamics explained in previous chapters, the following statement will be obvious to the reader: "If a single mutation is able to confer resistance to an antiviral agent, and the mutation does not cause a significant selective disadvantage to the virus (fitness decrease) in the considered environment, a drug-resistant virus mutant will be present in most, if not all, virus populations" (Domingo, 1989). If a virus replicates in such a way that a population size of 10^4 can never be achieved in a single population, it is extremely unlikely that any drug-resistance mutation (or any type of mutation associated with a phenotypic change) that is generated at a frequency of 10^{-4} or lower will be present in that viral population (Perales et al., 2011).

Selection of escape mutants depends on the replicative load and the concentration of inhibitor attained at the sites of virus replication. Consider different cell or tissue compartments in which an antiviral inhibitor reaches different concentrations (exerts different intensity of selection) (Figure 8.1). In each compartment, there are multiple replication complexes. A mutation conferring resistance to the inhibitor will occur at the same rate in each of them, assuming that the mutation rate is independent of the presence of the inhibitor. However, after its occurrence, the proportion of viral RNAs harboring the mutation will decrease depending on the inhibitor concentration. The time at which the effect of the inhibitor will be manifested depends on the inhibitor target. In the example of Figure 8.1

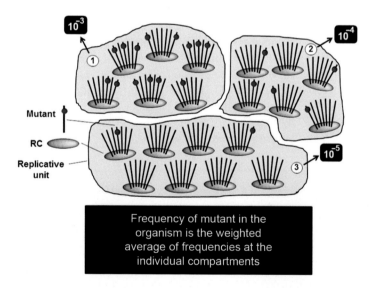

Frequency of mutant in the organism is the weighted average of frequencies at the individual compartments

FIGURE 8.1

Frequency of a drug-resistant mutant in different compartments (subcellular site, cell, tissue, or organ) of the same organism. Three compartments labeled 1, 2, and 3 are drawn. Replication complexes are depicted as ellipses, and replicating genomes as lines. A replicative unit is defined here as a set of replication complexes. The inhibitor-resistant mutant is represented by a red circle in a genome. Compartments 1, 2, and 3 reach increasing concentration of the inhibitor, rendering inhibitor-resistant mutant frequencies of 10^{-3}, 10^{-4}, and 10^{-5}, respectively. See text for the difference between occurrence and presence of the resistant mutant, and implications of compartmentalization.

we assume that the concentration of inhibitor-resistant mutants will decrease in the replication complexes, reaching resistant mutant frequencies of 10^{-3}, 10^{-4}, and 10^{-5} in compartments 1, 2, and 3, respectively. The frequency of inhibitor-resistant mutants in the entire cell, tissue, or organism at that time will be given by the weighted average of mutation frequencies at the individual compartments. In the case of a virus producing viremia, assuming no bottleneck effects or differential selection for other traits, the frequency of resistant mutants calculated for the virus in blood should reflect the average frequency in all compartments that supply virus to blood. Low inhibitor concentration in a compartment will favor selection of the resistant mutant that can either be archived as an adaptive reservoir or penetrate other compartments, depending on the sequence of events of virus spread to other compartments.

If two or more independent mutations can confer resistance to an inhibitor, the probability of occurrence of an inhibitor-resistant mutant is equal to the sum of probabilities of occurrence of each individual mutation. For multiple mutations, the probability will be the sum of probabilities of the different mutations, a frequent case in viruses since they often display several evolutionary pathways to drug resistance. The probability of finding a viral genome resistant to two or more inhibitors directed to different targets is given by the product of probabilities of resistance to each of the individual inhibitors. The basic probability considerations regarding the frequency of occurrence of inhibitor-resistant viral mutants are summarized in Box 8.2. When two or more mutations occur in the same genome, they may be subjected to epistatic effects, meaning either increase (positive epistasis) or decrease (negative epistasis) of viral fitness (see Section 2.3 of Chapter 2 for the concept of epistasis).

The diversity of chemical structures of the antiviral compounds that can select for escape mutants is illustrated in Figures 8.2 and 8.3 with the formulae of some antiviral agents in current or historical use. They include relatively simple organic molecules, nucleoside analogs, and complex heterocyclic compounds with a variety of residues (CH_3-, $C=O$, NH, NH_2, F, and Cl) that may contribute to interactions with viral proteins or alter the electronic structure of neighbor bonds thus modifying the interaction behavior of some atoms. For all of them, resistant viral mutants have been identified, despite barriers imposed upon the virus to reach a drug-resistance phenotype.

BOX 8.2 PROBABILITY OF SELECTION OF INHIBITOR-ESCAPE MUTANTS

- If there are two or more different mutations that produce the same inhibitor-resistance phenotype, and once one of the mutations is present additional mutations are no longer necessary to produce the phenotype, the probability of achieving the phenotypic change is equal to the sum of probabilities of finding each mutation individually.
- If two or more independent mutations must happen to produce resistance to an inhibitor, the probability of occurrence of the necessary mutations is equal to the product of probabilities of occurrence of each mutation individually.
- If a virus is inhibited by an inhibitor combination, and the mutations that confer resistance to each inhibitor are independent (no cross-resistance is involved), the probability of a combination-resistant mutant to arise is equal to the product of probabilities of resistance to the individual mutations.
  These probability calculations are applicable to other mutation-dependent virus variations.

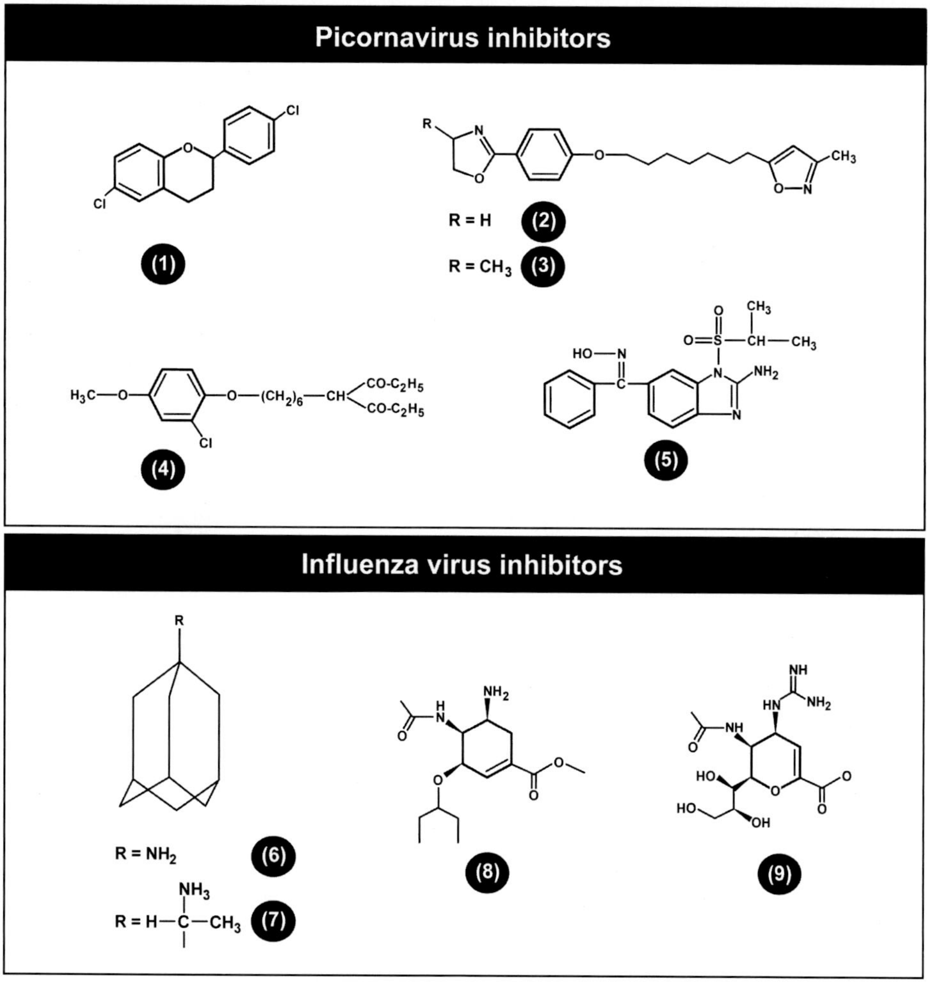

FIGURE 8.2

Some inhibitors of picornaviruses and influenza virus. The inhibitors are (1) Dichloroflavan (4′,6-dichloroflavan). (2) Disoxaril, 5-[7-[4-(4,5 dihydro-2-oxazolyl) phenoxyl] heptyl]-3-methyl-isoxazole (WIN 51711). (3) WIN 52084. (4) Arildone, 4-[6-(2-chloro-4-methoxyphenoxy) hexyl]-3,5-heptanedione. (5) Enviroxime, anti-6-[(hydroxyimino)-phenyl]-1-[(−methylethyl)sulfonylimidazol-2-amine]. (6) Amantadine, (1-amino-adamantane). (7) Rimantadine, (α-methyl-1-adamantane methylamine). (8) Oseltamivir (trade name Tamiflu®), ethyl (3R,4R,5S)-5-amino-4-acetamido-3-(pentan-3-yloxy)-cyclohex-1-ene-1-carboxylate. (9) Zanamivir (trade name Relenza®), (2R,3R,4S)-4-guanidino-3-(prop-1-en-2-ylamino)-2-((1R,2R)-1,2,3-trihydroxypropyl)-3,4-dihydro-2H-pyran-6-carboxylic acid.

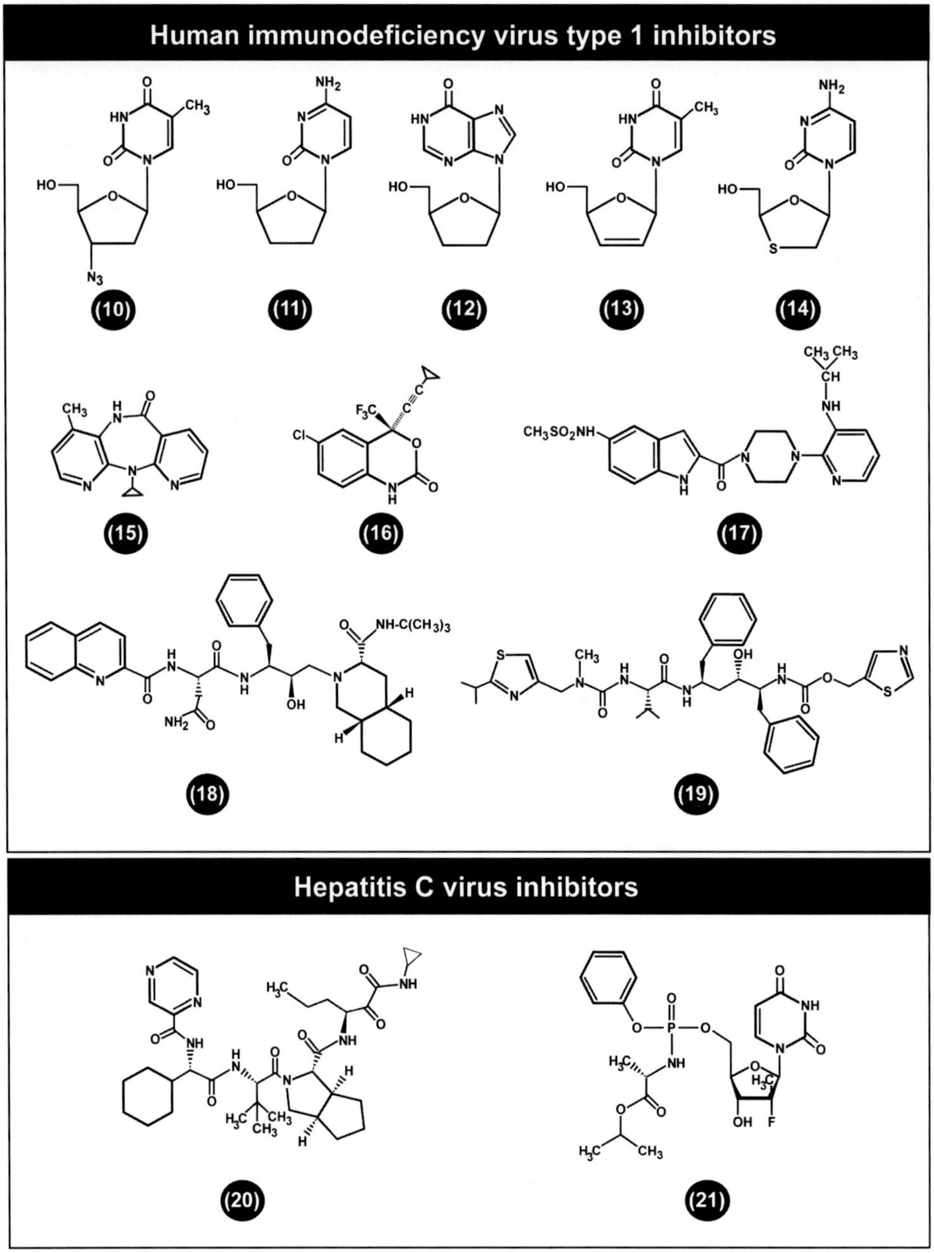

FIGURE 8.3

Some inhibitors of human immunodeficiency virus type 1 (antiretroviral agents) and hepatitis
C virus. The inhibitors are (10) Zidovudine (AZT), 1-[(2R,4S,5S)-4-Azido-5(hydroxymethyl)
oxolan-2-yl]-5-methylpyrimidine-2,4-dione. (11) Zalcitabine (ddC), 4-amino-1-((2R,5S)-5-
(hydroxymethyl)tetrahydrofuran-2-yl)pyrimidin-2(1H)-one. (12) Didanosine (ddI), 9-((2R,5S)-
5-(hydroxymethyl) tetrahydrofuran-2-yl)-3H-purin-6(9H)-one. (13) Stavudine (d4T),
1-[(2R,5S)-5-(hydroxymethyl)-2,5-dihydrofuran-2-yl]-5-mehyl-1,2,3,4-tetrahydropyrimidine-2,4-dione.

8.4.2 BARRIERS TO DRUG RESISTANCE

The impediments for a virus to attain resistance to an inhibitor are divided into genetic, phenotypic, and mutant swarm (population) barriers to resistance (Box 8.3).

The genetic barrier to resistance to a specific inhibitor is not a universal value for a virus group, since it may be affected by genetic differences among natural viral isolates. The diversification of HCV into genotypes 1a and 1b influenced the genetic barrier to resistance to the NS3/4A protease inhibitor telaprevir [formula (20) in Figure 8.3]. One of the amino acid substitutions that confer resistance to telaprevir is R155K in NS3. In genotype 1a, the triplet encoding R155 is AGA; therefore, a single nucleotide transition G → A can yield the triplet AAA which encodes K. In genotype 2b, the triplet encoding R-155 is CGA; therefore, two nucleotide changes (transversion C → A and transition G → A) are required to reach AAA, the triplet encoding K. Reaching the alternative AAG codon for K would require the same or a larger number of mutations (see Section 4.3.1 in Chapter 4 for another example of how the synonymous codon usage can influence an evolutionary outcome). The requirement of transitions versus transversions will affect the genetic barrier to antiviral resistance. Most viral polymerases tend to produce transition mutations more readily than transversions presumably because in the course

---

BOX 8.3 BARRIERS TO DRUG RESISTANCE IN VIRUSES

- Genetic barrier: Number and types of mutations needed to acquire the resistance trait.
- Phenotypic barrier: Fitness cost imposed by the resistance mutations. The cost may be due to effects at the RNA level, protein level, or both. A high fitness cost may result in reversion of the relevant mutation in the absence of the drug or incorporation of compensatory mutations that increase viral fitness.
- Mutant swarm barrier: Suppressive effect of mutant spectra may impede dominance of resistant mutants.
- The combined effect of the different barrier classes determine the ease of dominance of drug-resistant mutants in populations subjected to the selective pressure of one or multiple inhibitors.

FIGURE 8.3–CONT'D

(14) Lamivudine (3TC), 4-amino-1-[(2R,5S)-2-(hydroxymethyl)-1,3-oxathiolan-5-yl]-1,2-dihydropyrimidin-2-one. (15) Nevirapine, 11-cyclopropyl-4-methyl-5,11-dihydro-6H-dipyrido[3,2-*b*:2',3'-e][1,4] diazepin-6-one. (16) Efavirenz, (4S)-6-chloro-4-(2-cyclopropylethynyl)-4-(trifluoromethyl)-2,4-dyhidro-1*H*,3,1-benzoxazin-2-one. (17) Delavirdine, N-[2-({4-[3-(propan-2-ylamino) pyridin-2-yl] piperazin-1-yl} carbonyl)-1*H*-indol-5-yl] methanesulfonamide. (18) Saquinavir, (2S)-N-[(2S,3R)-4-[(3S)-3-(*tert*-butylcarbamoyl)-decahydroisoquinolin-2-yl]-3-hydroxy-1-phenylbutan-2-yl]-2-(quinolin-2-ylformamido)butanediamide. (19) Ritonavir, 1,3-thiazol-5-ylmethyl *N*-[(2S,3S,5S)-3-hydroxy-5-[(2S)-3-methyl-2-{[methyl({[2-(propan-2-yl)-1,3-thiazol-4-yl] methyl})carbamoyl]amino}butanamido]-1,6-diphenylhexan-2-yl]carbamate. (20) Telaprevir, (1S,3a*R*,6aS)-2-[(2S)-2-[[(2S)-2-cyclohexyl-2-(pyrazine-2-carbonylamino)acetyl]amino]-3,3-dimethylbutanoyl]-N-[(3S)-1-1(cyclopropylamino)-1,2-dioxohexan-3-yl]-3,3a,4,5,6,6a-hexahydro-1H-cyclopenta[c]pyrrole-1-carboxamide. (21) Sofosbuvir, isopropyl(2S)-2[[[(2*R*,3*R*,4*R*,5*R*)-5-(2,4-dioxopyrimidin-1-yl)-4-fluoro-3-hydroxy-4-methyl-tetrahydrofuran-2-yl]methoxy-phenoxy-phosphoryl]amino]propanoate. Additional drugs currently used in antiviral therapy can be found in the references quoted in the text and in Chapter 9.

of RNA elongation it is easier to misincorporate a purine by another purine than by a pyrimidine, and the same for pyrimidine misincorporations (Chapter 2). Mutation preference is one of several factors that determine the frequency of drug-escape mutants. Thus, evolution may diversify viruses to display different genetic barriers to the same drugs. To complicate matters even more, since in many cases several independent mutations may confer resistance to the same drug, it has to be considered also that the genetic barrier to one inhibitor may be affected by the presence of other inhibitors (Beerenwinkel et al., 2005).

The phenotypic barrier to drug resistance is equivalent to the fitness cost inflicted upon the virus by the mutations and corresponding amino acid substitution(s) required for resistance. Fitness cost is treated in Chapter 4 (Section 4.6) and in Chapter 7 (Section 7.4.2) as it may affect the frequency of monoclonal antibody- or cytotoxic T-cell-escape mutants in viral populations. When a drug-resistance mutation inflicts a high fitness cost, a likely result is reversion of the mutation when the virus replicates in the absence of the drug. An alternative outcome is that compensatory mutations are introduced in the genome so that viral fitness increases while maintaining the inhibitor-resistance mutation. The two outcomes are not mutually exclusive and may contribute to the multiple, transient selection pathways observed by the application of deep sequencing to monitor the response of a viral population to specific selective force (Tsibris et al., 2009; Fischer et al., 2010; Cale et al., 2011; Kortenhoeven et al., 2015) (see Section 6.3 in Chapter 6). A high fitness cost may prevent or delay selection of escape mutants. Sofosbuvir [formula (21) in Figure 8.3] is a very effective NS5B (viral polymerase) inhibitor of HCV. Amino acid substitution S282T in NS5B has been associated with sofosbuvir resistance, and the substitution has been detected in patients and in some natural isolates of HCV. In one of several clinical studies on sofosbuvir efficacy, the mutant spectrum composition of HCV genotype 2b in an infected patient treated with the drug was followed by NGS at baseline (prior to initiation of treatment), in the course of treatment, and posttreatment. The frequency of S282T was 0.05% at baseline, indicating preexistence of resistance mutations despite no exposure of the virus to the drug (Section 8.6). Two days after initiation of sofosbuvir treatment the level of S282T decreased to 0.03%, and viral breakthrough was detected 4 weeks later when 99.8% of the viral population included S282T. During the posttreatment period, genomes with the wild-type S282 amino acid regained dominance that was attributed to true reversion of mutant genomes rather than outgrowth of baseline wild-type genomes (Hedskog et al., 2015). This result suggests a high phenotypic barrier for sofosbuvir, but that HCV has mechanisms to overcome this barrier. The complexities of virus-host interactions render the elucidation on the pathways exploited by a virus to overcome the phenotypic barrier to a drug a highly empirical endeavor. The hope is that by combining high phenotypic barrier inhibitors, the forced reversion of the resistance mutations for survival may drive the virus to extinction (Chapter 9).

The mutant swarm barrier to resistance is a consequence of the interfering interactions that operate within quasispecies, and that are described in Chapter 3 (Section 3.8). It is a particular case of interference that can delay or impede the increase of frequency of a resistance mutation (Crowder and Kirkegaard, 2005). The possible contribution of mutant swarms to facilitate or impede the dominance of drug-resistant mutants in infected patients is still largely unexplored.

It is difficult to anticipate how the three types of barrier listed in Box 8.3 may result in a level of drug resistance for a particular virus, in a particular host individual, in a particular target organ, at a given time. Additional influences are drug pharmacokinetics, drug penetration into different cells, tissues, and organs where the virus replicates (see Figure 8.1), and prior history of virus replication in the infected host. It is not surprising that the study of drug resistance in viruses remains fundamentally descriptive.

8.4.3 DRUG EFFICACY, MUTANT FREQUENCIES, AND SELECTION OF ESCAPE MUTANTS

The genetic barrier, as defined in Box 8.3, can be anticipated from the number of point mutations that, according to the genetic code, are needed to convert an amino acid associated with drug sensitivity into another amino acid that confers drug resistance. When independent amino acid substitutions can lead to resistance to the same drug, alternative evolutionary pathways may be followed depending on tRNA abundances, mutational preferences, and relative nucleotide substrate concentrations at the virus replication sites. If resistance requires two or more amino acid substitutions, the genetic barrier will be correspondingly increased (Section 8.4.1 and Box 8.2).

Quantification of barriers to resistance in experiments in cell culture requires a prior characterization of the drug to be tested when acting on the cell culture-adapted virus as it infects a specific cell line. The two basic parameters to be determined are the toxicity of the drug for the host cell, and its capacity to inhibit the production of infectious virus. Toxicity is quantified by the concentration of drug that kills a given percentage (generally 50%, but sometimes another value) of cells under the conditions used in the infection. It is expressed as the cytotoxic concentration 50 (CC_{50}), as depicted in Figure 8.4. Toxicity may depend on the cell concentration, the extent of confluence in a cell monolayer, and the metabolic state of the cell (resting vs. actively dividing). The capacity of inhibition is quantified by the concentration of inhibitor that reduces the infectious progeny production by a given percentage (generally 50%, but sometimes another value) under the defined conditions of the infection, including a multiplicity of infection (MOI). It is expressed as the inhibitory concentration 50 (IC_{50}), as depicted in Figure 8.4. The therapeutic index (TI) is given by the quotient CC_{50}/IC_{50}, and although generally used for *in vivo* experiments of drug efficacy testing, it can be also applied to cell culture measurements.

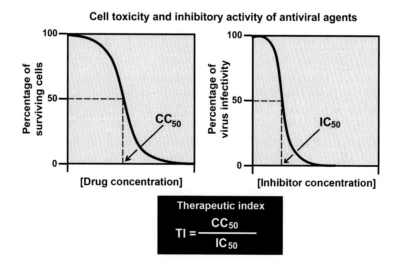

FIGURE 8.4

Schematic representation of two experiments to determine the concentration of an inhibitor needed to kill 50% of cells in culture (CC_{50} value, left) and the concentration of inhibitor that reduces the viral production to 50% (IC_{50} value, right). The therapeutic index is the quotient between CC_{50} and IC_{50} (box at the bottom). Similar tests can be performed with tissue explants or animals, under controlled environmental conditions. See text for pharmacological implications.

The three parameters, CC_{50}, IC_{50} and TI are not universal for a virus and a drug since they may be influenced by the composition of the viral population and environmental influences, as repeatedly expressed for other features of viruses in the present book. As a guide, TI values of 100 or more suggest excellent performance of an antiviral agent, values higher than 10 are acceptable, but values lower than 10 predict limited efficacy. The quantitative effects of a drug may vary when analyzing a single round of infection versus multiple rounds in serial passages, or when comparing *in vivo* versus cell culture experiments. CC_{50} and IC_{50} values serve as a guide to decide range of the drug concentration to be used in serial passage experiments to evaluate the possible selection of inhibitor-resistant mutants and to estimate the genetic barrier.

The possibility to overcome a genetic barrier depends on the virus population size. For viruses that replicate in cell culture, it is possible to estimate the minimal viral population size needed to select a drug-resistant mutant which is generally positively correlated with the genetic barrier (Figure 8.5). In the hypothetical example of the figure, a viral population is composed of inhibitor-sensitive viruses (blue spheres), and a low level of inhibitor-resistant viruses (red spheres). The proportion of inhibitor-resistant viruses is given by the mutational pressure (e.g., at a frequency of 10^{-4}, which is increased in the picture for clarity). Passage of a small amount of virus (e.g., 10^2 infectious virus in the small circle at the upper part of the figure) will exclude the mutant virus (red spheres) that will be maintained at the basal level dictated by mutational pressure in the course of passages (limited to two in the figure for simplicity). Selection of escape mutants is precluded by the limited population size at each transfer. In contrast, if the population size used for the successive infections is sufficiently large ($>10^4$, larger circles at the bottom that surround both sensitive and resistant viruses), the resistant mutant can become gradually dominant and can be isolated for further studies. To give another example, a single amino acid

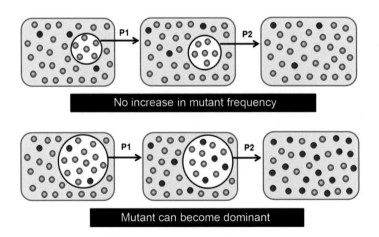

FIGURE 8.5

The effect of sampled virus population size on the dominance of an inhibitor-resistant mutant. The upper left population is composed of inhibitor-sensitive (blue spheres) and inhibitor-resistant (red spheres) viruses. If during passages P1 and P2, the amount of sampled virus is insufficient (upper three successive populations), the resistant virus mutant will not be enriched in the population. If during passages P1 and P2, the amount of sampled virus exceeds a critical value (bottom three successive populations), the population will be enriched in the resistant mutant. See text for numerical examples.

replacement that requires two mutations (the change from CAG to AUG to attain substitution Q151M in HIV-1 reverse transcriptase, associated with resistance to multiple antiretroviral nucleosides) will occur at lower frequency than replacements that require a single mutation. If each of the two mutations reach a frequency of 2×10^{-4}, the expected frequency of the drug-resistant genomes (ignoring fitness effects) will be $(2 \times 10^{-4}) \times (2 \times 10^{-4}) = 4 \times 10^{-8}$. Thus, at least 4×10^{8} viral genomes must undergo one round of copying (or a lower number of genomes a proportionally higher number of rounds of copying) to approach a good probability to obtain a drug-resistant genome in that viral population. Population size limitation of a drug selection event is a specific example of how random events may intervene in a process of positive selection (compare with Section 3.2 and Figure 3.2 in Chapter 3; in that figure, the random event that excludes the positively selected population is conceptually equivalent to the insufficient population size depicted in the upper infection series of Figure 8.5).

When two or more mutations are needed to confer the resistance phenotype, drug resistance will be less likely not only due to the lower probability of generating the two required mutations, but also because of the increased chances of two mutations entailing a fitness cost. A virus that requires three or more mutations to overcome a selective constraint may occur at a frequency in the range of 10^{-12} or lower which will often be insufficient for the mutant to be present in the mutant spectrum of the infected host (Figure 8.6).

Failure to select for a drug-resistant mutant in cell culture does not necessarily mean that the resistant mutant is not present in the population. It may mean that due to a high genetic barrier, the selection

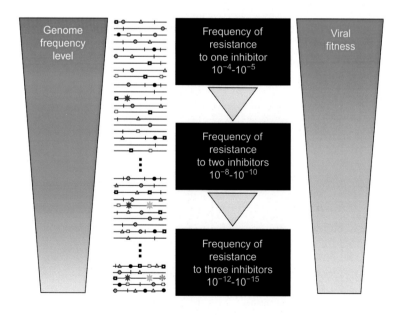

FIGURE 8.6

Decreased frequency and fitness of mutant genomes resistant to one, two, or three inhibitors. The genome frequency level decreases by several orders of magnitude when resistance to one inhibitor (red asterisk in the upper mutant spectrum), or two inhibitors (red and green asterisks in the middle-mutant spectrum), or three inhibitors (red, green, and blue asterisk in the bottom mutant spectrum) must occur in the same genome (left part of the figure and numerical values at the center). Increased number of mutations generally implies fitness decrease (right part of the figure). See text for implications.

experiment was designed to infect with an insufficient amount of virus. Similar, even more accentuated problems are encountered in selection experiments *in vivo*, since not only population size but also the source of the virus (blood and other organ) for the next infection may determine the inclusion or exclusion of the relevant mutation (compare Figures 8.1 and 8.6).

8.4.4 PHENOTYPIC BARRIER AND SELECTIVE STRENGTH

The phenotypic barrier or fitness cost inflicted by a drug-resistance mutation (Box 8.3) is often estimated empirically from the frequency of the relevant substitution in patients treated with the drug or in cell culture assays. An adequate procedure to quantify the phenotypic barrier to resistance is to determine the fitness of the virus expressing the wild-type amino acid (the one that confers drug sensitivity) relative to the virus expressing the substitution that confers resistance; fitness is measured in the absence and presence of the drug (double assay). This is an extension of the determination of fitness vectors described in Section 5.1.1 of Chapter 5, as depicted in Figure 8.7; the assays are best performed in cell culture, although use of explants or *in vivo* assays are also feasible. Two parameters can be calculated: the fitness cost inflicted by the amino acid substitution associated with resistance in the absence of the drug, and the selective advantage conferred by the substitution in the presence of the drug. In the presence of the drug, the mutant will yield a fitness value $f_{+DRUG} > 1$ relative to the wild type (necessarily if the mutation confers resistance and virus viability is preserved). In the absence of the drug, $f_{-DRUG} \leq 1$ relative to the wild type quantifies the fitness cost of the resistance mutation; the lower the value of f_{-DRUG}, the higher the fitness cost. We define the selective strength of the resistance mutation as f_{+DRUG}/f_{-DRUG}. For example, if we put arbitrary numbers (unrelated to values shown in ordinate) to the fitness values in the first graph of Figure 8.7, $f_{+DRUG} = 1.4$ and $f_{-DRUG} = 0.8$, we obtain a selective

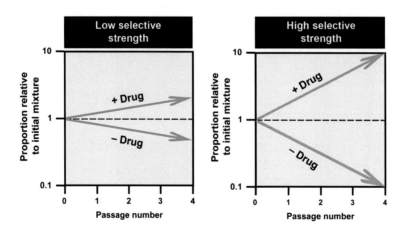

FIGURE 8.7

Selective strength of an inhibitor-resistance mutant. A mutation may confer a different degree of resistance at a different fitness cost for the virus. The fitness vectors in the left panel denote limited resistance at low fitness cost whereas the vectors in the right panel denote high resistance despite a considerable fitness cost inflicted by the resistance mutation. The selective strength is defined as the ratio of fitness in the presence and absence of the drug, as explained in the text.

strength of 1.7. For the vectors in the second graph, if $f_{+DRUG} = 3.6$ and $f_{-DRUG} = 0.1$, the selective strength is 36. High selective strength means an important selective advantage conferred by the amino acid substitution for the virus replication in the presence of the drug despite a high fitness cost inflicted by the substitution (compare with Section 4.2 in Chapter 4 for the *trade-off* and "no free lunch" concepts). If the substitution does not entail any fitness cost ($f_{-DRUG} = 1$), the fitness value in the presence of the drug equals the selective strength. Selective strength can be calculated for a mutation or group of mutations that confer resistance to a drug used at a given concentration in a defined environment. The limitations of fitness measurements (environment dependence, etc.) described in Section 5.1.2 of Chapter 5 apply here. Since viral genomic sequences may vary in the course of fitness assays, a limited number of passages and triplicate parallel assays are recommended. If a substitution entails a high fitness cost, direct reversion of the substitution or incorporation of compensatory mutations may occur. Nucleotide sequence monitoring in the course of the assay should reinforce the conclusions.

8.4.5 MULTIPLE PATHWAYS AND EVOLUTIONARY HISTORY IN THE ACQUISITION OF DRUG RESISTANCE

Most drug-resistance mutations inflict a fitness cost upon the virus and yet very rarely (not to say never) drug resistance represents an unsurmountable barrier to maintain viral infectivity. Several possibilities can account for the pertinacious occurrence and selection of drug-resistant, viable viral mutants. One possibility, supported by some experimental and clinical observations, is that a drug-resistant phenotype may be achieved through a number of alternative genetic modifications. Even if a specific amino acid substitution—that would serve as the most direct and effective determinant of drug resistance—were highly detrimental or lethal for the virus, alternative mutations can often be found that lead to a similar resistance phenotype, or at least a sufficient resistance to permit finding compensatory mutations. Not only the connectivity among points of sequence space plays a role but also the fact that several points in sequence space map into the same (or similar) drug-resistance phenotype. This phenotypic redundancy applies to both standard nonmutagenic inhibitors and to mutagenic inhibitors. The cascade of mutations that confer resistance of picornaviruses to the mutagenic purine analog ribavirin illustrates how alternative amino acid substitutions in the viral polymerase (some being genuine resistance mutations and others acting as compensatory substitutions to maintain polymerase function) can lead to the ribavirin-resistance phenotype (discussed in detail in Chapter 9).

A speculative interpretation of the systematic occurrence of drug-resistant viral mutants is that the majority of the chemicals used in antiviral therapy (Figures 8.2 and 8.3) have a structure which may be related to natural compounds that viruses and their ancestral replicative machineries encountered in their continuous struggle to survive. In this view, drug resistance would have been gradually built as a consequence of coevolution (Section 4.5 of Chapter 4) between virus replicative and gene expression machineries and the "space" of chemical compounds that interacted with them. Mechanisms of drug resistance might have had their roots in molecular events repeatedly experienced as viruses evolved in an interactive manner with protocellular and cellular metabolites in our biosphere. Discrimination in favor of small molecule substrates compatible with a flow of genome replication and gene expression, and avoidance of perturbing intruders that could alter catalytic activities, should have been positively selected. Unfortunately, this is a possibility we will never be able to test. Whatever the reasons behind, the unfortunate reality is that drug resistance is an extremely frequent event that complicates enormously the control of viral disease.

8.5 MOLECULAR MECHANISMS OF ANTIVIRAL RESISTANCE

The great majority of inhibitor-resistance mechanisms involve amino acid substitutions in viral proteins. The substitutions may render the drug ineffective through the following mechanisms:

- Substitutions in the protein targeted by the drug that decrease the affinity of the protein for the drug. To this group belong mutations that modify nucleotide selectivity in viral polymerases.
- If the inhibitor acts on a viral protein that itself has some other viral protein or genomic structure as a target, amino acid substitutions or mutations that affect that target may also contribute to resistance. Correlated mutations in the first and second target may also yield the resistance phenotype. This is the case of some protease inhibitor-resistance mutations in HIV-1.
- In the case of viral polymerases, some mutations permit the excision of a chain-terminating nucleotide at the 3′-end of the primer. It is achieved through phosphorolysis mediated by a pyrophosphate donor, probably ATP.

8.5.1 SOME EXAMPLES WITH HIV-1

Resistance to nucleoside/nucleotide reverse transcriptase (RT) inhibitors (NRTIs) is achieved by one of at least two mechanisms: (i) discrimination against the incorporation of the triphosphate form of the NRTI and (ii) excision of the chain-terminating nucleotide once incorporated at the 3′-end of the growing DNA chain. This occurs with thymidine analog resistance mutation (or TAMs) that are typically selected under treatment with AZT or d4T [formulae (10) and (13) in Figure 8.3]. Groups of amino acid substitutions may yield a multidrug-resistance phenotype. A well-studied example in HIV-1 is the Q151M complex in the reverse transcriptase which includes substitutions A62V, V75I, F77L, F116Y, and Q151M. The phenotype consists in the limitation of incorporation of several nucleotide analogs. These and other mutants are characterized by a decrease in the catalytic rate constant (k_{pol}) of incorporation of the analog or analogs relative to standard nucleotides (see Section 2.6 in Chapter 2 for the basic kinetic parameters for polymerase activity). The combination of enzymological and structural studies has provided a molecular interpretation of the mechanism of inhibition of HIV-1 by NRTIs in use or under preclinical development (reviews in Menendez-Arias, 2010, 2013; for a predictive model that includes nucleotide levels, see von Kleist et al., 2012).

Nonnucleotide RT inhibitors (NNRTIs) bind to a RT pocket 10Å away (a considerable distance) from the catalytic site composed of residues of the p66 and p51 subunits of the enzyme. The hydrophobic nature of NNRTIs is illustrated in Figure 8.3 with the structures of nevirapine, efavirenz, and delavirdine [formulae (15), (16), and (17), respectively]. Several mechanisms have been proposed to explain their inhibitory activity, including alteration of the catalytic amino acids YMDD at the RT active site, distortion of the nucleotide binding site, or modification of the position of the primer that receives the incoming nucleotides (Menendez-Arias, 2013). Amino acid substitutions that confer resistance to NNRTIs block the access of the inhibitors to their binding sites or alter the conformation and volume of the binding pocket.

The essential viral proteases are a target for the development of specific antiviral inhibitors. Examples are saquinavir and ritonavir for the HIV-1 [formulae (18) and (19), respectively, in Figure 8.3]. Multiple resistance mutations have been described for protease inhibitors. They can affect the substrate binding site or neighbor positions, often accompanied of compensatory substitutions at distant positions including sites of the target viral protein [e.g., the Gag cleavage sites in HIV-1 (Fun et al., 2012; Flynn et al., 2015)]. Some HIV-1 protease inhibitor combinations display a high genetic barrier to resistance but, significantly for the capacity of viruses to explore sequence space, resistant mutants with 20-30 amino acid substitutions in the protease-coding region have been isolated (Rhee et al., 2010).

The reviews by Menendez-Arias (2010 and 2013) provide excellent and detailed accounts of mechanisms of resistance to HIV-1 inhibitors that include virus entry and integrase inhibitors in addition to the RT and protease inhibitor briefly described here. Despite variations in the detailed mechanisms involved, resistance to inhibitors of other major viral pathogens such as HCV and HBV is a general phenomenon, and the number of resistance mutations keeps increasing as new treatments become available.

8.5.2 MUTATION SITE AND FUNCTIONAL BARRIER

Because of the distance effects that can be exerted among amino acids in viral proteins, substitutions that confer resistance to inhibitors of viral enzymes may lie far from the catalytic site. The effect of drug-resistance mutations on the general catalytic efficiency of a viral enzyme is one of the determinants of the functional barrier to resistance. When an enzymological assay *in vitro* is available, the effects of specific drug-resistance amino acid substitutions on enzyme activity can be tested, although the observed alteration may not be the only influence on the fitness modification of the corresponding mutant virus. The reason is that most viral proteins (including viral enzymes) are multifunctional and an enzyme activity assay may not capture the range of influences exerted by the enzyme. Regarding the new directly acting antiviral (DAA) agents for HCV, the inhibitors that target the HCV polymerase (NS5B) generally display a higher functional barrier to resistance than the protease inhibitors, and manifest a broader genotype coverage. Fitness decreases entail reductions in viral load and, consequently, decreased probability of viral breakthrough (treatment failure). Nucleotide analogs that bind to conserved residues at or near the active site of the viral polymerase tend to show subtype-independent antiviral activity. Because amino acid substitutions at or near the active site of viral enzymes are likely to inflict a fitness cost, such substitutions are less likely to preexist in treatment-naive patients (Margeridon-Thermet and Shafer, 2010; Sarrazin and Zeuzem, 2010). Interestingly, in the case of HIV-1, mutations conferring resistance to RT inhibitors inflict a lower fitness cost than mutations that confer resistance to protease inhibitors (reviewed in Martinez-Picado and Martinez, 2008). Thus, enzymes that perform similar functions for different viruses may have evolved to display different tolerance to amino acid substitutions. It is not possible to generalize which types of resistance mutations will display high or low functional barriers.

8.5.3 ADDITIONAL CONSIDERATIONS ON ESCAPE MUTANT FREQUENCIES

There is a broad range of frequencies of antibody- and drug-escape mutants in viral populations, although a frequent range is 10^{-4} to 10^{-6} mutants per infectious unit for many RNA and DNA viruses (see Table 7.2 in Chapter 7 for antibody-escape mutants). A few estimates for drug-escape mutants are listed in Table 8.1; several observations and characterization of escape mutants have not been accompanied with frequency measurements. A point worth emphasizing is that the quantification of viruses harboring biologically relevant mutations has been possible because an adequate biological

Table 8.1 The Frequency of Some Drug-Escape Mutants

Drug[a]	Virus[b]	Frequency	Comments	Reference
HBB[c]	CAV9	10^{-4}	Reversion from HBB dependence to HBB independence	Eggers and Tamm (1965)
Amantadine and rimantadine	IV	1×10^{-3} to 4×10^{-4}	Measurements in cell culture	Appleyard (1977), Lubeck et al. (1978)
Rimantadine[a]	IV	27%	Percentage of children treated with rimantadine that shed resistant IV	Belshe et al. (1988)
Disoxaril[a]	HRV14	1×10^{-3} to 4×10^{-4}	Low-level resistance in cell culture	Heinz et al. (1989)
		4×10^{-5}	High-level resistance in cell culture	
Guanidine[d]	PV	1.8×10^{-5} to 4×10^{-8}	Measurements in cell culture	Pincus and Wimmer (1986)

[a] The formula of these drugs is included in Figure 1 with the following number in parenthesis: amantadine (6); rimantadine (7); disoxaril (2).
[b] The virus abbreviations are CAV9, coxsackievirus type A9; IV, influenza virus; HRV14, human rhinovirus type 14; PV, poliovirus.
[c] HBB is 2-(alpha-hydroxybenzyl)-benzimidazole
[d] Guanidinium is carbamimidoylazanium.

assay is available. An antiviral agent or a neutralizing antibody measures the proportion of infectious viral particles that differ from the majority of the population in the relevant resistance trait. There is no reason to suspect that the viral amino acid residues (that are the target of an inhibitor or an antibody) that are substituted to confer the resistance phenotype are on average more prone to variation than other amino acids in viral proteins. This means that if we had additional selective agents to probe other viral sites, we would probably obtain a similar range of variant amino acids than using inhibitors or antibodies. This quite straightforward prediction is another way to state that there is general agreement in the mutation rates and frequencies for viruses being in the range of 10^{-3} to 10^{-5} substitutions per nucleotide (s/nt), calculated using a variety of biochemical and genetic methods (Chapter 2).

The calculated frequencies of resistant mutants in viral populations may be lower than the rate at which they originate by mutation due to fitness cost of the mutation (Section 8.4.4). The argument is parallel to the one used to justify why mutation rates and frequencies differ due to fitness effects of mutations (Chapter 2). Another prediction derived from the above considerations is that mutations conferring resistance to antiviral agents are expected to be detected in viral populations never exposed to the relevant drugs. In other words, the basal level of mutational pressure may be sufficient to provide a detectable level of escape mutants without the need of selection by the selective agent. This is an important aspect of antiviral therapy which is addressed next.

8.6 ANTIVIRAL RESISTANCE WITHOUT PRIOR EXPOSURE TO ANTIVIRAL AGENTS

The first demonstration that the baseline mutation level in viral quasispecies can include a detectable level of mutations that confer resistance to inhibitors in the absence of selection by the inhibitors, was obtained by D.D. Ho, I. Nájera, C. López-Galíndez, and their colleagues working with HIV-1 (Mohri et al., 1993; Nájera et al., 1994, 1995). One of the studies examined the *pol* gene of 60 HIV-1 genomes obtained directly from lymphocytes of infected patients. Mutation frequencies for independent viral isolates were in the range of $1.6.10^{-2}$ to $3.4.10^{-2}$ s/nt, while for mutant spectrum components of individual isolates the values were $3.6.10^{-3}$ to $1.1.10^{-2}$ s/nt. In the virus from these patients, mutation frequencies at the codons for amino acids involved in antiretroviral resistance were very similar to the average mutation frequency for the entire *pol* gene. Consistently with the mutation frequency values, several mutations that led to amino acid substitutions that conferred resistance to reverse transcriptase inhibitors were identified in patients not subjected to therapy. At the time of the study, the number of antiretroviral agents was still limited and a considerable number of patients were not treated. The authors gave convincing epidemiological arguments that the background of mutations related to antiretroviral resistance was a consequence of high mutation rates and quasispecies dynamics, and not of transmission of resistant virus from individuals that had been subjected to therapy (primary resistance) (Nájera et al., 1995).

The presence of inhibitor-resistance mutations in viral populations never exposed to the corresponding inhibitor has been confirmed for HIV-1 and for several other viruses, including HCV (Havlir et al., 1996; Lech et al., 1996; Cubero et al., 2008; Johnson et al., 2008; Toni et al., 2009; Tsibris et al., 2009, among other studies). Ample support has come also from NGS analyses of mutant spectra, opening a point of debate on the basal frequency of inhibitor-resistance mutations that constitutes an indication to avoid the use of the corresponding inhibitors in therapy. The data underline the relevance of mutant spectra as phenotypic reservoirs to confront selective constraints before constraints are in operation. For treatments including a drug that has already been administered to a patient in the past, the influence of quasispecies memory should also be considered (Section 5.5.1 in Chapter 5). Mutant spectra can be viewed as an anticipatory reservoir of phenotypes.

8.7 FITNESS OR A FITNESS-ASSOCIATED TRAIT AS A MULTIDRUG-RESISTANCE MECHANISM

The major mechanism of drug resistance in viruses is based on amino acid substitutions that render the drug ineffective through the several molecular mechanisms summarized in Section 8.5. Despite being the most common, the presence of specific resistance mutations is not the only mechanism of drug resistance. The cell culture system of HCV replication in human hepatoma cells (Lindenbach et al., 2005; Wakita et al., 2005; Zhong et al., 2005) permitted addressing the important issue of HCV resistance to interferon-alpha (IFN-α). IFN-α and ribavirin were the two components of the standard of care treatment against HCV infections until the advent of new therapies based on directly acting antiviral agents in 2014. Natural HCV isolates differ in IFN-α sensitivity and the molecular basis of the difference is largely unknown. The study in cell culture consisted in subjecting a clonal population of HCV (termed HCVp0, prepared by electroporation of hepatoma cells with RNA encoding the viral genome, transcribed from a plasmid) to 100 serial passages (of the type described in Section 6.1 of Chapter 6) in the absence or presence of increasing concentrations of IFN-α added to the culture medium. Several mutations scattered throughout the HCV genome were associated with IFN-α resistance (Perales et al., 2013). The selection of multiple alternative mutations is most likely due to the fact that IFN-α evokes a multicomponent antiviral response which is not focused toward a single viral protein (Perales et al., 2014). Unexpectedly, even the control HCV populations (those passaged 100 times in the absence of IFN-α) displayed a partial (but statistically significant) resistance to IFN-α, that could not be attributed to endogenous IFN production by the hepatoma cells (Perales et al., 2013).

In view of this intriguing result, the initial HCVp0 population and the HCV population passaged 45 and 100 times in the absence of IFN-α (termed HCVp45 and HCVp100, respectively) were tested for their resistance to other inhibitors of HCV replication: the protease inhibitor telaprevir [formula (20) in Figure 8.3], the NS5A inhibitor daclatasvir, the cellular protein cyclophilin A inhibitor cyclosporin A, and the mutagenic purine nucleoside ribavirin. HCVp45 and HCVp100 displayed significant increased resistance to all inhibitors tested, as compared with the parental population HCVp0 (Sheldon et al., 2014) (Figure 8.8). Passage of HCV entailed an increase of viral fitness and a broadening of the mutant spectrum, as expected (Section 5.4 of Chapter 5). Therefore, a clear possibility was that the broadening of the mutant spectrum increased the frequency of mutations associated with drug resistance, thus explaining the behavior of the multiply passaged HCV populations. The search for the resistant mutations was easier for telaprevir, daclatasvir, and cyclosporin A than for the other drugs because amino acid substitutions in the target protein had been previously identified as responsible for drug resistance. In the case of cyclosporin A resistance, substitutions map in NS5A and NS5B, because the drug binds to cyclophilin A which in turn interacts with NS5A. Analysis of the mutant spectra of HCVp45 and HCVp100 by molecular cloning and Sanger sequencing and by NGS failed to identify specific drug-resistance mutations. Since it could not be excluded that the broadening of the mutant spectrum might have increased the frequency of resistance mutations still to be characterized, two additional tests were performed. One was to determine the kinetics of viral production over a 1000-fold range of MOI in the absence and presence of telaprevir. Both the unpassaged and multiply passaged HCV displayed parallel kinetics at the different MOIs, which excludes that drug resistance was due to the presence of resistance mutations in minority components of the mutant spectrum (Figure 8.8). To further substantiate the findings, biological clones obtained by end-point dilution of the corresponding HCVp0 and HCVp100 populations were tested regarding drug resistance. A biological clone should have eliminated minority genomes that harbored drug-resistance mutations since biological cloning is the most severe form of bottleneck event (Sections 6.2 and 6.5 in Chapter 6). The biological clones did not display any decrease in drug resistance as compared with their corresponding parental, uncloned populations (Sheldon et al., 2014).

The above observations have established viral fitness, or some trait associated with viral fitness, as a multidrug-resistance determinant in HCV, that may also apply to other viruses. One possible molecular mechanism may consist in a competition between replicative complexes and inhibitory molecules inside the infected cells. This model implies that fitness increase is reflected either in more replicating molecules per each replicative unit or in an increase in the number of replicative units per cell, without any influence in the number of inhibitor molecules that reach the replication sites. Exploration of this model and other possible models, and the extension to other viral-host systems, are important challenges in the field of antiviral research.

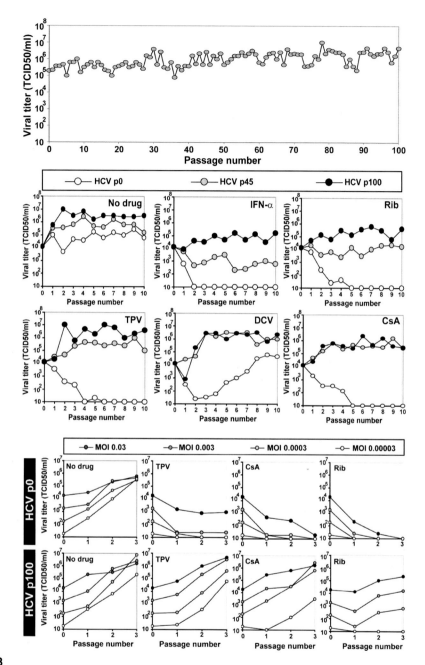

FIGURE 8.8

Multidrug resistance of hepatitis C virus passaged in human hepatoma cells. Top: Evolution of virus infectivity in the course of 100 serial passages. Middle: HCVp0, HCVp45, and HCVp100 are the initial hepatitis C virus, the HCV passaged 45 times and the virus passaged 100 times in the absence of any drug (samples from the experiment described at the top). The individual panels give the virus titer in the course of 10 passages either in the absence of any drug (No drug) or the presence of IFN-α, ribavirin (Rib), telaprevir (TPV), daclatasvir (DCV), or cyclosporin A (CsA). Note the resistance of HCVp45 and HCVp100 as compared to HCVp0 to several inhibitors. Bottom: Virus titer in the course of three serial passages of HCVp0 and HCVp100 (written in the filled boxes on the left) at the indicated multiplicity of infection (box). See text for the interpretation of these experiments and Sheldon et al. (2014) for details.

The figure has been modified from Sheldon et al. (2014), with permission from the American Society for Microbiology, Washington DC, USA.

To sum up, mutant spectra and quasispecies dynamics can mediate antiviral resistance by at least two mechanisms: (i) by increases in the proportion of resistance mutations in the mutant spectra and (ii) by a fitness increase promoted by continued viral replication in the same environment. Both mechanisms may act conjointly during viral infections *in vivo*. Some studies with HCV have documented drug-resistance phenotypes in infected patients, in the absence of specific drug-resistance mutations (Sullivan et al., 2013; Sato et al., 2015). In fact, prolonged chronic HCV infections represent an adequate scenario for fitness increase due to extended rounds of infections in the same host liver. As a consequence, chronic infections may be prone to display fitness-associated multidrug-resistance phenotypes, in the absence of drug-resistance mutations. The multiple mechanisms of drug resistance related to quasispecies dynamics justify even further the need of new antiviral strategies as presented in Chapter 9.

8.8 VIRAL LOAD, FITNESS, AND DISEASE PROGRESSION

High viral loads are predictors of disease progression. For HIV-1 and other lentiviruses, an efficient early control of virus replication by the host immune response is generally associated with limited disease severity. In HIV-1, the viral load that follows after the initial immune response to the virus is referred to as the "set point." In the absence of early therapy, low set points in HIV-1 are generally attributed to a strong cellular immune response, likely influenced by host and viral factors. [This and other aspects of HIV-1 replication and pathogenesis have been reviewed in excellent monographs by Levy (2007)]. A low set point predicts an asymptomatic outcome, and this is generally the case for viruses that establish persistent rather than acute infections. Again, viral population numbers, derived from the virus-host interaction, play a critical role in the result of an infection. Given a host environment, an initial high viral fitness during the early stages of viral replication can promote disease manifestations. An example was provided by the progression toward disease of a cohort of individuals that were infected during blood transfusion with an HIV-1 containing a large deletion in Nef. After more than 15 years, some of the infected individuals showed clinical signs, probably as a result of accumulation of mutations in the HIV-1 genome that compensated for the lack of Nef, an adaptor protein that mediates replication and pathogenesis (reviewed in Arien and Verhasselt, 2008). More generally, fitness-decreasing (but not lethal) genetic lesions in a viral genome may be compensated by additional genomic mutations that become increasingly dominant in the course of further viral replication. The kinetics of fitness gain will depend on the nature of the lesion and the functional implications of the altered protein or genomic regulatory region (Chapter 5).

Fitness, replicative capacity, and viral load are directly interconnected parameters, and they affect disease progression (Domingo et al., 2012) (Figure 8.9). Fitness gain will be more effective, the higher is the load of actively replicating virus in the infected organism. High replicative capacity and fitness sustain high viral loads. The reason for this basic feature of viral population dynamics is that given a basal mutation rate, a large number of replicating genomes entails a correspondingly higher probability that a required mutation for fitness gain can be produced. The events involved are a specific case of search for adaptive mutations in terms of exploration of sequence space as discussed in Section 3.7 of Chapter 3. While active viral replication, high load, and high fitness favor progression of the infection and disease manifestations, the fourth parameter included in the large arrow of Figure 8.9, mutant spectrum diversity, has an optimal range. Too low or too high intrapopulation diversity is detrimental

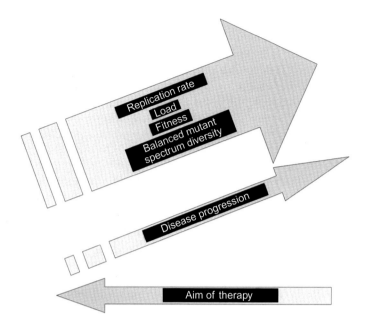

Replication rate
Load
Fitness
Balanced mutant
spectrum diversity

Disease progression

Aim of therapy

FIGURE 8.9

A schematic representation of interconnected parameters of viral replication that often relate to disease progression. An understanding of quasispecies dynamics has made it evident that the aim of antiviral therapy need not be only to diminish directly the viral load, but to affect other parameters that then can reduce the viral load. See text for justification and references, and Chapter 9 for new antiviral strategies that follow the concept expressed in this figure.

to virus adaptability. Insufficient diversity limits adaptability to complex environments (Pfeiffer and Kirkegaard, 2005; Vignuzzi et al., 2006), while excess diversity may force the virus to cross an extinction threshold, and this is the basis of lethal mutagenesis as an antiviral therapy (Chapter 9). An additional implication of the parameters shown in Figure 8.9 for antiviral interventions is that fitness decrease is recognized as an alternative to inhibition of viral replication to control viral infections (Clementi, 2008; Clementi and Lazzarin, 2010, reviewed in Domingo et al. (2012)). Thus, key features of quasispecies dynamics have a direct implication on the management of viral infections.

8.9 LIMITATIONS OF SIMPLIFIED REAGENTS AND SMALL MOLECULES AS ANTIVIRAL AGENTS

External interventions that are applied or have been envisaged to limit or suppress virus infection include not only vaccination and administration of antiviral agents as described in previous sections, but also passive immunotherapy, antisense RNAs, or oligonucleotides with various chemical modifications, interfering RNAs, ribozymes, or their combinations. Biotechnological developments have favored the design of chemically defined vaccines (consisting of expressed immunogenic proteins, synthetic peptides, or peptide arrays), without the need to handle or administer live virus. One of the most naive manifestations of trust in biotechnology in the middle of the twentieth century was the belief that a catalog of plasmids encoding the antigenic proteins of the circulating types of a pathogenic virus would suffice to prepare the required vaccine as needed. Concerning influenza vaccines, W.I. Beveridge wrote the following: "The first objective would be to capture the full range of influenza A subtypes. Their antigens would be studied by specialists at central laboratories and made available for the preparation of particular vaccines if and when required. It might be feasible to stockpile some vaccine against all the principal hemagglutinin antigens to be used in a fire brigade type of action as soon as an incipient pandemic is spotted" (Beveridge, 1977). Naivety is also perceived in current designs of universal vaccines based on conserved antigens. It is remarkable how our present perception of viral populations differs from the views expressed in the W.I. Beveridge book and in other writings at the boom of implementation of DNA recombinant techniques. With a conceptual similarity, in medical practice, monotherapy with an antiviral agent was traditionally preferred over drug mixtures. (The change of paradigm was largely a consequence of the AIDS epidemic, and it was publicly expressed by the pioneer hepatologist S. Sherlock in a summary address of an International Symposium on Viral Hepatitis held in Madrid in 1998.) The change of perspective is clear.

Some antiviral strategies such as antisense nucleic acids or virus-directed ribozymes were intensely investigated decades ago. It is unlikely that when used in isolation, they can be converted into useful antiviral therapies, because resistant mutants are likely to be selected. Yet, they could be part of combinations with other antiviral inhibitors to provide a larger antiviral barrier (Chapter 9). A similar fate is likely for interfering RNAs (Boden et al., 2003; Gitlin et al., 2005; Herrera-Carrillo and Berkhout, 2015; McDonagh et al., 2015). To a large extent, the failures of defined chemical entities (oligonucleotides, ribozymes, small molecule inhibitors, etc.) to control virus replication and spread are a consequence of their targeting a very defined viral genomic sequence combined with the adaptability of viral populations. Combination of multiple such elements have been envisaged and tested, but off target effects and the adaptive potential of viruses are likely to limit their efficacy. Unfortunately, biotechnological developments that have been so positive for many research areas and practical applications tend to simplify the types of agents to prevent disease or inhibit virus replication, ignoring the inherent complexity of the object to be controlled. Success is unlikely when "complexity" is combated with "simplicity." An increased understanding of viral population dynamics over the last decades has changed the picture dramatically, by providing an interpretation of "virus escape" as a general and largely unavoidable phenomenon. Such an awareness has pushed the development of new antiviral designs, which are fundamentally centered in two strategies: combination of multiple, independently acting elements or fitness decrease through excess mutations (Chapter 9).

8.10 "HIT EARLY, HIT HARD"

The full implications of quasispecies-mediated adaptation of viruses for antiviral therapy were expressed by D.D. Ho in an influential article entitled "Time to hit HIV, early and hard" (Ho, 1995). The article title captures what is needed to prevent adaptation of a virus in the infected host. Any opportunity to replicate is exploited by the virus to increase its fitness and to become less vulnerable to internal (intrahost) or external interventions such as antiviral therapy. Treatment interruptions during chronic infections, such as "drug holidays" that in the case of HIV-1-infected patients were justified to alleviate

---

BOX 8.4 SOME RECOMMENDATIONS FOR THE USE OF ANTIVIRAL AGENTS

- Avoid monotherapy. Ideally, use two or more antiviral agents which do not share a mechanism of action (Figure 8.6).
- Treat as soon as possible after virus diagnosis, to avoid virus adaptability associated with high virus population size and to minimize transmission of inhibitor-resistant mutants.
- Individual patients should be treated only during the time at which the drug proves effective. When viral load rebounds, treatment should be discontinued.
- Use NGS methodology to determine mutant spectrum composition for adequate choice of inhibitors. The aim is to design personalized treatments that consider probability of drug combination efficacy with minimal side effects.
- Consider temporary shelving of effective drugs when resistant mutants acquire epidemiological relevance.

---

side effects associated with administration of antiretroviral agents, provided an opportunity for the virus to gain fitness. In principle, given our current understanding of viral quasispecies dynamics, the proposal of D.D. Ho is applicable to other viral pathogens.

One argument that tones down the strength of the "hit early and hit hard" proposal is that some infected patients may not progress to disease, but maintain an asymptomatic lifelong persistent infection. This is the case with elite controllers in the case of HIV-1 infection, and individuals infected with HCV who will not progress toward liver disease. In the cases in which such nonprogression can be anticipated by viral and host parameters, it may be justified to exclude some patients from aggressive interventions (Suthar and Harries, 2015). As a general rule, however, the potential benefits of early treatment are obvious not only to avoid disease on an individual basis, but also to diminish the chances of virus transmission (reviewed in Suthar and Harries, 2015; see also Sections 7.1 and 7.7 in Chapter 7 regarding the relevance of viral population numbers in transmission). Restricting the number of treated patients for economic reasons will result in more expensive public health interventions when the infected individuals develop disease.

Box 8.4 includes recommendations for the use of antiviral agents, and recapitulates concepts explained in this and preceding sections.

8.11 INFORMATION AND GLOBAL ACTION

Despite emphasis on evolutionary aspects, prevention and treatment of viral disease has many other angles some of which were considered in Chapter 7 in connection with factors of disease emergence (Smolinski et al., 2003). Two of them should be mentioned here because they are as important as the adequate treatment designs described in this chapter: (i) public information about virus sources and means of contagion and (ii) need of global actions. Information to the public should aim at limiting the spread of disease, that is, undertaking personal and collective actions to reduce the R_0 value for a given virus (Chapter 7, Section 7.2). As an illustration of this key point, there was a quite extensive information campaign on HIV-1 and AIDS in developed countries during the early decades of HIV-1 spread, while the information about other potentially threatening viruses such as Ebola or the severe acute or Middle East respiratory syndrome (SARS and MERS, respectively) coronaviruses was more limited.

The need of a global response to limit the extension of disease episodes at the sites where they are initiated has been recognized for a long time, but it became obvious with the 2014-2015 West African Ebola epidemic (see Siedner et al., 2015). There is a need for international organizations and governments of developed countries to provide health-care work force to assist low- and middle-income countries to control viral episodes at an early stage. "Help early, help effectively" is the recognized need at a global scale, which is the parallel to "Hit early, hit hard" for the treatment of infected patients.

Global early action and adequate information can be as important as an adequate treatment design to control viral disease. It can restrict viral replication rounds and consequent adaptability. Information is thought to have been critical for the control of Ebola epidemic in Nigeria (Siedner et al., 2015). However, information must be also planned to reach the target population in a convincing manner, as learnt from the poliovirus vaccination and eradication campaign (Renne, 2010). The uncertainties regarding whether an initial, limited episode of viral disease will expand or die out does not help in decision making. However, the best choice in the case of emerging and reemerging infections is to act assuming the worst scenario.

8.12 OVERVIEW AND CONCLUDING REMARKS

Medical interventions represent a totally new set of selective constraints that viruses are facing only since decades ago, an infinitesimal time of their existence as biological entities. Yet, the evolutionary mechanisms available to viruses have successfully coped with many selective pressures, notably vaccines that do not evoke a broad immune response or treatment with antiviral agents. A common scenario with obvious commercial and sometimes even political influences is to test a new vaccine with an animal host, be it the authentic host or an animal model, and obtain full protection when the animal is challenged with a virus that matches the antigenic composition of the vaccine. Following the initial excitement, very often the vaccine displays only partial protection when tested in the natural environment. Somewhat parallel arguments can be made about clinical trials for antiviral agents, usually performed initially with selected groups of patients. This chapter has attempted to emphasize how the complexity of viral populations is a serious (often underestimated) difficulty to prevent and treat viral disease. Examination of the molecular mechanisms exploited by viruses to persist despite interventions suggests two major lines of action: first, a more judicious use of existing tools, that should consider the complexity of viral populations and their dynamics. Complexity cannot be combated with simplicity. Second, the need to design new antiviral strategies, a topic addressed in Chapter 9.

Several interconnected parameters determine the probability of success of an antiviral intervention. Most of them follow from the general concepts of Darwinian evolution explained in preceding chapters. It is important, however, to quantify as much as possible the evolutionary events that determine therapy success or failure. For this reason, the importance of viral population size, basic probability calculations of developing resistance, and the selective strength of mutations, have been analyzed with numerical examples. Hopefully, these simple quantifications will permit a higher awareness of when and why a treatment may succeed or fail.

We live in a very unequal society. The chapter closes with the recognition that there are many social economic issues that are as important as scientific planning to combat the pathogenic viruses around us (see Summary Box).

SUMMARY BOX

- Medical interventions represent a new class of selective constraints acting on viral populations.
- Viral evolution affects antiviral preventive and treatment strategies in two different ways: through the long-term diversification of viruses in nature, and through the molecular mechanisms of short-term response in treated individuals.
- Ineffective vaccines can contribute to selection of antigenic viral variants.
- Selection of viral mutants resistant to antiviral agents is a general phenomenon. Selection is favored by suboptimal treatments, and is delayed by the combined administration of multiple inhibitors. Resistance may also occur in the absence of specific resistance mutations, and it is associated with viral fitness or a fitness-related trait.
- The aim of therapy should be to increase the functional barrier to resistance and to give no opportunity to the virus to pursue replication that increases its replicative fitness.
- Replication rate, viral load, fitness, and mutant spectrum complexity are interconnected parameters that may tip the balance toward either control of the infection or disease progression. Each of these parameters can be targeted in an antiviral design.
- Firm political action and adequate public information is as important as antiviral designs to control virus infections at a global level.

REFERENCES

Appleyard, G., 1977. Amantadine-resistance as a genetic marker for influenza viruses. J. Gen. Virol. 36, 249–255.

Arien, K.K., Verhasselt, B., 2008. HIV Nef: role in pathogenesis and viral fitness. Curr. HIV Res. 6, 200–208.

Bailey, J., Blankson, J.N., Wind-Rotolo, M., Siliciano, R.F., 2004. Mechanisms of HIV-1 escape from immune responses and antiretroviral drugs. Curr. Opin. Immunol. 16, 470–476.

Beerenwinkel, N., Daumer, M., Sing, T., Rahnenfuhrer, J., Lengauer, T., et al., 2005. Estimating HIV evolutionary pathways and the genetic barrier to drug resistance. J. Infect. Dis. 191, 1953–1960.

Belshe, R.B., Smith, M.H., Hall, C.B., Betts, R., Hay, A.J., 1988. Genetic basis of resistance to rimantadine emerging during treatment of influenza virus infection. J. Virol. 62, 1508–1512.

Beveridge, W.I.B., 1977. Influenza: The Last Great Plague; and Unfinished Story of Discovery. Prodist, New York.

Bloom, B.R., Lambert, P.-H., 2003. The Vaccine Book. Academic Press, Elsevier, San Diego.

Boden, D., Pusch, O., Lee, F., Tucker, L., Ramratnam, B., 2003. Human immunodeficiency virus type 1 escape from RNA interference. J. Virol. 77, 11531–11535.

Burnet, M., 1966. Natural History of Infections Disease. Cambridge University Press, London.

Cale, E.M., Hraber, P., Giorgi, E.E., Fischer, W., Bhattacharya, T., et al., 2011. Epitope-specific CD8+ T lymphocytes cross-recognize mutant simian immunodeficiency virus (SIV) sequences but fail to contain very early evolution and eventual fixation of epitope escape mutations during SIV infection. J. Virol. 85, 3746–3757.

Cheng, B.Y., Ortiz-Riano, E., Nogales, A., de la Torre, J.C., Martinez-Sobrido, L., 2015. Development of live-attenuated arenavirus vaccines based on codon deoptimization. J. Virol. 89, 3523–3533.

Chumakov, K., Kew, O., 2010. The poliovirus eradication initiative. In: Ehrenfeld, E., Domingo, E., Roos, R.P. (Eds.), The Picornaviruses. ASM Press, Washington, DC, pp. 449–459.

Ciurea, A., Klenerman, P., Hunziker, L., Horvath, E., Senn, B.M., et al., 2000. Viral persistence *in vivo* through selection of neutralizing antibody- escape variants. Proc. Natl. Acad. Sci. U. S. A. 97, 2749–2754.

Ciurea, A., Hunziker, L., Martinic, M.M., Oxenius, A., Hengartner, H., et al., 2001. CD4+ T-cell-epitope escape mutant virus selected in vivo. Nat. Med. 7, 795–800.

Clementi, M., 2008. Perspectives and opportunities for novel antiviral treatments targeting virus fitness. Clin. Microbiol. Infect. 14, 629–631.

Clementi, M., Lazzarin, A., 2010. Human immunodeficiency virus type 1 fitness and tropism: concept, quantification, and clinical relevance. Clin. Microbiol. Infect. 16, 1532–1538.

Coleman, J.R., Papamichail, D., Skiena, S., Futcher, B., Wimmer, E., et al., 2008. Virus attenuation by genome-scale changes in codon pair bias. Science 320, 1784–1787.

Crowder, S., Kirkegaard, K., 2005. Trans-dominant inhibition of RNA viral replication can slow growth of drug-resistant viruses. Nat. Genet. 37, 701–709.

Cubero, M., Esteban, J.I., Otero, T., Sauleda, S., Bes, M., et al., 2008. Naturally occurring NS3-protease-inhibitor resistant mutant A156T in the liver of an untreated chronic hepatitis C patient. Virology 370, 237–245.

Domingo, E., 1989. RNA virus evolution and the control of viral disease. Prog. Drug Res. 33, 93–133.

Domingo, E., Holland, J.J., 1992. Complications of RNA heterogeneity for the engineering of virus vaccines and antiviral agents. Genet. Eng. (N.Y.) 14, 13–31.

Domingo, E., Biebricher, C., Eigen, M., Holland, J.J., 2001a. Quasispecies and RNA Virus Evolution: Principles and Consequences. Landes Bioscience, Austin.

Domingo, E., Mas, A., Yuste, E., Pariente, N., Sierra, S., et al., 2001b. Virus population dynamics, fitness variations and the control of viral disease: an update. Prog. Drug Res. 57, 77–115.

Domingo, E., Sheldon, J., Perales, C., 2012. Viral quasispecies evolution. Microbiol. Mol. Biol. Rev. 76, 159–216.

Eggers, H.J., Tamm, I., 1961. Spectrum and characteristics of the virus inhibitory action of 2-(alpha-hydroxybenzyl)-benzimidazole. J. Exp. Med. 113, 657–682.

Eggers, H.J., Tamm, I., 1965. Coxsackie A9 virus: mutation from drug dependence to drug independence. Science 148, 97–98.

Evans, A.S., Kaslow, R.A., 1997. Viral Infections of Humans. Epidemiology and Control. Plenum Medical Book Company, New York and London.

Fischer, W., Ganusov, V.V., Giorgi, E.E., Hraber, P.T., Keele, B.F., et al., 2010. Transmission of single HIV-1 genomes and dynamics of early immune escape revealed by ultra-deep sequencing. PLoS One 5, e12303.

Flynn, W.F., Chang, M.W., Tan, Z., Oliveira, G., Yuan, J., et al., 2015. Deep sequencing of protease inhibitor resistant HIV patient isolates reveals patterns of correlated mutations in gag and protease. PLoS Comput. Biol. 11, e1004249.

Fun, A., Wensing, A.M., Verheyen, J., Nijhuis, M., 2012. Human immunodeficiency virus gag and protease: partners in resistance. Retrovirology 9, 63.

Gandon, S., Mackinnon, M., Nee, S., Read, A., 2003. Imperfect vaccination: some epidemiological and evolutionary consequences. Proc. Biol. Sci. 270, 1129–1136.

Gitlin, L., Stone, J.K., Andino, R., 2005. Poliovirus escape from RNA interference: short interfering RNA-target recognition and implications for therapeutic approaches. J. Virol. 79, 1027–1035.

Golan, D.E., Tashjian, A.H. (Eds.), 2011. Principles of Pharmacology: The Pathophysiologic Basis of Drug Therapy. Lippincott Williams & Wilkins, Baltimore, Philadelphia.

Hagan, T., Nakaya, H.I., Subramaniam, S., Pulendran, B., 2015. Systems vaccinology: enabling rational vaccine design with systems biological approaches. Vaccine, in press.

Havlir, D.V., Eastman, S., Gamst, A., Richman, D.D., 1996. Nevirapine-resistant human immunodeficiency virus: kinetics of replication and estimated prevalence in untreated patients. J. Virol. 70, 7894–7899.

Hedskog, C., Dvory-Sobol, H., Gontcharova, V., Martin, R., Ouyang, W., et al., 2015. Evolution of the HCV viral population from a patient with S282T detected at relapse after sofosbuvir monotherapy. J. Viral Hepat., in press.

Heinz, B.A., Rueckert, R.R., Shepard, D.A., Dutko, F.J., McKinlay, M.A., et al., 1989. Genetic and molecular analyses of spontaneous mutants of human rhinovirus 14 that are resistant to an antiviral compound. J. Virol. 63, 2476–2485.

Herrera-Carrillo, E., Berkhout, B., 2015. The impact of HIV-1 genetic diversity on the efficacy of a combinatorial RNAi-based gene therapy. Gene Ther. 22, 485–495.

Herrmann Jr., E.C., Herrmann, J.A., 1977. A working hypothesis–virus resistance development as an indicator of specific antiviral activity. Ann. N. Y. Acad. Sci. 284, 632–637.

Ho, D.D., 1995. Time to hit HIV, early and hard. N. Engl. J. Med. 333, 450–451.

Hsu, H.Y., Chang, M.H., Liaw, S.H., Ni, Y.H., Chen, H.L., 1999. Changes of hepatitis B surface antigen variants in carrier children before and after universal vaccination in Taiwan. Hepatology 30, 1312–1317.

Ji, W., Niu, D.D., Si, H.L., Ding, N.Z., He, C.Q., 2014. Vaccination influences the evolution of classical swine fever virus. Infect. Genet. Evol. 25, 69–77.

Johnson, J.A., Li, J.F., Wei, X., Lipscomb, J., Irlbeck, D., et al., 2008. Minority HIV-1 drug resistance mutations are present in antiretroviral treatment-naive populations and associate with reduced treatment efficacy. PLoS Med. 5, e158.

Kekarainen, T., Gonzalez, A., Llorens, A., Segales, J., 2014. Genetic variability of porcine circovirus 2 in vaccinating and non-vaccinating commercial farms. J. Gen. Virol. 95, 1734–1742.

Kew, O., Morris-Glasgow, V., Landaverde, M., Burns, C., Shaw, J., et al., 2002. Outbreak of poliomyelitis in Hispaniola associated with circulating type 1 vaccine-derived poliovirus. Science 296, 356–359.

Kilbourne, E.D., 1987. Influenza. Plenum Medical Book Company, New York.

Kirkwood, C.D., 2010. Genetic and antigenic diversity of human rotaviruses: potential impact on vaccination programs. J. Infect. Dis. 202 (Suppl), S43–S48.

Kortenhoeven, C., Joubert, F., Bastos, A., Abolnik, C., 2015. Virus genome dynamics under different propagation pressures: reconstruction of whole genome haplotypes of west nile viruses from NGS data. BMC Genomics 16, 118.

Lech, W.J., Wang, G., Yang, Y.L., Chee, Y., Dorman, K., et al., 1996. In vivo sequence diversity of the protease of human immunodeficiency virus type 1: presence of protease inhibitor-resistant variants in untreated subjects. J. Virol. 70, 2038–2043.

Levy, J.A., 2007. HIV and the Pathogenesis of AIDS. ASM Press, Whashington, DC.

Lindenbach, B.D., Evans, M.J., Syder, A.J., Wolk, B., Tellinghuisen, T.L., et al., 2005. Complete replication of hepatitis C virus in cell culture. Science 309, 623–626.

Loh, L., Petravic, J., Batten, C.J., Davenport, M.P., Kent, S.J., 2008. Vaccination and timing influence SIV immune escape viral dynamics in vivo. PLoS Pathog. 4, e12.

Lubeck, M.D., Schulman, J.L., Palese, P., 1978. Susceptibility of influenza A viruses to amantadine is influenced by the gene coding for M protein. J. Virol. 28, 710–716.

Margeridon-Thermet, S., Shafer, R.W., 2010. Comparison of the mechanisms of drug resistance among HIV, hepatitis B, and hepatitis C. Viruses 2, 2696–2739.

Martinez-Picado, J., Martinez, M.A., 2008. HIV-1 reverse transcriptase inhibitor resistance mutations and fitness: a view from the clinic and ex vivo. Virus Res. 134, 104–123.

McDonagh, P., Sheehy, P.A., Norris, J.M., 2015. Combination siRNA therapy against feline coronavirus can delay the emergence of antiviral resistance in vitro. Vet. Microbiol. 176, 10–18.

Melnick, J.L., Crowther, D., Barrera-Oro, J., 1961. Rapid development of drug-resistant mutants of poliovirus. Science 134, 557.

Menendez-Arias, L., 2010. Molecular basis of human immunodeficiency virus drug resistance: an update. Antiviral Res. 85, 210–231.

Menendez-Arias, L., 2013. Molecular basis of human immunodeficiency virus type 1 drug resistance: overview and recent developments. Antiviral Res. 98, 93–120.

Menendez-Arias, L., Richman, D.D., 2014. Editorial overview: antivirals and resistance: advances and challenges ahead. Curr. Opin. Virol. 8, iv–vii.

Mitchison, D.A., 1950. Development of streptomycin resistant strains of tubercle bacilli in pulmonary tuberculosis; results of simultaneous sensitivity tests in liquid and on solid media. Thorax 5, 144–161.

Mohri, H., Singh, M.K., Ching, W.T., Ho, D.D., 1993. Quantitation of zidovudine-resistant human immunodeficiency virus type 1 in the blood of treated and untreated patients. Proc. Natl. Acad. Sci. U. S. A. 90, 25–29.

Muylkens, B., Meurens, F., Schynts, F., Farnir, F., Pourchet, A., et al., 2006. Intraspecific bovine herpesvirus 1 recombinants carrying glycoprotein E deletion as a vaccine marker are virulent in cattle. J. Gen. Virol. 87, 2149–2154.

Nájera, I., Richman, D.D., Olivares, I., Rojas, J.M., Peinado, M.A., et al., 1994. Natural occurrence of drug resistance mutations in the reverse transcriptase of human immunodeficiency virus type 1 isolates. AIDS Res. Hum. Retroviruses 10, 1479–1488.

Nájera, I., Holguín, A., Quiñones-Mateu, M.E., Muñoz-Fernández, M.A., Nájera, R., et al., 1995. Pol gene quasispecies of human immunodeficiency virus: mutations associated with drug resistance in virus from patients undergoing no drug therapy. J. Virol. 69, 23–31.

Pawlotsky, J.M., 2006. Hepatitis C virus population dynamics during infection. Curr. Top. Microbiol. Immunol. 299, 261–284.

Perales, C., Agudo, R., Manrubia, S.C., Domingo, E., 2011. Influence of mutagenesis and viral load on the sustained low-level replication of an RNA virus. J. Mol. Biol. 407, 60–78.

Perales, C., Beach, N.M., Gallego, I., Soria, M.E., Quer, J., et al., 2013. Response of hepatitis C virus to long-term passage in the presence of alpha interferon: multiple mutations and a common phenotype. J. Virol. 87, 7593–7607.

Perales, C., Beach, N.M., Sheldon, J., Domingo, E., 2014. Molecular basis of interferon resistance in hepatitis C virus. Curr. Opin. Virol. 8, 38–44.

Perez-Sautu, U., Costafreda, M.I., Cayla, J., Tortajada, C., Lite, J., et al., 2011. Hepatitis a virus vaccine escape variants and potential new serotype emergence. Emerg. Infect. Dis. 17, 734–737.

Pfeiffer, J.K., Kirkegaard, K., 2005. Increased fidelity reduces poliovirus fitness under selective pressure in mice. PLoS Pathog. 1, 102–110.

Pincus, S.E., Wimmer, E., 1986. Production of guanidine-resistant and -dependent poliovirus mutants from cloned cDNA: mutations in polypeptide 2C are directly responsible for altered guanidine sensitivity. J. Virol. 60, 793–796.

Pircher, H., Moskophidis, D., Rohrer, U., Burki, K., Hengartner, H., et al., 1990. Viral escape by selection of cytotoxic T cell-resistant virus variants *in vivo*. Nature 346, 629–633.

Read, A.F., Baigent, S.J., Powers, C., Kgosana, L.B., Blackwell, L., et al., 2015. Imperfect vaccination can enhance the transmission of highly virulent pathogens. PLoS Biol. 13 (7): e1002198.

Renne, E.P., 2010. The Politics of Polio in Northern Nigeria. Indiana University Press, Bloomington.

Rhee, S.Y., Taylor, J., Fessel, W.J., Kaufman, D., Towner, W., et al., 2010. HIV-1 protease mutations and protease inhibitor cross-resistance. Antimicrob. Agents Chemother. 54, 4253–4261.

Ribeiro, R.M., Bonhoeffer, S., 2000. Production of resistant HIV mutants during antiretroviral therapy. Proc. Natl. Acad. Sci. U. S. A. 97, 7681–7686.

Richman, D.D., 1994. Resistance, drug failure, and disease progression. AIDS Res. Hum. Retroviruses 10, 901–905.

Richman, D.D. (Ed.), 1996. Antiviral Drug Resistance. John Wiley and Sons Inc., New York.

Richman, D.D., Wrin, T., Little, S.J., Petropoulos, C.J., 2003. Rapid evolution of the neutralizing antibody response to HIV type 1 infection. Proc. Natl. Acad. Sci. U. S. A. 100, 4144–4149.

Rowlands, D.J., Minor, P.D., 2010. Vaccine strategies. In: Ehrenfeld, E., Domingo, E., Roos, R.P. (Eds.), The Picornaviruses. ASM Press, Washington, DC, pp. 431–448.

Sarrazin, C., Zeuzem, S., 2010. Resistance to direct antiviral agents in patients with hepatitis C virus infection. Gastroenterology 138, 447–462.

Sato, M., Maekawa, S., Komatsu, N., Tatsumi, A., Miura, M., et al., 2015. Deep Sequencing and phylogenetic analysis of variants resistant to interferon-based protease inhibitor therapy in chronic hepatitis induced by genotype-1b hepatitis C virus. J. Virol. 89: 6105–6116.

Schat, K.A., Baranowski, E., 2007. Animal vaccination and the evolution of viral pathogens. Rev. Sci. Tech. 26, 327–338.

Sheldon, J., Beach, N.M., Moreno, E., Gallego, I., Pineiro, D., et al., 2014. Increased replicative fitness can lead to decreased drug sensitivity of hepatitis C virus. J. Virol. 88, 12098–12111.

Siedner, M.J., Gostin, L.O., Cranmer, H.H., Kraemer, J.D., 2015. Strengthening the detection of and early response to public health emergencies: lessons from the West African Ebola epidemic. PLoS Med. 12, e1001804.

Smolinski, M.S., Hamburg, M.A., Lederberg, J. (Eds.), 2003. Microbial Threats to Health. Emergence, Detection and Response. The National Academies Press, Washington DC.

Sullivan, J.C., De Meyer, S., Bartels, D.J., Dierynck, I., Zhang, E.Z., et al., 2013. Evolution of treatment-emergent resistant variants in telaprevir phase 3 clinical trials. Clin. Infect. Dis. 57, 221–229.

Suthar, A.B., Harries, A.D., 2015. A public health approach to hepatitis C control in low- and middle-income countries. PLoS Med. 12, e1001795.

Swayne, D.E., Kapczynski, D., 2008. Strategies and challenges for eliciting immunity against avian influenza virus in birds. Immunol. Rev. 225, 314–331.

Taboga, O., Tami, C., Carrillo, E., Núñez, J.I., Rodríguez, A., et al., 1997. A large-scale evaluation of peptide vaccines against foot-and-mouth disease: lack of solid protection in cattle and isolation of escape mutants. J. Virol. 71, 2606–2614.

Tami, C., Taboga, O., Berinstein, A., Nuñez, J.I., Palma, E.L., et al., 2003. Evidence of the coevolution of antigenicity and host cell tropism of foot-and-mouth disease virus in vivo. J. Virol. 77, 1219–1226.

Thiry, E., Muylkens, B., Meurens, F., Gogev, S., Thiry, J., et al., 2006. Recombination in the alphaherpesvirus bovine herpesvirus 1. Vet. Microbiol. 113, 171–177.

Toni, T.A., Asahchop, E.L., Moisi, D., Ntemgwa, M., Oliveira, M., et al., 2009. Detection of human immunodeficiency virus (HIV) type 1 M184V and K103N minority variants in patients with primary HIV infection. Antimicrob. Agents Chemother. 53, 1670–1672.

Tsibris, A.M., Korber, B., Arnaout, R., Russ, C., Lo, C.C., et al., 2009. Quantitative deep sequencing reveals dynamic HIV-1 escape and large population shifts during CCR5 antagonist therapy in vivo. PLoS One 4, e5683.

Valarcher, J.F., Schelcher, F., Bourhy, H., 2000. Evolution of bovine respiratory syncytial virus. J. Virol. 74, 10714–10728.

Vignuzzi, M., Stone, J.K., Arnold, J.J., Cameron, C.E., Andino, R., 2006. Quasispecies diversity determines pathogenesis through cooperative interactions in a viral population. Nature 439, 344–348.

Vignuzzi, M., Wendt, E., Andino, R., 2008. Engineering attenuated virus vaccines by controlling replication fidelity. Nat. Med. 14, 154–161.

von Kleist, M., Metzner, P., Marquet, R., Schutte, C., 2012. HIV-1 polymerase inhibition by nucleoside analogs: cellular- and kinetic parameters of efficacy, susceptibility and resistance selection. PLoS Comput. Biol. 8, e1002359.

Wakita, T., Pietschmann, T., Kato, T., Date, T., Miyamoto, M., et al., 2005. Production of infectious hepatitis C virus in tissue culture from a cloned viral genome. Nat. Med. 11, 791–796.

Weidt, G., Deppert, W., Utermohlen, O., Heukeshoven, J., Lehmann-Grube, F., 1995. Emergence of virus escape mutants after immunization with epitope vaccine. J. Virol. 69, 7147–7151.

Zhong, J., Gastaminza, P., Cheng, G., Kapadia, S., Kato, T., et al., 2005. Robust hepatitis C virus infection in vitro. Proc. Natl. Acad. Sci. U. S. A. 102, 9294–9299.

TRENDS IN ANTIVIRAL STRATEGIES

Virus as Populations. http://dx.doi.org/10.1016/B978-0-12-800837-9.00009-5

ABBREVIATIONS

AIDS	acquired immunodeficiency syndrome
APOBEC3G	apolipoprotein B mRNA editing complex 3G
AZA-C	azacytidine
FMDV	foot-and-mouth disease virus
FU	5-fluorouracil
FUTP	5-fluorouridine-triphosphate
GTP	guanosine-5′-triphosphate
HAART	highly active antiretroviral therapy
HCV	hepatitis C virus
HIV-1	human immunodeficiency virus type 1
IFN	interferon
IMPDH	inosine monophosphate dehydrogenase
LCMV	lymphocytic choriomeningitis virus
NGS	next generation sequencing
NLS	nuclear localization signal
PV	poliovirus
Rib	ribavirin
RMP	ribavirin-monophosphate
VSV	vesicular stomatitis virus

9.1 THE CHALLENGE

There is no general procedure that can effectively prevent or control viral infections, at least based on the strategies developed over the last century and a half: vaccination, immunotherapy, chemotherapy, or their combinations. One of the main reasons is that each pathogenic virus to be controlled has some unique features in its interaction with the host cell. Not only for viruses but also for cellular pathogens (bacteria, fungi, and protozoa) it is amply recognized that current practices for prevention and treatment have clear limitations. The difficulties arise also from the evolutionary potential of viral and cellular pathogens that can jeopardize control strategies. For many decades there has been limited awareness of the adaptive potential of pathogens, and preventive and therapeutic designs were implemented ignoring evolution. Different experts have expressed the need to search for new paradigms to approach infectious disease, with the incorporation of Darwinian principles together with concepts from evolutionary ecology, and considering also re-implementation of old methods such as passive antibody therapy (Casadevall, 1996; Stearns, 1999; Williams, 2009; Casadevall and Pirofski, 2015). Only very recently the adaptive potential of pathogens, in particular viral quasispecies dynamics, and its extensions to other pathogenic entities (Chapter 10) have been considered an aspect of the disease control problem.

The challenge to be confronted is symbolically portrayed in Figure 9.1, using one of the ways to depict viral quasispecies throughout this book (lines to depict genomes and symbols on the lines to represent mutations). About 100 million sheets (!) similar to the one shown in the figure are necessary to represent the hepatitis C virus (HCV) genomes found in an acutely infected human liver at a given time point. The precise number, types, and distribution of mutations among genomes varies not only among patients but also as a function of time in each patient. A few minutes later a slightly different image will be produced, with relative proportions of different genomes, and the location of several mutations

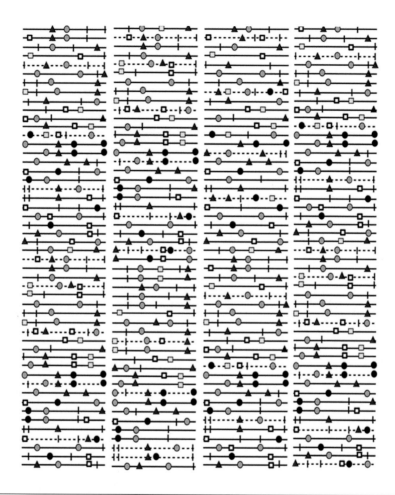

FIGURE 9.1

A scheme of quasispecies intended to stress the challenge of controlling replication of an immense and diverse entity. Each horizontal line represents a viral genome and symbols on the lines symbolize mutations, in a typical depiction of quasispecies used throughout the book. The discontinuous lines represent genomes with five or six mutations considered in this example a sufficient number to severely decrease fitness. Their fate is to attain very low frequency levels or be eliminated from the evolving population. About 100 million of displays similar to the one shown in this page are needed to depict the hepatitis C viral genomes present in an individual acutely infected with the virus.

modified. The dynamic mutant cloud here drawn on the basis of estimated HCV population numbers (and probably with underrepresentation of the number of mutations per genome!) is equally applicable to the majority of RNA viruses and error-prone DNA viruses. It should be quite obvious to the reader of previous chapters that there is a nonzero probability that mutations present in the mutant spectrum may contribute to drug resistance, immune evasion, or fitness enhancement. Even if they are hidden as a tiny minority, they are ready to be selected. The challenge is clear.

9.1.1 VIRUS AS MOVING TARGETS

The potential deleteriousness of mutational load (excessive number of mutations in the same genome) is symbolized in Figure 9.1 by the genomes drawn as discontinuous lines that contain five mutations, here assumed to be a sufficiently large number to decrease fitness and, therefore, the frequency of their potential descendants in subsequent replication rounds. Quasispecies dynamics imposes, however, that those genomes that move from a high frequency to a low frequency level be replaced by newly arising mutants that display higher fitness. The mutant cloud is highly dynamic with newly arising genomes incessantly exposed to the scrutiny of selection and the lottery of random drift. The image here portrayed through mutations is even more complex if we add recombination and genome segment reassortment in the case of viruses with segmented or multipartite genomes (Chapters 2 and 3). From the point of view of antiviral interventions, viruses are true "moving targets," in the sense that the repertoire of variants that we should inhibit at one time point is not exactly the same to be inhibited at a subsequent time point. The difference may be irrelevant regarding the efficacy of an antiviral treatment, or it may not. The impact of the dynamic change is unpredictable. Applied to clinical practice, it means that a specific inhibitor combination may be effective for many infected individuals for a long time, or only for a few individuals for a short time, with a range of possible intermediate outcomes. The new antiviral strategies discussed in this chapter take into consideration the moving target feature of the viruses to be controlled, and they have been proposed by an increasing number of experts aware of the quasispecies challenge.

9.2 PRACTICED AND PROPOSED STRATEGIES TO CONFRONT THE MOVING TARGET CHALLENGE WITH ANTIVIRAL INHIBITORS

The basic statistical considerations that justify the need to use multiepitopic vaccines to protect against variable viruses characterized by quasispecies behavior (keep in mind Figure 9.1) were discussed in Section 8.3 of Chapter 8. In this chapter, we focus on antiviral therapies; immunotherapy is only mentioned as a potential ingredient of combination therapies with antiviral inhibitors.

The systematic selection of drug (or multidrug)-resistant mutants in viral populations has encouraged the design of antiviral strategies intended to avoid viral breakthrough and treatment failure. The main options in clinical practice or under investigation using nonmutagenic and mutagenic antiviral inhibitors are summarized in Box 9.1. None of the listed strategies is totally free of the problem of selection of drug-resistant mutants, but the proposals are intended to avoid or delay their selection. Next, we discuss the major features of each of the six suggestions listed in Box 9.1.

9.2.1 COMBINATION TREATMENTS

Studies on the advantage of the combined administration of two drugs were pioneered by H.J. Eggers and I. Tamm (Eggers and Tamm, 1963; Eggers, 1976). Combination therapy applied to human immunodeficiency virus type 1 (HIV-1) has been the great success for the control of acquired immunodeficiency syndrome (AIDS) that has drastically reduced AIDS-related mortality. This type of treatment for HIV-1 is termed highly active antiretroviral therapy (HAART) and it is ideally implemented with three different antiretroviral agents. HAART efficacy has steadily improved due to availability of new inhibitors directed at different HIV-1 targets. Success is only partial due to side effects (that may be

---

BOX 9.1 SOME ANTIVIRAL STRATEGIES TO CONTROL VIRAL QUASISPECIES

Based on Nonmutagenic Inhibitors
- Combination treatments. Use of two or more inhibitors directed to independent viral targets.
- Splitting the treatment into an induction and a different maintenance regimen.
- Targeting cellular functions.
- Use of drugs that stimulate the host innate immune system.
- Combined use of immunotherapy and chemotherapy.

Based on Mutagenic Agents
- Lethal mutagenesis in their two modes of sequential and combined administration of inhibitors and mutagens.

---

derived from off-target activities of the drugs), incomplete patient adherence to the treatment, selection of multidrug-resistant viral variants and, above all, to the retroviral nature of HIV-1 that renders virus extinction from the organism a difficult endeavor. In retroviruses and hepadnaviruses, the viral DNA is not only hidden from antiviral drugs but it can serve as an archive of genomes that can reactivate replication upon treatment discontinuation. If combination therapy could be coupled with an efficient elimination of the proviral reservoirs in host DNA (De Crignis and Mahmoudi, 2014), HIV-1 could probably be eliminated from the infected organism, as achieved in the case of HCV infections in some patients.

The general advantage of combination therapy over monotherapy is a consequence of quasispecies dynamics, and it has been amply evidenced by clinical practice, and supported by straightforward statistical considerations and theoretical models of virus dynamics (Domingo, 1989; Domingo and Holland, 1992; Ho, 1995; Bonhoeffer et al., 1997; Pol et al., 1999; Ribeiro and Bonhoeffer, 2000; Le Moing et al., 2002; Van Vaerenbergh et al., 2002; Domingo et al., 2008; Müller and Bonhoeffer, 2008; Nijhuis et al., 2009) (see Section 8.4 in Chapter 8). Furthermore, use of combination therapies conforms to the "hit early, hit hard" dictum of D.D. Ho, justified in Section 8.10 of Chapter 8. There is the misconception (among some experts, but mainly among politicians!) that only patients in advanced phases of a viral infection (i.e., only precirrhotic or cirrhotic patients infected with HCV) should be treated aggressively with drug combinations. This is not what evolutionary virology teaches us: patients should be treated as strongly as possible and as early as possible, provided side effects can be controlled. Viruses should be given no chance to walk in sequence space in search of adaptive pathways (Chapter 3). Adequate combination treatments can fulfill such purpose.

9.2.2 SPLIT TREATMENTS

A second proposed strategy is to divide an antiviral treatment in two steps: an induction and a maintenance step (von Kleist et al., 2011). It is based on a theoretical model developed for the treatment of HIV-1 infections. The key argument is that when a treatment has to be changed after confirmation of treatment failure (virus rebound), mutants resistant to the ineffective drug have had the opportunity to replicate in the patient, increasing the viral load and supplying a proviral archive with inhibitor-resistant latent viruses, thus excluding the drug as a component of future combinations. A treatment with an induction regimen should be followed by a shift to a maintenance regimen at a point in time in which the second drug finds a low viral load and limited numbers of mutants resistant to the initial treatment. Critical issues for the implementation of this proposal are the timing of treatment switch, the decrease of viral load as a result of the induction regimen, and the genetic and phenotypic barriers to the maintenance regimen. Clinical trials are needed to explore this interesting proposal.

9.2.3 TARGETING CELLULAR FUNCTIONS

Many cellular functions are needed to complete any step of a virus replication cycle. A good deal of research in virology has as its main objective to identify and characterize cellular functions that participate in virus entry into the cell, intracellular multiplication, or release from cells. Since cellular proteins cannot vary in response to the presence of an inhibitor (at least within the time frame of a viral infection), an obvious thought is to administer inhibitors of those cellular functions (often proteins) that are needed to sustain viral replication; such inhibitors should suppress viral replication without selection of inhibitor-resistant mutants (Geller et al., 2007; Hopkins et al., 2010; Garbelli et al., 2011; Kumar et al., 2011; Vidalain et al., 2015). Despite potential benefits, two major problems may be encountered with antiviral agents directed to cellular proteins: (i) toxic effects derived from suppression or alteration of activities in which the target protein is involved and (ii) selection of viral mutants that are insensitive to the presence of the inhibitor, despite the inhibitor not being directed to the virus. Insensitivity may come about by at least three different mechanisms. If the cellular protein which is the target of the inhibitor forms a complex with a viral protein in the course of viral replication, amino acid substitutions in the viral protein may permit progression of the infection in the presence of the inhibitor. This is the case of HCV resistance to the nonimmunosuppressive cyclophilin inhibitor SCY-635. NS5A interacts with cyclophilins and with NS5B; amino acid substitutions in NS5A (T17A, E295K, and V44A) and in NS5B (T77K and I432V) decrease the sensitivity of the virus to SCY-635, although the precise molecular mechanism of resistance has been debated (Chatterji et al., 2010; Sarrazin and Zeuzem, 2010; Delang et al., 2011; Kwong et al., 2011; Vermehren and Sarrazin, 2011).

When inhibitors are targeted to the cellular receptor for a virus, mutants may be selected that can enter cells through an alternative receptor. This has been documented with HIV-1 that can use coreceptor CCR5, CXCR4, or both (dual-tropic viruses). A modified RANTES [*r*egulated on *a*ctivation, *n*ormal *T* cell *e*xpressed, and *s*ecreted; also termed chemokine (C–C motif) ligand 5 or CCL5] selected coreceptor switch variants in a SCID (*s*evere *c*ombined *i*mmuno*d*eficiency) mouse model (Mosier et al., 1999). The mutants with amino acid substitutions in loop V3 of Env protein were selected to use CXCR4 rather than CCR5 as coreceptor (see also Section 4.4 in Chapter 4).

An alternative mechanism to overcome the inhibition of a cellular protein is that viral mutants are selected that can utilize the cellular protein in complex with the inhibitor. This is the case of HIV-1 mutants resistant to small coreceptor CCR5-binding inhibitors; mutants with amino acid substitution in the V3 loop region or elsewhere in Env can use either free or inhibitor-bound CCR5 to enter cells, with an efficiency that depends on the CCR5 expression level and the host cell type (Pugach et al., 2007, 2009). These examples illustrate the multiple pathways that viruses can exploit to overcome inhibitors directed to cellular proteins. They constitute additional evidence that "abundant sources of genetic variation exist for viruses to learn new tricks, …" emphasized by J. Lederberg in connection with viral disease emergence and reemergence (quoted also in Section 7.7.1 of Chapter 7). Since many viruses can use alternative receptors for entry into cells (Section 4.4 in Chapter 4), inhibitors directed to viral receptors may promote selection of virus subpopulations that can use a different receptor.

9.2.4 USE OF DRUGS THAT STIMULATE THE HOST INNATE IMMUNE SYSTEM

Some inhibitors of enzymes of the *de novo* pyrimidine biosynthesis pathway (DD264, bequinar, and A3) stimulate the innate immune response, and behave as broad-spectrum antiviral inhibitors (Lucas-Hourani et al., 2013; Munier-Lehmann et al., 2015; Vidalain et al., 2015). The observed effect is one among other connections that have been established between nucleotide and DNA metabolism and immune stimulation (Motani et al., 2015). For some of the pyrimidine biosynthesis inhibitors, additional mechanisms of antiviral activity might be involved, as in the inhibition of lymphocytic choriomeningitis virus (LCMV) replication and transcription by A3 (Ortiz-Riano et al., 2014). Stimulation of the innate immune response may restrict the selection of escape mutants because the virus must mutate at several sites to overcome the different branches of the response, as is the case with interferon (IFN) resistance (Perales et al., 2014). Multifactorial antiviral responses increase the genetic and phenotypic barriers to resistance (Section 8.4.2 in Chapter 8).

9.2.5 COMBINED USE OF IMMUNOTHERAPY AND CHEMOTHERAPY

An extension of the advantage of combination therapy to decrease the selection of antiviral-resistant mutants consists in the combined use of immunotherapy (administration of neutralizing antibodies or other means of immune stimulation such as vaccination) together with antiviral inhibitors. The concept was pioneered by R.G. Webster and colleagues, and was proposed as a strategy for the control of influenza viruses (Webster et al., 1985). The authors showed that the simultaneous administration of inactivated H5N2 vaccine and the inhibitor amantadine conferred protection against H5N2 influenza virus A/Chick/Pennsylvania/83 in chickens (Webster et al., 1986). Related notions have been investigated with other viruses and different components of the immune response (Seiler et al., 2000; Li et al., 2005). However, additional model *in vivo* experiments with animals and clinical trials with patients are necessary to investigate the effectiveness of combined immunotherapeutic approaches.

As a general outlook on the strategies summarized in previous sections, the potential of combining two (or even more) of the proposals listed in Box 9.1 is encouraging, provided off-target effects of drugs or immune interventions can be controlled and side effects minimized. A trend toward "complex" treatment protocols is the expected response to the adaptive capacity of viral quasispecies. The introduction of mutagenic agents in antiviral designs is an important departure that exploits one of the corollaries of quasispecies behavior: the error threshold relationship (introduced in Section 3.6.3 of Chapter 3). It is the basis of lethal mutagenesis discussed next.

9.3 LETHAL MUTAGENESIS AND THE ERROR THRESHOLD

Lethal mutagenesis is defined as the process of viral extinction due to an excess of mutations in the viral genome. J.J. Holland and colleagues pioneered studies on the adverse effects of several chemical mutagens on the yield of infectious poliovirus (PV) and vesicular stomatitis virus (VSV) in cell culture (Holland et al., 1990). The term lethal mutagenesis was first proposed by L. Loeb, J.I. Mullins, and colleagues in a study of the loss of replicative potential of HIV-1 upon multiplication in human CEM cells in the presence of the deoxynucleoside analog 5-hydroxydeoxycytidine (5-OH-dC) (Loeb et al., 1999). The experimental design of lethal mutagenesis was inspired in the error threshold relationship derived from the basic equation of quasispecies dynamics (Eigen and Schuster, 1979; Swetina and Schuster,

1982; Nowak and Schuster, 1989; Schuster, 2016). Initially elaborated on simple (single peak) fitness landscapes, quasispecies theory has as one of its main corollaries that the stability of genetic information during a replicative process is dependent on two parameters: the error rate during replication and the amount of genetic information to be maintained. The basic equations that describe the relationship between the maximum tolerable mutation rate (μ_{max}) to ensure the transfer of genetic information to next generations of genomes with length ν, are included in Figure 9.2. They are the most relevant to an application of the error threshold concept to virology (Eigen, 2002; Schuster, 2016). In the equations, an important variable is the superiority of the master sequence over its surrounding mutant spectrum, denoted by σ_m. Consistent with σ_m being in the numerator of the equations, virologists will intuitively understand that a high superiority of the dominant sequence means a strong settlement of the virus with a well-adapted master sequence and its surrounding cloud in that environment. In consequence, a higher mutational input is necessary to destabilize the distribution.

There are additional factors that can modify the stability of a mutant distribution and the position of the error threshold; the mathematical justification and practical implications for virology have been detailed (Schuster, 2016). Of particular relevance to virus population stability is the influence of the fitness landscape in maintenance of genetic information. The first noteworthy result of the theoretical studies is that an error threshold is present in realistic rugged fitness landscapes, as those proposed to best describe viruses on the basis of experimental evidence (Section 5.3 in Chapter 5). The mathematical derivation of this important conclusion has been obtained by Schuster (2016). Second, the position of the error threshold moves toward lower mutation rates when the ruggedness of the fitness landscape (represented by a parameter that consists of a band of fitness values) is increased. How replication of viruses in variable fitness landscapes (e.g., under variable environmental conditions) versus a constant environment may affect the ease of extinction by increased mutagenesis is a largely unexplored question.

9.3.1 RECONCILIATION OF THEORY AND EXPERIMENT: A PROPOSAL

Several theoretical models have been presented to explain lethal mutagenesis of viruses (review in Tejero et al., 2016). The models are conceptually diverse, and at times with remarkably counterintuitive proposals. Some deny a connection between the error threshold of quasispecies theory and the extinction of viruses by enhanced mutagenesis. In one of the models discussed by H. Tejero and colleagues, it was suggested that error catastrophe could not occur in the presence of lethal genotypes, a proposal that was considered peculiar by experimentalists, and that it was proven incorrect by Takeuchi and Hogeweg (2007) [see also references in that publication and in Tejero et al. (2016)]. Other models that have suggested that lethal mutagenesis is unrelated to the error threshold have defined an extinction threshold to mean the mutation rate at which a viral population goes extinct. A rather counterintuitive proposal is that the error threshold is caused by the "survival of the flattest," which means dominance of genomes with low replicative capacity and high tolerance to mutations (robustness) that would hinder virus extinction (Tejero et al., 2011) (see Section 5.7 in Chapter 5 on the advantage of the flattest in a fitness landscape).

C. Perales and I have carefully reviewed the main experimental results obtained in our and other laboratories on the molecular events that accompany the transition toward virus extinction, and have tried to harmonize the experimental observations with the most realistic and significant theoretical models reviewed by Tejero et al. (2016). The main stream of experimental results can be summarized as follows. The first studies, that were an extension of those carried out in J.J. Holland's laboratory,

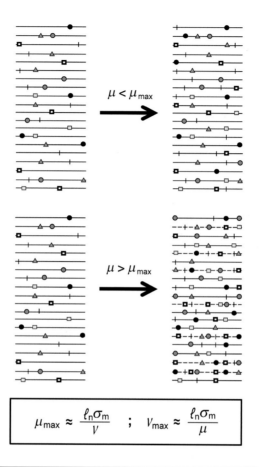

$$\mu_{max} \approx \frac{\ell_n \sigma_m}{V} \quad ; \quad V_{max} \approx \frac{\ell_n \sigma_m}{\mu}$$

FIGURE 9.2

A representation of virus entry into error catastrophe and the error threshold relationships of quasispecies theory. The mutant spectrum at the top replicates with a mutation rate (μ) below the maximum compatible with maintenance of genetic information (μ_{max}), and the population remains stable. The mutant spectrum below the first one replicates with a mutation rate (μ) that exceeds the highest tolerable (μ_{max}) for maintenance of the genetic information. As a consequence, there is a progressive deterioration of viral functions and genome replication, represented here by genomes drawn as discontinuous lines. Their number increases as replication proceeds under conditions of $\mu > \mu_{max}$, until the system collapses into total loss of information (loss of virus identity and transition into replication-incompetent sequences). The box below the mutant distributions includes the basic mathematical formulae of the error threshold, in which σ_m is the selectivity or superiority of the master sequence over its mutant spectrum, ν is the chain length of the replicating genome, and ν_{max} is the maximum length whose information can be maintained when replicating with mutation rate μ. For the mathematical derivation of the error threshold relationships, see Schuster (2016). See text for the connection between the error threshold and lethal mutagenesis.

confirmed a connection between a mutagenic activity and decrease of viral infectivity and showed that low viral load and low viral fitness favored virus extinction (Loeb et al., 1999; Sierra et al., 2000). Then, following the important discovery that the nucleoside analog ribavirin (1-β-D-ribofuranosyl-1-H-1,2,4-triazole-3-carboxamide) is mutagenic for PV (Crotty et al., 2000) it was observed that ribavirin decreased PV-specific infectivity (the ratio of infectivity to the amount of viral RNA) (Crotty et al., 2001). A decrease of specific infectivity, together with increase of mutation frequency is now considered the standard way to distinguish lethal mutagenesis from mere inhibition (Box 9.2). [This is not to mean that decreases of specific infectivity are exclusive of lethal mutagenesis. They have been also documented as a result of codon deoptimization (Section 4.3 of Chapter 4) and in viral clones subjected to many plaque-to-plaque transfers (Section 6.5.2 in Chapter 6)].

In a subsequent study, A. Grande-Pérez and colleagues documented that treatment of cells persistently infected with LCMV with 5-fluorouracil [5-Fluoro-1H,3H-pyrimidine-2,4-dione (FU)] resulted in a decay of infectivity that preceded the decay of viral RNA (see Section 9.4 for the mutagenic mechanisms of ribavirin and FU). This result suggested that a class of replication-competent RNA that did not lead to infectious virus was generated during the FU-mediated transition toward viral extinction. A theoretical model developed by S. Manrubia predicted that such defective genomes that were termed defectors were required to achieve LCMV extinction with a limited mutagenic intensity by FU (Grande-Pérez et al., 2005b). The proposal that loss of viral infectivity is due to the action of defective genomes produced by the mutagenic agent is termed the lethal defection model of virus extinction, and it is consistent with several observations on extinction of other viruses (reviewed in Domingo et al., 2012). A diagnostic test is that loss of infectivity precedes loss of viral RNA (Figure 9.3). Lethal defection can be regarded as an extreme outcome of interfering interactions that are exerted among components of the mutant spectrum when their mutational load increases (compare with Section 3.8 of Chapter 3).

The studies by A. Grande-Pérez, S. Manrubia, and colleagues distinguished two pathways that viruses can follow when subjected to mutagenesis: lethal defection at low mutagenic intensities, and overt lethality at high mutagenic intensities. Figure 9.4 recapitulates our understanding of the steps involved in mutagenesis-driven virus extinction, with the important qualification that no sharp boundary exists

---

BOX 9.2 MAIN OBSERVATIONS ON VIRAL EXTINCTION BY MUTAGENIC AGENTS

- Low viral load and low viral fitness favor extinction.
- During the transition toward extinction, the viral population:
    Decreases its specific infectivity.
    Increases its mutant spectrum complexity (movement toward usually unfavored regions of sequence space).
    Maintains an invariant consensus sequence.
- Reduction in viral load *per se* is not the mechanism of virus extinction. A load decrease by a mutagenic agent may extinguish the virus while the same decrease with an inhibitor may not.
- Viruses displaying different replicative features respond to mutagenic agents in a very similar way: the strategy is of general applicability provided a virus-specific mutagenic agent is available.
- The use of mutagens as antiviral agents has been validated *in vivo*.

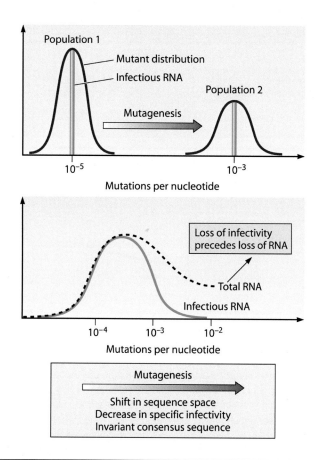

FIGURE 9.3

A summary of the main experimental observations upon mutagenesis of a viral population. Top: mutagenesis results in an expansion of mutant spectrum (red Gaussian distribution, that may be skewed in real populations), a 100-fold increase in the average number of mutations per nucleotide, and a decrease of infectious RNA (blue thin column) (specific values depend on each experimental system). Bottom: the increase in the average number of nucleotides leads to maintenance of viral RNA (discontinuous curve) despite loss of infectivity (blue curve). The box at the bottom underlines the major changes underwent by the viral genomes as the result of mutagenesis; these points are also included in Box 9.2.

Figure reproduced from Domingo et al. (2012), with permission from the American Society for Microbiology, Washington DC, USA.

between the lethal defection and the overt lethality phases. Studies with foot-and-mouth disease virus (FMDV) have provided further support to the lethal defection phase of extinction. Defective FMDV RNAs inhibited replication of standard FMDV RNA in a specific manner when co-electroporated into cells (Perales et al., 2007). Specificity means that the defective FMDV RNAs did not inhibit replication of the related encephalomyocarditis viral RNA. Specificity was also evidenced by the loss of interfering activity of a defector genome by the introduction of a mutation that prevented RNA replication. The requirement of replication of the defective RNAs suggests that a sufficient amount of expressed

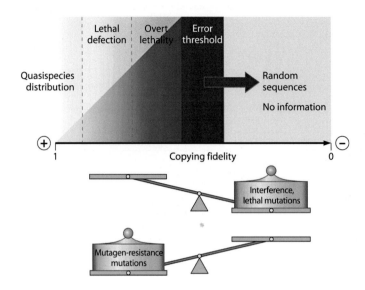

FIGURE 9.4

Scheme of the steps involved in virus extinction by enhanced mutagenesis. The graph includes a basic element of quasispecies fidelity: the horizontal black arrow that spans the entire possible range of copying fidelity values: from value 1 (perfect copying fidelity, no mutations; incompatible with evolution of life) to 0 (complete infidelity, mutation as the norm; incompatible with maintenance of genetic information). On top of the horizontal axis, four successive recognized steps, that are reached with decreasing copying fidelity, are represented: the region of maintenance of a dynamic quasispecies, the region of lethal defection, the region of overt lethality and, finally, the crossing of the error threshold (red arrow). The very last transition can be visualized as the viral genomic sequences having degenerated into random sequences; that is, a transition from information to no-information. The balances drawn at the bottom symbolize influences that may either favor or prevent virus extinction.

Figure reproduced from Domingo et al. (2012), with permission from the American Society for Microbiology, Washington DC, USA.

proteins is needed for interference (Perales et al., 2007). These results are consistent with enhanced and also specific interfering activity exerted by mutagenized FMDV RNA (González-López et al., 2004). Thus, the available evidence suggests that mutant RNAs—whose frequency and mutational load increase as the mutagenesis proceeds—adversely affect the replication of nonmutated RNAs that coexist in the same replicative ensemble. This is the major molecular event that has been recognized in the lethal defection phase of Figure 9.4.

The infectious FMDV RNA that can be retrieved from a mutagenized population by dilution and plaque isolation display decreases in replicative fitness (up to 200-fold lower infectivity) relative to the parental, nonmutagenized virus (Arias et al., 2013) associated with eightfold increase in mutation frequency. This result reinforces the complexity of events during lethal mutagenesis, confirms that fitness may be impaired in clones that survive transiently the mutagenic activity, and suggests an overlap between the lethal defection and overt lethality phases drawn in Figure 9.4.

We can now consider theoretical models in the light of the experimental results. First, the proposed interference by substituted, *trans*-acting proteins is in line with the early descriptions of fitness deterioration of cells due to the collapse of interdependent nucleic acid and protein *trans* networks in connection with the process of aging (Orgel, 1963). In the case of lethal defection, the effects of mutations have to be calibrated keeping in mind the multifunctional nature of viral proteins (Section 3.8.1 in Chapter 3). When a protein is defective it can jeopardize the activities of any other proteins that interact with it: a *trans* network can collapse by a domino effect. The possible influence of the topology of the network of interactions among genomes for maintenance of population stability is a largely unexplored possibility which is briefly addressed in the closing Chapter 10.

The notion that viral mutagenesis promotes drift in sequence space was shown by a direct amplification of A, U-rich genomic sequences of FMDV subjected to ribavirin mutagenesis (Perales et al., 2011b). The main effect of ribavirin was to accelerate the occupation of A, U-rich regions of sequence space, presumably due to the tendency of this purine analog to produce an excess of $G \rightarrow A$ and $C \rightarrow U$ transitions (Section 9.4.1). Analysis of the numbers and types of mutations suggests that the A, U-enriched portion of sequence space is detrimental to viral fitness. Movements toward unfavorable regions of sequence space are also suggested by mutant spectrum analyses of FMDV subjected to FU mutagenesis and other viruses subjected to other mutagenic agents (Grande-Pérez et al., 2002, 2005a; Agudo et al., 2008; Ortega-Prieto et al., 2013).

In view of the above evidence, any theoretical model of lethal mutagenesis that proposes a delocalization of the genome population in sequence space fits the experimental results of extinction. Specifically, models based on the advantage of the flattest that predict absence of extinction (Tejero et al., 2016) in reality predict the extinction of a real virus. This is because the mutagenesis-driven, astray walk in sequence space in the absence of a dominant master sequence should produce increased number of defective genomes (lethal defection) in unfavorable regions of sequence space (such as the A, U-rich regions promoted by ribavirin). The net result should be not only lethal defection but also an increasingly frequent hitting of lethal portions of the space (overt lethality phase in Figure 9.4). Thus, any theoretical models based on genome sequence delocalization fit with the experimental observations (Perales and Domingo, 2016). How such delocalization can be turned into an antiviral strategy is discussed in the next sections.

9.4 VIRUS EXTINCTION BY MUTAGENIC AGENTS

The pioneer experiments by J.J. Holland and colleagues demonstrated the adverse effects of mutagenic agents—including the base analog FU and the nucleoside analog 5-azacytidine [4-amino-1-β-D-ribofuranosyl-1,3,5-triazin-2(1*H*)-one (5-AZA-C)]—on the production of infectious PV and VSV progeny (Holland et al., 1990; Lee et al., 1997). These investigations were followed by several others that examined the effect of base and nucleoside analogs on virus survival. A few studies with animal viruses are summarized in Table 9.1 to illustrate that viruses displaying diverse replicative strategies (positive and negative strand RNA viruses and retroviruses) are vulnerable to increases in mutation rate. The investigations in cell culture or animal hosts employ mutagenic base or nucleoside analogs which are converted intracellularly into the nucleoside-triphosphate forms. The latter can be incorporated into RNA that in the next round of template copying will give rise to misincorporations (point mutations), and in some cases to RNA chain termination. Mutagenesis is mainly due to ambiguous pairing between the mutagenic base and standard nucleotides, with the formation of Watson-Crick or wobble base pairs. Some base pair structures of FU with A or G are drawn in Figure 9.5 (Sowers et al., 1988;

Table 9.1 Some Studies on Lethal Mutagenesis of Viruses

Virus[a]	Base or Nucleoside Analog[b]	Main Observations	Reference
FMDV	FU 5-AZA-C	Extinction favored by low viral load and low viral fitness	Sierra et al. (2000), Pariente et al. (2001)
FMDV	Rib	Enhanced mutagenesis eliminated virus from persistently infected cells	Airaksinen et al. (2003)
PV	Rib	Evidence of Rib-mediated error catastrophe of PV	Crotty et al. (2001)
HCV	Rib	Rib increased the mutation frequency in conserved regions of a HCV replicon genome	Contreras et al. (2002)
GBvB	Rib	Evidence of error-prone replication induced by Rib in cell culture, but no significant reduction of viremia *in vivo* (tamarin model). RTP incorporated in viral RNA can induce misincorporations and an elongation block	Lanford et al. (2001), Maag et al. (2001)
WNV	Rib	Error-prone replication and transition to error catastrophe were dependent on the host cell line	Day et al. (2005)
Hantaan virus	Rib	Rib-induced high mutation frequency in S segment. A range of Rib concentrations revealed a nonlinear fit of mutation frequency and mutagen concentration	Severson et al. (2003), Chung et al. (2007)
LCMV	FU	Largest increases in mutation frequency did not predict virus extinction. Virus extinction occurred without modification of the consensus sequence	Grande-Pérez et al. (2002, 2005a)
LCMV	Rib	Mutagenic activity documented in cell culture	Moreno et al. (2011)
IV	T-705	First evidence that the broad-spectrum antiviral agent T-705 (favipiravir) can act as a lethal mutagen	Baranovich et al. (2013)
IV	FU 5-AZA-C Rib	Effective lethal mutagenesis of H3N2 and H1N1 IVs. Evidence of high barrier to resistance	Pauly and Lauring (2015)
HIV-1	5-AZA-C	Lethal mutagenesis during reverse transcription is the main antiviral effect of 5-AZA-C, following reduction to the deoxycytidine form	Dapp et al. (2009)

[a] *The virus abbreviations are FMDV, foot-and-mouth disease virus; GBvB, GB virus B; HCV, hepatitis C virus; HIV-1, human immunodeficiency virus type 1; LCMV, lymphocytic choriomeningitis virus; PV, poliovirus; WNV, West Nile virus.*
[b] *The abbreviations for drug names are 5-AZA-C, 5-azacytidine; FU, 5-fluorouracil; Rib, ribavirin; T-705, favipiravir.*

Yu et al., 1993), and of ribavirin and T-705 (favipiravir) (5-Fluoro-2-oxo-1*H*-pyrazine-3-carboxamide) in Figure 9.6 (Crotty et al., 2000; Jin et al., 2013) (see Section 2.2 in Chapter 2 for the standard Watson-Crick and some wobble base pairs).

The possible point mutations generated as a consequence of FU or ribavirin incorporation during viral genome replication are depicted in Figures 9.7 and 9.8, respectively. The figures represent the

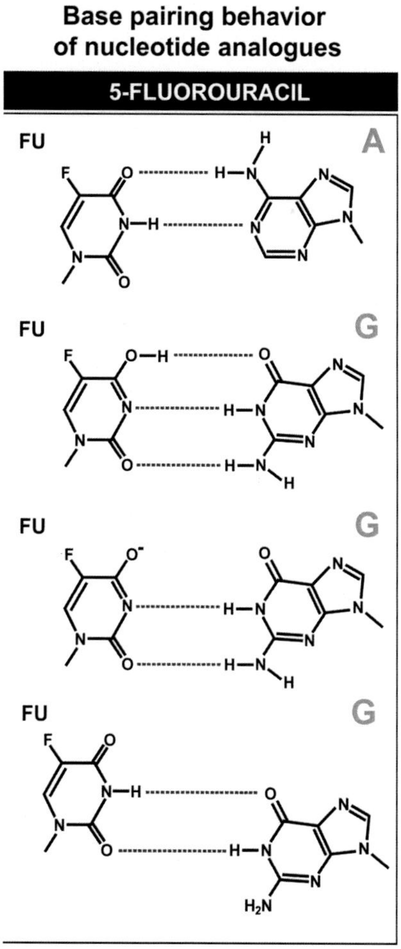

FIGURE 9.5

Some of the possible base pairs between 5-fluorouracil (FU) and the standard nucleotides A and G. Hydrogen bonds are indicated by discontinuous red lines; sugar and phosphate residues are not included.

most frequent case of the successive analog incorporation and subsequent events upon copying of positive and negative strand viral RNA, although the pathways are equally applicable to DNA in the case of deoxynucleotide analogs. The schemes can serve as a guide to anticipate the possible mutation types induced by other analogs provided their base pairing behavior has been investigated by physicochemical procedures, considering that mutation preferences may be modified by environmental factors.

The mutant repertoire produced by any nucleotide analog may be influenced by the structure of the viral polymerase [similar in general shape, but differing in critical molecular details (see Section 2.6 of Chapter 2 on polymerase structure and fidelity mutants, and Section 9.6 on subtle interactions in viral polymerases to confer resistance to mutagenic nucleotide analogs)]. For several viruses, the mutant

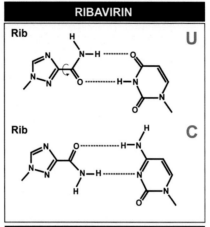

FIGURE 9.6

Some of the possible base pairs between ribavirin (Rib) or favipiravir (T-705) and the standard nucleotides U and C. Hydrogen bonds are indicated by discontinuous red lines; sugar and phosphate residues are not included. The rotation about the carboxamide bond that contributes to alternative pairing with U or C is indicated by the small circular arrow.

spectra produced upon replication in the presence of FU include an excess of A→G and U→C over G→A and C→U transitions (Sierra et al., 2000; Grande-Pérez et al., 2002; Ruiz-Jarabo et al., 2003; review on FU antiviral activity and mutagenesis in Agudo et al., 2009). The bias in favor of A→G and U→C mutations suggests that FU behaved as U rather than C when incorporated by the viral polymerases and that once FU was a template residue it tended to behave as C as rather than U (Figure 9.7). Studies with several viruses have indicated that ribavirin mutagenesis yields mutant spectra with an excess of G→A and C→U over A→G and U→C (Figure 9.8) (Crotty et al., 2001; Airaksinen et al.,

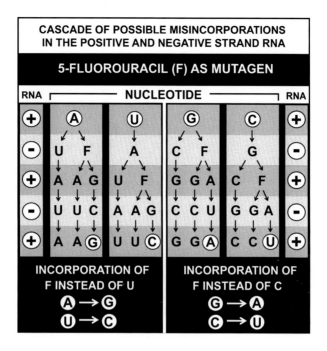

FIGURE 9.7

Mutations produced by 5-fluorouracil (F) as a result of its incorporation into viral RNA. The RNA polarity is indicated on the left (+, positive strand or genomic RNA in the case of positive strand RNA viruses; −, negative strand or complementary RNA; in the case of negative strand RNA viruses the genomic RNA is of negative polarity). Downward arrows indicate standard base copying and F incorporation. The block on the left explains the consequences of incorporation of F instead of U either in the minus strand (second row) or in the plus strand (third row). The block on the right indicates the consequences of incorporation of F instead of C either into the minus strand (second row) or into the plus strand (third row). The boxes at the bottom indicate the types of mutations expected in the mutant spectrum (see text for references).

2003; Chung et al., 2007; Agudo et al., 2010; Moreno et al., 2011; Dietz et al., 2013; Ortega-Prieto et al., 2013; among other studies; as reviews of the mutagenic activity of ribavirin see Crotty et al., 2002; Beaucourt and Vignuzzi, 2014).

The variation of mutational spectrum evoked by ribavirin depending on the polymerase and environmental factors is illustrated by a study of C.B. Jonsson and colleagues with Hantaan virus. They observed a much higher frequency of G → A than C → U mutations induced by ribavirin in this virus (Chung et al., 2007), as detected also with West Nile virus (Day et al., 2005). The effect of a mutagenic nucleotide analog may be influenced by the position it occupies in the template. This was shown in the early site-directed mutagenesis experiments of bacteriophage Qβ RNA performed by R. Flavell, C. Weissmann, and colleagues (Chapter 3). When the pyrimidine analog N⁴-hydroxy CMP was present at the extracistronic position 15, it directed the incorporation of GMP slightly more efficiently than AMP, while at position 39 incorporation of AMP was threefold higher than GMP (compare Flavell et al., 1974 and Domingo et al., 1976).

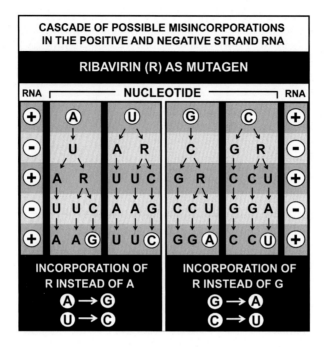

FIGURE 9.8

Mutations produced by ribavirin (R) as a result of its incorporation into viral RNA. The RNA polarity is indicated on the left (+, positive strand or genomic RNA in the case of positive strand RNA viruses; −, negative strand or complementary RNA; in the case of negative strand RNA viruses the genomic RNA is of negative polarity). Downward arrows indicate standard base copying and R incorporation. The block on the left explains the consequences of incorporation of R instead of A either into the minus strand (second row) or into the plus strand (third row). The block on the right indicates the consequences of incorporation of R instead of G either into the minus strand (second row) or into the plus strand (third row). The boxes at the bottom indicate the types of mutations expected in the mutant spectrum (see text for references).

9.4.1 THE SEARCH FOR NEW MUTAGENIC NUCLEOTIDE ANALOGS

New nucleotide analogs are currently being investigated as potential lethal mutagens for viruses (Harki et al., 2002, 2006, 2007; Graci and Cameron, 2004; Beach et al., 2014; Dapp et al., 2014; Vivet-Boudou et al., 2015; among other studies). There is active research to apply drugs (or its derivatives) used in antibacterial or anticancer therapy to lethal mutagenesis of viruses, in a strategy known as drug repositioning or drug repurposing, quite extended in current pharmacology. L.M. Mansky and his associates have pioneered such efforts for the search of new antiretroviral agents, as well as the study of new combination therapies based on lethal mutagenesis (Dapp et al., 2012, 2013; Bonnac et al., 2013; Rawson and Mansky, 2014). M.J. Dapp, L.M. Mansky, and colleagues studied the joint effect of 5-AZA-C and apolipoprotein B mRNA editing complex 3G (APOBEC3G) on the mutational spectrum of HIV-1. The results revealed unexpected changes in the mutational trend, particularly an increase in G → A transitions and a decrease of G → C transversions observed with 5-AZA-C alone (Dapp et al., 2009).

The complexities of the interaction among mutagens and between inhibitors and mutagens in infected cells are still poorly understood. A few studies have provided evidence that some sequential inhibitor-mutagen treatments may have an advantage over the corresponding combinations, fundamentally because they avoid the simultaneous presence of a mutagen and inhibitor during viral replication (Section 9.8).

9.5 LETHAL MUTAGENESIS *IN VIVO*: COMPLICATIONS DERIVED FROM MULTIPLE MECHANISMS OF DRUG ACTION—THE CASE OF RIBAVIRIN

Most of the investigations summarized in the previous section were carried out in cell culture. Some animal experiments and a clinical trial have provided the proof of principle of the feasibility of antiviral interventions based on lethal mutagenesis *in vivo*. J.C. de la Torre and colleagues documented that administration of FU to mice prevented the establishment of a persistent LCMV infection in the animals (Ruiz-Jarabo et al., 2003). J.I. Mullins and colleagues carried out the first phase II clinical trial using lethal mutagenesis by administering KP1461 (N4-heptyloxycarbonyl-5,6-dihydro-5-aza-2′-deoxycitidine), the prodrug of KP1212 (5,6- dihydro-5-aza-2′-deoxycitidine), to HIV-1 infected volunteers previously treated with antiretroviral agents (Mullins et al., 2011). No reduction in viral load or increase in average mutation frequencies was noted in the treated patients. However, mutations that likely occurred in HIV-1 in the course of the treatment were predominantly A → G and G → A transitions, as expected from the base pairing behavior of the pyrimidine analog. The study validated lethal mutagenesis as an antiviral approach for human disease.

The purine analog T-705 (favipiravir) (base pairing behavior shown in Figure 9.6) acted as an effective antinorovirus agent in a mouse model, with features diagnostic of lethal mutagenesis: increases of mutation frequency and decreases of specific infectivity (Arias et al., 2014) (Box 9.2). T-705 is an interesting compound because it shares with ribavirin a broad antiviral spectrum of activity. It has proven effective against several viruses *in vivo*, including West Nile infection of mice and hamsters (Morrey et al., 2008), lethal inhalation of Rift Valley fever virus in rats (Caroline et al., 2014), and highly pathogenic H5N1 IV in mice (Kiso et al., 2010). It is still an open question if T-705 acts as a lethal mutagen in these *in vivo* systems as it does with IV in cell culture (Baranovich et al., 2013) and norovirus in cell culture and *in vivo* (Arias et al., 2014).

An interesting possibility is that the broad-spectrum activity of ribavirin and T-705 is associated with the mutagenic properties of these analogs since the main requirement is that their nucleoside-triphosphate forms be incorporated into replicating RNA. Viral polymerases from different viruses share some features of nucleotide recognition and polymerization so that several of them may be vulnerable to the same nucleotide analogs.

Ribavirin has been used for many years in combination with pegylated IFN-α as the standard of care treatment for HCV infections (McHutchison et al., 1998; Cummings et al., 2001; Di Bisceglie et al., 2001). It is not clear whether lethal mutagenesis is part of the anti-HCV activity of ribavirin, with some studies favoring a mutagenic activity on the virus and others not (Gerotto et al., 1999; Querenghi et al., 2001; Sookoian et al., 2001; Dixit et al., 2004; Asahina et al., 2005; Chevaliez et al., 2007; Lutchman et al., 2007; Perelson and Layden, 2007; Cuevas et al., 2009; Dietz et al., 2013). A mutagenic activity of ribavirin has been documented in cell culture with HCV and subgenomic replicons (Contreras et al., 2002; Zhou et al., 2003; Kanda et al., 2004; Ortega-Prieto et al., 2013), albeit with exceptions

(Kato et al., 2005; Mori et al., 2011). The main problem to interpret the mechanism of anti-HCV activity of ribavirin *in vivo* is that this antiviral agent can act through several nonexclusive mechanisms, including: (i) immunomodulatory activity with enhancement of the Th1 antiviral response (Hultgren et al., 1998; Ning et al., 1998); (ii) upregulation of expression of genes related to IFN signaling pathways (Zhang et al., 2003; Feld et al., 2007); (iii) depletion of intracellular guanosine-5′-triphosphate (GTP) levels associated with the inhibition of inosine-monophosphate dehydrogenase (IMPDH, the enzyme that converts inosine-monophosphate into xanthosine-monophosphate in the GTP biosynthesis pathway) by ribavirin-monophosphate (RMP) (Streeter et al., 1973); (iv) inhibition of mRNA cap formation (Goswami et al., 1979); (v) inhibition of viral polymerases, independent of a mutagenic activity (Eriksson et al., 1977; Wray et al., 1985; Toltzis et al., 1988; Fernandez-Larsson et al., 1989; Maag et al., 2001; Bougie and Bisaillon, 2003); (vi) lethal mutagenesis, as evidenced with several RNA viruses (some of the studies are listed in Table 9.1), but not with other viruses (Leyssen et al., 2005, 2006; Kim and Lee, 2013). As general reviews of the antiviral properties of ribavirin, see (Snell, 2001; Graci and Cameron, 2002; Parker, 2005; Beaucourt and Vignuzzi, 2014).

It is not easy to separate an inhibitory from a mutagenic activity exerted by mutagenic base or nucleotide analogs. Inhibition may be a consequence of mutagenesis or inhibition and mutagenesis may have two totally separate modes of action. In the case of the antiviral activity of FU against FMDV, it was possible to show inhibition of the uridylylation of primer protein VPg (the initial step of picornavirus RNA synthesis) by 5-fluorouridine-triphosphate (FUTP), in addition to a mutagenic activity of FUTP during RNA elongation (Agudo et al., 2008). In the case of ribavirin, the situation is more complex since the involvement of many cellular functions in HCV replication (in reality in the replication of any virus), does not allow excluding the participation of any of the six mechanisms listed in the previous paragraph in its anti-HCV activity. In particular, it would be interesting to reexamine the evidence for inhibition of viral polymerases to ascertain that the inhibition is independent of a mutagenic activity.

In my group we favor lethal mutagenesis as being part of the antiviral activity of ribavirin against HCV, without excluding the participation of other mechanisms *in vivo*. The main arguments on which our preference is based are: in the study by (Dietz et al., 2013) that involved next generation sequencing (NGS) analyses of HCV from patients subjected to ribavirin monotherapy, the virus showed the mutational bias expected from ribavirin mutagenesis. A second argument is the consistent ribavirin mutagenesis of HCV observed in cell culture that could not be accounted for by GTP depletion (see Ortega-Prieto et al., 2013 and references therein). It has been suggested that in cell culture experiments the concentrations of ribavirin used are not attainable *in vivo*. Until measurements of the concentration of ribavirin nucleotides at the HCV replication complexes are available, it cannot be concluded that ribavirin concentrations *in vivo* are incompatible with mutagenesis; also, a broad range of ribavirin concentration in human serum have been reported during treatment (discussion and references in Ortega-Prieto et al., 2013). Furthermore, there are several reasons to miss a mutagenic activity *in vivo* with current analytical procedures. Ribavirin produced transient expansions followed by compression of mutant spectrum complexity in model studies with FMDV in cell culture (Ojosnegros et al., 2008; Perales et al., 2011b) (Figure 9.9). A possible interpretation is that when the mutational load due to ribavirin mutagenesis surpasses some critical value, increasingly defective genomes cease to contribute to progeny, therefore, resulting in mutant spectrum compression. The compression can be viewed as a mutagenesis-mediated bottleneck-associated reduction of complexity. Additional variations in FMDV mutant spectrum complexity were observed upon other mutagenic treatments and their interruption (discussed in Ojosnegros et al., 2008).

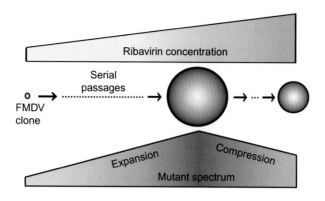

FIGURE 9.9

Expansion and compression of a FMDV mutant spectrum following ribavirin mutagenesis. A biological clone of FMDV (left) was subjected to serial passages in the presence of increasing concentrations of ribavirin (200 up to 5000 μM). The size of the three spheres quantifies in an approximate manner the complexity of the mutant spectrum, following the average distance parameter used to quantify genome subpopulations in the partition analysis of quasispecies (PAQ) clustering procedure of Baccam et al. (2001). The amplitude of the mutant spectrum increased first, but then it was compressed to a significant extent. The scheme is based on data reported in (Ojosnegros et al., 2008) and possible interpretations and implications are discussed in the text.

These model studies render unsurprising that no consistent expansions of HCV mutant spectra have been observed in patients subjected to ribavirin-based treatments or in chronically infected, untreated patients (compare, e.g. Farci et al., 2000; Duffy et al., 2002; Sullivan et al., 2007; Ramachandran et al., 2011; Palmer et al., 2014). Thus, even during ribavirin treatment alteration of mutant spectrum complexity may be missed depending on the time after treatment onset at which the viral sample is obtained. Also, it is necessary to quantify the amount of viral RNA to ensure that the sequence analyses either by standard molecular cloning-Sanger sequencing or NGS provide a faithful representation of the biological sample (see Section 3.6.4 in Chapter 3). Biases in the quantification of mutant spectrum complexity due to limitation in the initial amount of viral nucleic acid cannot be excluded unless sufficient methodological detail is described. Finally, in our view it is unlikely that a mutagenic activity readily observed in cell culture may be totally absent and bear no relationship with the activity of ribavirin *in vivo*, even though other mechanisms of ribavirin action can also be involved.

The question of participation of lethal mutagenesis in the *in vivo* antiviral activity of ribavirin, favipiravir, and other antiviral nucleotide analogs under development is expected to be solved soon, given the increasing application of NGS methodologies. If the answer were positive it would mean that inadvertently lethal mutagenesis might have been the mode of action of broad-spectrum antiviral agents used for decades, and traditionally considered standard nonmutagenic inhibitors.

9.6 VIRUS RESISTANCE TO MUTAGENIC AGENTS: MULTIPLE MECHANISMS AND EVIDENCE OF ABORTIVE ESCAPE PATHWAYS

It took more than one decade of use of ribavirin to obtain the first resistant mutants using Sindbis virus and mycophenolic acid (a nonmutagenic inhibitor of IMPDH); the selected mycophenolic acid-resistant mutants displayed cross-resistance to ribavirin (Scheidel et al., 1987; Scheidel and Stollar, 1991). A decade later the first ribavirin-resistant mutants of poliovirus selected in the laboratory (Pfeiffer and Kirkegaard, 2003), and of HCV from patients under ribavirin monotherapy were described (Young et al., 2003). Subsequent work has characterized viral mutants resistant to mutagenic agents, particularly ribavirin, FU, and 5-AZA-C (Pfeiffer and Kirkegaard, 2005b; Sierra et al., 2007; Arias et al., 2008; Agudo et al., 2010; Levi et al., 2010; Arribas et al., 2011; Feigelstock et al., 2011; Domingo-Calap et al., 2012; Sadeghipour et al., 2013; Zeng et al., 2013, 2014; among other studies). The major overall conclusion of these investigations is that viruses can develop resistance to mutagenic nucleotide analogs as they do to nonmutagenic inhibitors.

Several mechanisms of resistance to mutagenic nucleotide analogs have been described, and the major ones are listed in Box 9.3. They can be broadly divided into two main categories: modifications of the viral polymerase and modification of other proteins. The first substitution to be described that conferred ribavirin resistance was G64S in the PV polymerase (3D), that resulted in an increase of polymerase template-copying fidelity. Studies with PV harboring this mutation have been instrumental to show the relevance of mutant spectrum complexity in virus adaptability (Pfeiffer and Kirkegaard, 2005a; Vignuzzi et al., 2006) (see Section 2.6 in Chapter 2). These and other studies have established the important concept that to maintain an adequate fitness and a good survival probability, a virus population must keep its genome heterogeneity within a suitable range: too low a diversity impairs adaptability, and too high a diversity may approach the population to an extinction threshold (Smith et al., 2013; Smith and Denison, 2013; Zeng et al., 2014). Several studies have demonstrated an attenuation phenotype associated with low- or high-fidelity viral mutants (Borderia et al., 2016). In a clinical setting, maintaining mutant spectra of viruses within an optimal range is one of the predictors of viral survival and progression toward disease (Section 8.8 in Chapter 8).

BOX 9.3 MAIN MOLECULAR MECHANISMS OF RESISTANCE TO MUTAGENIC NUCLEOTIDE ANALOGS

- Amino acid substitutions in the viral polymerase:
  Substitutions that increase the general copying fidelity of the enzyme.
  Substitutions that specifically limit the incorporation of the mutagenic nucleotide.
  Substitutions that modulate the relative incorporation of the standard nucleotides.
- Amino acid substitutions in other viral proteins:
  Substitutions in nonstructural viral proteins that participate in viral replication or modify polymerase fidelity.
- Combinations of some of these mechanisms.
- Fitness-enhancing mutations (unrelated to resistance *per se*) can contribute to the expression and stability of the resistance trait.
  (Most virological studies have involved base or nucleoside analogs rather than nucleoside-triphosphates. It is assumed that at least for the mechanisms described to date, base and nucleosides are converted into the nucleoside-triphosphate derivatives which are responsible for the mutagenic activity. See text for references).

9.6.1 UNPREDICTABLE EFFECTS OF SOME POLYMERASE SUBSTITUTIONS

The same amino acid substitution in the polymerase of related viruses may have a totally different phenotypic effect. The ribavirin resistance, fidelity-enhancing PV substitution G64S in 3D was obtained independently in two different laboratories (Pfeiffer and Kirkegaard, 2003; Castro et al., 2005). This suggested that PV may have a very restricted number of mutations to attain ribavirin resistance, despite ample evidence that viruses generally display alternative pathways toward resistance to nonmutagenic inhibitors (Section 8.4.5 in Chapter 8). When FMDV whose polymerase is closely related to that of PV (Ferrer-Orta et al., 2004) was passaged in the presence of ribavirin, replacement G62S (the one equivalent to G64S in PV) was not selected. Instead, the 3D substitution selected was M296I (Sierra et al., 2007). A mutant FMDV encoding G62S in 3D that was constructed by site-directed mutagenesis, displayed a strong selective disadvantage relative to the standard virus, that was partially compensated by the presence of M296I. The mutant reverted upon passage in cell culture while FMDV with substitution M296I in 3D was stable (Ferrer-Orta et al., 2010). The comparison of the enzymological properties of the FMDV polymerase (3D) with either G62S, M296I, or both indicated that G62S impairs RNA binding, RNA polymerization, and the incorporation of RMP into RNA. Therefore, despite G62S being of potential benefit for the replication of FMDV in the presence of ribavirin, its selection would be abortive. Despite the sites of substitution G62S and M296I being separated by 13.1 angstroms, a network of interactions allowed a cross influence between the two sites, with an effect on the catalytic domain of the enzyme (Ferrer-Orta et al., 2010). Thus, due to distance effects and the subtle, sequence-dependent network connections among residues within the polymerase molecule, the same amino acid replacement may have disparate and unpredictable effects on the behavior of closely related polymerases.

9.6.2 POLYMERASE FIDELITY AND MODULATION OF NUCLEOTIDE INCORPORATION

Substitution M296I in 3D of FMDV offers an example of mutagen resistance that limited the incorporation of RMP in the viral RNA without a significant alteration of the general template-copying fidelity of the enzyme (Sierra et al., 2007; Arias et al., 2008) (Box 9.3). Fitness of the mutant FMDV relative to the standard virus was 3.8 in the presence of $800\,\mu M$ ribavirin, and 0.5 in the absence of ribavirin [these values represent a selective strength of 7.6 for substitution M296I ($f_{+Ribavirin}/f_{-Ribavirin}$), calculated as described in Section 8.4.4 of Chapter 8].

The selective strength conferred by substitution M296I was not sufficient to maintain this substitution in 3D as the only one for the virus to respond to higher ribavirin concentrations. Passage of the 3D M296I mutant virus in the presence of larger concentrations of ribavirin, resulted in selection of two additional substitutions, P44S and P169S, to yield a virus with the triple substitution P44S, P169S, and M296I in 3D; this triple mutant was termed SSI (Agudo et al., 2010). The selective strength of the triple mutant measured in the presence of $800\,\mu M$ ribavirin was 18.3, 2.4-fold higher that the selective strength conferred by M296I alone. Significantly, the most salient biochemical feature displayed by the mutant polymerase was that P44S restricted the incorporation of RMP more strongly opposite C than opposite U in a number of *in vitro* incorporation assays. As a consequence, during replication in the presence of ribavirin, mutant SSI could maintain a balance of the different transition types [measured as the ratio of $(G \rightarrow A)+(C \rightarrow U)$ to $(A \rightarrow G)+(U \rightarrow C)$ mutations] typical of the standard virus when replicating in the absence of mutagens (Agudo et al., 2010) (see Figure 9.8 for the ribavirin-mediated mutational pathways). The modulation mechanism associated with 3D substitution P44S in FMDV is one of the mutagen resistance mechanisms listed in Box 9.3. Modulation of transition types allows the virus to maintain its typical mutant spectrum complexity with its corresponding adaptability. Interestingly, substitution P169S contributed a selective advantage only in the presence of very high (5000 µM) ribavirin concentrations, once the transition-modulating phenotype through P44S had already been acquired. Box 9.3 lists also as mutations contributing to mutagen escape those that increase fitness without being *bona fide* resistance mutations. Comparison of the fitness-enhancing mutations described for antiviral resistance (Chapter 8) and the effect of P169S in 3D of FMDV suggests that such mutations can be divided in two classes: those that increase fitness generally and those that do so only in the presence of mutagen and even in the presence of some range of mutagen concentrations (Agudo et al., 2010).

The analyses of mutant spectra produced by FMDV with the standard and SSI 3D in the absence and presence of ribavirin suggest that a bias in favor of $G \rightarrow A$ and $C \rightarrow U$ is detrimental to the virus because it is associated with an increase of nonsynonymous mutations, and probably also of defective genomes, as evidenced by the presence of a stop codon in one of the clones of the mutagenized standard virus population (Agudo et al., 2010).

The three-dimensional structures of the ribavirin-resistant polymerases revealed alterations in the N-terminal region of 3D (Agudo et al., 2010), affecting sites that belong to a nuclear localization signal (NLS) present in FMDV 3D (Sanchez-Aparicio et al., 2013). Amino acid substitutions within the NLS that diminished the transport to the nucleus of 3D and 3D3C (a functional precursor intermediate of 3D and the protease 3C) modified also the template binding and nucleotide recognition properties of 3D (Ferrer-Orta et al., 2015). Interestingly, some replacements within the NLS increased the incorporation of RMP relative to standard substrates, suggesting that structural alterations in viral polymerases may enhance the vulnerability of viruses to nucleotide analogs. Despite uncertainties repeatedly exposed in this book (dependence of enzyme behavior on the protein and template sequence context, behavior modifications due to introduction of additional amino acid substitutions, etc.) it can be envisaged that through structure-based designs, new drugs could be found that enhance the incorporation of mutagenic nucleotides and contribute to new antiviral combinations. The results on the effect of amino acid substitutions within the NLS of FMDV 3D emphasize the multifunctional nature of this viral polymerase, in line with the recognized multifunctionality of many viral proteins, a feature that increases vulnerability to lethal mutagenesis.

The comparison of the several amino acid and groups of amino acid substitutions that in 3D of FMDV can be involved in ribavirin resistance (Sierra et al., 2007; Agudo et al., 2010; Zeng et al., 2014; Ferrer-Orta et al., 2015) suggests that there are multiple, in some cases even independent, evolutionary pathways for a virus to achieve resistance to mutagenic agents, as there are multiple pathways toward resistance to standard, nonmutagenic inhibitors. In the case of PV, different ribavirin resistance substitutions would be expected if independent viral lineages were subjected to a range of ribavirin concentrations.

Furthermore, the viral polymerases are not the only determinants of template-copying fidelity. The viral replicative machineries consist of a complex of several viral and host proteins, and it has been shown that proteins other than the polymerase can also contribute to polymerase fidelity and resistance to mutagenic agents (Box 9.3). The possibility that the read through protein of bacteriophage Qβ may contribute to 5-AZA-C resistance was suggested by a study of E. Lázaro and colleagues (Arribas et al., 2011), although limited amino acid substitutions in the viral replicase can confer resistance (Cabanillas et al., 2014). The small, nonenzymatic coronavirus nonstructural protein 10 is involved in maintaining the replication fidelity of the virus (Smith et al., 2015). Two substitutions in protein NS5A of an HCV replicon conferred low-level resistance to ribavirin (Pfeiffer and Kirkegaard, 2005b). In the case of the DNA bacteriophage φX174 that does not encode its own polymerase, FU resistance was achieved through substitutions in the virus-coded lysis protein (Pereira-Gomez and Sanjuan, 2014). It was proposed that delayed cell lysis, increase in the amount of progeny virus per cell, and limitation of the number of infectious cycles reduced the chances of mutagenesis. Therefore, current evidence anticipates multiple adaptive mechanisms of virus resistance to mutagenic nucleotides, not necessarily confined to the polymerase as documented for standard inhibitors. There is, however, an interesting possibility of an initial advantage of mutagenic agents over standard inhibitors to impede selection of resistant mutants. Before justifying this suggestion, current therapeutic alternatives based on lethal mutagenesis are discussed.

9.7 VIRUS EXTINCTION AS THE OUTCOME OF REPLACEMENT OF VIRUS SUBPOPULATIONS: TEMPO AND MODE OF MUTATION ACQUISITION

Some general considerations derived from the observations described in previous sections are worth commenting. The extinction of a virus is not only a consequence of the reduction of viral load (with insufficient R_0 value to ensure cell-to-cell transmission to sustain the infection; see Section 7.2 in Chapter 7 for the concept of R_0), or of a simple movement in sequence space, or even of an increase of mutational load in replicating genomes. It is the result of a combination of several influences under specific circumstances of the kinetics of accumulation of mutations in viral genomes. We consider these points in turn.

Studies by N. Pariente, C. Perales, and colleagues showed that a given reduction in viral load achieved through mutagenesis was sufficient to drive FMDV toward extinction, but that when the same reduction was achieved through inhibition, it did not lead to virus extinction (Pariente et al., 2001, 2003, 2005; Perales et al., 2011a). The main difference is that mutagenesis is a dynamic process of genome variation and reduction of viral load, while inhibition is basically a static process regarding incorporation of mutations. Only when a decrease in viral load has already taken place, the virus becomes more vulnerable to mutagenesis (Sierra et al., 2000).

Movements in sequence space are the norm during virus replication and evolution, independently of being subjected to increased mutagenesis or not. An A, U-rich genome subpopulation that became dominant upon passage of the virus in the presence of ribavirin existed also in the standard viral population, albeit at a lower level (Perales et al., 2011b). That is, the main effect of ribavirin was to shift the dominant part of the mutant spectrum into an A, U-rich region. This shift may increase the presence of defective genomes, including defectors (Section 9.3.1). We encounter here again one of the main departures introduced by quasispecies in the understanding of virus evolution: replacement of some genome subpopulations by others. In the case of extinction, the movement consists in the replacement of standard subpopulations by others lying in unfavorable portions of sequence space. Expressed in another manner: mutagenesis forces an uncontrolled, not a fitness gradient-mediated, relocalization in sequence space.

The mutational load, calculated as the number of mutations incorporated in an individual genome is not by itself a predictor of survival or extinction. It depends on the location of the mutations. Accumulation of mutations may reduce fitness and robustness to the effect of additional mutations (Section 6.7.1 in Chapter 6) but still allow replicative competence. A FMDV clone subjected to multiple plaque-to-plaque transfers accumulates mutations at a rate of 0.1-0.25 mutations per passage, and viable genomes can accumulate a number of mutations equivalent to a mutation frequency of

10^{-2} substitutions per nucleotide (Escarmís et al., 2008). A 10-fold lower mutation frequency is sufficient to extinguish the same viral clone upon passage in the presence of a mutagenic agent. The tempo and mode of accumulation of mutations is the determinant factor. E. Lázaro and colleagues showed that 5-AZA-C enhanced Qβ replication during plaque development, but drove the virus to extinction during growth in liquid culture medium, unless a mutagen-resistant Qβ was selected (Cases-Gonzalez et al., 2008; Arribas et al., 2011). The experimental design in the plaque transfers is such that at each transfer a viral particle capable of developing a visible plaque on the cell monolayer is rescued no matter how many companions are left behind (extinguished). The applied selection is powerful. In contrast, during mutagenesis without intervening plaque isolations, no means to recover viable minority survivors are included in the design: if a rare potential survivor with a cluster of mutations arises transiently, it will soon perish due to the next rounds of mutagenesis, or will be suppressed by the surrounding mutant spectrum. Thus, even if beneficial mutations are more likely to arise in low-fitness genomes, their contribution to survival is minimal under a mutagenic environment.

A critical point for prediction of virus extinction is how close to an error (or extinction) threshold virus replication takes place. The evidence is that DNA genomes whose replication is catalyzed by a high-fidelity DNA polymerase may resist transient increases of mutation rate and even benefit from them (Cupples and Miller, 1989; Solé and Deisboeck, 2004; Springman et al., 2010). Thus, viral extinction by enhanced mutagenesis is conditioned by several factors related mainly to basal replicative parameters of the virus, and the kinetics of mutagenesis.

9.8 THE INTERPLAY BETWEEN INHIBITORS AND MUTAGENIC AGENTS IN VIRAL POPULATIONS: SEQUENTIAL VERSUS COMBINATION TREATMENTS

As a test of the paradigm of the general advantage of combination therapies over monotherapy, initial experiments were designed to study the efficacy of a combination of a mutagenic agent and an antiviral inhibitor. The results documented that extinction of high-fitness FMDV or HIV-1 could be achieved by a combination of a mutagenic agent and a nonmutagenic inhibitor, but not with the mutagenic agent alone (Pariente et al., 2001, 2005; Tapia et al., 2005). Despite being an expected result, additional experiments pursued by C. Perales with FMDV demonstrated that the response of the virus to the combined action of an inhibitor and a mutagenic agent is a bit more complex than initially thought. The critical comparative experiment was performed by C. Perales, and it is shown in Figure 9.10. The production of infectivity and viral RNA was measured in the course of viral passages in the presence of the inhibitor guanidinium chloride alone, or ribavirin alone, or a combination of guanidinium chloride and ribavirin, or with an initial passage in the presence of guanidinium chloride, followed by passages in the presence of ribavirin alone (Figure 9.10a). The results of quantification of FMDV progeny and a sensitive RT-PCR test of virus extinction, documented that the design consisting of administering guanidine alone first, followed by additional passages in the presence of ribavirin alone was the most effective to reduce progeny production and to drive the virus toward extinction (Figure 9.10b–e) (Perales et al., 2009b). S. Manrubia developed a theoretical model whose main parameters were the concentration of standard and defective viruses sensitive and resistant to the inhibitor, the viral mutation rate, the rate of generation of inhibitor-resistant mutants, the number of standard and defective progeny genomes, and the number of infected cells and of infectious cycles per cell. The model (described in

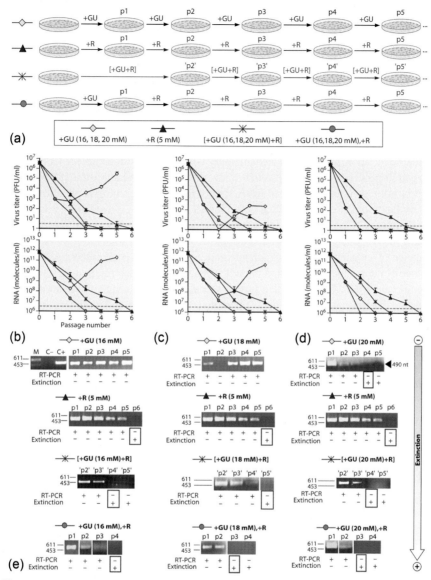

FIGURE 9.10

Alternative antiviral designs using an inhibitor and a mutagenic agent. This model study was carried out with FMDV replicating in BHK-21 cells. The inhibitor used was guanidine hydrochloride (GU), and the mutagen was ribavirin (R). (a) The four types of serial passages (p indicates passage number) are, from top to bottom: passages in the presence of guanidine alone (yellow diamond); passages in the presence of ribavirin (R) alone (red triangle); passages in the presence of a mixture of GU and R (denoted as [+GU+R]; red star; passage number is in quotes because the first passage with the mixture of GU and R was considered equivalent to the second passage in the presence of a single drug); finally, a first passage in the presence of GU, followed by four passages in the presence of R (blue circle). (b), (c), (d) Virus titer and viral RNA level in the course of six passages in the presence of 5 mM R and 16, 18, 20 mM GU in (b), (c), (d), respectively, with passage regimen code indicated in the box. Note that the highest reductions of FMDV progeny production are achieved using the sequential GU-R protocol (blue circles) and that infectivity and viral RNA are lost earlier at the highest GU concentration tested. (e) A highly sensitive RT-PCR amplification used as diagnostic of viral extinction confirmed the advantage of the sequential GU-ribavirin treatment. Symbols are as in (a).

Figure reproduced from Perales et al. (2009b) where additional experimental details can be found.

Perales et al., 2009b; Iranzo et al., 2011) explained the advantage of the sequential inhibitor-mutagen administration protocol over the corresponding combination. When an inhibitor and a mutagen are present together during viral replication, their joint influence on the viral RNA may jeopardize virus extinction. Some possible mechanisms of interaction between inhibitors and mutagenic agents are summarized in Box 9.4. The first of the points listed in Box 9.4 concerns us here: the presence of the mutagen may increase transiently the frequency of inhibitor-resistant mutants, therefore, favoring virus escape and treatment failure. A second mechanism by which the simultaneous presence of an inhibitor and mutagen can impede virus extinction is that in the presence of the inhibitor, the RNA defectors that are generated by mutagenesis of the viral RNA cannot interfere with viral replication. The reason is that defector replication is needed for lethal defection (Section 9.3.1), presumably because replication increases the amount of RNA that encodes *trans*-acting substituted proteins that are detrimental to the virus life cycle, and responsible for lethal defection.

The advantage of sequential inhibitor-mutagen administration has been evidenced also with LCMV, taking advantage of the double inhibitor and mutagenic character of ribavirin, that depends on its concentration (Moreno et al., 2012). The model that supports the advantage of sequential treatments predicts that such an advantage depends on set of replicative parameters of the virus and on the intensity of inhibition and mutagenesis (Iranzo et al., 2011; Perales et al., 2012). For each virus-host system, it will be necessary to carry out experiments to delimit the range of inhibitor and mutagen concentrations at which the sequential treatment is advantageous. The translation into clinical practice is of interest, but it will have the added complications derived from target compartmentalization and uneven inhibitor and mutagen concentrations in different compartments (Steinmeyer and Wilke, 2009). The model studies by S. Manrubia and J. Iranzo predict also that when therapy is based on the use of either two mutagenic agents or two inhibitors, a combination treatment is always preferred over the sequential. Only a few studies have investigated the consequences of using two mutagens together, with or without inhibitors (Perales et al., 2009a; Dapp et al., 2012). The availability of two or more virus-specific mutagen agents of different mutational preferences has the advantage that they will target different regions of sequence space (related to network connections discussed in Chapter 10), although the resulting mutational patterns may be difficult to interpret (Dapp et al., 2012). Also, a mutagen may extinguish a virus that has acquired resistance to another mutagen as in the extinction of a ribavirin-resistant FMDV mutant by FU (Perales et al., 2009a). According to the model, the worst option is to administer first a mutagen and then an inhibitor. Unless the mutagen achieves extreme reductions of viral load, it may generate an expanded mutant spectrum from which inhibitor-resistant mutants may be selected (Iranzo et al., 2011; Perales et al., 2012).

BOX 9.4 INTERACTIONS BETWEEN INHIBITORS AND MUTAGENIC NUCLEOTIDES THAT CAN AFFECT THE EFFICACY OF COMBINATION TREATMENTS

- A mutagen can increase the frequency of inhibitor-resistant mutants.
- An inhibitor can prevent replication of interfering mutants that contribute to lethal defection.
- The mutant spectrum can suppress inhibitor-resistant mutants that affect a *trans*-complementable protein.
- A mutagenized mutant spectrum can suppress high-fitness genomes.
  See text for justification and references.

9.9 **PROSPECTS FOR A CLINICAL APPLICATION OF LETHAL MUTAGENESIS**

Lethal mutagenesis constitutes an example of how a fundamental theoretical concept initially unrelated to virology can unfold into a potential application in the form of antiviral designs. This transition is exemplified by M. Eigen, who established the basis of quasispecies theory (Eigen, 1971) and commented a paper on virus extinction by mutagenic nucleotides 30 years later! (Eigen, 2002). Can you imagine a grant application to work on quasispecies with the aim of controlling viral infections? Some readers will probably object immediately: nothing has been achieved yet with lethal mutagenesis at the clinical level. True, unless ribavirin has been clearing HCV with participation of mutagenesis (see Section 9.5). To be able to talk about the prospect of an application is sufficient to make the point about the relevance of basic research, and it is the prospect that we address here.

Box 9.5 lists advantages and limitations of lethal mutagenesis treatments as compared with standard, nonmutagenic inhibitors. Quantifications of mutagen-escape mutant frequencies have not been performed. The derivation of some of the mutants that have been studied (Section 9.6) required a gradual increase of mutagen concentration in the course of selective passages. The difficulty to isolate mutants from a mutagen-treated viral population may stem from the suppressive environment created by the mutagenesis itself that may preclude or delay dominance of the resistant mutant.

The possibility of sequential inhibitor-mutagen treatments (Section 9.8) is clinically relevant because such designs may diminish the severity of side effects by avoiding the simultaneous presence of two drugs in the treated patients. Shorter treatment duration of sequential treatments is also a possibility.

Some natural cellular mechanisms of defense against genetic parasites are based on producing an excess of mutations in the invader. In Chapter 2 (Section 2.7), APOBEC (*a*polipoprotein *B* mRNA

BOX 9.5 ADVANTAGES AND LIMITATIONS OF ANTIVIRAL TREATMENTS BASED ON LETHAL MUTAGENESIS

Advantages
- A possible high barrier to resistance.
- Its mechanism of action favors suppression of possible resistant variants by the mutagenized mutant spectrum within infected cells.
- Lethal mutagenesis may be included in sequential or combination designs with other classes of antiviral agents.
- Natural mechanisms of resistance to genetic parasites include lethal mutagenesis-like strategies: APOBEC, ADAR, RIP, etc.

Limitations
- The mutagenic activity of the agents must be virus specific. It cannot mutagenize cellular nucleic acids.
- Possible off-target effects are still poorly understood.
- The number of available antiviral mutagenic agents is restricted.
- Additional experiments with animal models and preclinical and clinical trials are needed prior to possible therapy implementation.
- Resistance of expert panels to encourage the exploration of unconventional antiviral approaches both for scientific and commercial reasons.

editing *c*omplex) and ADAR (*a*denosine *d*eaminase *a*cting on double-stranded *R*NA) were discussed. They exert nucleic acid editing functions used by the cell that can be recruited as antiviral responses (Harris and Dudley, 2015). Editing is part of replication cycle of several viruses. It has been suggested that *Paramyxovirinae* might have evolved to possess a genome of polyhexameric length (known as the "rule of six") to avoid uncontrolled editing and error catastrophe of the virus (Kolakofsky et al., 2005). There are additional mutagenic-like activities that mimic lethal mutagenesis. One of them is termed RIP (*r*epeat-*i*nduced *p*oint mutations) that operates in some filamentous fungi to mutate genetic intruders, including transposable elements (Galagan and Selker, 2004; Clutterbuck, 2011; Braga et al., 2014; Amselem et al., 2015). Some experts regard as highly positive that an intended medical intervention resembles some natural process.

Box 9.5 lists also several limitations, some of which are obvious (need of specificity to mutagenize viral but not cellular nucleic acids, lack of information on off-target effects, and a necessity to explore treatment efficacy *in vivo*). It is likely that the number of virus-specific mutagenic agents will increase in coming decades. If their efficacy *in vivo* can be properly documented, expert panels may flexibilize their attitude toward these and other new antiviral approaches.

9.10 SOME ATYPICAL PROPOSALS

Decisions on the suitability of new treatments are mainly based on the three basic parameters CC_{50} (as a measure of toxicity), IC_{50} (as a measure of inhibitory potential), and the therapeutic index that they yield ($TI = CC_{50}/IC_{50}$) (explained in Section 8.4.3 of Chapter 8). There is an additional parameter that should also be considered: the spectrum of antiviral activity, regarding the number of unrelated viral pathogens that are effectively inhibited by the treatment. Although evidence is still lacking, the possibility that broad-spectrum in the above sense may predict a capacity to inhibit broad repertoires of quasispecies swarms is appealing. The broadness of antiviral activity is basically dependent on the mechanism of antiviral activity. In this chapter and in the preceding chapter, we have described two classes of broad-spectrum antiviral inhibitors: those that stimulate the innate immune response (notably inhibitors of pyrimidine biosynthesis) and mutagenic nucleotide analogs (notably ribavirin and favipiravir as the most relevant examples, likely to be followed by additional ones). Thus, it would be interesting to investigate the joint use of these two classes of compounds (stimulators of immune responses and lethal mutagens) in sequential or combination protocols of the type described in Section 9.8.

Drug repositioning may offer additional possibilities. F. Sobrino and colleagues demonstrated that valproic acid (2-propylpentanoic acid, VPA) displays a broad antiviral activity against many enveloped viruses (Martin-Acebes et al., 2011; Vazquez-Calvo et al., 2013). VPA is used to treat epilepsy and bipolar mania, among other disorders. It is listed among the essential medicines by the World Health Organization. Despite its antiviral potency being modest, its inclusion as part of combined therapies should be also considered in view of its broad-spectrum of activity. Triple, broad-spectrum drug combinations appear as an attractive possibility provided antagonistic interactions are avoided and side effects are tolerable.

There is a long way to go before lethal mutagenesis or other new designs based on the concepts listed in Box 9.1 can be applied to the treatment of viral disease. However, the hope is that the experience gained mainly with HIV-1 but also with other error-prone viruses, will favor explorations that regard the adaptive potential of viruses as the major challenge to be confronted.

9.11 **OVERVIEW AND CONCLUDING REMARKS**

It is uncertain whether the possibilities presented here as new trends in antiviral strategies will satisfy the demands of new paradigms to approach infectious disease (emphasized in the opening paragraphs of this chapter), or it will be necessary to wait for more innovative advance. The available possibilities have as a common trend that they respond to the quasispecies challenge, and fulfill requirements expressed repeatedly in this book: viral populations to be controlled should be denied any opportunity to replicate, because replication is synonymous with adaptation. Viruses have to be hit hard and as soon as possible both at the level of individual infections and at the epidemiological level. To the extent that new treatment designs can fulfill this requirement, they are worth being pursued.

Probably, combinations of broad-spectrum antiviral agents, irrespective of their nature (lethal mutagens, stimulators of the immune response, mixtures of highly neutralizing antibodies, or their combinations) are among the best options presently available. Despite the general preference for combination treatments, once a mutagenic agent enters the antiviral formulations, the possibility of sequential inhibitor-mutagen treatments should be considered. They may be as effective as the corresponding combinations and may alleviate the burden of side effects, one of the problems of current treatments.

Given the challenge of existing viral diseases and the emergent viral diseases unavoidably to come, what should not be done is to minimize the importance of the challenge and, as a consequence, withdraw the focus from the essential issue. Antiviral designs must be planned counting on quasispecies dynamics. Controversies about which theoretical models best describe extinction of viruses by lethal mutagenesis should be used to clarify antiviral mechanisms and find the adjustments for best efficacy. Controversies should not be presented as evidence against the real nature of dynamic quasispecies. With the application of deep sequencing to viral populations, the challenge has become dramatically evident. New antiviral targets, new drugs (mutagenic and nonmutagenic), and studies with animal models will hopefully contribute to reach the goal (see Summary Box).

SUMMARY BOX

- There is a need to develop new antiviral strategies to control viral pathogens characterized by quasispecies dynamics.
- Several new designs have been implemented or are under investigation. They aim at increasing the barrier to resistance or diminishing viral fitness.
- Lethal mutagenesis consists in virus extinction by excess mutations. Experimental evidence suggests that lethal mutagenesis is directly related to the error threshold concept of quasispecies theory.
- Mutagen-resistant viral mutants have been isolated, and their study has been instrumental to document the adaptive value conferred by a mutant spectrum of adequate complexity.
- Sequential inhibitor-mutagen treatments may have an advantage over the corresponding combination, depending on replicative parameters of the virus to be controlled and the inhibitory and mutagenic intensities.
- New treatments using broad-spectrum mutagenic and nonmutagenic drug combinations offer new prospects for the control of error-prone viruses.

REFERENCES

Agudo, R., Arias, A., Pariente, N., Perales, C., Escarmis, C., et al., 2008. Molecular characterization of a dual inhibitory and mutagenic activity of 5-fluorouridine triphosphate on viral RNA synthesis. Implications for lethal mutagenesis. J. Mol. Biol. 382, 652–666.

Agudo, R., Arias, A., Domingo, E., 2009. 5-Fluorouracil in lethal mutagenesis of foot-and-mouth disease virus. Future Med. Chem. 1, 529–539.

Agudo, R., Ferrer-Orta, C., Arias, A., de la Higuera, I., Perales, C., et al., 2010. A multi-step process of viral adaptation to a mutagenic nucleoside analogue by modulation of transition types leads to extinction-escape. PLoS Pathog. 6, e1001072.

Airaksinen, A., Pariente, N., Menendez-Arias, L., Domingo, E., 2003. Curing of foot-and-mouth disease virus from persistently infected cells by ribavirin involves enhanced mutagenesis. Virology 311, 339–349.

Amselem, J., Lebrun, M.H., Quesneville, H., 2015. Whole genome comparative analysis of transposable elements provides new insight into mechanisms of their inactivation in fungal genomes. BMC Genomics 16, 141.

Arias, A., Arnold, J.J., Sierra, M., Smidansky, E.D., Domingo, E., et al., 2008. Determinants of RNA-dependent RNA polymerase (in)fidelity revealed by kinetic analysis of the polymerase encoded by a foot-and-mouth disease virus mutant with reduced sensitivity to ribavirin. J. Virol. 82, 12346–12355.

Arias, A., Isabel de Avila, A., Sanz-Ramos, M., Agudo, R., Escarmis, C., et al., 2013. Molecular dissection of a viral quasispecies under mutagenic treatment: positive correlation between fitness loss and mutational load. J. Gen. Virol. 94, 817–830.

Arias, A., Thorne, L., Goodfellow, I., 2014. Favipiravir elicits antiviral mutagenesis during virus replication in vivo. eLife 3, e03679.

Arribas, M., Cabanillas, L., Lazaro, E., 2011. Identification of mutations conferring 5-azacytidine resistance in bacteriophage Qbeta. Virology 417, 343–352.

Asahina, Y., Izumi, N., Enomoto, N., Uchihara, M., Kurosaki, M., et al., 2005. Mutagenic effects of ribavirin and response to interferon/ribavirin combination therapy in chronic hepatitis C. J. Hepatol. 43, 623–629.

Baccam, P., Thompson, R.J., Fedrigo, O., Carpenter, S., Cornette, J.L., 2001. PAQ: partition analysis of quasispecies. Bioinformatics 17, 16–22.

Baranovich, T., Wong, S.S., Armstrong, J., Marjuki, H., Webby, R.J., et al., 2013. T-705 (favipiravir) induces lethal mutagenesis in influenza A H1N1 viruses in vitro. J. Virol. 87, 3741–3751.

Beach, L.B., Rawson, J.M., Kim, B., Patterson, S.E., Mansky, L.M., 2014. Novel inhibitors of human immunodeficiency virus type 2 infectivity. J. Gen. Virol. 95, 2778–2783.

Beaucourt, S., Vignuzzi, M., 2014. Ribavirin: a drug active against many viruses with multiple effects on virus replication and propagation. Molecular basis of ribavirin resistance. Curr. Opin. Virol. 8, 10–15.

Bonhoeffer, S., May, R.M., Shaw, G.M., Nowak, M.A., 1997. Virus dynamics and drug therapy. Proc. Natl. Acad. Sci. U. S. A. 94, 6971–6976.

Bonnac, L.F., Mansky, L.M., Patterson, S.E., 2013. Structure-activity relationships and design of viral mutagens and application to lethal mutagenesis. J. Med. Chem. 56, 9403–9414.

Borderia, A.V., Rozen-Gagnon, K., Vignuzzi, M., 2016. Fidelity variants and RNA quasispecies. Curr. Top. Microbiol. Immunol, in press.

Bougie, I., Bisaillon, M., 2003. Initial binding of the broad spectrum antiviral nucleoside ribavirin to the hepatitis C virus RNA polymerase. J. Biol. Chem. 278, 52471–52478.

Braga, R.M., Santana, M.F., Veras da Costa, R., Brommonschenkel, S.H., de Araujo, E.F., et al., 2014. Transposable elements belonging to the Tc1-Mariner superfamily are heavily mutated in Colletotrichum graminicola. Mycologia 106, 629–641.

Cabanillas, L., Sanjuan, R., Lazaro, E., 2014. Changes in protein domains outside the catalytic site of the bacteriophage Qbeta replicase reduce the mutagenic effect of 5-azacytidine. J. Virol. 88, 10480–10487.

Caroline, A.L., Powell, D.S., Bethel, L.M., Oury, T.D., Reed, D.S., et al., 2014. Broad spectrum antiviral activity of favipiravir (T-705): protection from highly lethal inhalational Rift Valley Fever. PLoS Negl. Trop. Dis. 8, e2790.

Casadevall, A., 1996. Crisis in infectious diseases: time for a new paradigm? Clin. Infect. Dis. 23, 790–794.

Casadevall, A., Pirofski, L.A., 2015. The ebola epidemic crystallizes the potential of passive antibody therapy for infectious diseases. PLoS Pathog. 11, e1004717.

Cases-Gonzalez, C., Arribas, M., Domingo, E., Lazaro, E., 2008. Beneficial effects of population bottlenecks in an RNA virus evolving at increased error rate. J. Mol. Biol. 384, 1120–1129.

Castro, C., Arnold, J.J., Cameron, C.E., 2005. Incorporation fidelity of the viral RNA-dependent RNA polymerase: a kinetic, thermodynamic and structural perspective. Virus Res. 107, 141–149.

Chatterji, U., Lim, P., Bobardt, M.D., Wieland, S., Cordek, D.G., et al., 2010. HCV resistance to cyclosporin A does not correlate with a resistance of the NS5A-cyclophilin A interaction to cyclophilin inhibitors. J. Hepatol. 53, 50–56.

Chevaliez, S., Brillet, R., Lazaro, E., Hezode, C., Pawlotsky, J.M., 2007. Analysis of ribavirin mutagenicity in human hepatitis C virus infection. J. Virol. 81, 7732–7741.

Chung, D.H., Sun, Y., Parker, W.B., Arterburn, J.B., Bartolucci, A., et al., 2007. Ribavirin reveals a lethal threshold of allowable mutation frequency for Hantaan virus. J. Virol. 81, 11722–11729.

Clutterbuck, A.J., 2011. Genomic evidence of repeat-induced point mutation (RIP) in filamentous ascomycetes. Fungal Genet. Biol. 48, 306–326.

Contreras, A.M., Hiasa, Y., He, W., Terella, A., Schmidt, E.V., et al., 2002. Viral RNA mutations are region specific and increased by ribavirin in a full-length hepatitis C virus replication system. J. Virol. 76, 8505–8517.

Crotty, S., Maag, D., Arnold, J.J., Zhong, W., Lau, J.Y., et al., 2000. The broad-spectrum antiviral ribonucleoside ribavirin is an RNA virus mutagen. Nat. Med. 6, 1375–1379.

Crotty, S., Cameron, C.E., Andino, R., 2001. RNA virus error catastrophe: direct molecular test by using ribavirin. Proc. Natl. Acad. Sci. U. S. A. 98, 6895–6900.

Crotty, S., Cameron, C., Andino, R., 2002. Ribavirin's antiviral mechanism of action: lethal mutagenesis? J. Mol. Med. 80, 86–95.

Cuevas, J.M., González-Candelas, F., Moya, A., Sanjuán, R., 2009. Effect of ribavirin on the mutation rate and spectrum of hepatitis C virus in vivo. J. Virol. 83, 5760–5764.

Cummings, K.J., Lee, S.M., West, E.S., Cid-Ruzafa, J., Fein, S.G., et al., 2001. Interferon and ribavirin vs interferon alone in the re-treatment of chronic hepatitis C previously nonresponsive to interferon: a meta-analysis of randomized trials. JAMA 285, 193–199.

Cupples, C.G., Miller, J.H., 1989. A set of lacZ mutations in Escherichia coli that allow rapid detection of each of the six base substitutions. Proc. Natl. Acad. Sci. U. S. A. 86, 5345–5349.

Dapp, M.J., Clouser, C.L., Patterson, S., Mansky, L.M., 2009. 5-Azacytidine can induce lethal mutagenesis in human immunodeficiency virus type 1. J. Virol. 83, 11950–11958.

Dapp, M.J., Holtz, C.M., Mansky, L.M., 2012. Concomitant lethal mutagenesis of human immunodeficiency virus type 1. J. Mol. Biol. 419, 158–170.

Dapp, M.J., Patterson, S.E., Mansky, L.M., 2013. Back to the future: revisiting HIV-1 lethal mutagenesis. Trends Microbiol. 21, 56–62.

Dapp, M.J., Bonnac, L., Patterson, S.E., Mansky, L.M., 2014. Discovery of novel ribonucleoside analogs with activity against human immunodeficiency virus type 1. J. Virol. 88, 354–363.

Day, C.W., Smee, D.F., Julander, J.G., Yamshchikov, V.F., Sidwell, R.W., et al., 2005. Error-prone replication of West Nile virus caused by ribavirin. Antiviral Res. 67, 38–45.

De Crignis, E., Mahmoudi, T., 2014. HIV eradication: combinatorial approaches to activate latent viruses. Viruses 6, 4581–4608.

Delang, L., Vliegen, I., Froeyen, M., Neyts, J., 2011. Comparative study of the genetic barriers and pathways towards resistance of selective inhibitors of hepatitis C virus replication. Antimicrob. Agents Chemother. 55, 4103–4113.

Di Bisceglie, A.M., Thompson, J., Smith-Wilkaitis, N., Brunt, E.M., Bacon, B.R., 2001. Combination of interferon and ribavirin in chronic hepatitis C: re-treatment of nonresponders to interferon. Hepatology 33, 704–707.

Dietz, J., Schelhorn, S.E., Fitting, D., Mihm, U., Susser, S., et al., 2013. Deep sequencing reveals mutagenic effects of ribavirin during monotherapy of hepatitis C virus genotype 1-infected patients. J. Virol. 87, 6172–6181.

Dixit, N.M., Layden-Almer, J.E., Layden, T.J., Perelson, A.S., 2004. Modelling how ribavirin improves interferon response rates in hepatitis C virus infection. Nature 432, 922–924.

Domingo, E., 1989. RNA virus evolution and the control of viral disease. Prog. Drug Res. 33, 93–133.

Domingo, E., Holland, J.J., 1992. Complications of RNA heterogeneity for the engineering of virus vaccines and antiviral agents. Genet. Eng. 14, 13–31.

Domingo, E., Flavell, R.A., Weissmann, C., 1976. In vitro site-directed mutagenesis: generation and properties of an infectious extracistronic mutant of bacteriophage Qβ. Gene 1, 3–25.

Domingo, E., Grande-Perez, A., Martin, V., 2008. Future prospects for the treatment of rapidly evolving viral pathogens: insights from evolutionary biology. Expert. Opin. Biol. Ther. 8, 1455–1460.

Domingo, E., Sheldon, J., Perales, C., 2012. Viral quasispecies evolution. Microbiol. Mol. Biol. Rev. 76, 159–216.

Domingo-Calap, P., Pereira-Gomez, M., Sanjuan, R., 2012. Nucleoside analogue mutagenesis of a single-stranded DNA virus: evolution and resistance. J. Virol. 86, 9640–9646.

Duffy, M., Salemi, M., Sheehy, N., Vandamme, A.M., Hegarty, J., et al., 2002. Comparative rates of nucleotide sequence variation in the hypervariable region of E1/E2 and the NS5b region of hepatitis C virus in patients with a spectrum of liver disease resulting from a common source of infection. Virology 301, 354–364.

Eggers, H.J., 1976. Successful treatment of enterovirus-infected mice by 2-(alpha-hydroxybenzyl)-benzimidazole and guanidine. J. Exp. Med. 143, 1367–1381.

Eggers, H.J., Tamm, I., 1963. Synergistic effect of 2-(alpha-hydroxybenzyl)-benzimidazole and guanidine on picornavirus reproduction. Nature 199, 513–514.

Eigen, M., 1971. Self-organization of matter and the evolution of biological macromolecules. Die Naturwissenschaften 58, 465–523.

Eigen, M., 2002. Error catastrophe and antiviral strategy. Proc. Natl. Acad. Sci. U. S. A. 99, 13374–13376.

Eigen, M., Schuster, P., 1979. The Hypercycle. A principle of natural self-organization, Springer, Berlin.

Eriksson, B., Helgstrand, E., Johansson, N.G., Larsson, A., Misiorny, A., et al., 1977. Inhibition of influenza virus ribonucleic acid polymerase by ribavirin triphosphate. Antimicrob. Agents Chemother. 11, 946–951.

Escarmís, C., Lazaro, E., Arias, A., Domingo, E., 2008. Repeated bottleneck transfers can lead to non-cytocidal forms of a cytopathic virus: implications for viral extinction. J. Mol. Biol. 376, 367–379.

Farci, P., Shimoda, A., Coiana, A., Diaz, G., Peddis, G., et al., 2000. The outcome of acute hepatitis C predicted by the evolution of the viral quasispecies. Science 288, 339–344.

Feigelstock, D.A., Mihalik, K.B., Feinstone, S.M., 2011. Selection of hepatitis C virus resistant to ribavirin. Virol. J. 8, 402.

Feld, J.J., Nanda, S., Huang, Y., Chen, W., Cam, M., et al., 2007. Hepatic gene expression during treatment with peginterferon and ribavirin: identifying molecular pathways for treatment response. Hepatology 46, 1548–1563.

Fernandez-Larsson, R., O'Connell, K., Koumans, E., Patterson, J.L., 1989. Molecular analysis of the inhibitory effect of phosphorylated ribavirin on the vesicular stomatitis virus in vitro polymerase reaction. Antimicrob. Agents Chemother. 33, 1668–1673.

Ferrer-Orta, C., Arias, A., Perez-Luque, R., Escarmis, C., Domingo, E., et al., 2004. Structure of foot-and-mouth disease virus RNA-dependent RNA polymerase and its complex with a template-primer RNA. J. Biol. Chem. 279, 47212–47221.

Ferrer-Orta, C., Sierra, M., Agudo, R., de la Higuera, I., Arias, A., et al., 2010. Structure of foot-and-mouth disease virus mutant polymerases with reduced sensitivity to ribavirin. J. Virol. 84, 6188–6199.

Ferrer-Orta, C., de la Higuera, I., Caridi, F., Sanchez-Aparicio, M.T., Moreno, E., et al., 2015. Multifunctionality of a picornavirus polymerase domain: nuclear localization signal and nucleotide recognition. J. Virol. 89, 6848–6859.

Flavell, R.A., Sabo, D.L., Bandle, E.F., Weissmann, C., 1974. Site-directed mutagenesis: generation of an extracistronic mutation in bacteriophage Q beta RNA. J. Mol. Biol. 89, 255–272.

Galagan, J.E., Selker, E.U., 2004. RIP: the evolutionary cost of genome defense. Trends Genet. 20, 417–423.

Garbelli, A., Radi, M., Falchi, F., Beermann, S., Zanoli, S., et al., 2011. Targeting the human DEAD-box polypeptide 3 (DDX3) RNA helicase as a novel strategy to inhibit viral replication. Curr. Med. Chem. 18, 3015–3027.

Geller, R., Vignuzzi, M., Andino, R., Frydman, J., 2007. Evolutionary constraints on chaperone-mediated folding provide an antiviral approach refractory to development of drug resistance. Genes Dev. 21, 195–205.

Gerotto, M., Sullivan, D.G., Polyak, S.J., Chemello, L., Cavalletto, L., et al., 1999. Effect of retreatment with interferon alone or interferon plus ribavirin on hepatitis C virus quasispecies diversification in nonresponder patients with chronic hepatitis C. J. Virol. 73, 7241–7247.

González-López, C., Arias, A., Pariente, N., Gómez-Mariano, G., Domingo, E., 2004. Preextinction viral RNA can interfere with infectivity. J. Virol. 78, 3319–3324.

Goswami, B.B., Borek, E., Sharma, O.K., Fujitaki, J., Smith, R.A., 1979. The broad spectrum antiviral agent ribavirin inhibits capping of mRNA. Biochem. Biophys. Res. Commun. 89, 830–836.

Graci, J.D., Cameron, C.E., 2002. Quasispecies, error catastrophe, and the antiviral activity of ribavirin. Virology 298, 175–180.

Graci, J.D., Cameron, C.E., 2004. Challenges for the development of ribonucleoside analogues as inducers of error catastrophe. Antivir. Chem. Chemother. 15, 1–13.

Grande-Pérez, A., Sierra, S., Castro, M.G., Domingo, E., Lowenstein, P.R., 2002. Molecular indetermination in the transition to error catastrophe: systematic elimination of lymphocytic choriomeningitis virus through mutagenesis does not correlate linearly with large increases in mutant spectrum complexity. Proc. Natl. Acad. Sci. U. S. A. 99, 12938–12943.

Grande-Pérez, A., Gómez-Mariano, G., Lowenstein, P.R., Domingo, E., 2005a. Mutagenesis-induced, large fitness variations with an invariant arenavirus consensus genomic nucleotide sequence. J. Virol. 79, 10451–10459.

Grande-Pérez, A., Lázaro, E., Lowenstein, P., Domingo, E., Manrubia, S.C., 2005b. Suppression of viral infectivity through lethal defection. Proc. Natl. Acad. Sci. U. S. A. 102, 4448–4452.

Harki, D.A., Graci, J.D., Korneeva, V.S., Ghosh, S.K., Hong, Z., et al., 2002. Synthesis and antiviral evaluation of a mutagenic and non-hydrogen bonding ribonucleoside analogue: 1-beta-D-Ribofuranosyl-3-nitropyrrole. Biochemistry 41, 9026–9033.

Harki, D.A., Graci, J.D., Galarraga, J.E., Chain, W.J., Cameron, C.E., et al., 2006. Synthesis and antiviral activity of 5-substituted cytidine analogues: identification of a potent inhibitor of viral RNA-dependent RNA polymerases. J. Med. Chem. 49, 6166–6169.

Harki, D.A., Graci, J.D., Edathil, J.P., Castro, C., Cameron, C.E., et al., 2007. Synthesis of a universal 5-nitroindole ribonucleotide and incorporation into RNA by a viral RNA-dependent RNA polymerase. Chembiochem 8, 1359–1362.

Harris, R.S., Dudley, J.P., 2015. APOBECs and virus restriction. Virology 479-480C, 131–145.

Ho, D.D., 1995. Time to hit HIV, early and hard. N. Engl. J. Med. 333, 450–451.

Holland, J.J., Domingo, E., de la Torre, J.C., Steinhauer, D.A., 1990. Mutation frequencies at defined single codon sites in vesicular stomatitis virus and poliovirus can be increased only slightly by chemical mutagenesis. J. Virol. 64, 3960–3962.

Hopkins, S., Scorneaux, B., Huang, Z., Murray, M.G., Wring, S., et al., 2010. SCY-635, a novel nonimmunosuppressive analog of cyclosporine that exhibits potent inhibition of hepatitis C virus RNA replication in vitro. Antimicrob. Agents Chemother. 54, 660–672.

Hultgren, C., Milich, D.R., Weiland, O., Sallberg, M., 1998. The antiviral compound ribavirin modulates the T helper (Th) 1/Th2 subset balance in hepatitis B and C virus-specific immune responses. J. Gen. Virol. 79 (Pt 10), 2381–2391.

Iranzo, J., Perales, C., Domingo, E., Manrubia, S.C., 2011. Tempo and mode of inhibitor-mutagen antiviral therapies: a multidisciplinary approach. Proc. Natl. Acad. Sci. U. S. A. 108, 16008–16013.

Jin, Z., Smith, L.K., Rajwanshi, V.K., Kim, B., Deval, J., 2013. The ambiguous base-pairing and high substrate efficiency of T-705 (Favipiravir) Ribofuranosyl 5'-triphosphate towards influenza A virus polymerase. PLoS One 8, e68347.

Kanda, T., Yokosuka, O., Imazeki, F., Tanaka, M., Shino, Y., et al., 2004. Inhibition of subgenomic hepatitis C virus RNA in Huh-7 cells: ribavirin induces mutagenesis in HCV RNA. J. Viral Hepat. 11, 479–487.

Kato, T., Date, T., Miyamoto, M., Sugiyama, M., Tanaka, Y., et al., 2005. Detection of anti-hepatitis C virus effects of interferon and ribavirin by a sensitive replicon system. J. Clin. Microbiol. 43, 5679–5684.

Kim, Y., Lee, C., 2013. Ribavirin efficiently suppresses porcine nidovirus replication. Virus Res. 171, 44–53.

Kiso, M., Takahashi, K., Sakai-Tagawa, Y., Shinya, K., Sakabe, S., et al., 2010. T-705 (favipiravir) activity against lethal H5N1 influenza A viruses. Proc. Natl. Acad. Sci. U. S. A. 107, 882–887.

Kolakofsky, D., Roux, L., Garcin, D., Ruigrok, R.W., 2005. Paramyxovirus mRNA editing, the "rule of six" and error catastrophe: a hypothesis. J. Gen. Virol. 86, 1869–1877.

Kumar, N., Liang, Y., Parslow, T.G., Liang, Y., 2011. Receptor tyrosine kinase inhibitors block multiple steps of influenza a virus replication. J. Virol. 85, 2818–2827.

Kwong, A.D., Najera, I., Bechtel, J., Bowden, S., Fitzgibbon, J., et al., 2011. Sequence and phenotypic analysis for resistance monitoring in hepatitis C virus drug development: recommendations from the HCV DRAG. Gastroenterology 140, 755–760.

Lanford, R.E., Chavez, D., Guerra, B., Lau, J.Y., Hong, Z., et al., 2001. Ribavirin induces error-prone replication of GB virus B in primary tamarin hepatocytes. J. Virol. 75, 8074–8081.

Le Moing, V., Chene, G., Carrieri, M.P., Alioum, A., Brun-Vezinet, F., et al., 2002. Predictors of virological rebound in HIV-1-infected patients initiating a protease inhibitor-containing regimen. AIDS 16, 21–29.

Lee, C.H., Gilbertson, D.L., Novella, I.S., Huerta, R., Domingo, E., et al., 1997. Negative effects of chemical mutagenesis on the adaptive behavior of vesicular stomatitis virus. J. Virol. 71, 3636–3640.

Levi, L.I., Gnadig, N.F., Beaucourt, S., McPherson, M.J., Baron, B., et al., 2010. Fidelity variants of RNA dependent RNA polymerases uncover an indirect, mutagenic activity of amiloride compounds. PLoS Pathog. 6, e1001163.

Leyssen, P., Balzarini, J., De Clercq, E., Neyts, J., 2005. The predominant mechanism by which ribavirin exerts its antiviral activity in vitro against flaviviruses and paramyxoviruses is mediated by inhibition of IMP dehydrogenase. J. Virol. 79, 1943–1947.

Leyssen, P., De Clercq, E., Neyts, J., 2006. The anti-yellow fever virus activity of ribavirin is independent of error-prone replication. Mol. Pharmacol. 69, 1461–1467.

Li, M.J., Kim, J., Li, S., Zaia, J., Yee, J.K., et al., 2005. Long-term inhibition of HIV-1 infection in primary hematopoietic cells by lentiviral vector delivery of a triple combination of anti-HIV shRNA, anti-CCR5 ribozyme, and a nucleolar-localizing TAR decoy. Mol. Ther. 12, 900–909.

Loeb, L.A., Essigmann, J.M., Kazazi, F., Zhang, J., Rose, K.D., et al., 1999. Lethal mutagenesis of HIV with mutagenic nucleoside analogs. Proc. Natl. Acad. Sci. U. S. A. 96, 1492–1497.

Lucas-Hourani, M., Dauzonne, D., Jorda, P., Cousin, G., Lupan, A., et al., 2013. Inhibition of pyrimidine biosynthesis pathway suppresses viral growth through innate immunity. PLoS Pathog. 9, e1003678.

Lutchman, G., Danehower, S., Song, B.C., Liang, T.J., Hoofnagle, J.H., et al., 2007. Mutation rate of the hepatitis C virus NS5B in patients undergoing treatment with ribavirin monotherapy. Gastroenterology 132, 1757–1766.

Maag, D., Castro, C., Hong, Z., Cameron, C.E., 2001. Hepatitis C virus RNA-dependent RNA polymerase (NS5B) as a mediator of the antiviral activity of ribavirin. J. Biol. Chem. 276, 46094–46098.

Martin-Acebes, M.A., Vazquez-Calvo, A., Rincon, V., Mateu, M.G., Sobrino, F., 2011. A single amino acid substitution in the capsid of foot-and-mouth disease virus can increase acid resistance. J. Virol. 85, 2733–2740.

McHutchison, J.G., Gordon, S.C., Schiff, E.R., Shiffman, M.L., Lee, W.M., et al., 1998. Interferon alfa-2b alone or in combination with ribavirin as initial treatment for chronic hepatitis C. Hepatitis Interventional Therapy Group. N. Engl. J. Med. 339, 1485–1492.

Moreno, H., Gallego, I., Sevilla, N., de la Torre, J.C., Domingo, E., et al., 2011. Ribavirin can be mutagenic for arenaviruses. J. Virol. 85, 7246–7255.

Moreno, H., Grande-Pérez, A., Domingo, E., Martín, V., 2012. Arenaviruses and lethal mutagenesis. Prospects for new ribavirin-based interventions. Viruses 4, 2786–2805.

Mori, K., Ikeda, M., Ariumi, Y., Dansako, H., Wakita, T., et al., 2011. Mechanism of action of ribavirin in a novel hepatitis C virus replication cell system. Virus Res. 157, 61–70.

Morrey, J.D., Taro, B.S., Siddharthan, V., Wang, H., Smee, D.F., et al., 2008. Efficacy of orally administered T-705 pyrazine analog on lethal West Nile virus infection in rodents. Antiviral Res. 80, 377–379.

Mosier, D.E., Picchio, G.R., Gulizia, R.J., Sabbe, R., Poignard, P., et al., 1999. Highly potent RANTES analogues either prevent CCR5-using human immunodeficiency virus type 1 infection in vivo or rapidly select for CXCR4-using variants. J. Virol. 73, 3544–3550.

Motani, K., Ito, S., Nagata, S., 2015. DNA-mediated cyclic GMP-AMP synthase-dependent and -independent regulation of innate immune responses. J. Immunol. 194, 4914–4923.

Müller, V., Bonhoeffer, S., 2008. Intra-host dynamics and evolution of HIV infections. In: Domingo, E., Parrish, C.R., Holland, J.J. (Eds.), Origin and Evolution of Viruses. second ed.. Elsevier, London, pp. 279–302.

Mullins, J.I., Heath, L., Hughes, J.P., Kicha, J., Styrchak, S., et al., 2011. Mutation of HIV-1 genomes in a clinical population treated with the mutagenic nucleoside KP1461. PLoS One 6, e15135.

Munier-Lehmann, H., Lucas-Hourani, M., Guillou, S., Helynck, O., Zanghi, G., et al., 2015. Original 2-(3-alkoxy-1H-pyrazol-1-yl)pyrimidine derivatives as inhibitors of human dihydroorotate dehydrogenase (DHODH). J. Med. Chem. 58, 860–877.

Nijhuis, M., van Maarseveen, N.M., Boucher, C.A., 2009. Antiviral resistance and impact on viral replication capacity: evolution of viruses under antiviral pressure occurs in three phases. Handb. Exp. Pharmacol, 189, 299–320.

Ning, Q., Brown, D., Parodo, J., Cattral, M., Gorczynski, R., et al., 1998. Ribavirin inhibits viral-induced macrophage production of TNF, IL-1, the procoagulant fgl2 prothrombinase and preserves Th1 cytokine production but inhibits Th2 cytokine response. J. Immunol. 160, 3487–3493.

Nowak, M., Schuster, P., 1989. Error thresholds of replication in finite populations mutation frequencies and the onset of Muller's ratchet. J. Theor. Biol. 137, 375–395.

Ojosnegros, S., Agudo, R., Sierra, M., Briones, C., Sierra, S., et al., 2008. Topology of evolving, mutagenized viral populations: quasispecies expansion, compression, and operation of negative selection. BMC Evol. Biol. 8, 207.

Orgel, L.E., 1963. The maintenance of the accuracy of protein synthesis and its relevance to ageing. Proc. Natl. Acad. Sci. U. S. A. 49, 517–521.

Ortega-Prieto, A.M., Sheldon, J., Grande-Pérez, A., Tejero, H., Gregori, J., et al., 2013. Extinction of hepatitis C virus by ribavirin in hepatoma cells involves lethal mutagenesis. PLoS One 8, e71039.

Ortiz-Riano, E., Ngo, N., Devito, S., Eggink, D., Munger, J., et al., 2014. Inhibition of arenavirus by A3, a pyrimidine biosynthesis inhibitor. J. Virol. 88, 878–889.

Palmer, B.A., Dimitrova, Z., Skums, P., Crosbie, O., Kenny-Walsh, E., et al., 2014. Analysis of the evolution and structure of a complex intrahost viral population in chronic hepatitis C virus mapped by ultradeep pyrosequencing. J. Virol. 88, 13709–13721.

Pariente, N., Sierra, S., Lowenstein, P.R., Domingo, E., 2001. Efficient virus extinction by combinations of a mutagen and antiviral inhibitors. J. Virol. 75, 9723–9730.

Pariente, N., Airaksinen, A., Domingo, E., 2003. Mutagenesis versus inhibition in the efficiency of extinction of foot-and-mouth disease virus. J. Virol. 77, 7131–7138.

Pariente, N., Sierra, S., Airaksinen, A., 2005. Action of mutagenic agents and antiviral inhibitors on foot-and-mouth disease virus. Virus Res. 107, 183–193.

Parker, W.B., 2005. Metabolism and antiviral activity of ribavirin. Virus Res. 107, 165–171.

Pauly, M.D., Lauring, A.S., 2015. Effective lethal mutagenesis of influenza virus by three nucleoside analogs. J. Virol. 89, 3584–3597.

Perales, C., Domingo, E., 2016. Antiviral strategies based on lethal mutagenesis and error threshold. Curr. Top. Microbiol. Immunol, in press.

Perales, C., Mateo, R., Mateu, M.G., Domingo, E., 2007. Insights into RNA virus mutant spectrum and lethal mutagenesis events: replicative interference and complementation by multiple point mutants. J. Mol. Biol. 369, 985–1000.

Perales, C., Agudo, R., Domingo, E., 2009a. Counteracting quasispecies adaptability: extinction of a ribavirin-resistant virus mutant by an alternative mutagenic treatment. PLoS One 4, e5554.

Perales, C., Agudo, R., Tejero, H., Manrubia, S.C., Domingo, E., 2009b. Potential benefits of sequential inhibitor-mutagen treatments of RNA virus infections. PLoS Pathog. 5, e1000658.

Perales, C., Agudo, R., Manrubia, S.C., Domingo, E., 2011a. Influence of mutagenesis and viral load on the sustained low-level replication of an RNA virus. J. Mol. Biol. 407, 60–78.

Perales, C., Henry, M., Domingo, E., Wain-Hobson, S., Vartanian, J.P., 2011b. Lethal mutagenesis of foot-and-mouth disease virus involves shifts in sequence space. J. Virol. 85, 12227–12240.

Perales, C., Iranzo, J., Manrubia, S.C., Domingo, E., 2012. The impact of quasispecies dynamics on the use of therapeutics. Trends Microbiol. 20, 595–603.

Perales, C., Beach, N.M., Sheldon, J., Domingo, E., 2014. Molecular basis of interferon resistance in hepatitis C virus. Curr. Opin. Virol. 8, 38–44.

Pereira-Gomez, M., Sanjuan, R., 2014. Delayed lysis confers resistance to the nucleoside analogue 5-fluorouracil and alleviates mutation accumulation in the single-stranded DNA bacteriophage varphiX174. J. Virol. 88, 5042–5049.

Perelson, A.S., Layden, T.J., 2007. Ribavirin: is it a mutagen for hepatitis C virus? Gastroenterology 132, 2050–2052.

Pfeiffer, J.K., Kirkegaard, K., 2003. A single mutation in poliovirus RNA-dependent RNA polymerase confers resistance to mutagenic nucleotide analogs via increased fidelity. Proc. Natl. Acad. Sci. U. S. A. 100, 7289–7294.

Pfeiffer, J.K., Kirkegaard, K., 2005a. Increased fidelity reduces poliovirus fitness under selective pressure in mice. PLoS Pathog. 1, 102–110.

Pfeiffer, J.K., Kirkegaard, K., 2005b. Ribavirin resistance in hepatitis C virus replicon-containing cell lines conferred by changes in the cell line or mutations in the replicon RNA. J. Virol. 79, 2346–2355.

Pol, S., Couzigou, P., Bourliere, M., Abergel, A., Combis, J.M., et al., 1999. A randomized trial of ribavirin and interferon-alpha vs. interferon-alpha alone in patients with chronic hepatitis C who were non-responders to a previous treatment. Multicenter Study Group under the coordination of the Necker Hospital, Paris, France. J. Hepatol. 31, 1–7.

Pugach, P., Marozsan, A.J., Ketas, T.J., Landes, E.L., Moore, J.P., et al., 2007. HIV-1 clones resistant to a small molecule CCR5 inhibitor use the inhibitor-bound form of CCR5 for entry. Virology 361, 212–228.

Pugach, P., Ray, N., Klasse, P.J., Ketas, T.J., Michael, E., et al., 2009. Inefficient entry of vicriviroc-resistant HIV-1 via the inhibitor-CCR5 complex at low cell surface CCR5 densities. Virology 387, 296–302.

Querenghi, F., Yu, Q., Billaud, G., Maertens, G., Trepo, C., et al., 2001. Evolution of hepatitis C virus genome in chronically infected patients receiving ribavirin monotherapy. J. Viral Hepat. 8, 120–131.

Ramachandran, S., Campo, D.S., Dimitrova, Z.E., Xia, G.L., Purdy, M.A., et al., 2011. Temporal variations in the hepatitis C virus intrahost population during chronic infection. J. Virol. 85, 6369–6380.

Rawson, J.M., Mansky, L.M., 2014. Retroviral vectors for analysis of viral mutagenesis and recombination. Viruses 6, 3612–3642.

Ribeiro, R.M., Bonhoeffer, S., 2000. Production of resistant HIV mutants during antiretroviral therapy. Proc. Natl. Acad. Sci. U. S. A. 97, 7681–7686.

Ruiz-Jarabo, C.M., Ly, C., Domingo, E., de la Torre, J.C., 2003. Lethal mutagenesis of the prototypic arenavirus lymphocytic choriomeningitis virus (LCMV). Virology 308, 37–47.

Sadeghipour, S., Bek, E.J., McMinn, P.C., 2013. Ribavirin-resistant mutants of human enterovirus 71 express a high replication fidelity phenotype during growth in cell culture. J. Virol. 87, 1759–1769.

Sanchez-Aparicio, M.T., Rosas, M.F., Sobrino, F., 2013. Characterization of a nuclear localization signal in the foot-and-mouth disease virus polymerase. Virology 444, 203–210.

Sarrazin, C., Zeuzem, S., 2010. Resistance to direct antiviral agents in patients with hepatitis C virus infection. Gastroenterology 138, 447–462.

Scheidel, L.M., Stollar, V., 1991. Mutations that confer resistance to mycophenolic acid and ribavirin on Sindbis virus map to the nonstructural protein nsP1. Virology 181, 490–499.

Scheidel, L.M., Durbin, R.K., Stollar, V., 1987. Sindbis virus mutants resistant to mycophenolic acid and ribavirin. Virology 158, 1–7.

Schuster, P., 2016. Quasispecies on fitness landscapes. Curr. Top. Microbiol. Immunol, in press.

Seiler, P., Senn, B.M., Klenerman, P., Kalinke, U., Hengartner, H., et al., 2000. Additive effect of neutralizing antibody and antiviral drug treatment in preventing virus escape and persistence. J. Virol. 74, 5896–5901.

Severson, W.E., Schmaljohn, C.S., Javadian, A., Jonsson, C.B., 2003. Ribavirin causes error catastrophe during Hantaan virus replication. J. Virol. 77, 481–488.

Sierra, S., Davila, M., Lowenstein, P.R., Domingo, E., 2000. Response of foot-and-mouth disease virus to increased mutagenesis: influence of viral load and fitness in loss of infectivity. J. Virol. 74, 8316–8323.

Sierra, M., Airaksinen, A., González-López, C., Agudo, R., Arias, A., et al., 2007. Foot-and-mouth disease virus mutant with decreased sensitivity to ribavirin: implications for error catastrophe. J. Virol. 81, 2012–2024.

Smith, E.C., Denison, M.R., 2013. Coronaviruses as DNA wannabes: a new model for the regulation of RNA virus replication fidelity. PLoS Pathog. 9, e1003760.

Smith, E.C., Blanc, H., Vignuzzi, M., Denison, M.R., 2013. Coronaviruses lacking exoribonuclease activity are susceptible to lethal mutagenesis: evidence for proofreading and potential therapeutics. PLoS Pathog. 9, e1003565.

Smith, E.C., Case, J.B., Blanc, H., Isakov, O., Shomron, N., et al., 2015. Mutations in coronavirus nonstructural protein 10 decrease virus replication fidelity. J. Virol. 89, 6418–6426.

Snell, N.J., 2001. Ribavirin-current status of a broad spectrum antiviral agent. Expert. Opin. Pharmacother. 2, 1317–1324.

Solé, R.V., Deisboeck, T.S., 2004. An error catastrophe in cancer? J. Theor. Biol. 228, 47–54.

Sookoian, S., Castano, G., Frider, B., Cello, J., Campos, R., et al., 2001. Combined therapy with interferon and ribavirin in chronic hepatitis C does not affect serum quasispecies diversity. Dig. Dis. Sci. 46, 1067–1071.

Sowers, L.C., Eritja, R., Kaplan, B., Goodman, M.F., Fazakerly, G.V., 1988. Equilibrium between a wobble and ionized base pair formed between fluorouracil and guanine in DNA as studied by proton and fluorine NMR. J. Biol. Chem. 263, 14794–14801.

Springman, R., Keller, T., Molineux, I.J., Bull, J.J., 2010. Evolution at a high imposed mutation rate: adaptation obscures the load in phage T7. Genetics 184, 221–232.

Stearns, S.C., 1999. Evolution in Health and Disease. Oxford University Press, Oxford.

Steinmeyer, S.H., Wilke, C.O., 2009. Lethal mutagenesis in a structured environment. J. Theor. Biol. 261, 67–73.

Streeter, D.G., Witkowski, J.T., Khare, G.P., Sidwell, R.W., Bauer, R.J., et al., 1973. Mechanism of action of 1- -D-ribofuranosyl-1,2,4-triazole-3-carboxamide (Virazole), a new broad-spectrum antiviral agent. Proc. Natl. Acad. Sci. U. S. A. 70, 1174–1178.

Sullivan, D.G., Bruden, D., Deubner, H., McArdle, S., Chung, M., et al., 2007. Hepatitis C virus dynamics during natural infection are associated with long-term histological outcome of chronic hepatitis C disease. J. Infect. Dis. 196, 239–248.

Swetina, J., Schuster, P., 1982. Self-replication with errors. A model for polynucleotide replication. Biophys. Chem. 16, 329–345.

Takeuchi, N., Hogeweg, P., 2007. Error-threshold exists in fitness landscapes with lethal mutants. BMC Evol. Biol. 7, 15 author reply 15.

Tapia, N., Fernandez, G., Parera, M., Gomez-Mariano, G., Clotet, B., et al., 2005. Combination of a mutagenic agent with a reverse transcriptase inhibitor results in systematic inhibition of HIV-1 infection. Virology 338, 1–8.

Tejero, H., Marin, A., Montero, F., 2011. The relationship between the error catastrophe, survival of the flattest, and natural selection. BMC Evol. Biol. 11, 2.

Tejero, H., Montero, F., Nuño, J.C., 2016. Theories of lethal mutagenesis: from error catasthophe to lethal defection. Curr. Top. Microb. Immunol, in press.

Toltzis, P., O'Connell, K., Patterson, J.L., 1988. Effect of phosphorylated ribavirin on vesicular stomatitis virus transcription. Antimicrob. Agents Chemother. 32, 492–497.

Van Vaerenbergh, K., Harrer, T., Schmit, J.C., Carbonez, A., Fontaine, E., et al., 2002. Initiation of HAART in drug-naive HIV type 1 patients prevents viral breakthrough for a median period of 35.5 months in 60% of the patients. AIDS Res. Hum. Retroviruses 18, 419–426.

Vazquez-Calvo, A., Martin-Acebes, M.A., Saiz, J.C., Ngo, N., Sobrino, F., et al., 2013. Inhibition of multiplication of the prototypic arenavirus LCMV by valproic acid. Antiviral Res. 99, 172–179.

Vermehren, J., Sarrazin, C., 2011. New HCV therapies on the horizon. Clin. Microbiol. Infect. 17, 122–134.

Vidalain, P.O., Lucas-Hourani, M., Helynck, O., Tangy, F., Munier-Lehmann, H., 2015. Stimulation of the antiviral innate immune response by pyrimidine biosynthesis inhibitors: a surprise of phenotypic screening. Med. Sci. (Paris) 31, 98–104.

Vignuzzi, M., Stone, J.K., Arnold, J.J., Cameron, C.E., Andino, R., 2006. Quasispecies diversity determines pathogenesis through cooperative interactions in a viral population. Nature 439, 344–348.

Vivet-Boudou, V., Isel, C., El Safadi, Y., Smyth, R.P., Laumond, G., et al., 2015. Evaluation of anti-HIV-1 mutagenic nucleoside analogues. J. Biol. Chem. 290, 371–383.

von Kleist, M., Menz, S., Stocker, H., Arasteh, K., Schutte, C., et al., 2011. HIV quasispecies dynamics during pro-active treatment switching: impact on multi-drug resistance and resistance archiving in latent reservoirs. PLoS One 6, e18204.

Webster, R.G., Kawaoka, Y., Bean, W.J., Beard, C.W., Brugh, M., 1985. Chemotherapy and vaccination: a possible strategy for the control of highly virulent influenza virus. J. Virol. 55, 173–176.

Webster, R.G., Kawaoka, Y., Bean, W.J., 1986. Vaccination as a strategy to reduce the emergence of amantadine- and rimantadine-resistant strains of A/Chick/Pennsylvania/83 (H5N2) influenza virus. J. Antimicrob. Chemother. 18, 157–164.

Williams, P.D., 2009. Darwinian interventions: taming pathogens through evolutionary ecology. Trends Parasitol. 26, 83–92.

Wray, S.K., Gilbert, B.E., Noall, M.W., Knight, V., 1985. Mode of action of ribavirin: effect of nucleotide pool alterations on influenza virus ribonucleoprotein synthesis. Antiviral Res. 5, 29–37.

Young, K.C., Lindsay, K.L., Lee, K.J., Liu, W.C., He, J.W., et al., 2003. Identification of a ribavirin-resistant NS5B mutation of hepatitis C virus during ribavirin monotherapy. Hepatology 38, 869–878.

Yu, H., Eritja, R., Bloom, L.B., Goodman, M.F., 1993. Ionization of bromouracil and fluorouracil stimulates base mispairing frequencies with guanine. J. Biol. Chem. 268, 15935–15943.

Zeng, J., Wang, H., Xie, X., Yang, D., Zhou, G., et al., 2013. An increased replication fidelity mutant of foot-and-mouth disease virus retains fitness in vitro and virulence in vivo. Antiviral Res. 100, 1–7.

Zeng, J., Wang, H., Xie, X., Li, C., Zhou, G., et al., 2014. Ribavirin-resistant variants of foot-and-mouth disease virus: the effect of restricted quasispecies diversity on viral virulence. J. Virol. 88, 4008–4020.

Zhang, Y., Jamaluddin, M., Wang, S., Tian, B., Garofalo, R.P., et al., 2003. Ribavirin treatment up-regulates antiviral gene expression via the interferon-stimulated response element in respiratory syncytial virus-infected epithelial cells. J. Virol. 77, 5933–5947.

Zhou, S., Liu, R., Baroudy, B.M., Malcolm, B.A., Reyes, G.R., 2003. The effect of ribavirin and IMPDH inhibitors on hepatitis C virus subgenomic replicon RNA. Virology 310, 333–342.

COLLECTIVE POPULATION EFFECTS IN NONVIRAL SYSTEMS

CHAPTER CONTENTS

ABBREVIATIONS

LTR long terminal repeat
NGS next generation sequencing
www World Wide Web

10.1 CONCEPT GENERALIZATION

The major concepts that support the population structure and dynamics of replicating viruses are of a general nature and can be applied to any biological system. They include extended Darwinian principles of genetic variation, competition, selection and random events, error-prone replication, and information theory and network connectivity. The requirement for extension is to adjust parameters to realistic

Virus as Populations. http://dx.doi.org/10.1016/B978-0-12-800837-9.00010-1
© 2016 Elsevier Inc. All rights reserved.

values for each particular system. Examination of these major concepts leads in quite a straightforward manner to the identification of several interesting implications for nonviral systems.

The scope of Darwinian principles was documented by prebiotic and biological activities summarized in Chapters 1 and 6. As a relevant modern example, the external (human-made) intervention consisting in administration of antiviral agents (Chapters 8 and 9) has a parallel in the use of antibiotics to control pathogenic bacteria. The differences are obviously not in the general principle of natural selection that pushes bacteria to evolve antibiotic resistance but in the molecular mechanisms of acquisition and spread of the resistance phenotype. Similarities and differences between antiviral and antibiotic resistance are described in Section 10.2.1.

The second major notion is error-prone replication which is the basis of the potential rapid evolution of RNA genetic elements, or any genetic system whose replication is catalyzed by a low fidelity nucleic acid polymerase (Chapters 2 and 3). It is based on imperfect template copying, unchecked by repair activities. Limited copying fidelity is ultimately due to atomic fluctuations (the consequence of quantum mechanical uncertainties associated with the electronic distributions among interacting atoms; see Chapter 2) that restrict recognition accuracies. Inexact recognition and errors during replication have been the basis of the origin of life and its subsequent evolution and diversification into multiple forms (Chapter 1). Therefore, quantum mechanical uncertainties constitute a second basic concept that can be generalized to all biological systems, even nongenetic systems (Section 10.6).

Moving to the third major concept, theory of information was one of the key ingredients in the formulation of quasispecies by M. Eigen and P. Schuster (Chapter 3). The connectivity among related genomes in the form of walks in sequence space is recognized as one of the basis of virus adaptability. A collection of related viral genomes such as those that constitute a mutant spectrum are connected through mutation and recombination. They are a specific case of links among many other classes of connected elements such as proteins within a cell, biological species in an ecosystem, or sites (nodes or hubs) in the World Wide Web (www or, simply, the web). Because of its connectivity and self-organized nature, the internet can be studied as a complex evolving system. The error threshold notion derived from quasispecies theory and relevant for the development of new antiviral strategies (Chapters 3, 8 and 9) is of broad applicability to systems consisting of connected elements. Identification of critical thresholds can define parameters of stability versus vulnerability of self-organized systems. While the web topology is known to affect connectivity, accessibility, and survival of its components, little is known of alternative topologies of connected genomes within viral quasispecies. Prospects for future research in this area prompted by application of next generation sequencing (NGS) are examined in section 10.8. Thus, network connectivity stands as a generalizable feature that permits comparing intra-population virus interactions with those of other genetic and nongenetic systems. The relationship among major notions and some of their practical implications are represented schematically in Figure 10.1. We examine some of them beginning with the most immediate: a comparison of viruses and cells, and of antiviral resistance in viruses with antibiotic resistance in bacteria, both events occurring as a consequence of positive selection in response to a medical intervention.

10.2 VIRUSES AND CELLS: THE GENOME SIZE-MUTATION-TIME COORDINATES REVISITED

High mutation rates and the dynamic mutant cloud nature of viral quasispecies −that renders use of consensus sequences insufficient to describe viruses− have as one of its consequence the efficient (almost hectic) exploration of sequence space. The relative exploration efficiency of other biological

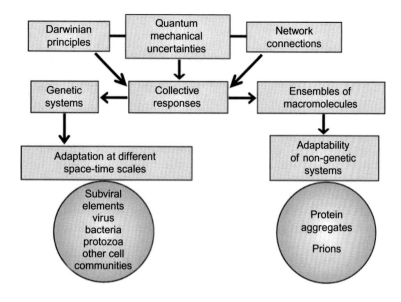

FIGURE 10.1

Some general concepts and their implications. The three top boxes indicate three generalizable notions that permit extending features of viral populations to non-viral systems. The major consequence of the interaction among the three notions is the collective responses observed in genetic systems and in macromolecular assemblages. Genetic systems display adaptability at different space and time scales. Collective responses in populations of macromolecules (non-genetic systems) display Darwinian behavior, such as in the case of prions. The behavior of cellular and macromolecular systems is analyzed and compared with virus behavior in several sections of the present chapter.

systems is largely dictated by genome size, mutation rate, and time frame in which events develop. Section 3.7.2 of Chapter 3 compares the orders of magnitude difference that exists in sequence space occupation and exploration between cells and viruses.

For the entire population of human genomes to be perceived as a dynamic mutant cloud, the time frame between the primitive humans (about 2.8 million years ago with the first members of *Homo habilis*) and us should be compressed by a factor of 10^3. Then your admired Egyptian pharaoh Ramses II and yourself would be separated by about 3 years, and those humans between Ramses II and you and surrounding relatives would appear as a dynamic mutant cloud with mutations and other genetic lesions coming and going in minutes. If human genome sequencing could have been applied retrospectively and interpreted with the knowledge we will likely have next century, you could even trace the origin of the set of mutations responsible for your uncle's propensity to suffer cancer or your in-law's friend's schizophrenia. The accelerated motion would, however, show an important difference with the viral clouds you can follow in infected hosts: in the course of human history, a great majority of points in sequence space could not be reached. The total number of human beings that have been born on Earth has been estimated in 1.08×10^{11}, less than the number of infectious HCV or HIV-1 particles in many infected patients. At most given times of our compressed human history, the number of mutants would be inferior to the total number of possible single mutants in the human genome (parameters in Box 3.3 of Chapter 3).

Bacterial genome size lies somewhere in between that of viruses and multicellular differentiated organisms (Chapter 1), and so is the capacity of bacteria to explore sequence apace through point mutations. Not surprisingly, bacterial adaptive mechanisms have been enriched with many possibilities of gene transfers, and some of this enrichment is manifested in the mechanisms of selection of antibiotic-resistant bacteria, the phenomenon parallel to antiviral resistance described in Chapter 8.

10.2.1 A COMPARISON OF ANTIVIRAL AND ANTIBIOTIC RESISTANCE

In addition to mutation and several forms of recombination (homologous, nonhomologous, mobile elements such as transposons and insertion elements with generation of insertions and deletions), bacteria can engage in conjugation, fusion, assimilation, infection, transduction, and transformation (review in Miller and Day, 2004; Read and Massey, 2014). These multiple mechanisms provide adaptive potential that can generate and spread antibiotic-resistant forms of different bacteria. Antibiotic resistance represents a growing health problem, as evidenced by the 25,000 people that die each year in the European Union as a consequence of sepsis (pathology derived from the presence of microorganisms in the blood and immune system reactions) due to antibiotic-resistant bacteria [November 2014 report of the European Academies Science Advisory Council (www.easac.eu)].

Antibiotic resistance has similarities and differences with antiviral resistance (Box 10.1). Some of the common features justify the extension to antibiotic use of the recommendations proposed to minimize the effects of antiviral resistance in clinical practice (Section 8.10 and Box 8.4 in Chapter 8). Specific suggestions are the occasional shelving of antibiotics to repopulate natural bacterial communities with sensitive forms, avoidance of the indiscriminate use of antibiotics, and the need to fund research for the discovery of new antibiotics to expand the available repertoire to control forms resistant to the classical antibiotics.

Most of the features that distinguish antiviral resistance from antibiotic resistance can be attributed to the variety of molecular mechanisms that bacteria have for the acquisition and transfer of genetic material. While antiviral resistance tends to be confined to the virus that develops it, antibiotic resistance can spread to cognate and to phylogenetically distant bacteria even beyond the clinical setting of the treated patients. Multidrug-resistant viruses can be selected by recombination and then spread to new hosts. However, recombination is relatively restricted in the field (Section 10.7), and this mode of antiviral resistance spread is less efficient than in the case of antibiotic resistance. Moreover, while the use of antiviral agents is restricted to treat infected patients, antibiotics have been used for agricultural purposes (treatment of farm animals and plants) so that resistance can evolve in nonpathogenic bacteria and then be transferred into bacterial pathogens. No parallel event in terms of environmental penetration has been yet described for antiviral agents.

The comparison between antiviral and antibiotic resistance mechanisms (Box 10.1) constitutes a very clear example of how the molecular mechanisms of genome variation and mobility available to a biological system influences the way selection events unfold. High mutation rates set the scene for the evolution of error-prone replicating viruses, and the general escape strategy or acquisition of mutation-dependent phenotypes is widespread among them. In contrast, for more complex genomes that must refrain from generalized high mutation rates to avoid lethality, mechanisms based on genome rearrangements and transfers are the molecular frame for evolutionary events. Complex DNA viruses exploit genome expansions, rearrangements, and exchanges as adaptive mechanisms, close to the strategies employed by bacteria.

BOX 10.1 SOME SIMILARITIES AND DIFFERENCES BETWEEN ANTIBIOTIC (AB) RESISTANCE IN BACTERIA AND ANTIVIRAL (AV) RESISTANCE IN VIRUSES

Common Features

- Resistant forms gradually replace sensitive forms upon use of ABs or AVs.
- Low effective concentrations of AB or AV either in an organism or at specific sites within the organism favor selection of resistant pathogens.
- Resistant pathogens often display lower fitness than their sensitive counterparts, when fitness is measured in the absence of AB or AV. In consequence, parental sensitive pathogens often replace resistant forms in the absence of AB or AV.
- Fitness decrease associated with resistance can be compensated by additional genetic change in the pathogen.
- The degree of resistance may increase with the continued presence of the AB or AV.
- Multidrug resistant forms can be selected.

Distinctive Features

Antibiotics	Antiviral Agents
Resistance maps in chromosomal genes, plasmids, or mobile elements	Resistance maps in the viral genome, often the gene encoding the target of the AV
Resistance is transmitted among bacterial species	Resistance is often restricted to the virus isolate that acquired the resistance and its progeny
AB use selects for resistance in nonpathogenic bacteria	AV resistant forms generally remain in the treated patients and their contacts. No selection of resistance in nonpathogenic viruses has been described yet
Resistance can be transferred from nonpathogenic to pathogenic bacteria	Transference of AV resistance from one virus to another is unlikely. Recombination-mediated acquisition of resistance necessitates the involved viruses to be closely related

Expanded from Domingo et al. (2001) and other studies quoted in the text.

10.3 DARWINIAN PRINCIPLES AND INTRAPOPULATION INTERACTIONS ACTING ON CELL POPULATIONS

The emphasis on mutation as being inherent to quasispecies theory versus other mathematical formulations of evolutionary dynamics (Section 3.1 in Chapter 3) justifies quasispecies as a descriptor of bacteria in at least three different but related aspects:

- The recognized genetic heterogeneity of bacterial populations which is accentuated in mutator bacteria, those that display mutation rates which are 10^2- to 10^3-fold higher than the standard values for bacteria. Virtually, any individual bacterium is likely to vary genetically from any other individual bacterium of the same species.
- The dynamics of genetic variation, competition, and selection that render bacterial populations highly dynamic and adaptable.
- Intrapopulation interactions by cellular signals that modify the behavior of bacteria when they are organized as a collectivity, as opposed to the same bacteria acting in isolation.

The first point listed above is a straightforward extension of mutant generation in bacteria which occurs with average rates that are 10^2- to 10^3-fold lower than those of RNA viruses, leading to quasispecies-like populations (Covacci and Rappouli, 1998). The second point is a statement of general adaptive potential, with the genome size-mutation-time frame implications for the limitation of sequence space exploration (Section 10.1). The third point is an extension to bacterial populations of the interactions established among components of a mutant spectrum in viruses extensively discussed in previous chapters, but with bacteria using different mechanisms. In the case of viruses, the interactions are mediated by *trans*-acting gene products that conferred a viral population the status of unit of selection. In the case of bacteria, the intrapopulation interactions are based on elaborate systems of intercellular communication which are increasingly recognized as relevant to the behavior of the microbial world generally (Foster, 2011; Gibbs and Federle, 2015).

The regulated, cell density-dependent response of bacterial collectivities is known as quorum sensing (Nealson and Hastings, 1979; Fuqua et al., 1994; Miller and Bassler, 2001). Quorum sensing is based on signal transduction schemes that allow bacterial population to produce a coordinated response to external stimuli. The result is the possibility to modulate bacterial virulence as shown also with complex viral populations following subpopulation differentiation as a result of serial viral infections (Rumbaugh et al., 2009; Ojosnegros et al., 2010a; Ojosnegros et al., 2010b; Vignuzzi and Andino, 2010). Bacterial biofilms are communities of microorganisms supported by some types of extracellular polymeric matrices (Costerton et al., 1978; Costerton et al., 1987; Costerton et al., 1999). Belonging to a biofilm modifies the cells physiologically. Biofilms may display mechanisms of antibiotic resistance additional to those that operate in isolated bacteria (Section 10.2), including a physical impairment to the diffusion of the antibiotic to reach the target bacteria (Stewart and Costerton, 2001). The presence or absence of bacteriophage in a biofilm may also affect bacterial virulence (Rice et al., 2009). Biofilm being a class of bacterial consortium, is a manifestation of social behavior, and constitutes a collective target of selection (Caldwell and Costerton, 1996), a feature shared with viral populations.

10.4 THE DYNAMICS OF UNICELLULAR PARASITES IN THE CONTROL OF PARASITIC DISEASE

The selection of drug resistance in viruses and bacteria summarized in Section 10.2.1 is also a common occurrence among pathogenic unicellular eukaryotes such as fungi, protozoans such as amoebas, the malaria parasite *Plasmodium falciparum*, or different *Leishmania* species, the causative agents of several forms of leishmaniasis and kala-azar. The molecular mechanisms of resistance range from point mutations in the drug target genes to activation of the efflux pumps that prevent the transport into the cells or changes in the "transportome" that affect one or multiple drugs (multidrug resistance) (Wahlgren et al., 1999; Koenderink et al., 2010; Naidoo and Roper, 2010; Sanchez et al., 2010; Munday et al., 2015). Drug resistance has been associated with a defective mismatch repair system in *Plasmodium falciparum* (Ahmad and Tuteja, 2014). Natural populations of unicellular parasites are highly polymorphic (Escalante et al., 1998; Rottschaefer et al., 2011; Nkhoma et al., 2012). Extensive intra-strain genomic heterogeneity in *Leishmania*, termed mosaic aneuploidy, is essential for parasite adaptability and selection of drug resistance mutations (Mannaert et al., 2012; Sterkers et al., 2012; Lachaud et al., 2014). Alterations of growth conditions may perturb the population equilibrium reflected in the replacement of some parasite subpopulations by others (Andrews et al., 1992; Valadares et al., 2012; Duncan et al., 2013). As in the case of viruses, cellular parasites can escape antibodies or CTLs (De La Cruz et al., 1989; Baltz et al., 1991; al-Khedery et al., 1999), in particular, immune evasion through recombination-mediated switching of components of the variant surface glycoprotein coat (Horn and McCulloch, 2010). Subpopulations of *Plasmodium falciparum* with decreased sensitivity to monoclonal antibodies have been selected *in vitro* (Iqbal et al., 1997). The mechanism involved may be selection of preexisting parasite subpopulations with constitutive reduced expression of the relevant antigens, or antibody-induced downregulation of antigen expression. Induction of complement resistance *in vivo* and *in vitro* has been described for some amoebas (Hamelmann et al., 1993).

The trend in the last decades has been to unveil an increasing level of genetic heterogeneity in protozoan populations, with similarities in the evolution of concepts in viruses that occurred a few decades earlier. Except for the differences in genome size-population size relationship with viruses (emphasized in general terms in Section 10.2), intrapopulation heterogeneity, coexistence of different subpopulations, and replacement of some subpopulations by others described in protozoa are clearly parallel events to those detailed for viruses in preceding chapters. Application of whole genome sequencing and new bioinformatic procedures to *Plasmodium falciparum* is currently used to obtain accurate genomic data sets, and to identify positive selection, in particular, *loci* involved in immune evasion and drug resistance (Samad et al., 2015). As an outcome of the implications of population complexity, the advantage of double and triple drug therapies is also increasingly recognized (Shanks et al., 2015). Therefore, endowed with mutation and recombination as major molecular mechanisms of genome variation, cellular parasites display features of quasispecies dynamics (Domingo et al., 2001) (see Section 10.7 for a discussion of clonality in the microbial world).

10.5 CANCER DYNAMICS: HETEROGENEITY AND GROUP BEHAVIOR

Cancer cells are extremely heterogeneous entities that can be regarded as cellular parasites of differentiated organisms that have acquired autonomous behavior. Already several decades ago, tumor heterogeneity was considered an important adaptive trait relevant to cancer metastasis (secondary growth in a part of the organism different from the original tumor site) (Hansemann, 1890; Boveri, 1914; Nowell, 1976; Poste et al., 1982; Nicolson, 1987; Chin et al., 2011; Hanahan and Weinberg, 2011). Cell heterogeneity and modified behavior of cancer cells are mediated by point mutations, chromosomal rearrangements, and epigenetic modifications (trait modifications not due to changes in DNA sequence; examples are DNA methylation and histone modifications such as phosphorylation). Most tumors are characterized by high mutation rates, although the necessity of high mutation rates for tumorigenesis is a debated issue. In a study by W.F. Bodmer and colleagues, the dual participation of mutation and selection in tumor progression was considered. It was argued that the standard cellular mutation rates (in the order of 10^{-8} per *locus* per cell per generation) may be sufficient to explain the key mutations associated with the transformed phenotype, although in the course of tumor progression, mutation rates can increase up to 10^4-fold (Tomlinson et al., 1996). These authors consider cell selection more important than an initial high mutation rate to trigger tumor initiation. The rate of tumor progression may be favored by progressive increases in the mutation rate as tumor cells replicate and diversify.

Tumor progression presents a situation of genome instability which has not yet been described during virus infection: an increase in mutation rate as the virus replicates in an infected host. This is a possibility that cannot be excluded given the evidence that multiple viral proteins, not only the viral polymerases, can be involved in template-copying fidelity (Section 9.6 in Chapter 9). The difficulty, encountered in viruses as well as in tumors (or other cellular collectivities), can be reduced to the distinction between mutation rate (occurrence of mutations) and mutation frequency (the combined outcome of occurrence and selection; Section 2.5 in Chapter 2). Two main possibilities can be envisaged: an increase in mutation rate is necessary either to reach a phenotypic trait required for tumor progression, or not to reach the trait but only to increase the rate of progression. In the latter scenario, the trait would be equally accessible with the standard cellular mutation rate. A mutator phenotype would act as an evolutionary motor to increase the chances of finding the most advantageous traits for tumor growth and dispersal (Nicolson, 1987; Loeb, 2001; Bielas et al., 2006). The mutator phenotype in some types of cancer may be linked to APOBEC-mediated deamination (Vartanian et al., 2008; Caval et al., 2014).

The contribution of multiple mechanisms of DNA damage for the cell to reach a level of transformation is subject to population size constraints similar to those that may limit selection of multidrug resistance in viruses (Chapter 8). Alternative phenotypes may afford tumor progression and metastatic potential to different degrees. If approximation to the most effective phenotype is considered akin to fitness increase, attaining the required constellation of genetic alterations will depend on the tumor size (number of cells) on which genetic variation acts. Variation in this case includes point mutations, recombination and rearrangements, genetic modifications mediated by cellular factors and epigenetic changes. Mutations affecting repair genes, and proteins that bind to DNA, transcriptional regulators, or proteins involved in signaling pathways, among others, have been implicated in the gain of an oncogene or loss of a tumor suppressor (Futreal et al., 2004; Greenman et al., 2007; Chin et al., 2011; Shlien et al., 2015). These observations suggest that evolving tumor cells display features of complex adaptive systems sustained by cellular heterogeneity and clonal evolution (Nowell, 1976; Deisboeck and Couzin, 2009) (see also Section 10.8). This dynamic cellular system evolves, fueled by genetic and epigenetic change, following the general Darwinian principles of competition and selection (Maley and Forrest, 2000; Maley et al., 2006; Merlo et al., 2006). Cell populations may also be subjected to complementation or suppression interactions that may modulate tumor characteristics.

The similarity between cancer and RNA virus evolution has long been recognized (Gonzalez-Garcia et al., 2002; Brumer et al., 2006; Tannenbaum et al., 2006; Fox and Loeb, 2010); reviewed in (Ojosnegros et al., 2011; Domingo and Schuster, 2016). The parallels between the two systems extend to the proposal of an error threshold in cancer (Solé and Deisboeck, 2004; Fox and Loeb, 2010). Cells, as viruses, have a maximum tolerable error rate above which maintenance of information for cell viability is no longer possible (Figure 10.2). Genomic mutations (in general, genomic lesions) in cancer cells are several orders of magnitude higher than in normal cells (10^3-fold according to the data of Fox and Loeb, 2010 used in Figure 10.2). In parallel with viruses, relatively modest increases in mutation rate may drive cancer cells into error catastrophe (Fox and Loeb, 2010; Loeb, 2011). This important conceptual parallelism opens anticancer treatment options based on administration of nucleotide analogs.

10.5.1 THE TWO-COMPONENT THEORY OF CANCER: SIMILARITIES WITH OTHER BIOLOGICAL SYSTEMS

Some models distinguish two components regarding the genetic information in cancer cells: a robust component that ensures stability, and a variable component that permits phenotypic explorations that confer adaptability (Solé et al., 2008). The distinction is again parallel to one found in viruses: the core information that preserves virus identity versus flexible information that generates diversity to cope with environmental demands. This two-component theory is reminiscent of the "pan-genome" concept in bacteria. It proposes that individuals of a given bacterial species share a core component, while different representatives of the same species differ in part of their genetic information

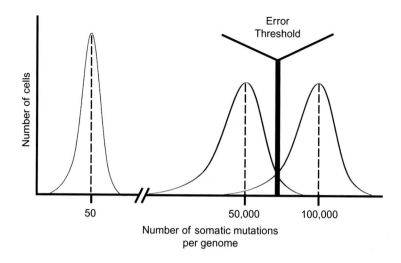

FIGURE 10.2

Diagram depicting error threshold and lethal mutagenesis in cancer. The average number of somatic mutations per genome is taken as 50 for normal cells (distribution on the left). This number increases to 50,000 for cancer cells (second distribution). If the number of mutations is increased twofold, an error threshold is crossed that results in loss of information and cancer extinction.

Data copied and figure adapted from Fox and Loeb (2010), with permission from the authors.

(Whitaker and Banfield, 2006; Mira et al., 2010). The genome complement of the bacterial species as a whole is larger than the genome portion present in individual strains within the species. Events such as genetic exchanges among bacteria contribute to enrich the flexible component while preserving bacterial identity as well as the minimal ingredients for survival.

It is expected that the phenotypic pluripotency of cancer cells will rapidly select for cell mutants resistant to chemotherapeutic agents, particularly during monotherapy (Luo et al., 2009). Tumor cell heterogeneity represents also a limitation to the efficacy of oncolytic viruses (Forbes et al., 2013). The objective is to achieve highly active anti-cancer treatments similar to the highly active treatments used for virus infections. Designs to suppress chemoresistant populations have been proposed (Gatenby et al., 2009) Cancer therapies require addressing different targets in a personalized manner. In parallel with viral therapies, lethal mutagenesis has been proposed for the treatment of solid tumors (Fox and Loeb, 2010). The requirement of combination treatments to control viruses emphasized in Chapters 8 and 9 is recognized as a valid approach also for cellular pathogens (Avner et al., 2012).

We reach the unifying notion for viruses and cellular collectivises that, as proposed in the quasispecies theory, the unit of selection can be a collectivity of individuals. Interestingly, a collective behavior may also be manifested in protein aggregates such as prions, with Darwinian behavior despite not being a genetic system.

10.6 COLLECTIVE BEHAVIOR OF PRIONS

The extensions of quasispecies theory outlined in previous sections have been applied to systems in which inheritance in the form of template copying is the key ingredient, common to viral and cellular life styles. Yet, interestingly, C. Weissmann and colleagues have provided direct evidence that prions (protein-only pathogenic entities) display features of quasispecies. Prions are the causative agents of transmissible spongiform encephalopathies (Alper et al., 1967; Griffith, 1967; Bolton et al., 1982; Prusiner, 1982; Prusiner, 1991; Walker and Jucker, 2015). The prion-associated diseases include Creutzfeldt-Jakob disease, fatal familial insomnia, and Gerstmann-Sträussler-Scheinker disease in humans, as well as scrapie in sheep, bovine spongiform encephalopathy in cattle, and chronic wasting disease in cervids. L.C. Walker and M. Jucker have proposed to define prions as "proteinaceous nucleating particles" and in fact "nucleation" may be one of the events responsible of their quasispecies behavior as discussed in coming paragraphs. The basis of prion disease is the propagation of PrP^{Sc} protein aggregates. PrP^{Sc} differs in conformation from a normal host glycoprotein termed PrP^{C} (C stands for *c*ommon or *c*ellular, Sc for *sc*rapie). Protein aggregation is increasingly viewed as a normal physiological process so that beneficial aggregation should be distinguished from pathologic aggregation often associated with neurodegeneration (Tyedmers et al., 2008; Jarosz et al., 2010; Tuite and Serio, 2010; Falsone and Falsone, 2015).

Protein conformation defines the strain characteristics and host range of prions. Different conformations can be regarded as "prion mutants." In a strikingly parallel behavior of RNA viruses, prions cloned by end-point dilution in cell culture can become heterogeneous by increasing the repertoire of conformers in the prion population (Li et al., 2010). Expressed in a virus-centric fashion, passage of prions increases their mutation frequency (frequency of different conformations). Selective pressures result in the selection of variants, including drug-resistant mutants (Ghaemmaghami et al., 2009; Li et al., 2010; Mahal et al., 2010; Shorter, 2010; Weissmann et al., 2011). Prion behavior can be modified depending on the population size (large populations versus bottleneck events produced by limiting dilution) in the protein misfolding cyclic amplification assay (the parallel to serial viral passages described in Section 6.2 of Chapter 6). Conformational variants are identified only after large population passages, and a decline in variant amplification rates is observed upon bottleneck passages (Vanni et al., 2014). Thus, Darwinian behavior and a fitness effect of passage regime are properties of prions despite being a nongenetic macromolecular population system.

The events that underlie the quasispecies behavior of prions, reflected in a dynamics of conformation heterogeneity, competition, and selection, are largely unknown. They may involve differences in glycosylation or transitions among conformations triggered by environmental factors, and perhaps with participation of truncated forms of the protein (Helmus et al., 2008; Notari et al., 2008). We have proposed that, irrespective of likely effects of the molecular composition of the prion proteins and environmental factors in modifying conformational preferences, some basic mechanisms acting on molecular fluctuations may also play a role (Ojosnegros et al., 2011). Protein conformation is the result of multiple amino acid-amino acid interactions, and atomic modifications of the type described to explain the origin of mutations through nonstandard base pairs among nucleotides (Chapters 2 and 9).

In the case of protein aggregates, the adoption of one versus alternative conformations may depend on atom ionization, and ionic and hydrophobic contacts, all of which depend in turn on bond torsion angles subjected to thermal fluctuations. The different influences at play may also determine the presence of minority conformations at different frequencies. Structural heterogeneities and transitions among conformations are widespread in protein populations (Smith et al., 2015). Such transitions may play physiological roles, and only occasionally they may be involved in pathological conditions. The relevance of alternative conformations in prions became apparent because they can give rise to disease, but in other cases they may remain unnoticed. A possibility is that a specific conformation variant can nucleate the conversion of additional molecules into the same conformation to yield aggregates enriched in the new variants (see Bernacki and Murphy, 2009; Kastelic et al., 2015 and references therein for some theoretical aspects of protein aggregation). It would be extremely interesting to develop a theory of quasispecies for macromolecular populations (nongenetic systems). It would serve as framework to understand prion-related disease, as well as other pathologies that may be related to altered aggregation behavior of cellular proteins, as the molecular quasispecies of replicative systems helped to interpret virus behavior (Mas et al., 2010; Ojosnegros et al., 2011).

10.7 MOLECULAR MECHANISMS OF VARIATION AND CLONALITY IN EVOLUTION

In this and previous chapters, it has been emphasized that while point mutations are a universal means of genetic variation, changes based on loss, gain, or rearrangements of genomic portions or segments are used unevenly by different biological systems. For bacteria, protozoa, and cancer cells, large genome modifications are widely exploited as adaptive mechanisms. This poses the question if their evolution is predominantly clonal; in this context, clonality means that individual lineages evolve separately and that either sex or recombination plays no role or only a minimal role in their mode of life and survival strategies. The debate bears on the historical or present advantage of sex as a reproductive strategy (Barton and Charlesworth, 1998; Agrawal, 2006). Possible advantages of sex to counteract the fitness-reducing effects of Muller's ratchet were discussed in Section 6.5.1 of Chapter 6.

The controversy on clonality has been quite vivid for protozoa, with interesting studies and controversial proposals by many authors. The clonality model has been championed by M. Tibayrenc and F.J. Ayala, and is suggested to be applicable not only to protozoa but also to viruses and bacteria (Tibayrenc and Ayala, 2012; Tibayrenc and Ayala, 2014). The concept of predominant clonal evolution proposed by these authors is based on significant linkage disequilibrium (nonrandom association of genotypes at different loci), and maintenance of the identity of different lineages or genetic subdivisions in the form of recognized clades (or near-clades) that are stable in space and time, among other arguments. The absence of recombination could be either because of absence of molecular mechanisms to perform it or because of scarcity of double infections during which two genetically distinct protozoans from the same species could exchange genetic material (the "starving sex" hypothesis). Other authors consider that there is also evidence for non-clonality and an active role of recombination during the evolution of protozoa and other cellular entities (Heitman, 2010; Ramirez and Llewellyn, 2014; Rogers et al., 2014). In other cases, such as in some yeasts, there is evidence of both clonality and recombination in the field (Lott et al., 2010). Rather than going deeper in the controversy with cellular organisms, here we will examine clonality in viral systems. Viruses may provide adequate systems to evaluate the evidence and the extent of clonality during their evolution. In particular, viruses may shed light on why clonality is their main mechanism of evolution despite the presence of an active recombination machinery.

The term clonality applied to virus populations and evolution has at least three different meanings. The one more distant from the concept under discussion is that it may refer to biological clones of viruses, to describe populations that are the progeny of a single genome. We sometimes speak as a clonal population of virus to signify a population derived from a single genome (obtained from a single viral plaque or end-point dilution series). Biological cloning is the most severe form of genetic bottleneck, and the effects of bottlenecks are extensively discussed in other chapters of this book. A second meaning is the clonal expansion of viral genomes integrated in cellular DNA, accompanying the expansion of the cells that harbor them. Some retroviruses may integrate their genomes for considerable time periods so that they follow the pattern of variation typical of the host cell rather than the error-prone replication of the virus in the process of progeny production. This is the case of the human T-cell lymphotropic virus types 1 and 2 (HTLV-1 and HTLV-2). They have two routes for expanding their genetic material: to produce viral particles by activating viral expression from the latent reservoir, or by duplication of their genetic material as part of cellular DNA (see Melamed et al., 2014 and references therein). The latter has been termed the mitotic stage of the replication cycle of these viruses. As evidence that cell expansion confers a selective advantage to the virus, there are viral functions that drive cellular proliferation. Such proliferation should be distinguished from virus-driven malignancy that involves other mechanisms. In this second meaning of clonality, the viruses undergo clonal expansion through proliferation of their carrying cells.

The third meaning of clonality in viruses that do not integrate DNA copies of their genome in the cellular DNA corresponds closely to the meaning of clonal evolution of cellular parasites. Evidence of clonality in viruses is provided by the existence of near-clades and consistent linkage disequilibrium. However, there is extensive evidence of recombination in most viruses, including RNA viruses (Section 2.9 in Chapter 2). The question is how can clonality be maintained and considered the predominant mode of evolution in the face of extensive recombination. The proposal to solve this issue has been to distinguish unproductive or inconsequential recombination from biologically meaningful recombination (Perales et al., 2015). Current evidence is that only a minority of all the recombination events undergone by viruses have biological consequences. Despite the recombination machinery (which is closely ingrained into the replication machinery) remaining intact and active, most recombination events go unnoticed due to lack of appropriate markers to distinguish parental from recombinant genomes, and, more relevant biologically, absence of opportunity of fitness increase through recombination. Recombination in HIV-1 illustrates these points. Recombination is probably a very frequent occurrence in HIV-1 (Shriner et al., 2004) yet its biological effects regarding fitness gains of mosaic genomes were only noticed when HIV-1 lineages had previously diversified through point mutations (Thomson et al., 2002).

A study of recombination intermediates in poliovirus-infected cells using NGS illustrates an additional level of unproductive recombination (Lowry et al., 2014). The study by D.J. Evans and colleagues unveiled multiple imperfect recombination events during infection, in what was termed the step of generation of recombinant genomes. It was followed by the resolution step in which a few successful recombinants gave rise to viral progeny. This is yet another example of negative selection acting on nascent variant molecules to allow survival of only a few out of many genetic variants (Section 3.4 in Chapter 3). This example emphasizes the distinction between "occurrence" and "biological consequences" of recombination in viruses. There are many examples of biologically relevant recombination in viruses and some of them were reviewed in Chapter 2, including the establishment of epidemiologically relevant vaccine-derived poliovirus recombinants or well-established HIV-1 recombinants, as well as a salient transition toward genome segmentation. Since recombination requires coinfection of the same cells with two (or more) virus types that can recombine, the "starving sex hypothesis" may apply due to low probability of cell coinfections in differentiated organisms. In addition, coinfection may be prevented by mechanisms of superinfection exclusion by which active replication of a virus blocks a second infection by a related virus (Webster et al., 2013 and references therein). In favor of the possibility of recombination is the observation that in host organisms, some cell subsets may be more prone than average to capture viral particles (Cicin-Sain et al., 2005; Chohan et al., 2005).

In view of the above evidence, we have suggested a model for virus clonality whose main ingredients are summarized in Box 10.2, and it is diagrammatically represented in Figure 10.3 (Perales et al., 2015).

BOX 10.2 MAINTENANCE OF CLONALITY IN VIRUS EVOLUTION

- Genetic recombination is frequent but biologically meaningful recombination is limited.
- Recombination events are probably frequent during intracellular virus replication, with many recombination intermediates subjected to negative selection.
- Biologically meaningful recombination is limited by:
  - The need of prior diversification of parental viruses by mutation.
  - The requirement of coinfection of the same cell by the two distinguishable viruses.
  - The fitness requirements of the recombinants that must coexist with or outcompete the parental viruses.

A critical point is that despite recombination offering the virus a means to explore distant regions of sequence space (essential for occupation of new niches and long-term survival), it has not been established as a necessity to complete the replication cycle of viruses. Strand transfers during reverse transcription, or discontinuous RNA synthesis mechanisms during RNA replication or transcription, can be regarded as mechanistically akin to recombination, but they do not affect the genetic composition of progeny except in the case of coinfections which are restricted as indicated in Box 10.2. In contrast, recombination is an intrinsic necessity of all organisms in which sexual mechanisms have been established. The scarcity of biologically meaningful recombination maintains stable near-clades in viruses, as documented with many phylogenetic analyses. As indicated in Figure 10.3, occasional discontinuity points (highlighted in the figure by selection of mosaic genomes) may be biologically very relevant but they are not part of the way of life of viruses. The scheme of Figure 10.3 does not imply any space or temporal frame. It may refer to intra-host or inter-host events, or to a single infection or multiple infections (host-to-host transmissions). The concept applies equally to both situations. Thus, clonality can coexist with recombination in the evolution of cells and viruses. What the model we propose for virus stresses is that recombination can play an evolutionary role at the discontinuity points (Figure 10.3) but, contrary to sexual reproduction, recombination is not a norm imposed upon viruses to complete their replication cycles.

10.8 GENOMES, CLONES, CONSORTIA, NETWORKS, AND POWER LAWS

There are several levels of organization and interactions in biological and nonbiological systems that affect their behavior and stability. Classical ones at the level of cells and organisms are symbiosis (including endosymbiosis), parasitism, commensalism, and mutualism. Reexamining the flow of increasing complexity in error-prone viral genomes, we have noted the following order: individual genome → multiple genomes → quasispecies as unit of selection → multiple interacting quasispecies. While the consideration of viral populations as interacting sets of individuals is now well established (Domingo et al., 2012; Villarreal and Witzany, 2013; Domingo and Schuster, 2016), the interactions among different viral quasispecies and their possible significance have to be considered.

Viral quasispecies can behave as a complex adaptive system as evidenced by properties discussed in this book, notably the presence of a molecular memory of past evolutionary events, akin to memory in the immune system (Section 5.5 in Chapter 5). The critical ingredient of quasispecies memory or a mutant ensemble directing behavior is a network of interactions established among members of a mutant

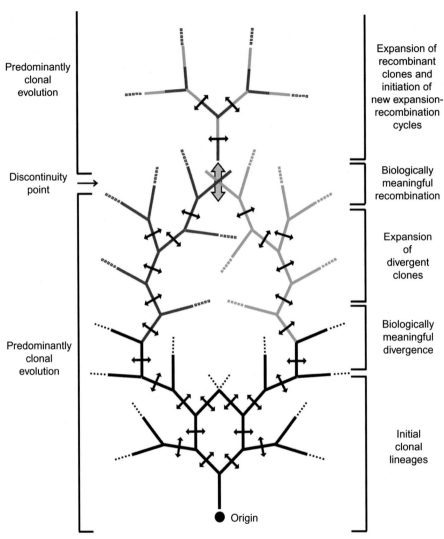

FIGURE 10.3

A schematic representation of clonal evolution of a virus despite continuous, active recombination. At an initial point (Origin), virus expands through replication (straight black lines), with continuous occurrence of inconsequential recombination (double arrows perpendicular to the replication lines). Discontinuous lines indicate that multiple divergent lineages are formed and pursue expansion. At a certain point, two sublineages diversify by mutation yielding viruses with potentially relevant differences (the red and blue lineages). In turn, each of the two new lineages proceeds with their diversification. At a critical or discontinuity point (thick vertical double arrow), a biologically meaningful recombination event occurs that gives rise to a mosaic virus (red-blue chimera) that initiates its clonal propagation. The major events in the process are indicated at the left and right margins of the diagram. See text for further clarification and implications.

Figure reproduced from Perales et al. (2015), with permission from the National Academy of Sciences USA.

spectrum. Two categories of interactions can be distinguished: those due to the mutations *per se*, and those due to gene expression products, in particular altered proteins that act in *trans* on other proteins or genomes of the same replicative unit. Mutation *per se* allows connections among component of the mutant spectrum, including mutational transitions between the master or dominant genome and its surrounding genomes located at 1, 2, etc. mutations distance in sequence space. We have described that an apparently disorganized mutant spectrum may in reality be composed of multiple subclasses of genomes, as revealed, for example, by partition methods that should be increasingly applied to analyze mutant spectra dissected by NGS (Section 3.6.4 in Chapter 3). A still largely unexplored subject is the extent of quasispecies substructuring in viral samples *in vivo* (a viral population from a cell, or a group of cells, an organ, etc.) and the network of interactions among them. Before addressing this point and its implications, we should briefly comment on networks.

Many natural and artificial systems have as one of their defining features being self-organized in a network manner. Two main points should be mentioned about such class of networks. First, they are widespread in our world, and second the network connections are often not random. Being widespread means that they describe disparate categories of interactions, some familiar to biologists but others not. Interacting networks are established among metabolites and metabolic reactions, or proteins within a cell, or organisms within an ecological system. There are also social networks, for example, the interactions among your relatives or your friends and their acquaintances, or the nodes and hubs that conform the www (for an overview that includes specific references, see Barabási, 2014).

The second point is that the connections in these types of self-organized networks follow some patterns. This can be intuitively understood: the proteins within a cell do not interact at random; some proteins have many partners while the majority has only a few partners. Nodes in the web have unequal weight in the interactions: some important hubs have many interactions while the majority has few interactions. This type of network topology is termed scale-free, and it is mathematically described as a power law, meaning that when two quantities are compared (number of proteins versus the number of their interactions in the cell, or number of nodes versus the number of their interactions in the web), one of the quantities varies as a power of the other quantity. Events described by a power law are frequent in nature (Bak, 1994; Bak, 1996; Newman, 2005). They describe apparently disparate events such as the intensity of earthquakes as a function of their numbers, or the number on inhabitants in a city versus the number of cities, among many other examples. A scale-free (nonrandom) organization described by a power law generally reflects an underlying dynamic process in their construction: how cellular evolution has adjusted the expression level and interaction among proteins or how dynamic sociological aspects have generated megacities in our planet. A power law is not the only statistical distribution that reflects an underlying mechanism. In Section 6.6 of Chapter 6, we encountered the Weibull distribution that characterized viral titers in successive plaque transfers, and that probably reflected the multiple cell-virus interactions that participate in the burst of viral progeny (Lázaro et al., 2003).

A few studies have described power laws in relation to viral dynamics. A. Moya and his colleagues described network dynamics of eukaryotic long terminal repeat (LTR) retroelements of eukaryotes. The connectivities among phenotypic markers were distributed according to a power law. This reflected a mode of evolution guided by an intrinsic capacity of LTR phenotypes to self-organize following a pattern characteristic of complex systems (Llorens et al., 2009). In an NGS analysis to examine the mutational load of HCV in patients subjected to peginterferon-alpha 2a-ribavirin therapy, A.M. Di Bisceglie and colleagues found that 36,818 mutations detected in 56 patients displayed a power-law distribution of mutation frequencies (Wang et al., 2014).

Self-organized systems display critical thresholds for spreading efficacy. For example, to be implemented, an innovation (scientific or technological) must spread among the network of potential recipients above a critical threshold of contacts (Abrahamson and Rosenkopf, 1997). We discussed the error threshold concept in Chapter 9 as a limit for maintenance of genetic information. In complex systems, there are critical thresholds needed to achieve a result such as the spread of an epidemics or the successful implementation of an innovation. Networks have additional features that may be relevant to virus population stability.

10.9 AN ADDITIONAL LEVEL OF VIRUS VULNERABILITY?

Self-organized networks have two additional features that may offer unexplored possibilities to guide viral populations toward extinction: substructuring and target-dependent vulnerability. An apparently homogeneous network of connections such as the www is in reality divided into subcompartments (also termed continents; see Barabási, 2014 for review and references). The stability of the ensemble depends not only on maintaining sufficient numbers of connections but mainly on maintaining some key connections: some minor nodes can be eliminated with no effect on the ensemble while some highly connected hubs must be preserved to ensure stability. Translated to viral populations, mutational links are established between the master (or dominant) genomic sequence and their neighbors in sequence space, and also among the minority neighbors. Not all points in sequence space have the same weight regarding quasispecies stability. If we could specifically hit the master sequence (by some "finely directed" mutagenesis procedure), we could destabilize the ensemble in a much more effective way than just hitting at random any genome from the population.

The issue of substructuring within viral populations is a largely unexplored one. The increasing application of NGS with powerful platforms that should allow determination of multitudes of whole genome sequences is expected to clarify the possible presence of substructures within viral populations as they replicate at different size-time scales (within single cells, sets of cells, cell cultures under various environmental influences, etc.). Some possibilities of absence or presence of different types of substructures are schematically presented in Figure 10.4. The assumption is often made that a viral quasispecies is dominated by a master sequence which is surrounded by a mutant cloud that influences the performance of the ensemble (Figure 10.4a). We know, however, that a viral population can be dominated by two or more master sequences. In these cases, there might be mutant spectra essentially in isolation because the mutational distance among them is too large (Figure 10.4b). Alternatively, the different quasispecies may be connected among them by accessible mutational distances (Figure 10.4c). We may envisage different topologies with various degrees of complexity. If they exist, it should be revealed by powerful NGS analyses applied to viral populations.

Two general outcomes can be anticipated: absence of substructuring or presence of multiple clusters ("nodes" or "hubs") of genomes with different numbers of connections and, thus, a different contribution to the stability of the system as a whole. A specific antiviral activity directed to debilitate the genomes with the maximum number of connections would be much more effective than a general antiviral activity directed to any genome in an indiscriminate manner. Again, debilitating the masters is more effective than debilitating minorities. The obvious difficulty is how to achieve such specific targeting. One possibility is to pursue designs of the type defined as treatment splitting into an induction and a maintenance regimen (Box 9.1 in Chapter 9) proposed by M. von Kleist and colleagues

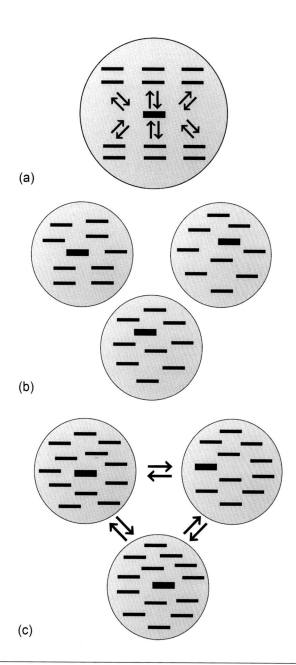

FIGURE 10.4

Possible interactions among quasispecies. Different mutant distributions (circles) are depicted in a simplified manner with genomes as horizontal lines (symbols that represent mutations have been omitted). The master or dominant genome is represented as a thicker line. (a) A single quasispecies in which interactions (mutational transitions) between the master sequence and surrounding members of the mutant spectrum are indicated by arrows. (b) Three quasispecies in which the interactions between the master sequence and mutant spectrum components have been omitted for clarity. In this case, the three mutant distributions are isolated by a mutational barrier. (c) Same as (b) except that in this case interactions (mutational transitions) are established among the three mutant distributions. See text for implications.

(von Kleist et al., 2011), and adapted to a second mutagenesis step. The induction phase should enrich the population in the most frequent and connected components of the mutant distributions, enhancing their exposure to mutagenic agents in the second treatment phase. Another largely unexplored possibility which is complementary to the previous one is, to choose the most adequate mutagenic agents to produce the highest frequency of deleterious mutations. It is known that different mutagenic agents produce different repertoires of mutations, shifting sequence space into different directions (Section 9.4 in Chapter 9). The base composition and codon usage in the genomes that conform the mutant distributions may suggest a mutagen-dependent efficacy of viral load reduction. At the risk of failing in scientific predictions (as many colleagues have), I end this section by suggesting that an increased number of viral-specific mutagenic agents, together with a deeper knowledge of the network organization of viral quasispecies (Figure 10.4), may reveal new vulnerabilities and help in the design of more effective antiviral strategies.

10.10 OVERVIEW AND CONCLUDING REMARKS

This chapter, as it also happens with chapter 1, deviates somehow from the main stream of the book. Several concepts relevant to the understanding of viruses as populations do not apply obviously solely to viruses. The reason is that they are of a very general nature. They can be translated to other biological systems, provided key parameters are adjusted. The comparative analysis has served to emphasize that different molecular mechanisms of genetic variation (with genomic transfers and rearrangements becoming more prominent relative to mutation as genome complexity increases) are used to achieve similar features and goals. Features are collective behavior (with several cell-to cell communication mechanisms) and selection of cellular subpopulations in the face of selective constraints. This has been illustrated with striking similar mechanisms of resistance to antibiotics and drugs displayed by bacteria and unicellular eukaryotes. If anything, the observations reinforce the conclusions reached for viruses in preceding chapters and indicate that control of viruses and pathogenic cells share the need of new paradigms.

A well-accepted parallelism is the one between RNA virus evolution and cancer development, both being Darwinian process of accelerated genetic variation, competition, and selection of the most suitable individuals in each environment. Cancer metastasis is the typical manifestation of tumor cell adaptability. This may be regarded as a restatement of the difficulties to control tumor progression. However, there is another side to it: the existence of an error threshold for maintenance of the information needed for cellular viability. Therefore, a new possible therapeutic design for cancer is on the horizon: lethal mutagenesis, following the strategy for RNA viruses.

A remarkable finding is that prions and possibly other proteins that function as macromolecular aggregates display features of quasispecies despite not being genetic systems. It means that Darwinian behavior does not require genetic modification to operate, as we already encountered in Chapter 1 with selective events in a prebiotic scenario. Quasispecies dynamics in prions necessitates a theoretical development because it may help approaching only-protein-based pathologies, as quasispecies theory helped to rationalize the understanding of pathogenic viruses.

We have also addressed an existing controversy on clonality in microbial evolution. Centered on viruses, the proposal made in this chapter is that clonality (absence of recombination as a way of life) can be the norm despite the continuous operation of recombination. The main point to solve this paradox is a distinction between unproductive or inconsequential recombination and biologically meaningful recombination. A deeper penetration into the replicating structures of viruses and cellular parasites should tell us soon if the model proposed for viruses is valid and if it is applicable also to cellular microbes.

Finally, this chapter (and the book) ends by acknowledging the importance of network interactions at different levels. It is worth noting that possible substructuring of viral quasispecies should be amenable to analysis during the coming years mainly as a result of massive whole genome sequencing. The feeling is that a new window of exciting information about virus population complexity will soon open. At the risk of falling victim to the inability of scientists to predict the future, I have adventured possible ways by which a detailed knowledge of quasispecies subpopulations (if such subpopulations exist and can be defined) might provide new approaches to combat viral disease. It looks still far, but unless some new paradigm takes the scene, it seems like a worth exploration, in line with what this book has aimed at transmitting (See Summary Box).

SUMMARY BOX

Extensions to Genetic and Nongenetic systems

- Darwinian principles, quantum mechanic uncertainties, and information theory are the pillars on which viral population dynamics rests. Being of a fundamental nature, these three concepts can be applied to a variety of genetic and nongenetic systems.
- Bacteria and cancer cells display collective features parallel to those described for viral populations: population heterogeneity, intrapopulation interactions, and collective responses to external stimuli.
- Cancer cells have a high mutational load and multiply close to an error threshold for maintenance of genetic information. Lethal mutagenesis by mutagenic nucleotides is a therapeutic option to treat cancer.
- Viruses evolve in a predominantly clonal way. Despite active recombination, biologically meaningful recombination is not the norm in the replication cycle of viruses.
- Scale-free, interacting networks describe many natural and artificial systems. Possible network connections within viral populations may unveil new points of virus vulnerability.

REFERENCES

Abrahamson, E., Rosenkopf, L., 1997. Social network effects on the extent of innovation diffusion: a computer simulation. Organ. Sci. 8, 289–309.

Agrawal, A.F., 2006. Evolution of sex: why do organisms shuffle their genotypes? Curr. Biol. 16, R696–R704.

Ahmad, M., Tuteja, R., 2014. Emerging importance of mismatch repair components including UvrD helicase and their cross-talk with the development of drug resistance in malaria parasite. Mutat. Res. 770, 54–60.

al-Khedery, B., Barnwell, J.W., Galinski, M.R., 1999. Antigenic variation in malaria: a 3' genomic alteration associated with the expression of a P. knowlesi variant antigen. Mol. Cell 3, 131–141.

Alper, T., Cramp, W.A., Haig, D.A., Clarke, M.C., 1967. Does the agent of scrapie replicate without nucleic acid? Nature 214, 764–766.

Andrews, R.H., Chilton, N.B., Mayrhofer, G., 1992. Selection of specific genotypes of Giardia intestinalis by growth in vitro and in vivo. Parasitology 105 (Pt 103), 375–386.

Avner, B.S., Fialho, A.M., Chakrabarty, A.M., 2012. Overcoming drug resistance in multi-drug resistant cancers and microorganisms: a conceptual framework. Bioengineered 3, 262–270.

Bak, P., 1994. Self-organized criticality: a holistic view of nature. In: Cowan, G.A., Pines, D., Meltzer, D. (Eds.), Complexity. Metaphors, Models and Reality. Addison-Wesley Publishing Co., Reading, MA. pp. 477–496.

Bak, P., 1996. How Nature Works. Springer-Verlag New York, Inc., New York.

Baltz, T., Giroud, C., Bringaud, F., Eisen, H., Jacquemot, C., et al., 1991. Exposed epitopes on a Trypanosoma equiperdum variant surface glycoprotein altered by point mutations. EMBO J. 10, 1653–1659.

Barabási, A.-L., 2014. Linked: How Everything Is Connected to Everything Else and What It Means for Business, Science, and Everyday Life. Basic Books, New York.

Barton, N.H., Charlesworth, B., 1998. Why sex and recombination? Science 281, 1986–1990.

Bernacki, J.P., Murphy, R.M., 2009. Model discrimination and mechanistic interpretation of kinetic data in protein aggregation studies. Biophys. J. 96, 2871–2887.

Bielas, J.H., Loeb, K.R., Rubin, B.P., True, L.D., Loeb, L.A., 2006. Human cancers express a mutator phenotype. Proc. Natl. Acad. Sci. U. S. A. 103, 18238–18242.

Bolton, D.C., McKinley, M.P., Prusiner, S.B., 1982. Identification of a protein that purifies with the scrapie prion. Science 218, 1309–1311.

Boveri, T., 1914. Zur Frage der Entstehung Maligner Tumoren. Verlag, Jena.

Brumer, Y., Michor, F., Shakhnovich, E.I., 2006. Genetic instability and the quasispecies model. J. Theor. Biol. 241, 216–222.

Caldwell, D.E., Costerton, J.W., 1996. Are bacterial biofilms constrained to Darwin's concept of evolution through natural selection? Microbiologia 12, 347–358.

Caval, V., Suspene, R., Shapira, M., Vartanian, J.P., Wain-Hobson, S., 2014. A prevalent cancer susceptibility APOBEC3A hybrid allele bearing APOBEC3B 3'UTR enhances chromosomal DNA damage. Nat. Commun. 5, 5129.

Chin, L., Hahn, W.C., Getz, G., Meyerson, M., 2011. Making sense of cancer genomic data. Genes Dev. 25, 534–555.

Chohan, B., Lavreys, L., Rainwater, S.M., Overbaugh, J., 2005. Evidence for frequent reinfection with human immunodeficiency virus type 1 of a different subtype. J. Virol. 79, 10701–10708.

Cicin-Sain, L., Podlech, J., Messerle, M., Reddehase, M.J., Koszinowski, U.H., 2005. Frequent coinfection of cells explains functional in vivo complementation between cytomegalovirus variants in the multiply infected host. J. Virol. 79, 9492–9502.

Costerton, J.W., Geesey, G.G., Cheng, K.J., 1978. How bacteria stick. Sci. Am. 238, 86–95.

Costerton, J.W., Cheng, K.J., Geesey, G.G., Ladd, T.I., Nickel, J.C., et al., 1987. Bacterial biofilms in nature and disease. Annu. Rev. Microbiol. 41, 435–464.

Costerton, J.W., Stewart, P.S., Greenberg, E.P., 1999. Bacterial biofilms: a common cause of persistent infections. Science 284, 1318–1322.

Covacci, A., Rappuoli, R., 1998. Helicobacter pylori: molecular evolution of a bacterial quasi-species. Curr. Opin. Microbiol. 1 (1), 96–102.

De La Cruz, V.F., Maloy, W.L., Miller, L.H., Good, M.F., McCutchan, T.F., 1989. The immunologic significance of variation within malaria circumsporozoite protein sequences. J. Immunol. 142, 3568–3575.

Deisboeck, T.S., Couzin, I.D., 2009. Collective behavior in cancer cell populations. Bioessays 31, 190–197.

Domingo, E., Schuster, P., 2016. Quasispecies: from theory to experimental systems. Curr. Top. Microbiol. Immunol, in press.

Domingo, E., Biebricher, C., Eigen, M., Holland, J.J., 2001. Quasispecies and RNA Virus Evolution: Principles and Consequences. Landes Bioscience, Austin.

Domingo, E., Sheldon, J., Perales, C., 2012. Viral quasispecies evolution. Microbiol. Mol. Biol. Rev. 76, 159–216.

Duncan, A.B., Gonzalez, A., Kaltz, O., 2013. Stochastic environmental fluctuations drive epidemiology in experimental host-parasite metapopulations. Proc. Biol. Sci. 280, 20131747.

Escalante, A.A., Lal, A.A., Ayala, F.J., 1998. Genetic polymorphism and natural selection in the malaria parasite Plasmodium falciparum. Genetics 149, 189–202.

Falsone, A., Falsone, S.F., 2015. Legal but lethal: functional protein aggregation at the verge of toxicity. Front. Cell. Neurosci. 9, 45.

Forbes, N.E., Abdelbary, H., Lupien, M., Bell, J.C., Diallo, J.S., 2013. Exploiting tumor epigenetics to improve oncolytic virotherapy. Front. Genet. 4, 184.

Foster, K.R., 2011. The sociobiology of molecular systems. Nat. Rev. 12, 193–203.

Fox, E.J., Loeb, L.A., 2010. Lethal mutagenesis: targeting the mutator phenotype in cancer. Semin. Cancer Biol. 20, 353–359.

Fuqua, W.C., Winans, S.C., Greenberg, E.P., 1994. Quorum sensing in bacteria: the LuxR-LuxI family of cell density-responsive transcriptional regulators. J. Bacteriol. 176, 269–275.

Futreal, P.A., Coin, L., Marshall, M., Down, T., Hubbard, T., et al., 2004. A census of human cancer genes. Nat. Rev. Cancer 4, 177–183.

Gatenby, R.A., Silva, A.S., Gillies, R.J., Frieden, B.R., 2009. Adaptive therapy. Cancer Res. 69, 4894–4903.

Ghaemmaghami, S., Ahn, M., Lessard, P., Giles, K., Legname, G., et al., 2009. Continuous quinacrine treatment results in the formation of drug-resistant prions. PLoS Pathog. 5, e1000673.

Gibbs, K.A., Federle, M.J., 2015. A social medium: ASM's 5th cell-cell communication in bacteria meeting in review. J. Bacteriol.

Gonzalez-Garcia, I., Sole, R.V., Costa, J., 2002. Metapopulation dynamics and spatial heterogeneity in cancer. Proc. Natl. Acad. Sci. U. S. A. 99, 13085–13089.

Greenman, C., Stephens, P., Smith, R., Dalgliesh, G.L., Hunter, C., et al., 2007. Patterns of somatic mutation in human cancer genomes. Nature 446, 153–158.

Griffith, J.S., 1967. Self-replication and scrapie. Nature 215, 1043–1044.

Hamelmann, C., Foerster, B., Burchard, G.D., Shetty, N., Horstmann, R.D., 1993. Induction of complement resistance in cloned pathogenic Entamoeba histolytica. Parasite Immunol. 15, 223–228.

Hanahan, D., Weinberg, R.A., 2011. Hallmarks of cancer: the next generation. Cell 144, 646–674.

Hansemann, D.V., 1890. Ueber asymmetrische Zelltheilung in epithel Krebsen und deren biologische Bedeutung. Virchow's Arch. Path. Anat. 119, 299–326.

Heitman, J., 2010. Evolution of eukaryotic microbial pathogens via covert sexual reproduction. Cell Host Microbe 8, 86–99.

Helmus, J.J., Surewicz, K., Nadaud, P.S., Surewicz, W.K., Jaroniec, C.P., 2008. Molecular conformation and dynamics of the Y145Stop variant of human prion protein in amyloid fibrils. Proc. Natl. Acad. Sci. U. S. A. 105, 6284–6289.

Horn, D., McCulloch, R., 2010. Molecular mechanisms underlying the control of antigenic variation in African trypanosomes. Curr. Opin. Microbiol. 13, 700–705.

Iqbal, J., Siripoon, N., Snounou, V.G., Perlmann, P., Berzins, K., 1997. Plasmodium falciparum: selection of parasite subpopulations with decreased sensitivity for antibody-mediated growth inhibition in vitro. Parasitology 114 (Pt 4), 317–324.

Jarosz, D.F., Taipale, M., Lindquist, S., 2010. Protein homeostasis and the phenotypic manifestation of genetic diversity: principles and mechanisms. Annu. Rev. Genet. 44, 189–216.

Kastelic, M., Kalyuzhnyi, Y.V., Hribar-Lee, B., Dill, K.A., Vlachy, V., 2015. Protein aggregation in salt solutions. Proc. Natl. Acad. Sci. U. S. A. 112, 6766–6770.

Koenderink, J.B., Kavishe, R.A., Rijpma, S.R., Russel, F.G., 2010. The ABCs of multidrug resistance in malaria. Trends Parasitol. 26, 440–446.

Lachaud, L., Bourgeois, N., Kuk, N., Morelle, C., Crobu, L., et al., 2014. Constitutive mosaic aneuploidy is a unique genetic feature widespread in the Leishmania genus. Microbes Infect. 16, 61–66.

Lázaro, E., Escarmis, C., Perez-Mercader, J., Manrubia, S.C., Domingo, E., 2003. Resistance of virus to extinction on bottleneck passages: study of a decaying and fluctuating pattern of fitness loss. Proc. Natl. Acad. Sci. U. S. A. 100, 10830–10835.

Li, J., Browning, S., Mahal, S.P., Oelschlegel, A.M., Weissmann, C., 2010. Darwinian evolution of prions in cell culture. Science 327, 869–872.

Llorens, C., Munoz-Pomer, A., Bernad, L., Botella, H., Moya, A., 2009. Network dynamics of eukaryotic LTR retroelements beyond phylogenetic trees. Biol. Direct 4, 41.

Loeb, L.A., 2001. A mutator phenotype in cancer. Cancer Res. 61, 3230–3239.

Loeb, L.A., 2011. Human cancers express mutator phenotypes: origin, consequences and targeting. Nat. Rev. Cancer 11, 450–457.

Lott, T.J., Frade, J.P., Lockhart, S.R., 2010. Multilocus sequence type analysis reveals both clonality and recombination in populations of Candida glabrata bloodstream isolates from U.S. surveillance studies. Eukaryot. Cell 9, 619–625.

Lowry, K., Woodman, A., Cook, J., Evans, D.J., 2014. Recombination in enteroviruses is a biphasic replicative process involving the generation of greater-than genome length 'imprecise' intermediates. PLoS Pathog. 10, e1004191.

Luo, J., Solimini, N.L., Elledge, S.J., 2009. Principles of cancer therapy: oncogene and non-oncogene addiction. Cell 136, 823–837.

Mahal, S.P., Browning, S., Li, J., Suponitsky-Kroyter, I., Weissmann, C., 2010. Transfer of a prion strain to different hosts leads to emergence of strain variants. Proc. Natl. Acad. Sci. U. S. A. 107, 22653–22658.

Maley, C.C., Forrest, S., 2000. Exploring the relationship between neutral and selective mutations in cancer. Artif Life 6, 325–345.

Maley, C.C., Galipeau, P.C., Finley, J.C., Wongsurawat, V.J., Li, X., et al., 2006. Genetic clonal diversity predicts progression to esophageal adenocarcinoma. Nat. Genet. 38, 468–473.

Mannaert, A., Downing, T., Imamura, H., Dujardin, J.C., 2012. Adaptive mechanisms in pathogens: universal aneuploidy in Leishmania. Trends Parasitol. 28, 370–376.

Mas, A., Lopez-Galindez, C., Cacho, I., Gomez, J., Martinez, M.A., 2010. Unfinished stories on viral quasispecies and Darwinian views of evolution. J. Mol. Biol. 397, 865–877.

Melamed, A., Witkover, A.D., Laydon, D.J., Brown, R., Ladell, K., et al., 2014. Clonality of HTLV-2 in natural infection. PLoS Pathog. 10, e1004006.

Merlo, L.M., Pepper, J.W., Reid, B.J., Maley, C.C., 2006. Cancer as an evolutionary and ecological process. Nat. Rev. Cancer 6, 924–935.

Miller, M.B., Bassler, B.L., 2001. Quorum sensing in bacteria. Annu. Rev. Microbiol. 55, 165–199.

Miller, R.V., Day, M.J., 2004. Microbial Evolution. Gene Establishment, Survival and Exchange. ASM Press, Washington, DC.

Mira, A., Martin-Cuadrado, A.B., D'Auria, G., Rodriguez-Valera, F., 2010. The bacterial pan-genome: a new paradigm in microbiology. Int. Microbiol. 13, 45–57.

Munday, J.C., Settimo, L., de Koning, H.P., 2015. Transport proteins determine drug sensitivity and resistance in a protozoan parasite, Trypanosoma brucei. Front Pharmacol. 6, 32.

Naidoo, I., Roper, C., 2010. Following the path of most resistance: dhps K540E dispersal in African Plasmodium falciparum. Trends Parasitol. 26, 447–456.

Nealson, K.H., Hastings, J.W., 1979. Bacterial bioluminescence: its control and ecological significance. Microbiol. Rev. 43, 496–518.

Newman, M.E.J., 2005. Power laws. Pareto distributions and Zipf's law. Contemp. Phys. 46, 323–351.

Nicolson, G.L., 1987. Differential growth properties of metastatic large-cell lymphoma cells in target organ-conditioned medium. Exp. Cell Res. 168, 572–577.

Nkhoma, S.C., Nair, S., Cheeseman, I.H., Rohr-Allegrini, C., Singlam, S., et al., 2012. Close kinship within multiple-genotype malaria parasite infections. Proc. Biol. Sci. 279, 2589–2598.

Notari, S., Strammiello, R., Capellari, S., Giese, A., Cescatti, M., et al., 2008. Characterization of truncated forms of abnormal prion protein in Creutzfeldt-Jakob disease. J. Biol. Chem. 283, 30557–30565.

Nowell, P.C., 1976. The clonal evolution of tumor cell populations. Science 194, 23–28.

Ojosnegros, S., Beerenwinkel, N., Antal, T., Nowak, M.A., Escarmis, C., et al., 2010a. Competition-colonization dynamics in an RNA virus. Proc. Natl. Acad. Sci. U. S. A. 107, 2108–2112.

Ojosnegros, S., Beerenwinkel, N., Domingo, E., 2010b. Competition-colonization dynamics: an ecology approach to quasispecies dynamics and virulence evolution in RNA viruses. Commun. Integr. Biol. 107, 2108–2112.

Ojosnegros, S., Perales, C., Mas, A., Domingo, E., 2011. Quasispecies as a matter of fact: viruses and beyond. Virus Res. 162, 203–215.

Perales, C., Moreno, E., Domingo, E., 2015. Clonality and intracellular polyploidy in virus evolution and pathogenesis. Proc. Natl. Acad. Sci. U. S. A. 112, 8887–8892.

Poste, G., Tzeng, J., Doll, J., Greig, R., Rieman, D., et al., 1982. Evolution of tumor cell heterogeneity during progressive growth of individual lung metastases. Proc. Natl. Acad. Sci. U. S. A. 79, 6574–6578.

Prusiner, S.B., 1982. Novel proteinaceous infectious particles cause scrapie. Science 216, 136–144.

Prusiner, S.B., 1991. Molecular biology of prion diseases. Science 252, 1515–1522.

Ramirez, J.D., Llewellyn, M.S., 2014. Reproductive clonality in protozoan pathogens–truth or artefact? Mol. Ecol. 23, 4195–4202.

Read, T.D., Massey, R.C., 2014. Characterizing the genetic basis of bacterial phenotypes using genome-wide association studies: a new direction for bacteriology. Genome Med. 6, 109.

Rice, S.A., Tan, C.H., Mikkelsen, P.J., Kung, V., Woo, J., et al., 2009. The biofilm life cycle and virulence of Pseudomonas aeruginosa are dependent on a filamentous prophage. ISME J. 3, 271–282.

Rogers, M.B., Downing, T., Smith, B.A., Imamura, H., Sanders, M., et al., 2014. Genomic confirmation of hybridisation and recent inbreeding in a vector-isolated Leishmania population. PLoS Genet. 10, e1004092.

Rottschaefer, S.M., Riehle, M.M., Coulibaly, B., Sacko, M., Niare, O., et al., 2011. Exceptional diversity, maintenance of polymorphism, and recent directional selection on the APL1 malaria resistance genes of Anopheles gambiae. PLoS Biol. 9, e1000600.

Rumbaugh, K.P., Diggle, S.P., Watters, C.M., Ross-Gillespie, A., Griffin, A.S., et al., 2009. Quorum sensing and the social evolution of bacterial virulence. Curr. Biol. 19, 341–345.

Samad, H., Coll, F., Preston, M.D., Ocholla, H., Fairhurst, R.M., et al., 2015. Imputation-based population genetics analysis of Plasmodium falciparum malaria parasites. PLoS Genet. 11, e1005131.

Sanchez, C.P., Dave, A., Stein, W.D., Lanzer, M., 2010. Transporters as mediators of drug resistance in Plasmodium falciparum. Int. J. Parasitol. 40, 1109–1118.

Shanks, G.D., Edstein, M.D., Jacobus, D., 2015. Evolution from double to triple-antimalarial drug combinations. Trans. R. Soc. Trop. Med. Hyg. 109, 182–188.

Shlien, A., Campbell, B.B., de Borja, R., Alexandrov, L.B., Merico, D., et al., 2015. Combined hereditary and somatic mutations of replication error repair genes result in rapid onset of ultra-hypermutated cancers. Nat. Genet. 47, 257–262.

Shorter, J., 2010. Emergence and natural selection of drug-resistant prions. Mol. Biosyst. 6, 1115–1130.

Shriner, D., Rodrigo, A.G., Nickle, D.C., Mullins, J.I., 2004. Pervasive genomic recombination of HIV-1 in vivo. Genetics 167, 1573–1583.

Smith, C.A., Ban, D., Pratihar, S., Giller, K., Schwiegk, C., et al., 2015. Population shuffling of protein conformations. Angew. Chem. Int. Ed. Engl. 54, 207–210.

Solé, R.V., Deisboeck, T.S., 2004. An error catastrophe in cancer? J. Theor. Biol. 228, 47–54.

Solé, R.V., Rodriguez-Caso, C., Deisboeck, T.S., Saldaña, J., 2008. Cancer stem cells as the engine of unstable tumor progression. J. Theor. Biol. 253, 629–637.

Sterkers, Y., Lachaud, L., Bourgeois, N., Crobu, L., Bastien, P., et al., 2012. Novel insights into genome plasticity in Eukaryotes: mosaic aneuploidy in Leishmania. Mol. Microbiol. 86, 15–23.

Stewart, P.S., Costerton, J.W., 2001. Antibiotic resistance of bacteria in biofilms. Lancet 358, 135–138.

Tannenbaum, E., Sherley, J.L., Shakhnovich, E.I., 2006. Semiconservative quasispecies equations for polysomic genomes: the haploid case. J. Theor. Biol. 241, 791–805.

Thomson, M.M., Perez-Alvarez, L., Najera, R., 2002. Molecular epidemiology of HIV-1 genetic forms and its significance for vaccine development and therapy. Lancet Infect. Dis. 2, 461–471.

Tibayrenc, M., Ayala, F.J., 2012. Reproductive clonality of pathogens: a perspective on pathogenic viruses, bacteria, fungi, and parasitic protozoa. Proc. Natl. Acad. Sci. U. S. A. 109, E3305–E3313.

Tibayrenc, M., Ayala, F.J., 2014. Cryptosporidium, Giardia, Cryptococcus, Pneumocystis genetic variability: cryptic biological species or clonal near-clades? PLoS Pathog. 10, e1003908.

Tomlinson, I.P., Novelli, M.R., Bodmer, W.F., 1996. The mutation rate and cancer. Proc. Natl. Acad. Sci. U. S. A. 93, 14800–14803.

Tuite, M.F., Serio, T.R., 2010. The prion hypothesis: from biological anomaly to basic regulatory mechanism. Nat. Rev. Mol. Cell Biol. 11, 823–833.

Tyedmers, J., Madariaga, M.L., Lindquist, S., 2008. Prion switching in response to environmental stress. PLoS Biol. 6, e294.

Valadares, H.M., Pimenta, J.R., Segatto, M., Veloso, V.M., Gomes, M.L., et al., 2012. Unequivocal identification of subpopulations in putative multiclonal Trypanosoma cruzi strains by FACs single cell sorting and genotyping. PLoS Negl. Trop. Dis. 6, e1722.

Vanni, I., Di Bari, M.A., Pirisinu, L., D'Agostino, C., Agrimi, U., et al., 2014. In vitro replication highlights the mutability of prions. Prion 8, 154–160.

Vartanian, J.P., Guetard, D., Henry, M., Wain-Hobson, S., 2008. Evidence for editing of human papillomavirus DNA by APOBEC3 in benign and precancerous lesions. Science 320, 230–233.

Vignuzzi, M., Andino, R., 2010. Biological implications of picornavirus fidelity mutants. In: Ehrenfeld, E., Domingo, E., Roos, R.P. (Eds.), The Picornaviruses. ASM Press, Washington DC, pp. 213–228.

Villarreal, L.P., Witzany, G., 2013. Rethinking quasispecies theory: from fittest type to cooperative consortia. World J. Biol. Chem. 4, 79–90.

von Kleist, M., Menz, S., Stocker, H., Arasteh, K., Schutte, C., et al., 2011. HIV quasispecies dynamics during pro-active treatment switching: impact on multi-drug resistance and resistance archiving in latent reservoirs. PLoS One 6, e18204.

Wahlgren, M., Fernandez, V., Chen, Q., Svard, S., Hagblom, P., 1999. Waves of malarial variations. Cell 96, 603–606.

Walker, L.C., Jucker, M., 2015. Neurodegenerative diseases: expanding the prion concept. Annu. Rev. Neurosci. 38, 87–103.

Wang, W., Zhang, X., Xu, Y., Weinstock, G.M., Di Bisceglie, A.M., et al., 2014. High-resolution quantification of hepatitis C virus genome-wide mutation load and its correlation with the outcome of peginterferon-alpha2a and ribavirin combination therapy. PLoS One 9, e100131.

Webster, B., Ott, M., Greene, W.C., 2013. Evasion of superinfection exclusion and elimination of primary viral RNA by an adapted strain of hepatitis C virus. J. Virol. 87, 13354–13369.

Weissmann, C., Li, J., Mahal, S.P., Browning, S., 2011. Prions on the move. EMBO Rep. 12, 1109–1117.

Whitaker, R.J., Banfield, J.F., 2006. Population genomics in natural microbial communities. Trends Ecol. Evol. 21, 508–516.

Further Reading

In this section a few key publications are listed in chronological order. They have been chosen because they either represent major developments or provide additional information on the topics that are main focus of this book.

Mills, D.R., Peterson, R.L., Spiegelman, S., 1967. An extracellular Darwinian experiment with a self-duplicating nucleic acid molecule. Proc. Natl. Acad. Sci. U. S. A. 58, 217–224.
The first study of RNA evolution in vitro.

Eigen, M., 1971. Self-organization of matter and the evolution of biological macromolecules. Naturwissenschaften 58, 465–523.
The first mathematical treatment of error-prone replication, and the basis of quasispecies theory. It links Darwinian evolution with information theory.

Flavell, R.A., Sabo, D.L., Bandle, E.F., Weissmann, C., 1974. Site-directed mutagenesis: generation of an extracistronic mutation in bacteriophage Q beta RNA. J. Mol. Biol. 89, 255–272.
The first site-directed mutagenesis experiment that marked the beginning of reverse genetics.

Domingo, E., Flavell, R.A., Weissmann, C., 1976. In vitro site-directed mutagenesis: generation and properties of an infectious extracistronic mutant of bacteriophage Qbeta. Gene 1, 3–25.

Batschelet, E., Domingo, E., Weissmann, C., 1976. The proportion of revertant and mutant phage in a growing population, as a function of mutation and growth rate. Gene 1, 27–32.
The first synthesis of an infectious virus mutant by site-directed mutagenesis, that permitted the calculation of a mutation rate for an RNA virus.

Domingo, E., Sabo, D., Taniguchi, T., Weissmann, C., 1978. Nucleotide sequence heterogeneity of an RNA phage population. Cell 13, 735–744.
Demonstration that bacteriophage Qβ populations evolve as dynamic mutant spectra.

Eigen, M., Schuster, P., 1979. The Hypercycle. A Principle of Natural Self-Organization. Springer, Berlin.
A compilation of the publications on the development of quasispecies theory and hypercyclic coupling. It describes concepts that are highly relevant to virology such as error-prone replication and the error threshold for maintenance of genetic information.

Holland, J.J., Grabau, E.A., Jones, C.L., Semler, B.L., 1979. Evolution of multiple genome mutations during long-term persistent infection by vesicular stomatitis virus. Cell 16, 495–504.
This study documents hundreds of mutations occurring in cloned vesicular stomatitis virus during persistence in cell culture. It includes the first statements of continuous and rapid evolution of an animal virus that drove the interest of John Holland towards viral dynamics.

Holland, J.J., Spindler, K., Horodyski, F., Grabau, E., Nichol, S., VandePol, S., 1982. Rapid evolution of RNA genomes. Science 215, 1577–1585.
It is an insightful exposure to the multiple biological implications of a dynamic RNA world in a DNA-based cellular biosphere.

Domingo, E., Holland, J.J., Ahlquist, P., 1988. RNA Genetics. CRC Press, Boca Raton, Florida.
These three volumes reviewed for the first time the impact of what became known as "RNA genetics". It describes key general concepts and specific findings with many viral groups. It includes the first detailed description of sequence space as it applies to virus evolution.

Meyerhans, A., Cheynier, R., Albert, J., Seth, M., Kwok, S., Sninsky, J., Morfeldt-Manson, L., Asjo, B., Wain-Hobson, S., 1989. Temporal fluctuations in HIV quasispecies in vivo are not reflected by sequential HIV isolations. Cell 58, 901–910.

The first demonstration of the implications of quasispecies dynamics for human immunodeficiency virus type 1, with emphasis on the difficulties to characterize viral populations in precise molecular terms.

Chao, L., 1990. Fitness of RNA virus decreased by Muller's ratchet. Nature 348, 454–455.

This study is the first experimental demonstration of the operation of Muller's ratchet in an RNA virus. It inaugurated studies linking concepts of population genetics with virus evolution. Subsequent studies by C. Escarmís and colleagues permitted the characterization of highly unusual mutants that populate mutant spectra of RNA viruses.

Holland, J.J., Domingo, E., de la Torre, J.C., Steinhauer, D.A., 1990. Mutation frequencies at defined single codon sites in vesicular stomatitis virus and poliovirus can be increased only slightly by chemical mutagenesis. J. Virol. 64, 3960–3962.

The first experimental test of the concept of error threshold. It documents negative effects of enhanced mutagenesis, opening the way to lethal mutagenesis as an antiviral strategy.

Eigen, M., 1992. Steps Towards Life. Oxford University Press.

It is a dissection of the main ingredients in the origin of life, with a molecular perspective.

Crotty, S., Maag, D., Arnold, J.J., Zhong, W., Lau, J.Y.N., Hong, Z., Andino, R., Cameron, C.E., 2000. The broad-spectrum antiviral ribonucleoside ribavirin is an RNA virus mutagen. Nat. Med. 6, 1375–1379.

It is the demonstration that the antiviral agent ribavirin is mutagenic for poliovirus. This result has exerted a lot of influence in the understanding of viral quasispecies, and the development of lethal mutagenesis as an antiviral strategy.

Solé, R., Goodwin, B., 2000. Signs of Life. How Complexity Pervades Biology. Basic Books, New York.

An interesting introduction to the concept of complexity as it affects biological systems.

Gago, S., Elena, S.F., Flores, R., Sanjuan, R., 2009. Extremely high mutation rate of a hammerhead viroid. Science 323, 1308.

A concise account of the relationship between mutation rate and genome complexity.

Cello, J., Paul, A.V., Wimmer, E., 2002. Chemical synthesis of poliovirus cDNA: generation of infectious virus in the absence of natural template. Science 297, 1016–1018.

It describes the chemical synthesis of a nucleic acid that can generate a virus. It is the ultimate demonstration that a polynucleotide chain has all the information to produce a virus.

Pfeiffer, J.K., Kirkegaard, K., 2005. Increased fidelity reduces poliovirus fitness under selective pressure in mice. PLoS Pathog. 1, 102–110.

This paper and the following one in this list inaugurated research on fidelity mutants of RNA viruses and opened fundamental studies on the consequences of altered template copying fidelity in virus adaptability.

Vignuzzi, M., Stone, J.K., Arnold, J.J., Cameron, C.E., Andino, R., 2006. Quasispecies diversity determines pathogenesis through cooperative interactions in a viral population. Nature 439, 344–348.

As the previous article in this list, this study revealed a number of biological implications of the error rate level during virus replication, and documented complementation among viral quasispecies in vivo.

Domingo, E., Parrish, C., Holland, J.J., 2008. Origin and Evolution of Viruses, second ed. Elsevier, Oxford.

It is a summary of theoretical and experimental observations related to virus evolution that spans population genetics concepts, and the modern focus on quasispecies and complexity.

Ehrenfeld, E., Domingo, E., Ross, R.P., 2010. The Picornaviruses. ASM Press, Washington D.C.

Picornaviruses have played major roles as model systems to understand basic features of viruses and virus-host interactions. This is the latest of a series of volumes that have periodically updated the contributions of picornavirology. It includes chapters dealing with mutation, recombination, fidelity mutants, lethal mutagenesis and antiviral agents.

Roossinck, M.J., 2011. The good viruses: viral mutualistic symbioses. Nat. Rev. Microbiol. 9, 99–108.

It describes the implication of viruses in several biological processes, unrelated to disease.

Domingo, E., Sheldon, J., Perales, C., 2012. Viral quasispecies evolution. Microbiol. Mol. Biol. Rev. 76, 159–216.

A review on quasispecies and its implications for antiviral interventions, including lethal mutagenesis approaches.

Eigen, M., 2013. From Strange Simplicity to Complex Familiarity. A Treatise on Matter, Information, Life and Thought. Oxford University Press, Oxford.

A journey from concepts of physics to information, error-prone replication, origin of life, complexity and self-organization. It includes several specific appendices, one by Peter Schuster.

Mateu, M.G., 2013. Structure and Physics of Viruses. An Integrated Textbook. Springer, Dordrecht, Heidelberg, New York, London.

This volume provides an excellent compilation of physicochemical procedures and concepts that have enriched our understanding of the structural basis of viral function, a discipline known as Physical Virology.

Ruiz-Mirazo, K., Briones, C., de la Escosura, A., 2014. Prebiotic systems chemistry: new perspectives for the origins of life. Chem. Rev. 114, 285–366.

An extensive and intensive account of prebiotic chemistry and processes leading to the origin and basic features of life.

Domingo, E., Schuster, P., 2016. Quasispecies: from theory to experimental systems. Curr. Top. Microbiol. Immunol. in press.

A review of theoretical and experimental aspects of quasispecies.

Author Index

Note: Page numbers followed by *b* indicate boxes, *f* indicate figures and *t* indicate tables.

367

Subject Index

Note: Page numbers followed by *f* indicate figures, *b* indicate boxes, and *t* indicate tables.

399

K

KP1461, in lethal mutagenesis, 317
K strategists, 126

L

Lamarckism, 75
Latency, 80, 232
Latent infection, 127
Latent reservoir, 7, 127, 349–350
Lateral gene transfers, 20, 23, 213
Lazarus effect, 101
Leishmania, 344–345
Lentiviruses, 23, 140, 142–143, 288
Lethal defection, 82, 308–310, 310*f*, 311, 324–326
Lethal mutagenesis, 49, 91, 327–328
    advantages and limitations, 327*b*
    and coronavirus, 49
    and error threshold, 305–311
    studies on, 312*t*
    theoretical models, 99–101, 306, 311, 329
    in vivo, 317–319
Life
    defined, 22–23
    origin, 7, 8–16, 17–18, 26–27, 340
    spontaneous generation, 9
Linkage disequilibrium, 349, 350
Lipid bilayer, 15
Lipid membrane, 21
Live-attenuated vaccine, 185, 269
Liver fibrosis, 26
Living matter, 13–14
Long-term coevolution, 20–21, 124–125, 134, 144
Long terminal repeat (LTR), 142–143, 353
Long terminal repeat (LTR) retroelements of eukaryotes, 353
Long-term virus evolution, 213, 269
    antigenic diversification, 238–242
    complexity revisited, 255
    extinction, 248–255
    interhost *vs.* intrahost rate of evolution, 235–237
    microbial disease, 252*b*
    monoclonal antibody-escape mutants, 242
    mutant cloud, 248–255
    phylogenetic relationships, 246–248
    rate discrepancies and clock hypothesis, 237–238
    reproductive ratio, 229–231
    serotypes, 239–241
    survival, 248–255
    time of sampling influence, 233–235
    viral emergence, 251–254
    virus evolution rate in nature, 232–238
    vs. viral genomes, 243–246
Luxury functions, 26

Lymphocytic choriomeningitis virus (LCMV), 26, 105, 140, 188, 305, 308, 312*t*, 317, 326
Lymphotropic minute virus of mice, 206, 207*t*

M

Mass extinction, 21–22, 27, 83
Master genome, 89–91, 179–180
Master sequence, 89, 90, 104–105, 212, 305–306, 307*f*, 311, 354, 355*f*
Maximum likelihood (ML), virus evolution, 232–233, 247
Maximum parsimony, 247
Mean pairwise diversity, 246
Measles virus (MV), 52, 105, 106–108, 127, 139*t*, 230, 242*t*
Mechanical transmission, 153
Medical interventions, 264–265, 292, 293, 327–328, 340
Megavirus, 6
Membrane, 6, 15, 21, 27, 46, 132, 133–134, 138, 151–152
Memory, 99, 180–184, 216–217
Memory, molecular, 99, 180, 182, 184, 190, 235, 351–353
Mendelian genetics, 75
Messenger RNAs (mRNAs), 6, 37, 38, 42
Metabolism, 15, 26–27, 305
Microbial evolution, 4, 19, 21, 75–76, 356–357
Micro-RNA (miRNA), 20–21, 42, 201–202
Middle East respiratory syndrome coronavirus (MERS), 146*t*, 291–292
Miller's approach, 9–10
Mimiviruses, 6–7
Mineral(s), 10–11
Mineral-organic complexes, 10–11
miRNA. *See* Micro-RNA (miRNA)
Misincorporation, 6, 37, 44–45, 46, 49, 50–52, 53, 91–92, 135, 188, 275–276, 311–312, 312*t*
Mobile genetic element, 23
Modern synthesis, 75, 77
Modulation, 104–105, 110, 147, 152, 155, 205–206, 321–323
Module
    archetypal, 20–21
    functional, 18–19, 20, 132
Molecular basis
    of fitness decrease, 213–215
    of mutation, 37–40
Molecular clock, 232–233, 237, 247, 255
Molecular clone, 78, 91–95, 180–182, 189
Monoclonal antibody (MAb), 47, 110, 111, 144–145, 146*t*, 170–171, 180, 218, 227, 240, 344–345
Monoclonal antibody-resistant mutants (MARMs), 82, 242, 242*t*
Monotherapy, 289–290, 302–303, 318, 319–320, 324–326, 347
Morphotype, 4–7, 239, 248
Mosaic genomes, 36–37, 56, 350–351
Mosquito vector, 153